1

4

5

2

3

6

Plate 1. Ancient litchi tree in Guangdong, China (photograph courtesy of Keith Chapman). **Plate 2.** Large litchi/longan orchard in southern China (photograph courtesy of Xuming Huang). **Plate 3.** Old litchi orchard in Guangxi, China. **Plate 4.** Mature litchi trees in southern China, with thin outer fruiting canopy (photograph courtesy of Zhan Wei Zhang). **Plate 5.** High-density longan orchard in Guangxi, China. **Plate 6.** Pruning of litchi in northern Queensland, Australia.

7

8

9

10

11

12

Plate 7. Heavy-yielding 'Tai So' ('Dazao') litchi after pruning in northern Queensland, Australia. **Plate 8.** Ten-year-old 'Mauritius' ('Tai So') litchi in South Africa. **Plate 9.** Wild litchi tree in southern China (photograph courtesy of Zhan Wei Zhang). **Plate 10.** Wild litchi fruit in Hanoi, Viet Nam. **Plate 11.** Vegetative growth in litchi.
Plate 12. Flowering in litchi, showing terminal inflorescences.

13

14

15

16

17

18

Plate 13. Fruit harvest in litchi. **Plate 14.** Male flower in litchi. **Plate 15.** Female flower in litchi. **Plate 16.** Vegetative growth in longan. **Plate 17.** Terminal inflorescence in longan. **Plate 18.** Fruit harvest in longan.

55

56

57

58

59

60

Plate 55. Algal spot, *Cephaleuros virescens*, in longan in Thailand (photograph courtesy of Chatree Sittigul). **Plate 56.** Black mildew of longan (photograph courtesy of Tony Cook). **Plate 57.** Phytophthora fruit rot in longan caused by *Colletotrichum* sp. (photograph courtesy of Tony Cook). **Plate 58.** Witches' broom in longan in Thailand (photograph courtesy of Chatree Sittigul). **Plate 59.** Harvesting litchis in China. **Plate 60.** Harvesting litchis in Thailand.

61

62

63

64

65

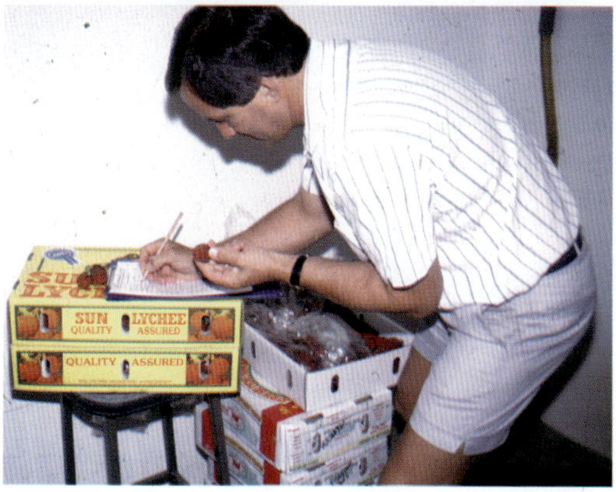

66

Plate 61. Sorting litchis in Mauritius. **Plate 62.** Sorting litchis in Viet Nam. **Plate 63.** Sorting litchis in Australia (photograph courtesy of Neil Greer). **Plate 64.** Hydrocooling of litchis in Australia (photograph courtesy of Neil Greer). **Plate 65.** Checking the quality of litchis in Australia (photograph courtesy of Neil Greer). **Plate 66.** Litchi cartons in China (photograph courtesy of Zhan Wei Zhang).

67

68

69

70

71

72

Plate 67. Litchi carton in Australia. **Plate 68.** Litchis in punnets in Australia. **Plate 69.** Tropical fruit market in South-east Asia, featuring litchi and the related rambutan. **Plate 70.** Litchi market in Chiang Mai, Thailand. **Plate 71.** Litchi market in Hanoi, Viet Nam. **Plate 72.** Fresh litchis in Mauritius.

Litchi and Longan

Botany, Production and Uses

Litchi and Longan

Botany, Production and Uses

Edited by

C.M. Menzel and G.K. Waite

Maroochy Research Station
Queensland Department of Primary Industries and Fisheries
Nambour
Queensland
Australia

CABI Publishing

CABI Publishing is a division of CAB International

CABI Publishing
CAB International
Wallingford
Oxfordshire OX10 8DE
UK

Tel: +44 (0) 1491 832111
Fax: +44 (0)1491 833508
E-mail: cabi@cabi.org
Web site: www.cabi-publishing.org

CABI Publishing
875 Massachusetts Avenue
7th Floor
Cambridge, MA 02139
USA

Tel: +1 617 395 4056
Fax: +1 617 354 6875
E-mail: cabi-nao@cabi.org

A catalogue record for this book is available from the British Library, London, UK.

Library of Congress Cataloging-in-Publication Data

Litchi and longan : botany, production, and uses / edited by C. Menzel and G. Waite.
p. cm.
Includes bibliographical references and index.
ISBN 0-85199-696-5 (alk. paper)
1. Litchi. 2. Longan. I. Menzel, C. (Chris) II. Waite, G. (Geoff) III. Title.

SB379.L8L58 2005
634.6--dc22

2004014693

ISBN 0 85199 696 5

Typeset by AMA DataSet Ltd, UK.
Printed and bound in the UK by Cromwell Press, Trowbridge.

Contents

Contributors

Ruth Ben-Arie, Fruit Storage Research Laboratory, Kiryat Shmona 10200, Israel
Lindy Coates, Department of Primary Industries and Fisheries, 80 Meiers Road, Indooroopilly, Queensland 4068, Australia
Tom L. Davenport, University of Florida, Tropical Research and Education Center, 18905 SW 280 Street, Homestead, FL 33031-3314, USA
Deirdre M. Holcroft, Dole Fresh Vegetables, PO Box 1759, Salinas, CA 93902, USA
Huibai Huang, College of Horticulture, South China Agricultural University, Guangzhou 510642, China
Xuming Huang, College of Horticulture, South China Agricultural University, Guangzhou 510642, China
Saichol Ketsa, Department of Horticulture, Kasetsart University, Chatuchak, Bangkok 10900, Thailand
Hetong Lin, College of Food Science and Technology, Fujian Agriculture and Forestry University, Fuzhou 350002, China
Richard E. Litz, Tropical Research and Education Center, University of Florida, 18905 SW 280 Street, Homestead, FL 33031-3314, USA
Chengming Liu, College of Horticulture, South China Agricultural University, Guangzhou 510642, China
Kazumitsu Matsumoto, Tropical Research and Education Center, University of Florida, 18905 SW 280 Street, Homestead, FL 33031-3314, USA
Christopher M. Menzel, Department of Primary Industries and Fisheries, PO Box 5083, SCMC, Nambour, Queensland 4560, Australia
Sisir K. Mitra, Faculty of Horticulture, Bidhan Chandra Krishi Viswavidyalaya, Mohanpur 741235, Nadia, West Bengal, India
Simon Raharjo, Tropical Research and Education Center, University of Florida, 18905 SW 280 Street, Homestead, FL 33031-3314, USA
Pravat K. Ray, Horticulture Department, Rajendra Agricultural University, Pusa 848125, Samastipur, India
Chatree Sittigul, Department of Plant Pathology, Chiang Mai University, Chiang Mai 50200, Thailand
Raphael A. Stern, MIGAL, Galilee Technology Center, PO Box 831, Kiryat Shmona 11016, Israel
Suranant Subhadrabandhu, Department of Horticulture, Faculty of Agriculture, Kasetsart University, Chatuchak, Bangkok 10900, Thailand

Geoff K. Waite, Department of Primary Industries and Fisheries, PO Box 5083, SCMC, Nambour, Queensland 4560, Australia

Witjaksono, Tropical Research and Education Center, University of Florida, 18905 SW 280 Street, Homestead, FL 33031-3314, USA

Erxun Zhou, Department of Plant Pathology, College of Resources and Environment, South China Agricultural University, Guangzhou 510642, China

About the Editors

Christopher Menzel is a Principal Horticulturist with the Queensland Department of Primary Industries and Fisheries at Maroochy Research Station, Nambour, on the Sunshine Coast of Australia. Chris completed his Bachelor of Agricultural Science at the University of Queensland in 1977, and then enrolled for a PhD, where he studied the growth of potato, *Solanum tuberosum*, in tropical climates. He has conducted research for the Australian Litchi Industry for 22 years and has written about 80 papers and book chapters. Some of these research papers have been translated for the benefit of litchi specialists in China.

Chris has been a guest speaker at national and international tropical fruit conferences, and has supervised several postgraduate students who have studied litchi physiology. Chris was leader of the canopy management project with CSIRO Plant Industry in Brisbane, which developed canopy management strategies for commercial litchi growers in Australia. His early work explored the factors affecting flowering and fruit set, suitability of cultivars for commercial production, along with recommendations for nutrition and water management.

In 1991, Dr Menzel reviewed the developing litchi and longan industries in northern Viet Nam for the United Nations Development Programme (UNDP) and in the following year spent 6 months at the Institute for Tropical and Subtropical Crops in Nelspruit in South Africa. In 1998 and 1999, Chris undertook consultancies for the Food and Agricultural Organization of the United Nations in China. He was co-organizer and editor of the First National Litchi Conference in Nambour in 1986, and the Fifth National Litchi Conference in Maroochydore in 1999. He was a keynote speaker and editor for the First International Litchi and Longan Symposium held in Guangzhou in 2000. In 2001, he represented Australian growers at an international workshop on litchi sponsored by the Food and Agricultural Organization of the United Nations in Bangkok. In 2002, he produced a technical guide for litchi producers in South-east Asia and the Pacific. Chris has also visited the litchi industries in Thailand, South Africa, Mauritius, Réunion, Hawaii, Florida and South America, and was senior contributor to the DPI & F's *Litchi Growing Guide* produced in 2002.

Dr Menzel has been an Associate Editor of the *Journal of Horticultural Science & Biotechnology*, produced in the UK since 1996. Chris also referees papers for *Tree Physiology*, *Australian Journal of Experimental Agriculture*, *Australian Journal of Plant Physiology*, *Australian Journal of Agricultural Research*, *Annals of Botany*, *New Zealand Journal of Crop and Horticultural Science*, *Journal of the American Society for Horticultural Science*, *HortScience*, *Journal of Tropical Science (Pertanika)*, *Journal of Southern African Society of Horticultural Science* and the *South African Journal of Botany*.

Geoff Waite is Principal Entomologist with the Queensland Department of Primary Industries and Fisheries at Maroochy. He commenced his professional career with the Department in Toowoomba in 1969, where he conducted research on lucerne, soybean and pasture pests. In 1973 he transferred to Emerald in central Queensland as the first entomologist to be stationed there, to manage an outbreak of the spur-throated and migratory locusts. At this time, the first farms in the Emerald Irrigation Scheme commenced production, and Geoff conducted research in cotton, developing an integrated pest management (IPM) system for the fledgling industry.

Geoff transferred to Maroochy in 1980, where he began research to develop IPM systems for several subtropical crops, especially litchi, longan, avocado, macadamia and strawberry. He has also conducted research on low-chill stonefruit, persimmon, mango, custard apple and pineapple. He has visited the tropical fruit industries in China, Thailand, Taiwan, Hawaii and South Africa, chiefly to survey and acquire natural enemies of litchi erinose mite, *Aceria litchii*, and macadamia nut borer, *Cryptophlebia ombrodelta*. Pest management in Queensland's strawberries relies almost exclusively on the IPM system Geoff developed, which uses the 'pest in first' strategy in conjunction with the predatory mite, *Phytoseiulus persimilis*, to control spider mites, the major pests of the crop. The subject of his current research is to find an effective, environmentally friendly, control for fruitspotting bugs, major pests of tropical and subtropical fruit in Queensland.

Maroochy Research Station
2 June 2004

Preface

Litchi and longan are indigenous to South-east Asia and make significant contributions to the lives and economic health of millions of people in the region. These fruit originated in southern China and northern Viet Nam, but their culture has now spread to most countries that experience a subtropical climate for part of the year. Litchi is most important in China, India, Viet Nam and Thailand, while longan is produced mainly in China, Viet Nam and Thailand. There is also interest in these two fruit in southern Africa, Madagascar, Réunion, Israel, Spain, the USA, Mexico, Brazil, Australia, Bangladesh, Nepal, the Philippines and Indonesia. Production in the Asia-Pacific region accounts for more than 95% of world cultivation, at about 3 million tonnes, more than world avocado production but less than citrus, banana, pineapple, mango and papaya.

Most of the fruit is produced by small landholders, with fewer than 100 trees each. Orchards with more than 1000 trees are rare, except in southern China, where there are single plantings of more than 10,000 trees, and in southern Africa and Australia. The fruit have a high value, and can significantly add to the income of small landholders, with a few trees doubling the income of some families in Asia.

Despite the long history of litchi and longan cultivation, many areas experience low productivity, with average yields being less than 5 t/ha due to a number of factors such as warm weather at flowering, reliance on inappropriate cultivars or lack of tree care. In Israel and some other countries, yields of up to 15 t/ha have been achieved, indicating that there is a large gap between actual and potential yields. Prospects for increasing production and marketing of these crops are high if some of the growing, postharvest handling and marketing issues are resolved. Inter- and intra-regional cooperation in this area would assist the development and status of these crops in many local economies.

Litchi requires temperatures of around 15°C or lower to flower, with a period of dry weather sometimes being beneficial. Once trees have set fruit, warm weather and good soil moisture are associated with heavy yields. Cropping is thus limited to areas that experience some cool weather before flowering. Production in litchi is very erratic in the true tropics where nights seldom fall below 25°C, with the majority of the industries based in environments where nights fall below 15°C. However, there are examples of industries such as that in central Thailand, which is based on cultivars that can flower at temperatures slightly higher than those normally considered ideal. These areas often supply early season fruit and return higher incomes than fruit from 'traditional' subtropical orchards, but can fail in some years. The quality of some of these cultivars is often inferior compared with the traditional types, indicating that new varieties that have better fruit quality need to be developed for these environments.

Longan has both subtropical and tropical ecotypes. The bulk of the crop in China and Thailand is still produced in cooler areas, whereas the main industry in Viet Nam is based in the Mekong Delta. More research is required to define the optimum temperatures for flowering in both species. There are also some growing techniques that can assist cropping in the warmer areas, but they have not been evaluated across many diverse environments. The other constraint related to climate is poor fruit set during cool, damp weather, and damage to trees and fruit after typhoons or hurricanes in China, Viet Nam, Madagascar, Mauritius, Australia, Hawaii and Florida.

Despite the long history of litchi and longan cultivation in Asia, there is a paucity of information concerning the yield of the major cultivars in different environments. The performance of some cultivars is disappointing, and production is unprofitable, with many industries based on only one or two cultivars. A lack of suitable cultivars limits production in many countries, because the existing varieties are low-yielding or are not well regarded in the market place.

Limited plant breeding and selection, facilitated by the exchange of germplasm, would increase the production of the crops. Some countries, such as India, China, Nepal and Viet Nam, have many seedling trees, which could form the basis of industry expansion. Breeding programmes are required in the long term to develop better cultivars, and this is best implemented with a regional focus. In the interim, the current gene pool should be more systematically evaluated. Standardization of cultivar names and descriptions would assist in this endeavour and encourage cooperation amongst producing countries. Based on the above, it can be concluded that there is a need for a much stronger varietal improvement programme.

Propagation is well described, most commercial orchards being planted with air-layers, with grafting and budding being popular in China and Viet Nam. It is reported that grafted trees are more drought- and wind-resistant, but little experimental evidence for this is available. Grafting also uses less planting material than air-layering. However, there are some disadvantages with grafting; it is not as easy to accomplish as air-layering, and requires the growing of seedling rootstocks. Grafted trees also take longer to develop to the stage where they can be planted out. There is little information on the graft compatibility between different cultivars, and the impact on production and fruit quality. Newly established young plants require regular watering, and many orchards in Asia experience serious losses at this stage due to a lack of irrigation facilities. The provision of irrigation, along with education of nursery workers and growers in tree care, would overcome this problem. It is apparent that further work is required in order to standardize nursery techniques.

Litchis and longans can be grown on a range of different soil types, including soils with a pH from 5 to 8. In very acid or alkaline soils, there can be problems with iron, zinc, boron and other nutrients. The soil must be freely draining, although the trees can tolerate a wet profile for part of the day. Tree health and production are best on sandy, sandy loam and clay loam soils, while heavy clay soils should be avoided.

Optimal production is achieved with irrigation, especially during the fruiting cycle, with rainfall varying from month to month across the different growing areas. Most of the orchards in Asia are not irrigated and are dependent on regular rainfall. Experiments in Australia and South Africa have shown that drought can reduce growth and fruit production, but the significance of this research for local producers in South-east Asia is not known. Most growers cannot afford the cost of irrigation and, in any case, irrigation is generally not available. Mulching and cover crops can assist water conservation in the absence of irrigation, although new orchards should be irrigated if possible.

Most growers apply fertilizers to their orchards. Tentative leaf and soil standards are available for both crops, but the tests are too expensive for small landholders. Local government extension staff could provide this service on a regional or district basis. Growers generally use a mixture of organic and chemical fertilizers, although the source of the fertilizer does not affect production. In contrast, the effect on cropping of the timing of fertilizer application is yet to be resolved. Crop nutrient removal data should be used for estimating fertilizer requirements.

High-density orchards are becoming popular, and should increase returns to growers, especially in the early years of a planting. There is evidence that these orchards have double the returns of traditional low-density plantings. Considerable experience has been acquired in China, with closer plantings expected to increase returns for both small and large landholders.

High-density plantings require canopy management to control the size of trees, and close attention to water and nutrient management is also essential. Experiments in China and Australia have shown that the optimal time to prune trees is during the first few weeks following harvest, and this research needs to be repeated in the other growing areas. Extension staff also need training in the various aspects and benefits of canopy management. Drought, girdling, pruning and chemical defoliation can improve flowering, and need to be evaluated across different environments.

Many insects and other pests affect the trees, leaves, flowers and fruit, and their impact on grower returns varies from orchard to orchard. Each country's local industry needs to develop its own system of integrated pest management. Pest management approaches, along with other methods of tree care, must be suited to the needs and abilities of small landholders. Diseases do not generally affect litchi production, apart from downy blight and anthracnose in China, and pepper spot in Australia. The main disorder affecting longan is witches' broom, endemic to much of South-east Asia.

Litchi and longan fruit are highly perishable and have short shelf-lives, which seriously limits marketing and expansion of the crops. Attempts have been made to reduce postharvest fruit browning and rotting, by heating and cooling the fruit, the use of various packages and packing materials, and application of fungicides and other chemicals. However, the protocols established for the use of these cannot guarantee fruit quality for more than a week or two. Some export industries are based on the treatment of the fruit with sulphur, but this chemical may soon be withdrawn. The development of new postharvest treatments is thus more urgent.

Most of the litchis and longans produced in Asia are marketed locally, with some exports to Hong Kong, Malaysia and Singapore, and to a lesser degree Europe and the Arab States. There are also exports of litchis from southern Africa, Madagascar, and Australia to Europe. Most of the fruit is sold fresh, with one-third of the Chinese crop being dried, and some being processed and canned.

Quarantine issues relating to fruit flies cause problems for exports from Australia and South Africa to Japan and the USA. Disinfestation protocols need to be established for the various markets. The market potential within Asia is strong because of the rising affluence in many of those countries. Good-quality fruit from the region is also highly regarded in Europe, although very little market intelligence has been collected. The preferred cultivars, packaging, etc. for the different markets are not known. The potential size and value of each market is also unknown. Restricted freight space is a limiting factor for countries such as Australia, which are some distance from overseas markets. Quality standards are not used in all countries, despite their demonstrated benefits in several markets.

This publication presents the current state of knowledge concerning the history, physiology, culture, and marketing of litchi and longan throughout the world. Although it contains a wealth of documented information from the literature, and the extensive personal experience of the authors within their respective disciplines in association with the crops, it can be concluded that further research into all aspects of these two tropical fruit is required.

Acknowledgements

We thank the authors and external reviewers who contributed to this publication. Special appreciation is given to Professor Sisir Mitra from India, who recognized the importance of a publication on these two crops. Geoff Waite, Lindy Coates and Neil Greer provided the funds for inclusion of the coloured plates.

Christopher Menzel and Geoff Waite
Maroochy Research Station
31 May 2004

Dedication

My first encounter with the tropical fruit litchi, *Litchi chinensis* Sonn., was in July 1977 when I visited northern Queensland during a study tour for fourth-year Agricultural Science students from the University of Queensland. I remember visiting an orchard owned by Ed Brittain and his family in the outer suburbs of Cairns. The trees were quite large at the time, all 'Mauritius' or 'Tai So'. Ed was one of the pioneers of the industry in northern Queensland, and now has a property on the Atherton Tableland, growing mostly 'Fay Zee Siu' ('Feizixiao'), which fruit much earlier than the other cultivars grown. Like many litchi growers, Ed has always had a thirst for information about the crop. To satisfy this, he has read widely, and has organized regular workshops for local enthusiasts.

It was not until the summer of 1980, when I was studying for my PhD, that I eventually experienced the delights of litchi. While on holiday at Noosa on the Sunshine Coast just north of Brisbane, I saw these strange-looking fruit, selling for what I thought was an outrageous 50 cents each. Curiosity got the better of me, and I purchased a few to share with friends, creating a desire to get closer to this little bit of heaven, and initiating a close relationship with the crop that has persisted to this day.

In May 1982, I was fortunate to obtain a research position with the Department of Primary Industries at the Subtropical Fruit Research Station at Nambour, also on the Sunshine Coast. Keith Chapman had left Maroochy to take up a temporary position in Sri Lanka to help the local sugarcane industry. Keith is well known to horticulturists throughout South-east Asia, and is currently based in the Food and Agricultural Organization of the United Nations (FAO) Regional Office in Bangkok, Thailand. In the period before he left, there was a change in the direction of research at Maroochy, with an expansion into exotic tropical fruit such as avocado, mango, litchi, longan, custard apple, kiwifruit, persimmon and macadamia. Up until the late 1970s, the research group had focused primarily on traditional crops such as pineapple, banana, citrus and papaya.

In my first 18 months with DPI, I worked with Brian Paxton, an experimentalist at Maroochy. Brian was very meticulous, and quite experienced at propagating litchis by air-layering. I quickly learned that not all cultivars were the same: 'Tai So' ('Mauritius', 'Dazao') had good fruit but didn't crop very regularly in the district. On the other hand, 'Bengal' looked attractive to the inexperienced litchi consumer, but had a large seed. It was at this time that the 'Salathiel' trees on the Research Station started to crop. We were lucky to have received two air-layers from the Department's Kamerunga Tropical Fruit Station near Cairns. The original tree at Kamerunga was quite large by this time, but had rarely fruited in the tropical environment. It was only when air-layers were brought to the Sunshine Coast that the commercial potential of the cultivar was realized. Andi and Annabel Flower, who owned a litchi orchard at Eudlo near Nambour, propagated and distributed trees to many of the new growers in the district.

The other cultivar gaining popularity at this time was 'Kwai May Pink', selected and promoted by Herb Bosworth and his family at Ingham in northern Queensland. This cultivar now dominates litchi plantings in Australia, and fruits regularly in most districts. There was quite a lot of activity in importing and evaluating new cultivars at this stage. Many varieties were introduced by Keith Chapman, Brian Watson from Kamerunga, Don Batten from New South Wales Agriculture at Alstonville, and private nurserymen such as Gordon Vallance. The Australian industry owes a considerable debt to the efforts of these pioneers. Many cultivars such as 'Salathiel' and 'Kwai May Pink' have now been exported to places as far afield as South Africa, the USA, Israel and Viet Nam.

Brian Paxton left the Department in 1983, and eventually moved to Hawaii, where he operated a very successful tropical fruit business. Brian was replaced at Maroochy by Don Simpson, who had been working with the Department's postharvest group in Brisbane. The success of the litchi programme at Maroochy was due, to a large degree, to Don's commitment and enthusiasm. At this time, we were very lucky to receive most of our research funds from the Government, whereas since the 1990s all of our operating and travel costs have been funded by industry levies. The programme was very active, with research into cultivars, flowering, fruit set, nutrition and water management. Geoff Waite has also been active in the litchi research programme at Maroochy, and has developed strategies for controlling the important insect pests.

I have been very lucky to have Geoff as my co-editor. Geoff has visited many countries in the search for predators and parasitoids to control pests such as macadamia nut borer, *Cryptophlebia ombrodelta*, and erinose mite, *Aceria litchii*. We were always keen to visit research centres and commercial farms in other districts, especially at the beginning and end of the litchi season, when the fruit was not available on the Sunshine Coast. The other member of the team at Maroochy was Neil Greer, who compiled the first growing guide for litchi producers. Neil was also instrumental in the establishment of '*Living Lychee*', the national industry newsletter, and edited the first issues.

In February 1986 the group at Maroochy, along with local producers, held the First National Litchi Conference. Key outcomes from this event were the formation of the Australian Lychee (Litchi) Growers' Association (ALGA), and the establishment of a research and promotion levy to assist industry expansion. This association was modelled on the work of the late Dr Lindsey Milne and his team from the South African Litchi Growers' Association (SALGA), which coordinates litchi R&D in southern Africa. This group holds an annual research symposium, regular grower meetings, and produces an annual yearbook. There are similar successful grower organizations in China, Thailand, Hawaii and Florida. We were very lucky to have the renowned litchi researcher, Professor Qiming Zhou from Guangzhou, attend our industry meeting in 1986, when he presented a paper on litchi-growing in China. Professor Zhou was one of the original students of George Weidman Groff, who wrote the famous litchi and longan text in 1921.

It was not until June 2000 that the First International Symposium on Litchi and Longan was held, and it was fitting that it was hosted by Drs Huibai Huang and Xuming Huang from the South China Agricultural University in Guangzhou. Both Professors have contributed chapters to this publication, along with Drs Chengming Liu (Cultivars and plant improvement), Erxun Zhou and Hetong Lin. The Second International Symposium was held in Chiang Mai, Thailand in 2003, and was expanded to include other members of the Sapindaceae such as rambutan, *Nephelium lappaceum*. The late Professor Suranant Subhadrabandhu from Kasetsart University in Thailand has contributed chapters on production and taxonomy, while his countrymen, Dr Chatree Sittigul and Saichol Ketsa, have assisted with reviews on diseases and postharvest physiology.

I have been fortunate to have had the opportunity to visit the litchi industries in China, Viet Nam, Thailand, Hawaii, Florida, South Africa, Mauritius, Réunion and Central America. These visits have contributed to the exchange of germplasm and ideas about litchi cultivation, between Australia and other countries. I thank my fellow litchi specialists for their collaboration and friendship, especially Trevor Olesen from Australia (external referee), Raffi Stern from Israel (chapters on production, taxonomy and flowering), Tom Davenport from Florida (flowering physiology), and Xuming Huang from Guangzhou (chapters on production, cultivars and fruit disorders). Special mention should be made of Professor Sisir Mitra from India, who had the original idea for a technical publication on litchi. I have been fortunate to meet Sisir at symposia in Guangzhou, Bangkok and Cairns, and with his colleague, Pravat K. Ray from Rajendra Agricultural University, he has contributed a chapter on propagation for this book. I have also appreciated the support of growers and other colleagues in Australia who have contributed to our research programme, and of PhD students such as Suwit Chaikiattiyos and Susanna Hieke, whose work is cited in this volume (see References).

Dr Deirdre Holcroft has no direct link with the litchi and longan industries, but has contributed an excellent review on postharvest physiology, in collaboration with Professor Hetong Lin from the Fujian Agriculture and Forestry University, and Professor Erxun Zhou from the South China Agricultural University. Deirdre has wide experience in the storage and handling of fruit, with part of her recent career spent at the University of Michigan, and in Stellenbosch, South Africa.

Dr Richard Litz has a long history of research in biotechnology, and has just edited a publication on the topic for CAB International. With his colleagues, Kazumitsu Matsumoto, Simon Raharjo and Witjaksono from Homestead in Florida, he has provided a comprehensive review of activities in litchi and longan.

Dr Lindy Coates is a Principal Plant Pathologist with the Queensland Department of Primary Industries and Fisheries in Brisbane, and has extensive experience in the management of diseases of tropical fruit. She has travelled to Asia, and presented the results of her research at workshops organized by the Australian Centre for International Agricultural Research (ACIAR). In 2003, she published a chapter on tropical fruit diseases in Randy Ploetz's CABI monograph, and, with colleagues from Thailand and China, has contributed to this volume with a review of litchi and longan diseases.

Dr Ruth Ben-Arie from the Fruit Storage Research Laboratory at Kiryat Shmona in Israel, with assistance from her colleague, Raffi Stern, has contributed to this publication by reviewing litchi and longan processing.

Chinese researchers have produced some excellent publications on litchi and longan, including Ling Yuen Fu's *An Album of Guangdong Litchi Varieties in Full Colour* (1985), and Zhan Wei Zhang's *Litchi – Pictorial Narration of Cultivation* (1997). Publications for growers have also been produced in Australia, South Africa, Mexico and India. Victor Galán Saúco and U.G. Menini produced a growing guide for FAO in 1989, but it did not detail the information being collected by litchi specialists in China, Thailand or Viet Nam. There has been a significant expansion in production and knowledge gained of these two tropical fruit in the past 20 years in South-east Asia. This publication is the first text to provide a global overview of litchi and longan cultivation.

References to litchi and longan in China are numerous, and many have been quoted by researchers and enthusiasts. Guangdong, Hainan, Fujian and Guangxi are well known for these fruit, which invariably evoke a Chinese connection for most Western horticulturists. According to Groff (1921), Chinese writers and poets have sung the praises of litchi and longan for centuries, and written about their medical and commercial value. Government officials have encouraged the cultivation of the fruit, protected ancient trees, and disseminated information about their propagation, cultivation, and insect pests and diseases. From the earliest times, travellers to China have reported on the merits of these fruit, and encouraged their introduction into Europe and the USA. It is only in the past 200 years that commercial production has spread to the rest of South-east Asia, India, southern Africa, the Mediterranean, Australia and America.

Sung Chio (Sung Ta-mo) from Fujian, in his 1608 treatise, details 33 auspicious occasions that add to the pleasure of eating litchis. These include 'the coming of agreeable friends', 'facing flowing water', 'examining treatises on the litchi' etc. Then, under 'Sombre happenings in eating litchis', he lists 34 unfavourable circumstances such as 'heavy rains', 'having people about who do not like to eat litchis', 'listening to bad poems or songs', 'appearance of the pine cone variety' (marking the end of the litchi season), etc.

Groff's (1921) publication also mentions a 'Litchi Club'. Part of its by-laws read, 'Each day, one member acts as Director and procures three thousand fruit as an average, but if there are more, then the pleasure is greater.' It was suggested that, 'At the meeting time, members will devote themselves to eating and drinking, and not occupy themselves with poems and songs, but each follow his own inclination, may either take the tripod for warming tea, play chess, recline upon a pillow or mat, enjoy fragrant incense, chat, laugh and not bother about anything else. Those who disturb our ideas and who shirk should be dealt with strictly, while the dilettanti (amateurs) who enter uninvited shall not be excluded.'

This book is a thank-you to litchi and longan specialists and producers throughout the world. I hope that it might lead to collaborations with friends and colleagues who appreciate the delights of these two tropical fruits. On a personal level, I dedicate this book to Sue Ruming. This publication would have not occurred without her love and support.

Christopher Menzel
Maroochy Research Station
7 June 2004

References

Chaikiattiyos, S., Menzel, C.M. and Rasmussen, T.S. (1994) Floral induction in tropical fruit trees: effects of temperature and water supply. *Journal of Horticultural Science* 69, 397–415.

Fu, L.Y., Wu, S.X. and Zhou, Q.M. (1985) *An Album of Guangdong Litchi Varieties in Full Colour.* Guangdong Province Scientific and Technology Commission, Guangzhou, China, 78 pp.

Galán Saúco, V. and Menini, U.G. (1989) *Lychee Cultivation.* FAO Plant Production and Protection Paper No. 83. Food and Agricultural Organization of the United Nations, Rome, 205 pp.

Groff, G.W. (1921) *The Lychee and Lungan.* Orange Judd Company, New York, 188 pp.

Hieke, S., Menzel, C.M., Doogan, V.J. and Lüdders, P. (2002a) The relationship between fruit and leaf growth in lychee (*Litchi chinensis* Sonn.). *Journal of Horticultural Science & Biotechnology* 77, 322–325.

Hieke, S., Menzel, C.M., Doogan, V.J. and Lüdders, P. (2002b) The relationship between yield and assimilate supply in lychee (*Litchi chinensis* Sonn.). *Journal of Horticultural Science & Biotechnology* 77, 326–332.

Hieke, S., Menzel, C.M. and Lüdders, P. (2002c) Shoot development, chlorophyll, gas exchange and carbohydrates in lychee seedlings (*Litchi chinensis*). *Tree Physiology* 22, 947–953.

Hieke, S., Menzel, C.M. and Lüdders, P. (2002d) Effects of leaf, shoot and fruit development on photosynthesis of lychee trees (*Litchi chinensis*). *Tree Physiology* 22, 955–961.

Hieke, S., Menzel, C.M. and Lüdders, P. (2002e) Effects of light availability on leaf gas exchange and expansion in lychee (*Litchi chinensis*). *Tree Physiology* 22, 1249–1256.

Zhang, Z.W., Yuan, P.Y., Wang, B.Q., Qiu, Y.P. and Li, J.S. (1997) *Litchi, Pictorial Narration of Cultivation.* Pomology Research Institute, Guangdong Academy Agricultural Science, China, 189 pp.

1 Origin, History, Production and Processing

Xuming Huang[1], Suranant Subhadrabandhu[2], Sisir K. Mitra[3], Ruth Ben-Arie[4] and Raphael A. Stern[5]

[1]College of Horticulture, South China Agricultural University, Guangzhou 510642, China; [2]Department of Horticulture, Faculty of Agriculture, Kasetsart University, Chatuchak, Bangkok 10900, Thailand; [3]Faculty of Horticulture, Bidhan Chandra Krishi Viswavidyalaya, Mohanpur 741235, Nadia, West Bengal, India; [4]Fruit Storage Research Laboratory, Kiryat Shmona 10200, Israel; [5]MIGAL, Galilee Technology Center, PO Box 831, Kiryat Shmona 11016, Israel

Introduction

Litchi, *Litchi chinensis* Sonn., and longan, *Dimocarpus longan* Lour., the two most popular members of the Sapindaceae, produce arillate fruit with sweet, translucent, juicy flesh. The fruit are high in sugar and contain several vitamins and minerals (Table 1.1), and can be eaten fresh, frozen, canned, dried, or processed into juice, wine, pickles, preserves, ice-cream and yoghurt. Both species have a long history of cultivation in China and throughout much of South-east Asia, but only litchi is well known in Europe, Africa and North America. The production of both crops is much less than that of the main tropical fruit such as banana, citrus and mango. However, each constitutes a very lucrative commodity and contributes significantly to the livelihood of several million people throughout South-east Asia.

Table 1.1. Nutritional value of litchi and longan per 100 g fresh weight (Deng *et al.*, 1999).

	Litchi	Longan
Water (g)	84	81
Protein (g)	0.7	1.2
Fat (g)	0.1	0.1
Carbohydrate (g)	15	16
Vitamin C (mg)	15	60
Thiamine (mg)	0.02	0.04
Niacin (mg)	1.1	1.0
Riboflavin (mg)	0.07	0.03
Phosphorus (mg)	32	26
Iron (mg)	0.7	0.4
Calcium (mg)	4	13

Litchi and longan share many similarities in origin, history and environmental requirements, as well as in the utilization of the fruit. Except for China, the two crops are relatively new in most countries. Commercial cultivation of litchi was limited to China and Viet Nam before the 17th century and has spread slowly over the past 400 years, whereas longan has developed outside China only in the last 250 years. There has been a very rapid development of both crops in South-east Asia since 1980.

Litchi and longan are adapted to the warm subtropics, cropping best in climates with hot, humid summers and dry, cool winters. Flower initiation in litchi is best below 20°C, while the optimal temperature for leaf and fruit growth is about 30°C (Menzel and Simpson, 1994).

Temperatures below 2°C damage new leaves, while those below –2°C can kill trees (Ni *et al.*, 1998). The limited environmental adaptability of both crops has confined commercial production mainly to the subtropics (Fig. 1.1).

About 95% of litchi production is in South-east Asia, with China, Viet Nam, Thailand, India, Bangladesh and Nepal being the most important (Table 1.2). Australia, South Africa and Madagascar are the major players in the southern hemisphere. Countries with smaller production include the Philippines, Indonesia, Israel, the USA, Brazil, Mexico, Canary Islands, Mauritius, Réunion, Zimbabwe and Mozambique. The production season for litchi in different regions is shown in Fig. 1.2. The industry has expanded rapidly in the past 20 years because of increasing interest in exotic fruit in Europe and increasing affluence in Asia, which in turn have provided lucrative returns to growers. However, productivity is low in many countries. Most industries are based on one or two major cultivars, which limits the production season. The fruit also have a relatively short shelf-life, which existing technologies have not been able to improve substantially.

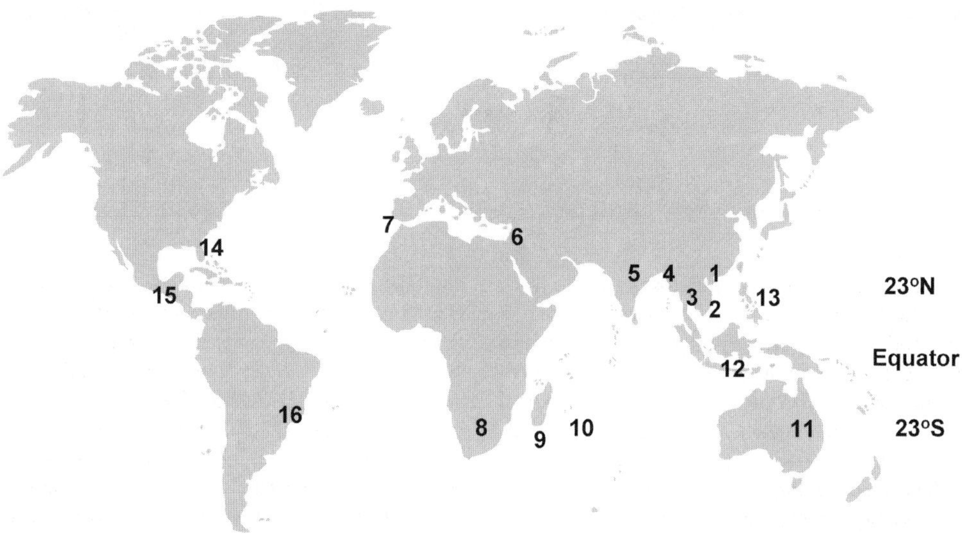

Fig. 1.1. Litchi- and longan-growing areas around the world. 1 = China, 2 = Viet Nam, 3 = Thailand, 4 = Nepal and Bangladesh, 5 = India, 6 = Israel, 7 = Spain, 8 = South Africa, 9 = Madagascar, 10 = Mauritius and Réunion, 11 = Australia, 12 = Indonesia, 13 = the Philippines, 14 = Florida, 15 = Mexico and Costa Rica, 16 = Brazil. Hawaii is not shown.

Table 1.2. Litchi acreage and production in different countries.

	Area (ha)	Production (t)	Reference
Mainland China	588,000	1,280,000	Huang (2002)
Taiwan	12,000	108,000	Mitra (2002)
Viet Nam	30,000	50,000	Hai and Dung (2002)
Thailand	23,000	81,000	Sethpakdee (2002)
India	56,200	429,000	Singh and Babita (2002)
Bangladesh	4,800	12,800	Abu Baker Siddiqui (2002)
Nepal	2,380	14,000	Budathoki (2002)
South Africa	1,500	8,000	Ghosh (Fisher 2004, personal communication)
Madagascar	3,000	20,000	Ghosh (2001)
Israel	300	2000	Goren *et al.* (2001)
Australia	1,500	5000	Menzel (2002)
Florida, USA	240	1000	Knight (2001)

Fig. 1.2. Harvest season for litchi in different countries.

Longan is less widely distributed than litchi and the fruit less popular outside Asia, with the demand for longan in Europe, North America and Australia mainly confined to ethnic Asian communities (Wong, 2000). Commercial activity is concentrated in China, Viet Nam and Thailand (Fig. 1.1), with small industries in Myanmar, India, Laos, Indonesia, Malaysia, Cambodia, Israel, Australia and the USA.

Origin of litchi

The litchi genus contains three subspecies: *Litchi chinensis* ssp. *chinensis*; *L. chinensis* ssp. *philippinensis*; and *L. chinensis* ssp. *javensis* (Leenhouts, 1978). Subspecies *philippinensis* is indigenous to the Philippines and is found growing wild from sea level to 500 m altitude (Sotto, 2002). The fruit are quite different from cultivated litchi and are long and oval-shaped, with long thorny protuberances and inedible flesh that partially covers the seed (Menzel, 1991). Subspecies *javensis* from the Malay Peninsula and Indonesia produces fruit similar to cultivated litchi, but with a thinner aril. Neither of these two subspecies is grown commercially.

Subspecies *chinensis*, the commercial form, originated in southern China and northern Viet Nam, from wild populations recorded in these regions (Wu, 1998; Hai and Dung, 2002). In China, abundant wild trees (Plate 9) have been found in moist forests from low elevations up to 1000 m in Hainan, below 500 m in hilly areas in western Guangdong and eastern Guangxi, and below 1000 m in valley or hilly regions of southern Yunnan (Wu, 1998). Litchi dominates several of these forests and may account for up to half of the virgin forest composition. Trees often grow in mixed stands with *Vatica astrotricha* (green plum), *Hopea hainensis*, *Heritiera parvifolia* of the Chinese parasol family, *Coelodepas hainanensis*, *Polyathia laui*, and *Diospyros hainanensis*, which belongs to the persimmon family. In Viet Nam, wild litchis have also been found growing at a low elevation in the Bavi Mountains and in forests in Tamdao (Vinhphuc Province) and Tuyenhoa (Quangbinh Province), although there are fewer pockets of natural rainforest there than in China (Hai and Dung, 2002).

The wild trees found in Hainan are similar in general appearance to cultivated litchi and produce fruit that are edible, although the flesh is relatively thin and sour (Plate 10). The fruit are variable and can be classified into three groups based on their shape, and arrangements of the skin segments and protuberances: Group 1 has sharp and pointed protuberances, Group 2 has protruded but obtuse skin segments, while Group 3 has flat skin segments. The wild types evolved in two directions, the skin segments becoming protruded and long as in 'Dazao' ('Tai So') and 'Guiwei' ('Kwai May'), or flattened as in 'Sanyuehong' ('Sum Yee Hong'), 'Shuidong' ('Souey Tung'), 'Nuomici' ('No Mai Chee') and 'Huaizhi' ('Wai Chee') (Wu, 1998). In Yunnan, a population of wild litchis was discovered that was quite different from most cultivated forms in that the trees required less cool weather to initiate flowering. These trees mature earlier and crop in warmer climates than traditional subtropical ecotypes. Flowers from these wild specimens have sepals with brownish hairs, and the species are known as 'Brown-hair litchi' or 'Hemaoli'.

Origin of longan

Longan is thought to have originated in the mountain chain that stretches from Myanmar through to southern China, and possibly down to the lowlands of south-west India and Sri Lanka (Tindall, 1994). Wild trees have been found dominating the tropical or monsoon rainforests in central and south-west Hainan, and in the west and south-west of Yunnan (Zhuang *et al.*, 1998). They are distributed mainly in the moist lowlands below 1200 m in Yunnan, and in the valley and hilly areas below 500 m in Hainan (Qiu and Zhang, 1996). Based on studies on the morphology of the pollen of longan cultivars and wild varieties in five ecological zones in China, Ke *et al.* (1994) suggested that Yunnan was the primary centre of origin, while Hainan, Guangdong and Guangxi were secondary centres.

Native longans that produce low-quality fruit are also found growing in the forests of Chiang Mai and Chiang Rai in northern Thailand (Subhadrabandhu and Yapwattanaphun, 2001a). Various tropical subspecies and varieties of longan that are similar to commercial longan grow throughout South-east Asia (especially in Malaysia), but they have a thin aril and a warty skin (Wong, 2001).

History of litchi (see Plates 1–4)

China has the longest history of litchi cultivation. Ancient documents, seeds unearthed from ancient tombs, and living millennium-old specimens all demonstrate that a significant litchi culture had developed by 200 BC in Hainan, Guangdong and Guangxi (Wu, 1998). The name 'litchi' or 'Li-zhi' in Chinese Pinying, originally meaning 'to be detached from the branch', first appeared in text in 200 BC. It was later found that the name referred to 'the rapid deterioration of the fruit once they were detached from the branch'. There were other documents that referred to litchi as 'to cut off the branches', because people needed to cut off the hard twigs to harvest the fruit.

According to Ge Hong (281–314), tropical and subtropical fruit including litchi and longan were presented to the Emperor by landlords in southern China as early as 200 BC. In 116 BC, the Emperor conquered Guangdong, and hundreds of trees were transplanted in the palace in Shanxi. This is the earliest recorded introduction of these crops from subtropical to temperate areas, although frosts killed all the trees.

Several monographs have provided details on cultivars, tree management, pest and disease control, postharvest handling and utilization of the fruit in ancient China (Dong, 1983). Growers chose seed from trees that produced large fruit with thick flesh, and paid special attention to trees that produced fruit with 'chicken-tongue' or aborted seeds. Crop improvement thus has a long history in China. The earliest known record of cultivars was made by Guo Yigong in his 'Guangzhi' Register published some 1800 years ago, with one of the four cultivars described as being shrivelled-seeded. In the 10th century, the 'Litchi Register in Guangzhong', compiled by Zheng Xiong, indicated 22 cultivars. Cai Xiang (1059) described 32 cultivars in Fujian, although a later document showed that most of these cultivars have been lost. Six hundred years later, Chen Ding registered 43 cultivars including 17 in Fujian, 6 in Sichuan, 3 in Guangxi, and 14 in Guangdong. The number in Guangdong increased to 58 in 1826, including the well known 'Guiwei', 'Gualu', 'Heiye', 'Huaizhi', 'Sanyuehong' and 'Feizixiao' (see Plates 20–29). The most recent register, published by Wu in 1998, recorded 222 cultivars, clones or individual trees with unique features, with detailed descriptions of their origin, distribution and morphology.

Litchi was propagated through seedlings before the 10th century. However, the long juvenile period and variations in performance associated with this practice eventually convinced growers to use vegetative material. It is estimated that air-layering or marcotting was first used in the 4th century AD, and grafting was first recorded in the 16th century (Anonymous, 1978). Grafting and air-layering used in litchi were first detailed in the Registers by Xu Bo in 1579 and by Deng Qingcai in 1628, respectively.

Litchi cultivation was confined to southern China and possibly northern Viet Nam until the late 17th century (Tindall, 1994; Hai and Dung, 2002). Distribution to other regions followed the route suggested by Galán Saúco and Menini (1989; see Fig. 1.3). Litchi spread to the part of Myanmar bordering Yunnan, China, in the late 17th century. Trees were taken to India in 1798 (Singh and Babita, 2002), and later to Nepal (Budathoki, 2002) and Bangladesh (Abu Baker Siddiqui, 2002). Litchi production has developed into a significant industry in these countries.

Litchi was first introduced into Thailand from China 300 years ago, by merchants who carried fruit with them. Some seedlings were adapted to the tropical conditions of the central region of the country and grew and fruited there. Many of these seedlings were selected because of their good quality and were named by local growers. They are known as lowland litchi or tropical litchi, as the trees do not require a long period of cold to initiate flowers (Subhadrabandhu, 1990). Plantings of litchi in Chiang Mai appeared around 1890, based on air-layers brought in by emigrants from Yunnan who migrated through Laos or Myanmar (Boonrat, 1984). These types are truly subtropical and need a longer period of low temperature for flowering. Many of them still retain their Chinese names such as 'O-Hia', 'Hong Huay' and 'Kim Cheng'; however, the Thai spellings and pronunciations are different (Subhadrabandhu and Yapwattanaphun, 2001a). Litchi was introduced from China to the Philippines earlier than 1916, but it never fruited at low altitudes (Sotto, 2002). Flowering was not recorded in the

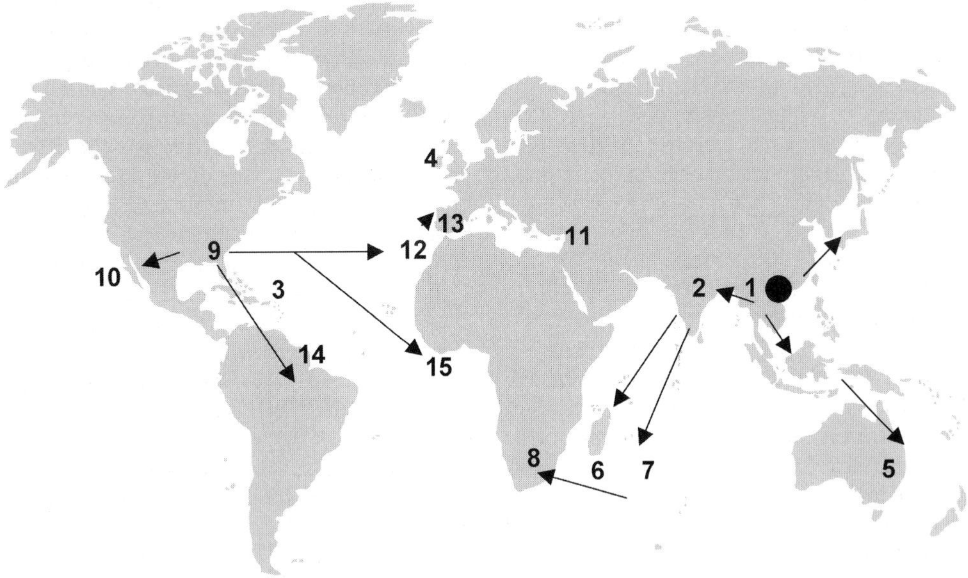

Fig. 1.3. Dissemination of litchi from China. Redrawn from Galán Saúco and Menini (1989). 1 = Thailand, 2 = India, 3 = Caribbean Islands, 4 = UK, 5 = Australia, 6 = Madagascar, 7 = Mauritius, 8 = South Africa, 9 = Florida, 10 = California, 11 = Israel, 12 = Canary Islands, 13 = Spain, 14 = South America, 15 = West Africa.

country until 1931. Later, trees from other sources were found to fruit in various places and gave hope for litchi-growing in the more elevated areas.

The crop reached Australia with Chinese migrants attracted by Gold Rushes around 1854 (Menzel *et al.*, 1988) and was introduced into southern Africa 50 years earlier (Milne, 1999). The introduction of litchi into Madagascar dates back to 1802, with many old plantations still in existence. A few relatively new plantations date back to the 1940s. The industry is currently based on trees older than this (Jahiel and Abraham, 2001). There is evidence that the first litchi trees were imported into South Africa from Mauritius in 1876, but there must have been earlier imports, as trees had already been observed in Natal in 1875 (Oosthuizen, 1991). From Natal, trees spread to the Transvaal Lowveld as well as to other suitable frost-free areas. The species arrived in Florida in the 1880s, but commercial cultivation did not start until the 1940s. The industry developed only after 'Brewster' ('Chenzi') was introduced, but the current industry is based mainly on 'Mauritius' ('Dazao', 'Tai So') (Knight, 2001). The first trees were transported to Hawaii in 1873 by Chinese merchants and were known as 'Afong' trees, later identified as being similar to 'Dazao' (Nakasone and Paull, 1998).

Litchi was introduced into Israel in the 1930s from three sources: South Africa ('Mauritius'), California ('Floridian') and India ('Bengal'), although a commercial industry did not develop until the late 1980s (Goren *et al.*, 2001).

History of longan

Longan, literally meaning 'the eye of the dragon' due to its round and bright black seed, has a documented history of more than 2200 years in China. Like litchi, longan was supplied to the Emperor as early as 200 BC during the Han Dynasty. There are fewer ancient monographs recording the history of longan, even though it is just as popular as litchi. Records detail the failure of plantings in Shanxi in 116 BC (Qiu and Zhang 1996), although introductions to Sichuan in south-west China a little later were more successful. Longan was sent to Fujian about 400 years later, where it thrived and proved to be a very profitable industry. The crop spread later from Fujian and Guangdong to Taiwan (see Plate 5).

It is believed that trees were brought to Thailand from China in the late 1800s, although longans were also found growing wild in northern Thailand over 200 years ago (Angasith *et al.*, 1999). The wild trees produce small fruit with a thin aril, and are called 'native longan' or 'ka-la longan'. In 1896, a Chinese merchant presented five trees to Queen Dara-Rasmee, wife of King Rama V. These trees were the ancestors of the present commercial varieties, which have been selected from several generations of seedlings.

Trees from China were sent to India, Sri Lanka, Myanmar, the Philippines, Cuba, the West Indies and Madagascar. The crop was taken to Australia in the mid-1800s, and to Hawaii and Florida by the early to mid-1900s (Menzel *et al.*, 1989).

Litchi in China

China accounts for nearly 80% of the world plantings, with production concentrated in Guangdong, Guangxi, Fujian, Hainan and Taiwan (Table 1.3). Smaller plantings are also found in Yunnan, Sichuan and Guizhou (see Fig. 1.4). Despite the long history of the crop, the fruit were rarely seen in northern markets until the late 1980s. There was a dramatic expansion during the 1990s, with plantings

Table 1.3. Area, production and productivity of litchi in China (Huang, 2002).

Province	Area (ha)	Production (t)	Average yield (t/ha)
Guangdong	303,080	793,200	2.6
Guangxi	210,000	310,000	1.5
Fujian	40,220	148,700	3.7
Hainan	18,600	15,000	0.9
Yunnan	2,700	2,700	1.0
Sichuan	8,620	2,000	0.2
Guizhou	430	500	1.2
Taiwan	11,961	108,668	9.1

Fig. 1.4. Major litchi-growing areas in southern China. 1 = Guangdong, 2 = Hainan Island, 3 = Fujian, 4 = Taiwan, 5 = Guangxi, 6 = Yunnan.

increasing from 161,700 ha to 592,000 ha. Over 60% of the plantings in mainland China have been made in the past 10 years, with many orchards opened in hilly areas that have access to irrigation (Figs 1.5 and 1.6). Production fluctuates according to seasonal weather, but has shown a sharp rise from 183,700 t in 1990 to 1,370,000 t in 1999 (Huang, 2002). In Taiwan, the crop is grown mostly in central and southern areas, with 11,580 ha producing 110,000 t. The pace of development is much slower than on the mainland, with fewer new orchards established. However, productivity is much higher (10 t/ha versus 2 t/ha), suggesting that there is potential for a large increase in production in Guangdong, Guangxi and Fujian (Table 1.3).

Over 200 cultivars, lines or individuals with unique features have been identified, although only about 20 are widely cultivated. The season starts with 'Sanyuehong' in late April in Hainan, and ends with 'Nanmuye' in mid-August in Sichuan. Significant expansion has occurred recently in Hainan, based on 'Feizixiao' ('Fay Zee Siu') (Plate 20), 'Sanyuehong' ('Sum Yee Hong') and 'Baitangying', which flower more readily under warm conditions. The early-season fruit produced in this area attracts a substantial premium in the market. However, 90% of production in mainland China is based on mid- to late-season cultivars and is concentrated in Guangdong, Guangxi and Fujian. The main season lasts for only about 50 days from late May to mid-July. 'Heiye' ('Haak Yip') (Plate 22) accounts for 90% of production in Taiwan, with some 'Yuhebao' in southern areas and 'Feizixiao' in central areas. The production season in Taiwan is usually from June to July.

Fig. 1.5. Commercial litchi and longan orchard in Guangdong, southern China (photograph courtesy of Xuming Huang).

Fig. 1.6. Commercial litchi orchard in Guangxi, southern China (photograph courtesy of Christopher Menzel).

The large crop produced over a relatively short harvest season has created a heavy burden on the industry in mainland China in terms of the concentration of labour required at this time. This is the main factor affecting expansion. The domestic market absorbs over 98% of the crop (Table 1.4). Slightly less than half of the production in Guangdong is consumed locally, half is transported to the northern provinces, and less than 2% is exported to Hong Kong, Japan, Singapore, the USA, Macau, Malaysia, Indonesia and Europe (Table 1.5). About 75% of the harvest is consumed fresh and 300,000 t is dried (Chen and Huang, 2001).

Litchi in Viet Nam

Although northern Viet Nam is considered to be part of the centre of origin of the crop, commercial production did not commence there until the 1980s. Since that time, there has been a rapid increase in plantings, with

Table 1.4. Import and export of litchis in China (Yi *et al.*, 2002).

	Export volume (t)	Export value (×10^4 US$)	Import volume (t)	Import value (×10^4 US$)
Fresh				
1995	3,711	683.1	569	31.1
1996	2,314	257.7	3,154	127.7
1997	4,787	621.6	1,827	69.9
1998	2,900	290.0	598	25.6
1999	12,762	672.0	6,752	244.4
2000	2,827	433.6	10,156	412.5
2001	1,193	160.0	7,177	340.3
Canned				
1995	11,579	1,496.0	73	0.1
1996	7,305	1,071.5	1,333	14.7
1997	7,559	930.9	448	9.0
1998	4,787	652.2	1,057	14.0
1999	14,995	1,254.8	21	1.2
2000	9,432	707.4	16	0.8
2001	4,029	302.7	27	0.5

Table 1.5. Average exports of litchi from China from 1995 to 2001 (Yi *et al.*, 2002).

Destination	Volume (t)
Hong Kong	21,453
Japan	2,832
Singapore	2,484
USA	802
Macau	442
Malaysia	409
Indonesia	356
England	82

30,000 ha of trees producing 50,000 t of fruit in 2000. Production is concentrated in an area within 40–200 km of Hanoi, with Bacgiang (20,250 t), Haiduong (11,600 t) and Quangninh (7000 t), being the most important centres. There are a number of local cultivars, but most of them are poor yielding and are of average quality. The industry is based on a single cultivar, 'Thieuthanhha'. The production season is relatively short, extending from late May to early June, with 70% of the crop consumed locally and the remainder exported to China, other parts of Asia and to Europe (Hai and Dung, 2002; Mitra, 2002).

Litchi in Thailand

Litchi ranks eleventh in value of fruit production in Thailand, whereas longan is ranked third. Most orchards are concentrated in the north of the country between Chiang Rai (8322 ha) and Chiang Mai (5763 ha), in a monsoon climate with a distinct dry season (Subhadrabandhu and Yapwattanaphun, 2001a; Sethpakdee, 2002). Plantings of more tropical cultivars have also been established in the warm, humid and high-rainfall areas of Chantaburi, Samut Songkram, Kanchanaburi and Nakhon Rajchasima; north, east and west of Bangkok (Sethpakdee, 2002). Flowering and production are more consistent in the cooler elevated areas, which account for 90% of the industry. The harvest in 2000 was a record 80,000 t from 23,000 ha. 'Hong Huay' ('Dazao', 'Tai So') occupies 70% of the area and 'Kom' 11% (see Fig. 1.7). Other important cultivars include 'O-Hia', 'Kim Cheng' ('Huaizhi', 'Wai Chee') and 'Chakrapad' ('Chacapat') (Subhadrabandhu and Yapwattanaphun, 2001a). Fruit are available from mid-March to mid-June across the range of climates and cultivars. Thailand exports about 10% of its production to Malaysia, Singapore, Hong Kong and Europe (Subhadrabandhu and Yapwattanaphun, 2001a; Sethpakdee, 2002).

Litchi in India

Like China, production in India has undergone substantial expansion in the past 50 years, with plantings increasing from 9400 ha in 1949 to 56,200 ha in 1998. The crop ranks seventh after mango, citrus, banana, apple, guava and papaya, and comprises 1.5% of the area under fruit (Singh and Babita, 2002). Production in 1998 was 429,000 t, with average yields of 7.5 t/ha. About 75% of the crop is produced in northern Bihar (310,000 t), followed by West Bengal (36,000 t), Tripura (27,000 t), Assam (17,000 t), Punjab (13,000 t), Uttar Pradesh (14,000 t) and Orissa (9000 t). India has 51 identified cultivars, of which 'Shahi', 'Bombai', 'China', 'Deshi', 'Calcuttia', 'Rose Scented' and 'Mazaffarpur' dominate (Ghosh *et al.*, 2001; Mitra, 2002). Fruit are available from early May

Fig. 1.7. Litchi market in Chiang Mai, northern Thailand (photograph courtesy of Christopher Menzel).

to early July in the different areas, with only a small quantity being exported in the last few years (Singh and Babita, 2002). Some of the crop is canned or juiced, whereas drying is relatively rare.

Litchi in Bangladesh and Nepal

The history of litchi in Bangladesh is unclear, although the species was probably introduced from Myanmar in the 1800s. Direct Chinese and Indian imports soon followed. The tree is cultivated over much of the country, with 4800 ha producing 13,000 t of fruit. Growing technology is poor, the average productivity of bearing orchards being only 1.7 t/ha in the different areas. The main cultivars are 'Bombai', 'Muzaffarpur', 'Bedana' and 'China-3' from India, and the season lasts for 2 months from May to June. The fruit are very popular locally, with none being exported (Abu Baker Siddiqui, 2002).

Although Nepal is a small country, variations in climate allow the production of a range of fruit, including litchi, on the plains and in the low hills at elevations of 60–950 m. Plantings are found across the country but there are more in the central and western regions. Production in 1999 was 13,875 t from 214,800 bearing trees on 1791 ha (total area of 2830 ha). Production is expected to increase steadily in the next few years as the 1039 ha of young trees begin to bear. Harvest extends from mid-May to late June and 60% of the fruit that are consumed locally are imported from India (Budathoki, 2002).

Litchi in the Philippines and Indonesia

Litchi is grown on only a small scale in the Philippines and Indonesia. In the Philippines, commercial plantings are based in Cordillera in northern Luzon, with 'Mauritius' ('Tai So', 'Dazao') and 'Sinco' the major cultivars harvested in May (Sotto, 2002). Since it is a very minor crop, there are no production data available. The Philippines is a net importer of litchi, with imports increasing by 8% each year. The current trade is 1500 t, worth US$500,000. In Indonesia, litchi is a minor fruit, with commercial activity limited to a few districts situated at 400–700 m in Bali (latitude 8°S). There are few commercial plantings and nearly all of the trees are grown in home gardens and along the roadside. Most fruit are consumed locally (Winarno, 2002).

Litchi in southern Africa

Litchi was introduced to Madagascar in 1802 and the crop is now extensively cultivated along 800 km of the eastern coast where the average rainfall is 2500–4000 mm. There are about 3000 ha of trees producing 30,000 t of fruit (Milne, 1999; Jahiel and Abraham, 2001). Few new plantations have been made and most orchards are at least 50 years old. Much of the industry is based on even older individual trees, some 100 years of age, which were planted near villages or in family fields rather than in well-structured orchards (Jahiel and Abraham, 2001). 'Mauritius' ('Dazao', 'Tai So') (Plates 7 and 8), the main variety grown, is harvested from November to December. The industry involves 20,000–30,000 producers and 5000 collectors. Exports have increased dramatically to the current 10,000 t, an increase of one-third since the late 1980s, with most fruit being sent to Europe (Milne, 1999; Jahiel and Abraham, 2001).

About 200,000 litchi trees are grown in South Africa. Nearly 50% of the plantings are in Onderberg, 31% in Tzaneen, 10% in Nelspruit and 6% in Natal and Levubu (Milne, 1999). 'Mauritius' accounts for more than 75% of production and 84% of exports, followed by 'McLean's Red' (16% of production and 13% of exports) (Milne, 1999; Ghosh, 2001). In 1992, The Institute for Tropical and Subtropical Crops at Nelspruit initiated a cultivar evaluation programme to determine the commercial suitability of local and imported cultivars. A breeding programme utilizing seedlings produced from self-, cross- and open-pollination commenced the following year. As a result of the programme, 'Huaizhi' ('Wai Chee') and 'Feizixiao' ('Fay Zee Siu') were recommended and released in 1999. There is a clear trend in the changing planting density from 70–150 trees/ha (10–12 m × 10–12 m) in the oldest plantings, to 200–280 trees/ha (6–7 m × 6–7 m) in recent plantings. Stassen (1999) recommended that high-density plantings of 300–600 trees/ha (8 m × 4 m) should be initiated, with trees being pruned and managed correctly. About 14% of growers have plantings of fewer than 100 trees, 31% have 101–500 trees, 13% have 501–1000 trees, 23% have 1001–3000 trees and 19% have more than 3000 trees (Milne, 1999). Production in 2000 was about 8000 t, with the fruit available from December to January. The fruit is handled by professional growers and about half of the crop is exported to France, the UK, Germany and the Netherlands.

The Manica Province of Mozambique has a climate suitable for litchi growing and has benefited from the building of the huge Chicamba-Real Dam. However, the country currently has only 60 ha of established orchards, based on 'Mauritius'. The industry has been constrained by poor access to capital, poor infrastructure and a poor economy after the long civil war. Mozambique currently exports litchis only to Zimbabwe and South Africa (Milne, 1999). Litchi production in Zimbabwe is conducted by a handful of growers in the Eastern Highlands near Mozambique. Total production is about 80 t, of which 20 t is exported by air to Europe. Again, 'Mauritius' is the main cultivar (Milne, 1999).

In Mauritius, there are 24,000 trees (mainly 'Mauritius') grown in backyards, and 8000 trees in commercial orchards covering an area of 55 ha and producing 1000 t. The fruit is very popular and only 10% is available for export (N. Ramburn, Mauritius, 2002, personal communication). Like Mauritius, Réunion, an island 800 km east of Madagascar, has traditionally grown litchi for export by air to France (Milne, 1999). The French research group CIRAD/IRFA is currently conducting experiments on tree spacing, cultivars, fertilization, irrigation, etc. There are about 1000 ha of trees producing 12,000 t of fruit (Ghosh, 2001).

Litchi in Australia, North America and Brazil

Litchi seeds were sent to Australia some 100 years ago (see Fig. 1.8), but commercial production expanded only in the late 1970s. The early industry was confined to the northern parts of Queensland, but has expanded to include much of the eastern subtropical coastline down to northern New South Wales (latitude 30°S). About 50% of production is in north Queensland, 40% in central and southern Queensland, and 10% in northern New South Wales (Menzel, 2002). Productivity is generally

Fig. 1.8. One-hundred-year-old seedling litchi tree at Bundaberg in southern Queensland, Australia (photograph courtesy of Christopher Menzel).

more reliable in central and southern districts. There are 300 growers under the well-organized Australian Lychee Growers' Association, producing 5000 t of fruit from 1500 ha. Important cultivars include 'Kwai May Pink' (Plates 6 and 25), followed by 'Feizixiao' ('Fay Zee Siu'), 'Shuidong' ('Souey Tung'), 'Huaizhi' ('Wai Chee') (Plate 28) and the locally selected 'Salathiel'. Fruit are available from October to March. About 35% of production is exported to South-east Asia, the Pacific and Europe. Improvements in fruit quality, grade standards, and quality assurance, and the formation of cooperative marketing groups have fostered successful exports. When properly grown and marketed, returns on a per hectare basis more than match those of other tropical crops such as avocado, mango and macadamia. Well-managed orchards can yield 10 t/ha.

In the USA, production is concentrated in Florida (95%), with minor cultivation in Hawaii and California. Total production was only 1044 t in 1998. Commercial cultivation in Florida commenced in the 1940s, although the crop arrived there in 1880. The Florida Lychee Growers' Association, formed in 1951, coordinated the small industry of 130 ha, based on the southern and central areas of the Florida peninsula. The industry suffered setbacks from disastrous frosts in 1958 and 1962, severe hurricanes, and relatively low productivity in the mainly 'Brewster' ('Chenzi') orchards. New plantings since 1992 have been based on the more productive 'Mauritius', with current cultivation standing at 240 ha (Campbell and Campbell, 2001a; Knight, 2001). However, the season is short, lasting from early June until mid-July.

In São Paulo State, Brazil, the high profitability of litchi has resulted in the expansion of cultivation throughout the State (Yamanishi *et al.*, 2001). There were only 347 ha with 102 growers in 1996. About 200 t was produced in 1998. The industry is based on 'Bengal', which is harvested in December and January.

Litchi in Mexico

Cultivation of litchi started in the early 1900s, when a grower in Sinaloa introduced trees from China. From 1960 to 1990, more stocks were introduced, with cultivation spreading to Veracruz, Nayarit and San Luis Potosi (SLP). Commercial activity was initially only successful in Sinaloa, because the fruit was relatively unknown to consumers and most new growers were not familiar with the crop and its agronomy. In 1976, the area under production was 182 ha, with only 2 ha outside Sinaloa. In 1985, a hurricane reduced the plantings to

143 ha. In the late 1980s, there was a downturn in agriculture in Mexico, with major crops, such as coffee, becoming less profitable. Growers were forced to look for alternative commodities, including tropical fruit such as litchi, which rapidly increased in popularity. In 1996, total plantings were 1025 ha, with commercial activity spread across ten states: Sinaloa (295 ha), Veracruz (203 ha), Campeche (180 ha), SLP (113 ha), Oaxaca (70 ha), Nayarit (66 ha), Pueblo (49 ha), Hidalgo (30 ha), Coahuila (15 ha) and Baja California Sur (4 ha). The fruit are marketed mainly in the areas where they are grown, chiefly to the Chinese community, with very little consigned to the large urban centres. Mexico is currently developing an export programme, with small volumes exported from Sinaloa and Nayarit to the USA, and fruit from Veracruz air-freighted to Japan and Europe.

Litchi in Israel

Litchi was introduced to Israel in the 1930s from South Africa ('Mauritius'), California ('Floridian') and India ('Bengali'), although a commercial industry was established only in the late 1970s (Goren *et al.*, 2001). In the past 25 years, 60 cultivars from around the world have been imported into Israel. The industry is based mainly on 'Mauritius' (80%) and 'Floridian' (20%), grown in orchards from 200 m below sea level up to 300 m above sea level in the north of the country. Total cultivation is 300 ha. Orchards are very productive, with average yields above 10 t/ha associated with close plantings, fertigation and controlled irrigation to induce flowering and strategic pruning. Growth regulators are also used to increase fruit retention and average fruit size (Gazit, 2001). The litchis mature from July to August, when the fruit is almost out of season in South-east Asia (Goren *et al.*, 2001), and therefore attract a premium price in Europe.

Longan in China

China has the largest longan industry in terms of cultivated area and production, with the crop having undergone a rapid expansion in parallel with litchi since 1980. In 2000, mainland China produced 608,500 t from 465,600 ha. Major producing regions include Guangdong (157,500 ha; 346,000 t), Guangxi (202,400 ha; 150,900 t) and Fujian (96,000 ha; 110,400 t). There are smaller plantings in Sichuan (2364 ha; 1200 t), Yunnan (3000 ha; 450 t) and Hainan (9400 ha; 890 t). As for litchi, average productivity is low, with great potential for improvement (Plate 5). Longan orchards in Taiwan are found mainly in the central and southern parts of the island, with a relatively stable 12,000 ha in recent years producing 53,000–130,000 t, depending on seasonal weather and growing conditions (Wong, 2000). The main cultivars in Taiwan are 'Fenke' ('Fenko'), 'Hongke' ('Hongko') and 'Qingke' ('Qingko'), with fruit available from mid-August to early September. Yields range from less than 5 t/ha in 'off' years to more than 10 t/ha in 'on' years; much higher than the 2 t/ha average yield on mainland China.

There are over 400 varieties or individual types with unique features (Liu and Ma, 2001), although commercial production is based only on a dozen cultivars (see Chapter 5, this volume), which provide fruit from early August to late September (Liu and Ma, 2001). The use of chemicals to induce off-season production in Thailand has persuaded growers in China to try some of these products; however, only limited success has been achieved. Mainland China has the largest consumption of longan in the world and imports fruit from Thailand during the off-season. In contrast, Taiwan is a net exporter, with fruit mainly in dried form being sent to Hong Kong, Singapore and the USA. In 1997, the export of dried longan totalled 1368 t, worth US$2.8 million (Wong, 2000).

Longan in Viet Nam

Viet Nam has a large longan industry, with 70–80% of plantings located in the Mekong Delta (Wong, 2000). Production in 1999 was 365,000 t of fruit from 41,600 ha. The most important cultivars are 'Longnhan', which can crop twice a year, 'Tieuhue', which can yield three crops in 2 years, and 'Longhungyen', which is more subtropical and produces one crop a year (Wong, 2000).

Longan in Thailand

Longan is one of the most important fruit crops in Thailand, with 358,420 t produced from 52,971 ha in 2000. This ranks Thailand equivalent to Viet Nam, but behind China (600,000 t). Production is confined mainly to the northern areas where the winters are cool. The main plantings are in Chiang Mai (34.8%), Lamphun (34.4%), Chiang Rai (8.5%), Lampang (3.3%), Payao (3.3%), Nan (3.2%) and Chantaburi (3.2%). The area under longan has increased from 17,757 ha in 1989 to 52,971 ha in 2000, reflecting the strong demand created by exports. Production increased dramatically from 44,661 t to 358,420 t over this period, but there were wide fluctuations from year to year, reflecting seasonal weather conditions and tree management strategies. Average productivity is greater in Chiang Mai and Lamphun than in the other areas (Table 1.6). 'E-Daw' ('Daw') is the most popular cultivar and accounts for 73% of plantings.

Longan is one of the top export earners for Thailand with a value of US$202 million in 2000 (Table 1.7). About 50% of production is exported as fresh (98,950 t), dried (55,900 t), frozen (977 t) or canned fruit (11,715 t), mainly to Hong Kong, Singapore, Malaysia and the Philippines (Subhadrabandhu and Yapwattanaphun, 2001a). Exports of dried fruit are increasing each year and exceed the volume of canned product. Frozen fruit comprises only a small section of the market.

The discovery that potassium chlorate can induce off-season flowering in longan has extended the production season to nearly all year round in Thailand (Subhadrabandhu and Yapwattanaphun, 2001b). The treatment has been used successfully in subtropical cultivars such as 'E-Daw' and 'Chompoo' ('Sri Chompoo'), which normally require cool winters for flower induction. Tropical cultivars such as 'Petchsakon Tawai' from Viet Nam, which are commonly grown in the central region of Thailand, flower naturally and need no chemical treatment, although fruit quality is inferior compared with the traditional cultivars. The success of the chlorate treatment has induced a 'longan mania' among some fruit growers in Thailand, as good prices are paid for the fruit (Sethpakdee, 2002). Exports are expected to continue to increase over the next few years.

Longan in Australia and America

Although longan trees were transported to Australia by Chinese immigrants in the mid-1800s, commercial orchards are mostly less than 20 years old. Plantings comprise about 72,000 trees, with most orchards established in the past 10 years (Fig. 1.9). Production is 300–1000 t, less than 20% of the Australian litchi crop (Nicholls, 2001). Cultivation is scattered along the eastern coast of Queensland and northern New South Wales, but is more concentrated near Mareeba in northern Queensland, at elevations of 400 m (Menzel and McConchie, 1998). Most of the older plantings were based on 'Kohala' introduced from Florida; however, new plantings usually

Table 1.6. Longan production in Thailand (The Department of Agriculture Extension, Bangkok, Thailand, 1999).

Province	Area (ha)	Production (t)	Yield (t/ha)
Chiang Mai	16,467	56,197	3.4
Lamphun	16,280	49,351	3.0
Chiang Rai	4,021	11,286	2.8
Lampang	1,567	4,084	2.6
Payao	1,551	4,266	2.7
Nan	1,522	3,721	2.4
Chantaburi	1,529	3,853	2.5

Table 1.7. Production and export of longan in Thailand. Value of exports in US$ million, in parentheses (Trade Statistics Center, Development of Business Economics, Bangkok, Thailand).

	Area (ha)	Production (t)	Export (t)
1996	28,303	236,428	104,266 (118)
1997	41,434	227,979	135,923 (201)
1998	41,504	238,000	–
1999	47,222	149,553	60,359 (84)
2000	52,971	358,420	167,546 (202)

Fig. 1.9. Ten-year-old longan in Brisbane in southern Queensland, Australia (photograph courtesy of Christopher Menzel).

consist of 'Chompoo', 'Haew' and 'Biew Kiew' from Thailand.

The production season extends from January in northern Queensland to April in northern New South Wales. The majority of the fruit are sold on the domestic market, and only a small amount is exported. The main domestic consumers are of Asian descent, comprising a relatively small percentage of the population. Consequently, current production has almost saturated the local market, with prices falling in years that produce heavy crops. Long-term profitability is dependent upon the market expanding to include non-traditional domestic consumers. Potential imports from South-east Asia could also weaken prices in the domestic market, with fewer prospects of exports into China, Viet Nam or Thailand.

Longan was introduced to Florida in 1903, but commercial cultivation became successful only after the introduction of 'Kohala' from Hawaii in 1954 (Campbell and Campbell, 2001b). The past 15 years has seen significant expansion of production due to the demand for the fruit from the increasing numbers of ethnic Asian people in the USA. Cultivar improvement and evaluation is being conducted in Florida through importation of cultivars, mostly from Asia. It is estimated that 150 ha are currently being cultivated in Florida, with fruit sold mostly in the fresh form in local ethnic markets (Campbell and Campbell, 2001b).

Significance of processing

Litchis and longans can be dried, canned, juiced or frozen (Subhadrabandhu and Yapwattanaphun, 2001a) (Fig. 1.10). The prospect of increasing the production of litchi and longan, and their relatively short postharvest life as fresh commodities, raises expectations for increased processing opportunities for these crops. A significant proportion of production from both the southern and northern hemispheres is exported to Europe and North America, where they have become popular exotic fruit. However, the harvest season is unduly short, and, in the absence of an effective postharvest technology, there is little fruit available from April to May, and from September to October (Ghosh, 2001). This situation would normally encourage efforts to develop new processing technologies; however, since Tindall's (1994) review of the Sapindaceae, few publications on processing these crops have appeared in the public domain.

Drying

Drying litchis to produce 'litchi nuts' is the oldest processing method known (Fig. 1.11), and was developed in China before other technologies for preserving the fruit became available (Chen

Fig. 1.10. Various products produced from processed litchis in southern China.

Fig. 1.11. Dried litchi in Guangxi, southern China (photograph courtesy of Christopher Menzel).

and Huang, 2001). Dried fruit are mainly consumed in South-east Asia, and are not well known in Europe and North America. However, consumption could be increased, since the dried fruit tastes quite different from fresh fruit (Chang *et al.*, 1998).

The fruits were traditionally dried in the sun, but the resulting 'nuts' varied greatly in quality. Galán Saúco and Menini (1989) have summarized the steps required to maintain the fruit in the best condition, including washing in boiling water, fumigation with sulphur dioxide, dipping in hydrochloric acid and, finally, washing in tap water. Alternatively, the fruit can be partially dried for 20 days while they are still attached to the panicles, and then fully dried for 10–20 days, when they turn yellow-brown (Zee *et al.*, 1998).

Oven-drying under controlled-temperature conditions following bleaching at 95°C takes approximately 100 h (Lin, 1992). In order to obtain consistent results, Zhao *et al.* (1996) studied the drying characteristics of the skin, flesh and seed under different temperatures, relative humidities and pressures. They found that the fruit membrane prevents water from moving from the flesh or seed to the skin. Although decompression assisted the removal of water from the exocarp, it was less effective on the endocarp, which took a long time to heat up. In similar studies in longan, Achariyaviriya *et al.* (2000) examined the effects of airflow on drying, and calculated the specific energy consumption for different drying conditions. The optimum specific energy consumption of 3.3 MJ/kg H_2O occurred with recycling 95% of the air, a flow rate of 28 kg dry air/h for each kilogram of dry longans, a temperature of 75°C, and a drying time of 33 h.

Canning

Canning comprises a small component of the litchi and longan industries in China, Taiwan, Viet Nam and Thailand (Figs 1.12 and 1.13). In Europe and North America, canned litchis are more popular than the dried fruit; however, there have been few studies on canning since Chakraborty *et al.* (1980) categorized different cultivars with respect to their suitability for canning. Zee *et al.* (1998) summarized the optimum procedure, and suggested the addition of sulphur dioxide to the syrup to prevent a pink tinge developing, which discoloured the final product.

Discoloration was originally thought to be due to the formation of a tin–anthocyanin complex (Garg *et al.*, 1974). However, Wu and Fang (1993) showed it to be primarily a problem in over-ripe fruit, with ripening associated with an increase in the concentration of a

Fig. 1.13. Canned litchis in Chiang Mai in northern Thailand (photograph courtesy of Christopher Menzel).

Fig. 1.12. Canning litchis in Guangdong, southern China.

flavanone precursor, and the activity of two enzymes (flavanone-3-hydroxylase and dihydroquercetin-4-reductase) involved in the biosynthesis of leucoanthocyanins, leading to higher concentrations of leucocyanidin. The conversion of leucocyanidin to a red cyanidin-containing compound occurs when the fruit is heated during processing. Longans retain their flavour better than litchis because their higher sugar content enables them to be canned in their own juice (Zee *et al.*, 1998).

Juice and nectar

Industrial uses of litchi in China include the preparation of juices and nectars (Li, 1999). Problems such as sedimentation, turbidity and losses of flavour and colour, which are attributed to pectins, cellulose, starch, proteins and tannins, are encountered in the manufacture of commercial fruit juices. Enzymes have been used to clarify juice extracted from apple and citrus fruit (Dietrich, 1998), and this approach has led to a trial of the technology in litchi. Pectinase, cellulase, α-amylase and papain at specific pHs, temperatures and reaction times have been shown to improve the quality of litchi juice (Chen *et al.*, 2001). Although a clear, stable, nutritional juice with an acceptable taste was produced, additional work is required in order to optimize quality. Heat-inactivated enzymes also produce a juice that retains acceptable colour and quality for 6 months when stored at room temperature, and for 12 months when stored at lower temperatures (Alex *et al.*, 2003). This method involves the addition of citric acid, potassium metabisulphite and ascorbic acid, after heating.

Frozen fruit

Frozen products do not constitute a significant proportion of the litchi and longan trade, with only 2% of longans frozen in Thailand in 1997 (Subhadrabandhu and Yapwattanaphun, 2001a). Fruit can be stored from one season to the next, but need to be consumed immediately they are thawed and turn brown. Recent studies indicate that citric acid, ascorbic acid and sugar help to retain the red colour and delay browning (Uthaibutra *et al.*, 2003).

World trade

There is little recent published information on the litchi and longan trade. An overview of litchi marketing was published by Ghosh (2001) and an earlier report was published on the Internet (*World Market for Litchi, RAP Market Information Bulletin No. 6*). Wong (2000) provided limited information on the longan trade within Asia and the Pacific.

Apart from small exports of litchi to Europe, the Middle East and North America, most of the fruit grown in South-east Asia and the Pacific is sold within or close to the areas in which it is produced. Trade has increased steadily in all major markets in the past decade. While demand for litchi and longan is strongest in Asian communities, litchi has successfully crossed over to the more mainstream markets in Europe and America, whereas longan consumption is essentially confined to people of Asian descent (Wong, 2000). The fruit is particularly popular with people from China and Thailand.

China has the largest production and market for both crops. Currently, about half of the litchi crop is sold in the southern provinces, about the same proportion is supplied to the large cities in the north, and less than 2% is exported. Mainland China exports 10,000–20,000 t to Hong Kong, Japan and Singapore (Tables 1.4 and 1.5). Inefficient post-harvest handling, as well as the huge domestic demand, limits exports. Recent advances in prolonging the shelf-life of the fruit have allowed some exports to Europe, the Middle East and North America. In 1999, mainland China earned US$6.7 million and US$12.5 million, respectively, from exports of 12,762 t of fresh litchis and 14,995 t in cans (Table 1.4). Most of the dried litchis are sold locally, with some exported to other countries in South-east Asia. Most frozen and canned fruit is exported to the USA, Japan, Korea and Australia.

During the early part of the season and during the off-season winter months, China imports fresh fruit (6752 t in 1999) from Thailand, Viet Nam and Australia. Taiwan also exports litchis to the Philippines (2000 t), Japan (1500 t), Singapore (500 t), the USA (1200 t) and Canada (1000 t). The demand for longan is so strong in China that the country is a net importer. Highlighting this, 7000 t of fresh longan fruit and 15,500 t of dried fruit were imported into China from Thailand in 1997. Fresh fruit imports have increased dramatically in recent years due to the success of off-season production in Thailand. Taiwan exports 500–3000 t of dried longans each year.

India, the second largest producer of litchi, has only recently ventured into the export trade. The development of marketing cooperatives, together with improvements in postharvest technology in the past few years, have fostered the development of limited exports to the Middle East (Ghosh, 2001).

Thailand has significant export industries, with 10% of its litchis and 50% of longans exported (Table 1.7). The fruit are sent by road to Malaysia and Singapore, and by air to Hong Kong and Europe. The major part of the Hong Kong market is for fresh litchis (9000 t), while Malaysia and the USA import canned fruit (6000 t). About 26,700 t of litchis worth US$38 million were exported in 1997. In contrast, the longan export trade of 81,600 t fresh, 38,000 t dried and 16,000 t canned or frozen is worth five times this amount (Subhadrabandhu and Yapwattanaphun, 2001a). The major destinations are Hong Kong, China, Singapore, Indonesia, Malaysia, Canada, the UK and France. Hong Kong is the biggest market, absorbing over 70% of the fresh fruit and nearly 40% of the dried fruit. Thailand has an advantage in Asia because it produces fruit earlier than China or India does. Nearly all of the export longans and some of the litchis are treated with sulphur.

The Australian litchi industry is relatively small but has a strong export focus. About one-third of the crop is exported by air without sulphur treatment to Hong Kong, Singapore, Europe, the Pacific and several Arab states. Because production is counter-seasonal to the northern hemisphere (Fig. 1.2), Australia has a sound potential for increased exports. The major competition comes from other southern-hemisphere producers such as South Africa and Madagascar. Exports are organized by marketing groups, which were established in the major growing areas in the early 1990s. These groups have a strong commitment to grade standards, postharvest treatment and quality assurance. China has recently applied to export litchi to Australia during the off-season. If the application is successful, Australia will apply to send litchis directly into China. Most of the fruit currently enters China via Hong Kong. In the domestic market, nearly all of the crop is sold fresh, with processing being virtually non-existent. Canned and frozen products are imported into Australia from Taiwan and Thailand.

Litchis are popular in Europe, and the fruit is imported from Madagascar, South Africa, Réunion, Mauritius (Fig. 1.14) and Australia during the winter, and from Thailand, China and Israel during the summer. The bulk of the crop from South Africa is treated with sulphur and shipped in refrigerated containers. Trade volume totalled 14,112 t in 1993. France is the largest market (8000 t per year) for fresh fruit in

Fig. 1.14. Two-kilogram export pack of litchis from Mauritius (photograph courtesy of Christopher Menzel).

Europe, taking more than 75% of direct EU imports from overseas suppliers, followed by the Netherlands, the UK and Germany. Madagascar captures the largest market share (73% in 1993) in Europe, followed by South Africa (22%). Thailand and Israel supplied a combined 3% of European imports, while Mauritius and Réunion provided less than 1%. Other minor suppliers include Taiwan, mainland China, Australia, Zimbabwe, Malaysia, Viet Nam, Indonesia and India.

The Middle East markets include Bahrain, Kuwait, the United Arab Emirates, Oman and Saudi Arabia (Ghosh, 2001). The major suppliers of fresh litchis to this market are Madagascar, South Africa, Mauritius, Thailand and Australia.

In North America, Florida has the only significant industry and a large proportion of the crop is sent outside the State. The USA also imports fresh litchis from fruit fly-free zones in Mexico, Israel, South Africa and China. The USA imported about 2500 t of processed litchi and longan valued at US$4.6 million in 1994, mainly from Thailand (72%), Taiwan (19%) and mainland China (8%). Canada imports fresh litchis from China, Thailand, Malaysia, Madagascar, South Africa, Mexico, Florida and the Dominican Republic.

Little information is available concerning cultivars, packaging or postharvest treatment in the different markets. The general preference is for large, well-coloured fruit with sweet flesh and small seeds. There are barriers to exports into Japan and the USA because of quarantine issues associated with some species of fruit fly. Appropriate disinfestation protocols need to be developed for these markets. There are also concerns about sulphur residues in fumigated fruit, especially in Europe and North America. Sulphur can preserve the red colour of litchis for more than 4 weeks and has been used to treat fruit sent by sea from Madagascar and South Africa (Jahiel and Abraham, 2001). In recent years there have been some concerns about excessive sulphur residues in the fruit, and the treatment is to be phased out. Sulphur-free treatments have been developed and applied in some countries, notably Australia and Israel. The Food and Agricultural Organization of the United Nations has developed CODEX standards for exports of fresh litchi (Ghosh, 2001). These include fruit uniformly red with minimum green areas, total soluble solids (TSS) above 18%, fruit greater than 25 mm in diameter, and sulphur concentrations below 10 mg/kg in the flesh and below 250 mg/kg in the peel.

Conclusions

Litchi and longan are indigenous to southern China and northern Viet Nam and are popular in South-east Asia, but are less well known in Africa, Europe and America. Litchi production is greatest in China, Viet Nam, Thailand and India, with smaller industries throughout South-east Asia, Australia, Africa, the Mediterranean and tropical America. In contrast, longan production is confined mainly to China, Viet Nam and Thailand. Production of the two crops is steadily rising, especially in Asia, where increasing affluence has led to greater demand. Total longan production is about half that of litchi.

In the northern hemisphere, litchis are available from March to August, while in the southern hemisphere they are harvested from October to March. Peak production of longan in South-east Asia is from July to September, although some sections of the industries in Viet Nam and Thailand can supply fruit nearly all year round. Production in most countries is based on one or two main cultivars. Irregular bearing and difficulties in postharvest handling limit expansion of the industries.

Most of the litchis produced in South-east Asia are consumed in the country in which they are produced, with only limited exports to Europe and America. In contrast, Thailand has a significant trade in fresh, dried and canned longans. There is also a growing interest in litchi in Europe, with significant imports of fresh fruit from Madagascar, South Africa and a few other producers.

Acknowledgements

We thank Minas Papademetriou for providing copies of *Lychee Production in the Asia-Pacific Region* and *Longan Production in Asia*, Zeng

Lian for data on litchi and longan production in China, and N. Ramburn for data on litchi production in Mauritius. Rita Schwentesius and Jonathan Crane supplied information on litchi in Mexico.

References

Abu Baker Siddiqui, S.B.M. (2002) Lychee production in Bangladesh. In: Papademetriou, M.K. and Dent, F.J. (eds) *Lychee Production in the Asia-Pacific Region.* Food and Agricultural Organization of the United Nations, Bangkok, Thailand, pp. 28–40.

Achariyaviriya, A., Soponronnarit S. and Tiansuwan J. (2000) Mathematical simulation of longan fruit drying. *Kasetsart Journal, Natural Sciences* 34, 300–307.

Alex, L., Suresh, C.P., Kabir, J. and Dhua, R.S. (2003) Litchi juice processing and preservation. *Advances in Plant Sciences* 16, 215–218.

Angasith, P., Na-Lampang, D. and Apichartpongchai, A. (1999) *Longan: An Important Economic Crop for Industry Development.* Ming Muang Publisher, Chiang Mai, Thailand, 137 pp.

Anonymous (1978) *Guangdong Litchi Register.* Guangdong Science Press, Guangzhou, China, 156 pp.

Boonrat, S. (1984) *Lychee Cultivation.* Horticultural Research Institute, Department of Agriculture, Bangkok, Thailand, 86 pp.

Budathoki, K. (2002) Lychee production in Nepal. In: Papademetriou, M.K. and Dent, F.J. (eds) *Lychee Production in the Asia-Pacific Region.* Food and Agricultural Organization of the United Nations, Bangkok, Thailand, pp. 81–93.

Campbell, R.J. and Campbell, C.W. (2001a) Evaluation and introduction of lychee cultivars in south Florida, USA. *Acta Horticulturae* 558, 119–123.

Campbell, R.J. and Campbell, C.W. (2001b) Longan evaluation and selection in Florida, USA. *Acta Horticulturae* 558, 125–127.

Chakraborty, S., Saha, N.K., Pradash, K. and Bisht, H.C. (1980) Studies on the varietal suitability of litchi (*L. chinensis*) for canning. *Indian Food Packer* 34, 17–22.

Chang, C.H., Yu, T.H., Lin, L.Y. and Chang, C.Y. (1998) Studies on the important volatile flavor compounds of fresh and dried longan fruits. *Journal of the Chinese Agricultural Chemical Society* 36, 521–532.

Chen, G.L., Chen, S., Sun, Y.M., Huang, X.Y., Yang, M. and Xie, Z.H. (2001) Studies on the application of composite enzyme in clarification of litchi juice. *Acta Horticulturae* 558, 393–398.

Chen, H.B. and Huang, H.B. (2001) China litchi industry: development, achievements and problems. *Acta Horticulturae* 558, 31–39.

Deng, X.M., Han, Z.H. and Li, S.H. (1999) *Fruit Tree Biology.* Higher Education Press, Beijing, China, 141 pp.

Dietrich, H. (1998) Enzymes in fruit juice processing. *Fruit Processing* 3, 105–107.

Dong, P.Y. (1983) *Historic Stories of Fruits.* Agriculture Press, Beijing, China, 362 pp.

Galán Saúco, V. and Menini, U.G. (1989) *Lychee Cultivation.* FAO Plant Production and Protection Paper No. 83. Food and Agricultural Organization of the United Nations, Rome, 205 pp.

Garg, R.C., Srivastava, R.K. and Ram, H.B. (1974) Effect of certain food additives on the pink discoloration in canned litchi (*Litchi chinensis* Sonn.). *Progressive Horticulture* 6, 93–103.

Gazit, S. (2001) How the problem of poor litchi productivity was solved in Israel. *Acta Horticulturae* 558, 98.

Ghosh, S.P. (2001) World trade in litchi: past, present and future. *Acta Horticulturae* 558, 23–30.

Ghosh, B., Mitra, S.K. and Sanyal, D. (2001) Litchi cultivars of West Bengal, India. *Acta Horticulturae* 558, 107–113.

Goren, M., Tomer, E. and Gazit, S. (2001) Litchi cultivars in Israel. *Acta Horticulturae* 558, 129–133.

Hai, V.M. and Dung, N.V. (2002) Lychee production in Viet Nam. In: Papademetriou, M.K. and Dent, F.J. (eds) *Lychee Production in the Asia-Pacific Region.* Food and Agricultural Organization of the United Nations, Bangkok, Thailand, pp.114–119.

Huang, X.M. (2002) Lychee production in China. In: Papademetriou, M.K. and Dent, F.J. (eds) *Lychee Production in the Asia-Pacific Region.* Food and Agricultural Organization of the United Nations, Bangkok, Thailand, pp. 41–54.

Jahiel, M. and Abraham, D. (2001) The growth of Malagasy litchi field face to the hardening of European control in the matter of sulphured residue. *Acta Horticulturae* 558, 63–71.

Ke, G.W., Wang, C.C. and Tang, Z.F. (1994) Palynological studies on the origin of longan cultivation. *Acta Horticulturae Sinica* 21, 323–328.

Knight, R.J., Jr (2001) The lychee's history and current status in Florida. *Acta Horticulturae* 558, 41–44.

Leenhouts, P.W. (1978) Systematic notes on the Sapindaceae-Nephelieae. *Blumea* 24, 395–403.

Li, M. (1999) Current state, problems and policy in litchi storage and processing in China. *Tropical Agricultural Science in China* 1, 50–54.

Lin, C.T. (1992) Studies on drying litchi fruit with solar energy. *Journal of Chinese Agricultural Engineering* 38, 68–77.

Liu, X.H. and Ma, C.L. (2001) Production and research of longan in China. *Acta Horticulturae* 558, 73–82.

Menzel, C.M. (1991) *Litchi chinensis* Sonn. In: Verheij, E.W. and Coronel, R.E. (eds) *Plant Resources of South East Asia No.2: Edible Fruit and Nuts.* Pudoc, Wageningen, The Netherlands, pp. 191–195.

Menzel, C.M. (2002) Lychee production in Australia. In: Papademetriou, M.K. and Dent, F.J. (eds) *Lychee Production in the Asia-Pacific Region.* Food and Agricultural Organization of the United Nations, Bangkok, Thailand, pp. 14–27.

Menzel, C.M. and McConchie, C.A. (1998) Lychee and longan. In: Hyde, K.W. (ed.) *The New Rural Industries.* Rural Industries Research and Development Corporation, Canberra, Australia, pp. 288–295.

Menzel, C.M. and Simpson, D.R. (1994) Lychee. In: Schaffer, B. and Andersen, P.C. (eds) *Handbook of Environmental Physiology of Fruit Crops, Vol. II: Sub-tropical and Tropical Crops.* CRC Press, Boca Raton, Florida, pp. 123–145.

Menzel, C.M., Watson, B.J. and Simpson, D.R. (1988) The lychee in Australia. *Queensland Agricultural Journal* 114, 19–27.

Menzel, C.M., Watson, B.J. and Simpson, D.R. (1989) Longan: a place in Queensland's horticulture? *Queensland Agricultural Journal* 115, 251–265.

Milne, D.L. (1999) Lychee production and research in Southern Africa. In: Menzel, C.M, Waite, G. and Bronson, R. (eds) *Proceedings of the Fifth National Lychee Conference.* Australian Lychee Growers' Association, Twin Waters, Queensland, pp. 25–31.

Mitra, S.K. (2002) Overview of lychee production in the Asia-Pacific region. In: Papademetriou, M.K. and Dent, F.J. (eds) *Lychee Production in the Asia-Pacific Region.* Food and Agricultural Organization of the United Nations, Bangkok, Thailand, pp. 5–13.

Nakasone, H.Y. and Paull, R.E. (1998) *Tropical Fruits.* CAB International, Wallingford, UK, 445 pp.

Ni, Y.Y., Lu, L.X., Lin, W.Z. and Hu, Z.Y. (1998) Litchi. In: Liang, Y.G. (ed.) *Fruits in Tropical and South Subtropical China.* China Agriculture Press, Beijing, China, pp. 69–90.

Nicholls, B.G. (2001) A perspective of the Australian longan industry. *Acta Horticulturae* 558, 45–47.

Oosthuizen, J.H. (1991) Lychee cultivation in South Africa. *Yearbook of the Australian Lychee Growers' Association* 1, 51–55.

Qiu, W.L. and Zhang, H.Z. (1996) *Encyclopedia of China Fruits: Longan.* China Forestry Press, Beijing, China, 85 pp.

Sethpakdee, R. (2002) Lychee production in Thailand. In: Papademetriou, M.K. and Dent, F.J. (eds) *Lychee Production in the Asia-Pacific Region.* Food and Agricultural Organization of the United Nations, Bangkok, Thailand, pp. 106–113.

Singh, H.P. and Babita, S. (2002) Lychee production in India. In: Papademetriou, M.K. and Dent, F.J. (eds) *Lychee Production in the Asia-Pacific Region.* Food and Agricultural Organization of the United Nations, Bangkok, Thailand, pp. 55–67.

Sotto, R.C. (2002) Lychee production in the Philippines. In: Papademetriou, M.K. and Dent, F.J. (eds) *Lychee Production in the Asia-Pacific Region.* Food and Agricultural Organization of the United Nations, Bangkok, Thailand, pp. 94–105.

Stassen, P.J.C. (1999) Are higher density litchi orchards practically and economically justified? *Yearbook of the South African Litchi Growers' Association* 10, 8–11.

Subhadrabandhu, S. (1990) *Lychee and Longan Cultivation in Thailand.* Rumthai Publishers, Bangkok, Thailand, 40 pp.

Subhadrabandhu, S. and Yapwattanaphun, C. (2001a) Lychee and longan production in Thailand. *Acta Horticulturae* 558, 49–57.

Subhadrabandhu, S. and Yapwattanaphun, C. (2001b) Regulation of off-season flowering of longan in Thailand. *Acta Horticulturae* 588, 193–198.

Tindall, H.D. (1994) Sapindaceous fruits: botany and horticulture. *Horticultural Reviews* 16, 143–195.

Uthaibutra, J., Saengnil, K., Boonkorn, P. and Kunagoon, B. (2003) Influence of pericarp color retention procedure on physical and chemical changes of fresh and frozen lychees. In: Subhadrabandhu, S. and Sukhvibul, N. (eds) *The Second International Symposium on Lychee, Longan, Rambutan and Other Sapindaceae Plants* (Abstract). Horticulture Research Institute, Department of Agriculture, Bangkok, Thailand.

Winarno, M. (2002) Lychee production in Indonesia. In: Papademetriou, M.K. and Dent, F.J. (eds) *Lychee Production in the Asia-Pacific Region.* Food and Agricultural Organization of the United Nations, Bangkok, Thailand, pp. 68–80.

Wong, K.C. (2000) *Longan Production in Asia.* Food and Agricultural Organization of the United Nations, Bangkok, Thailand, 44 pp.

Wong, K.C. (2001) Variations of local longan in Malaysia. *Acta Horticulturae* 558, 103–106.

Wu, M.C. and Fang, T.T. (1993) Studies on pink discoloration of litchi flesh. *Memoirs of the College of Agriculture National Taiwan University* 33, 275–289.

Wu, S.X. (1998) *Encyclopedia of China Fruits: Litchi.* China Forestry Press, Beijing, China, 221 pp.

Yamanishi, O.K., Machado Filho, J.A. and Kavati, R. (2001) Overview of litchi production in São Paulo State, Brazil. *Acta Horticulturae* 558, 59–62.

Yi, G.J., Wang, X.B. and He, H.Q. (2002) The litchi industry status in China and its export strategies. *Journal of Fruit Science* 19, 188–190.

Zee, F.T.P., Chan, H.T., Jr and Yen, C.R. (1998) Lychee, longan, rambutan and pulasan. In: Shaw, P.E., Chan, H.T., Jr and Nagy, S. (eds) *Tropical and Subtropical Fruits.* Agscience, Auburndale, Florida, pp. 290–335.

Zhao, H.H., Li, C.Y. and Guan, Z.J. (1996) Experimentation and research on litchi drying under atmospheric pressure and decompression. *Transactions of the Chinese Society of Agricultural Engineering* 12, 191–195.

Zhuang, Y.M., Ke, G.W., Zeng, W.X. and Pan, X.W. (1998) Longan. In: Liang, Y.G. (ed.) *Fruits in Tropical and South Subtropical China.* China Agriculture Press, Beijing, China, pp. 97–107.

2 Taxonomy, Botany and Plant Development

Suranant Subhadrabandhu[1] and Raphael A. Stern[2]
[1]*Department of Horticulture, Faculty of Agriculture, Kasetsart University, Chatuchak, Bangkok 10900, Thailand;* [2]*MIGAL, Galilee Technology Center, PO Box 831, Kiryat Shmona 11016, Israel*

Introduction

Litchi or lychee (*Litchi chinensis* Sonn., or the synonyms *Euphoria litchi* Desf., *Euphoria litchi* Juss., *Dimocarpus litchi* Willd., and *Sapindus litchi* Roxb.) and longan (*Dimocarpus longan* Lour., or the synonyms *Nephelium longan* Lam. and *Euphoria longan* Lam.) are the most important members of the Sapindaceae family, which contains more than 1000 species and at least 125 genera, widely distributed throughout the warm subtropics and tropics (Merrill, 1923; Bailey, 1949; Leenhouts, 1971, 1974, 1978, 1986; Chapman, 1984; Watson, 1984; van Welzen *et al.*, 1988; Menzel *et al.*, 1990, 1993; Menzel, 1991; Tindall, 1994).

The vernacular names of litchi are *linchi* in Thailand, *letsias* in the Philippines, *litsi* in Indonesia, *laici* in Malaysia, *litchia* in Portugal, *litchi de Chine* in France and *vai* and *cay vai* in Viet Nam (Tindall, 1994). The name 'longan', 'long yan' or 'lungngan' comes from China and means 'dragon eye', which is an apt description of the fruit after the skin has been removed (Wong, 2000). Other vernacular names include *lam-yai* in Thailand, *leng-keng* in Malaysia and Indonesia, *kyet mouk* in Myanmar, *mien* in Cambodia, *lam-nhai* or *nam-nhai* in Laos and *nhan* in Viet Nam (Choo and Ketsa, 1991).

Taxonomy of the Sapindaceae

The majority of species in the Sapindaceae are trees or shrubs that are native to Asia, although there are a few species in South America, Africa and Australia (Bailey, 1949; Leenhouts, 1971). The family name is derived from the soapberry, *Sapindus saponaria*, whose fruit are used as a soap substitute in the tropics (Nakasone and Paull, 1998). Related fruit from South-east Asia include rambutan, *Nephelium lappaceum*, and pulasan, *Nephelium mutabile*. These species are similar to litchi with red or yellow skin, but with long hairs or spinterns replacing the protuberances. Rambutan and pulasan are strictly tropical, while litchi and longan crop best in the warm subtropics or at elevation in the tropics. Litchi and longan are long-lived, evergreen trees that produce new leaves, flowers and fruit on terminal shoots. The inflorescences produce many hundreds of functionally male and female flowers, which carry from five to 80 attractive fruit at harvest. The fruit contain a single seed surrounded by a juicy sweet aromatic aril or flesh. Cultivars with large fruit, small seeds and a distinctive flavour are sought after in the market-place.

Other species of Sapindaceae in the subfamily Sapindoideae of local significance in tropical areas include tuan, dawa or Fiji longan, *Pometia pinnata*, from South-east Asia and the Pacific (tribe Nephelieae), mamoncillo,

 Litchi and Longan: Botany, Production and Uses (eds C.M. Menzel and G.K. Waite)

Meliococcus bijugatus, from the Caribbean (tribe Cupanieae), akee, *Blighia sapida*, from West Africa (tribe Sapindeae) and guarana, *Paullinia cupana*, from the Amazon basin (tribe Paullinieae) (Yeap, 1987; Menzel *et al.*, 1993). Other minor species worthy of evaluation for their fruit include: *Cubilia bancoi*, kubili, from low to medium elevations in the Philippines; *Diploglottis cunninghamii*, native tamarind, from subtropical Australia; *Talisia olivaeformis*, talisia, from tropical America; *Alectryon macrococcus*, mahoe, from Hawaii; and *Chrysanthus macrobotrys*, ndgulu, from Central Africa (Leenhouts, 1971, 1978).

The Sapindaceae were originally described by Cambessedes in 1828, although the first detailed systematic study was not published until the beginning of the 20th century. Radlkofer (1932) based his classification on a wide range of evidence including the presence or absence of a terminal leaflet, the number of ovules per carpel, the structure of the fruit, presence or absence of an aril, and pollen morphology.

There have been several revisions of the Sapindaceae but the scheme of Radlkofer's is essentially accepted, with only minor modification. According to plant characteristics, pollen morphology and geography, the Sapindaceae are split into two subfamilies – Dodonaeoideae (Austral distribution) and Sapindoideae. The latter can be separated into three main groups centered around Sapindeae (pantropical) or Cupanieae (pantropical) and a third group separating into Thiouinieae and Paullinieae, both predominantly American (Leenhouts, 1971, 1978, 1986).

The cultivated species in the Sapindaceae mainly belong to *Litchi*, *Nephelium*, *Dimocarpus* and *Blighia*, with their horticultural classification based largely on fruit characteristics (Tindall, 1994). Less attention has been paid to leaf and flower structures, although these vary significantly. The taxonomy of litchi and longan has been reviewed by Leenhouts (1971, 1978, 1986) and Choo and Ketsa (1991).

Taxonomy of litchi

Leenhouts (1978) indicated that there are three subspecies of *Litchi chinensis* based on the thickness of the twigs, arrangement of the flowers, number of stamens and fruit characteristics.

Litchi chinensis subspecies *chinensis* is the commercial litchi, which grows wild in southern China, northern Viet Nam and Cambodia (Groff, 1921) (Plates 9 and 10). The tree has slender twigs with the flowers borne in lax cymules. The flowers usually have six stamens. The fruit are smooth or with protuberances up to 2 mm high (Groff, 1921; Singh and Singh, 1954; Menzel and Simpson, 1990; Tindall, 1994).

The two other subspecies are not commercialized (Menzel *et al.*, 1993). *Litchi chinensis* subspecies *philippinensis* is known in the Philippines and Papua New Guinea as alupag, arupag or mamata, where it is rather common, but rarely cultivated. The tree has slender twigs and flowers borne in lax cymules with six to seven stamens. The fruit are long oval-shaped with thorn-like protuberances up to 3 mm high. The aril only partially covers the seed and is not edible. *Litchi chinensis* subspecies *javensis* is known only in cultivation in Malaysia and Indonesia and is called lengkeng or kalengkeng. It is suited to the humid tropics. The tree has thick twigs and flowers that have seven to 11 stamens carried in sessile clusters. The fruit are smooth or with protuberances up to 1 mm high, and have a thin aril.

Taxonomy of longan

The genus *Dimocarpus* from the subfamily Sapindodeae and the tribe Nepheliaea, contains six species of trees and shrubs (Leenhouts, 1971, 1978). *D. longan*, *D. dentatus*, *D. gardneri*, *D. foveolatus* and *D. fumatus* are from Sri Lanka, India and eastern Malaysia, while *D. australianus* comes from northern Australia. *D. longan* has two subspecies and five varieties, distinguished mainly by the structure of the leaflets (Leenhouts, 1971, 1978; Choo and Ketsa, 1991; Menzel *et al.*, 1993; Santisuk and Larsen, 1999). *D. longan* ssp. *longan* var. *longan* grows wild in Malaysia, Borneo, Sumatra and the Celebes, while *D. longan* ssp. *longan* var. *longepetiolulatus* occurs in southern Viet Nam. *D. longan* ssp. *longan* var. *obtusus*, *D. longan* ssp.

Fig. 2.1. Litchi tree characteristics (from Menzel *et al.*, 2002, with permission).

malesianus var. *malesianus*, and *D. longan* ssp. *malesianus* var. *echinatus* occur in northern Borneo and the southern Philippines.

D. longan ssp. *longan* var. *longan*, which originated near subtropical China, Myanmar and India, is the only one substantially grown for its edible fruit (Menzel *et al.*, 1990).

Wild trees have been found growing amongst the rainforests of Hainan Island, south of Guangdong, at a latitude of 22°N (Plate 9). Subhadrabandhu (1990) classified three edible longan types in Thailand. The native longan or *lamyai kradook* was found in northern areas, growing as an erect tree producing small fruit with large seeds. Forest longans were large trees producing small fruit with thin arils and were possibly of interest for breeding new cultivars. In contrast, commercial longan or *lamyai kraloke* produced large fruit with small seeds.

Botany and plant development in litchi

The botany of litchi has been reviewed by Groff (1921), Singh and Singh (1954), Joubert (1985, 1986), Menzel and Simpson (1986, 1990), Menzel *et al.* (1993) and Stern and Gazit (2003). The species is a medium to large evergreen tree, which can grow up to 10–12 m or even 20 m in very old specimens (Zhang, 1997). Its crown is generally round, dense, compact and symmetrical. In some cultivars, the branches are tightly curved or twisted and hang down to the ground (Figs 2.1 and 2.2). Some cultivars are more erect than others. Trees normally have a thick, straight, short trunk and dark brown-grey bark; however, branches often have V-shaped crotches and are easily broken off by strong winds.

The tree produces a number of shoots each year as flushes of growth, either vegetative (summer) or floral (winter). Each vegetative flush comprises several leaves and internodes (Plate 11), while the floral flush is a terminal inflorescence compound dichasium. Successive flushes are separated by quiescent periods, during which the expanded leaves darken and thicken. The interval between the start of successive flushes can be as short as 6 weeks, but is twice as long at low temperatures (Batten and Lahav, 1994). A new flush begins only when the leaves of the previous flush have matured or been removed (Menzel *et al.*, 2000; Olesen *et al.*, 2002). New flushes are bright red or brown, turning dark green when mature. At harvest, bunches of bright red fruit hang from the outside of the tree. Litchi is known for its longevity, with trees older than 1200 years still flowering and

Fig. 2.2. Litchi trunk and root system.

bearing fruit in China (Tieng, 1980; Zhang, 1997).

The leaves are pinnately compound with 4–7 leaflets about 7 cm long (Fig. 2.3). They are glossy dark green on the upper surface and grey-green on the under surface. The leaflets are arranged in opposite or slightly oblique order along the rachis on short petioles of their own. Mature leaflets are usually 5–15 cm long and 2.5–4.0 cm wide, and elliptical to lance-shaped. The bases of the blades are wedge-shaped or rounded.

The inflorescence is determinate and composed of several panicles produced on current-season shoots (Plate 12). The panicles are normally produced terminally in clusters of ten or more, but in some trees a high proportion of subterminals or axillaries may be produced. Panicles are generally in mixed form with the lowest buds producing leaves only, the middle buds producing floral buds in the axils of the leaves, and the topmost buds producing only floral branches and sometimes very small leaves that do not persist (Joubert, 1985). Panicles are 10–40 cm long and produce hundreds of small, white, green or yellow flowers (Fig. 2.4), which produce a distinctive scent when the tree is in full bloom.

The flowers are 3–6 mm wide when fully open and rest on 1.5 mm pedicles (Fig. 2.5). They possess a cup-shaped calyx with 4–5 short, serrated sepals, but have no petals. Each flower has 6–10 stamens (Plates 14 and 15). Flowers

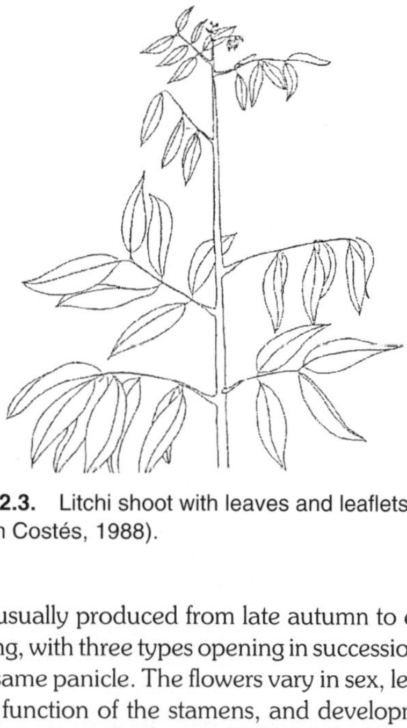

Fig. 2.3. Litchi shoot with leaves and leaflets (from Costés, 1988).

are usually produced from late autumn to early spring, with three types opening in succession on the same panicle. The flowers vary in sex, length and function of the stamens, and development and function of the pistil (Mustard *et al.*, 1953; Mustard, 1960; Stern *et al.*, 1996, 1997; Stern and Gazit, 1998). Type I flowers lack ovules and are functionally male (Male 1 = M_1). They have 6–8 stamens, which produce much pollen. Type II flowers are hermaphrodite but function as female with a well-developed pistil (two carpels) and stigma (two-lobed), and 5–8 stamens, which do not dehisce. Type III flowers are male (Male 2 = M_2), but have a rudimentary pistil lacking

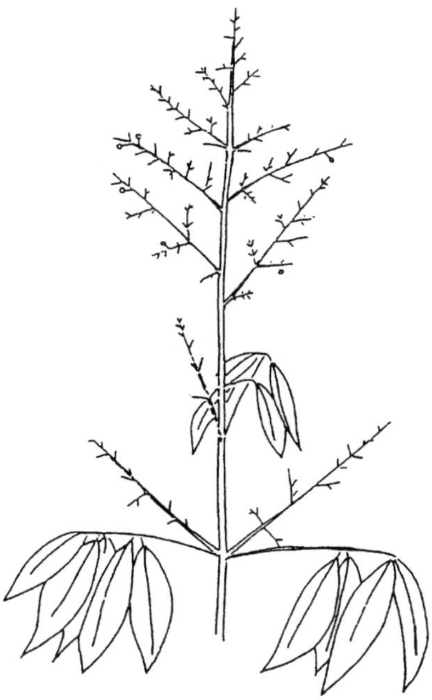

Fig. 2.4. A typical litchi inflorescence (from Costés, 1988).

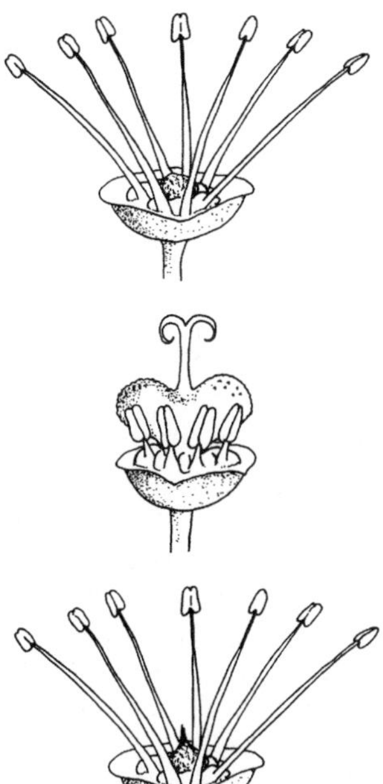

Fig. 2.5. The three litchi flower types in order of appearance (from top to bottom): Male 1, Female, and Male 2 (from Menzel *et al.*, 2002, with permission).

style and stigma. They have 6–8 stamens, which produce plentiful and viable pollen (Stern and Gazit, 1998). The ratio of male to female flowers varies with cultivar and environment (Stern and Gazit, 2003). In extreme cases, there may be no female flowers.

Floral anthesis occurs in overlapping cycles, normally of 10 days for Type I, 7–10 days for Type II and 7–10 days for Type III (Groff, 1921; Mustard *et al.*, 1953; Mustard, 1954; Stern and Gazit, 1996). The length of the cycle varies with cultivar and weather, and is much shorter under warm temperatures (Stern and Gazit, 2003).

Small insects usually pollinate the flowers (Du Toit and Swart, 1995; Singh and Chopra, 1998; Stern and Gazit, 2003), although sometimes fruit-set can occur in their absence (Batten and McConchie, 1992). In such cases the pollen is possibly carried by the wind. The most important pollinators are the honeybees, *Apis dorsata* and *A. mellifera*, along with *A. cerana* and *A. florae* (Kitroo and Abrol, 1996; Kumar *et al.*, 1996; Stern and Gazit, 1996; Abrol, 1999).

Fruit take 80–112 days to mature depending on the cultivar and weather (Groff, 1921). The fruit are drupes and may be round, ovoid or heart-shaped, varying up to 5 cm long and 4 cm wide (Plate 13). The skin or pericarp is thin, tough, hard and frangible. Fruit are green when immature and various shades of red when mature. The skin has sharp protuberances in some cultivars, but is smooth in others (Fig. 2.6).

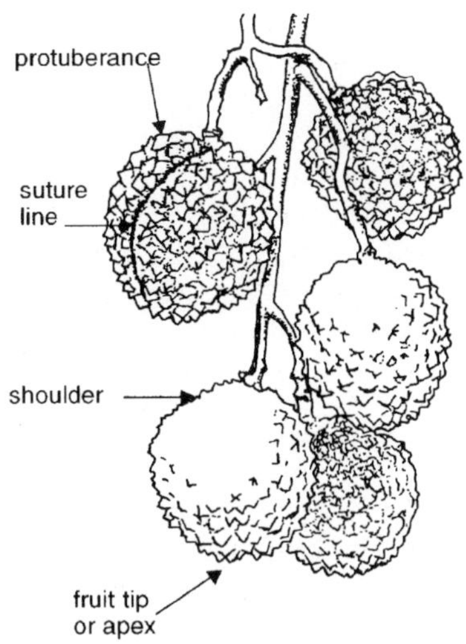

Fig. 2.6. A typical bunch of litchis.

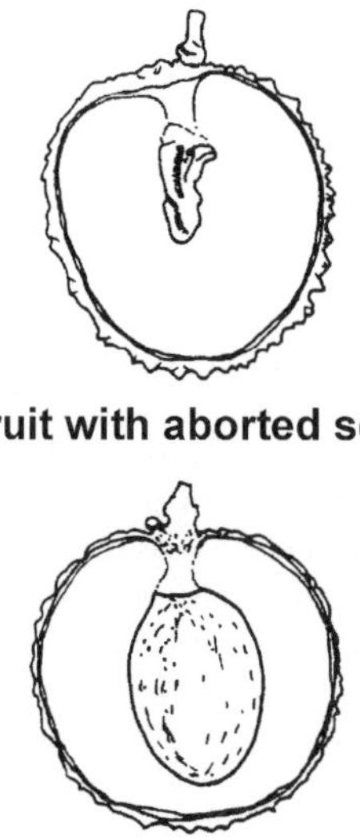

Fig. 2.7. Longitudinal sections of mature litchis showing normal and aborted seeds (from Menzel *et al.*, 2002, with permission).

The skin turns brown and becomes brittle when the fruit are exposed to dry air after harvest.

The edible part of the fruit is an aril formed from cells in the outer seed coat. The aril grows continuously from about 4 weeks after flowering and separates easily from the seed and pericarp at harvest. The aril contains little material from the cell membranes and is white to off-white, translucent, slightly acid, juicy and sweet, and similar in texture to a grape.

The fruits normally contain one chestnut-brown to dark-brown, ovoid to oblong seed, 1.0–3.3 cm long and 0.6–1.2 cm wide. In some cultivars, a high proportion of seeds may be abortive. The abortive seeds are small and shrivelled and are known as 'chicken tongues' (Fig. 2.7). Fruit with abortive seeds are preferred and often attract a high price, for although they are somewhat smaller than fruit with normal seeds, they usually contain a higher proportion of flesh. The proportion of small or shrivelled seeds is an important characteristic of a cultivar and varies from season to season and orchard to orchard. Fruit with chicken-tongue seeds are more susceptible to drought and other stresses and are usually shed before fruit containing normal seeds. Average fruit weight is 16–35 g and flesh recovery 50–70% (Menzel *et al.*, 1993).

Botany and plant development in longan

The botany of longan has been reviewed by Groff (1921), Leenhouts (1971, 1978), Watson (1984), Subhadrabandhu (1990) and Menzel *et al.* (1990), and is similar to that of litchi (Plates 16–19).

Longan is also an evergreen tree that can grow up to 20 m high, and has a spreading or erect habit, depending on the cultivar (Fig. 2.8). The trunk is brittle, with the branches having a corky bark that splits and peels, unlike litchi,

Fig. 2.8. A longan tree in southern Queensland, Australia (photograph courtesy of Christopher Menzel).

which is smooth. The compound leaves are alternate and paripinnate with 6–9 pairs of leaflets, which are dark glossy green on the upper surface and paler green on the lower surface. The young leaves are red-brown, becoming light to dark green when mature. Like litchi, the inflorescences are compound dichasia borne on terminal shoots, 8–60 cm long and with many branches (Fig. 2.9). They are borne on shoots produced over summer or autumn. The flowers are normally in cymules with only the central flower out of five developing into a fruit (Subhadrabandhu, 1990). The flowers are small and yellow-brown, on a lobed calyx 2–5 × 1–3 mm and, unlike litchi, they have five white and woolly to glabrous petals, 1.5–6 × 0.6–2 mm.

Fig. 2.9. Longan inflorescence (1) and fruit cluster (2).

Like litchi, there are three types of flowers. The male flowers (Type I) have about eight hairy stamens arranged in a single row on a light-brown disc (Fig. 2.10). Each stamen has a two-lobed anther. The female flowers (Type II) contain bicarpellate hairy ovaries with a bilobed stigma (Fig. 2.11). Normally only one locule in each female flower develops. The anthers have short filaments and are sterile. The second male flowers (Type III) have eight stamens with sessile filaments and produce viable pollen.

Type I flowers open first, followed by the female flowers and then the second male flowers (Type III). Flowering lasts 4–6 weeks depending on the cultivar and weather. Pollination is carried out by many insects, especially by the honeybee, *Apis mellifera*, which is the most important pollinator in commercial orchards in Thailand (Wongsiri and Chen, 1995; Waite and Hwang, 2002). However, *Trigona iridipennsis* may also contribute to pollination in Thailand and India (Boonithee *et al.*, 1991). Fruit normally take 120–150 days to mature, much longer than litchi. In Thailand, a panicle may carry up to 80 individual fruit, each weighing from 5 to 20 g, with premium fruit normally 14–18 g (1.5–3.5 cm in diameter). The skin is thin, but tough and leathery, and changes from green-yellow to yellow-brown with advancing maturity. The tubercles are typically flattened or

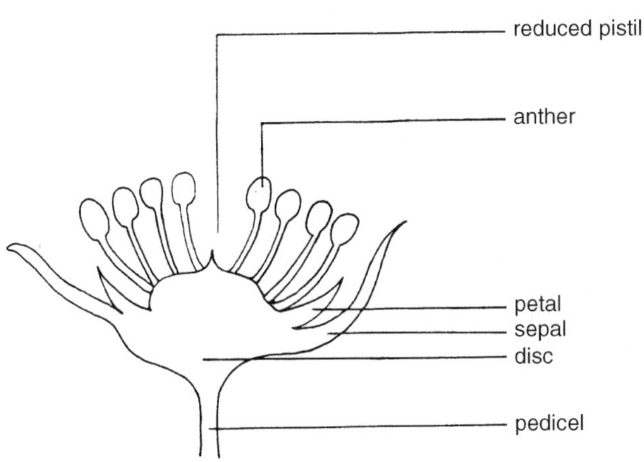

Fig. 2.10. Longan male 1 flower.

indistinct (Fig. 2.12). The aril is translucent white to off-white and contains 15–25% total soluble solids. It constitutes 60–75% of fruit fresh weight. Fruit are juicy to very crisp, bland to sweet and aromatic, but seldom acidic. The seeds are small (2–3 g), round to ovoid, glossy red-brown to black, and easily detach from the aril. A few cultivars in China and Thailand have small seeds weighing less than 1 g.

Conclusions

Litchi and longan have similar botany and share common characteristics such as three different flower types and have arillate fruit. However, they differ in fruit morphology, environmental adaptation, fertility, fruit size, colour and flavour. Litchi fruit are very attractive, with bright red skin covered by angular or conical protuberances. Longan fruit resemble those of litchi but are smaller, smoother and yellow to brown in colour. The fruit are also milder in flavour and less acidic. The nutrient requirements of longan are usually much higher than those of litchi. Bunches of longan have 50–100 fruit while those of litchi have only 5–30 fruit. McConchie *et al.* (1994) found that it was possible to produce litchi and longan hybrids when litchi was used as the female parent. This suggests the potential for incorporating some longan characteristics into the litchi gene pool. Other members of the Sapindaceae might also be exploited. In this regard, isozyme analysis (Degani *et al.*, 1995) or molecular and DNA marker technology (Degani *et al.*, 2003) might assist plant improvement by screening potential parents in future breeding programmes.

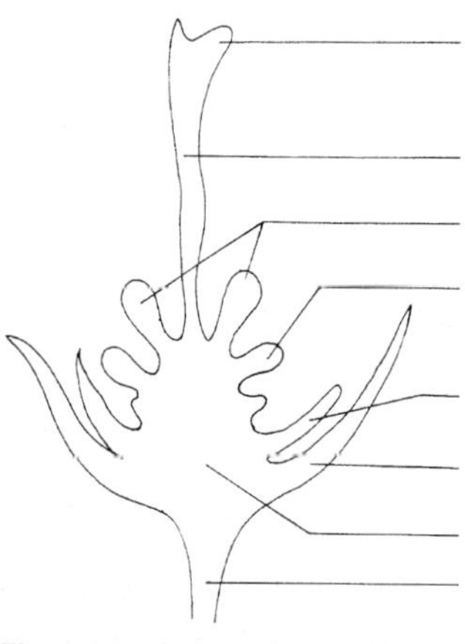

Fig. 2.11. Longan female flower.

Fig. 2.12. A typical bunch of longans (photograph courtesy of Christopher Menzel).

References

Abrol, D.P. (1999) Pollination importance value of bees. *Insect Environment* 5, 140–141.

Bailey, L.H. (1949) *Manual of Cultivated Plants.* Macmillan, New York, 1116 pp.

Batten, D.J. and Lahav, E. (1994) Base temperatures for growth processes of lychee, a recurrently flushing tree, are similar but optima differ. *Australian Journal of Plant Physiology* 21, 589–602.

Batten, D.J. and McConchie, C.A. (1992) Pollination in lychee. In: *Proceedings of the Third National Lychee Seminar*. Australian Lychee Growers' Association, Bundaberg, Australia, pp. 23–28.

Boonithee, A., Juntawong, N., Pechhacker, H. and Huttinger, E. (1991) Floral visits to select crops by four *Apis* species and *Trigona* sp. in Thailand. In: van Heemert, C. and de Ruijter, A. (eds) *Proceedings of the Sixth International Symposium on Pollination.* Tilburg, The Netherlands, pp. 74–80.

Chapman, K.R. (1984) Lychee (*Litchi chinensis* Sonn.). In: Page, P.E. (ed.) *Tropical Tree Fruits for Australia.* Queensland Department of Primary Industries, Brisbane, Australia, pp. 179–191.

Choo, W.C. and Ketsa, S. (1991) *Dimocarpus longan* Lour. In: Verheij, E.W.M. and Coronel, R.E. (eds) *Plant Resources of South-East Asia No. 2: Edible Fruits and Nuts.* Pudoc, Wageningen, The Netherlands, pp. 146–151.

Costés, E. (1988) Analyze architectural et modelisation du litchi (*Litchi chinensis*

Sonn.). PhD dissertation, University of Montpellier, France.

Degani, C., Beiles, A., El-Batsri, R., Goren, M. and Gazit, S. (1995) Identifying lychee cultivars by isozyme analysis. *Journal of the American Society for Horticultural Science* 120, 307–312.

Degani, C., Deng, J., Beiles, A., El-Batsri, R., Goren, M. and Gazit, S. (2003) Identifying lychee (*Litchi chinensis* Sonn.) cultivars and their genetic relationships using intersimple sequence repeat (ISSR) markers. *Journal of the American Society for Horticultural Science* 128, 838–845.

Du Toit, A.P. and Swart, D.J. (1995) Pollination and fruit set in Bengal litchi during the 1993 flowering season at Politsi: second report. *Yearbook of the South African Litchi Growers' Association* 7, 31–32.

Groff, G.W. (1921) *The Lychee and Lungan.* Orange Judd Company, New York, 188 pp.

Joubert, A.J. (1985) *Litchi chinensis.* In: Halevy, A.H. (ed.) *Handbook of Flowering.* CRC Press, Boca Raton, Florida, pp. 204–210.

Joubert, A.J. (1986) Litchi. In: Monselise, S.P. (ed.) *Handbook of Fruit Set and Development.* CRC Press, Boca Raton, Florida, pp. 233–246.

Kitroo, R. and Abrol, D.P. (1996) Studies on pollen carrying capacity and pollination efficiency of honey bees visiting litchi flowers. *Indian Bee Journal* 58, 55–57.

Kumar, R., Chaudhury, O.P. and Lenin, J.K. (1996) Studies on pollination of litchi (*Nephilium litchi* Camb.) in Dehradunn, India. *Indian Bee Journal* 58, 9–12.

Leenhouts, P.W. (1971) A revision of *Dimocarpus* (Sapindaceae). *Blumea* 19, 113–131.

Leenhouts, P.W. (1974) A new species of *Dimocarpus* (Sapindaceae) from Australia. *Blumea* 21, 377–380.

Leenhouts, P.W. (1978) Systematic notes on the Sapindaceae–Nephelieae. *Blumea* 24, 395–403.

Leenhouts, P.W. (1986) A taxonomic revision of *Nephelium* (Sapindaceae). *Blumea* 31, 373–436.

McConchie, C.A., Vithanage, V. and Batten, D.J. (1994) Intergeneric hybridisation between litchi (*Litchi chinensis* Sonn.) and longan (*Dimocarpus longan* Lour.). *Annals of Botany* 74, 111–118.

Menzel, C.M. (1991) *Litchi chinensis* Sonn. In: Verheij, E.W.M. and Coronel, R.E. (eds) *Plant Resources of South-East Asia No. 2: Edible Fruits and Nuts.* Pudoc, Wageningen, The Netherlands, pp. 191–195.

Menzel, C.M. and Simpson, D.R. (1986) Description and performance of major lychee cultivars in subtropical Queensland. *Queensland Agricultural Journal* 112, 125–136.

Menzel, C.M. and Simpson, D.R. (1990) Performance and improvement of lychee cultivars: a review. *Fruit Varieties Journal* 44, 197–215.

Menzel, C.M., Watson, B.J. and Simpson, D.R. (1990) Longan. In: Bose, T.K. and Mitra, S.K. (eds) *Fruits: Tropical and Subtropical.* Naya Prokash, Calcutta, India, pp. 522–546.

Menzel, C.M., Simpson, D.R. and Watson, B.J. (1993) Fruits of tropical climates: fruits of the Sapindaceae. In: Caballero, B., Trugo, L. and Finglas, P (eds) *Encyclopedia of Food Science, Food Technology and Nutrition.* Academic Press, London, pp. 2114–2120.

Menzel, C.M., Olesen, T., McConchie, C., Wiltshire, N., Diczbalis, Y. and Wicks, C. (2000) *Lychee, longan and rambutan: optimizing canopy management.* Rural Industries Research and Development Corporation, Canberra, Australia, 92 pp.

Menzel, C.M., Bagshaw, J., Campbell, T., Greer, N., Noller, J., Olesen, T. and Waite, G.K. (2002) *Lychee Information Kit.* Queensland Department of Primary Industries, Nambour, Australia.

Merrill, E.D. (1923) *An Enumeration of Philippine Flowering Plants.* Bureau of Printing, Manila, The Philippines, 67 pp.

Mustard, M.J. (1954) Fundamentals of panicle differentiation. *Proceedings of the Florida Lychee Growers' Association* 1, 7–8.

Mustard, M.J. (1960) Megagametophytes of the lychee (*Litchi chinensis* Sonn.). *Proceedings of the American Society for Horticultural Science* 75, 292–304.

Mustard, M.J., Liu, S. and Nelson, R.Q. (1953) Observation of floral biology, and fruit setting in lychee varieties. *Proceedings of the Florida State Horticultural Society* 66, 212–220.

Nakasone, H.Y. and Paull, R.E. (1998) Litchi, longan and rambutan. In: *Tropical Fruits.* CAB International, Wallingford, UK, pp. 173–207.

Olesen, T., Menzel, C.M., Wiltshire, N. and McConchie, C.A. (2002) Flowering and shoot elongation of lychee in eastern Australia. *Australian Journal of Agricultural Research* 53, 977–983.

Radlkofer, L. (1932) *Sapindaceae.* Pflanzenreich, Weinheim, Germany, 1539 pp.

Santisuk, T. and Larsen, K. (1999) *Flora of Thailand, Vol. 7, Part I (Apocynaceae, Primulaceae, Sapindaceae).* Diamond Printing Company, Bangkok, Thailand, 205 pp.

Singh, L.B. and Singh, U.P. (1954) *The Litchi.* Lucknow Superintendent of Printing and Stationary, Uttar Pradesh, India, 87 pp.

Singh, R. and Chopra, S.K. (1998) Flower visitors of litchi (*Litchi chinensis* Sonn.) and their role in pollination and fruit production. *Pest Management and Economic Zoology* 6, 1–5.

Stern, R.A. and Gazit, S. (1996) Lychee pollination by the honeybee. *Journal of the American Society for Horticultural Science* 121, 152–157.

Stern, R.A. and Gazit, S. (1998) Pollen viability in lychee. *Journal of the American Society for Horticultural Science* 123, 41–46.

Stern, R.A. and Gazit, S. (2003) The reproductive biology of the lychee. *Horticultural Reviews* 28, 393–453.

Stern, R.A., Eisenstein, D., Voet, H. and Gazit, S. (1996) Anatomical structure of two day old litchi ovules in relation to fruit set, and yield. *Journal of Horticultural Science* 71, 661–671.

Stern, R.A., Eisenstein, D., Voet, H. and Gazit, S. (1997) Female 'Mauritius' litchi flowers are not fully mature at anthesis. *Journal of Horticultural Science* 72, 19–25.

Subhadrabandhu, S. (1990) *Lychee and Longan Cultivation in Thailand.* Rumthai Publishers, Bangkok, Thailand, 40 pp.

Tieng, S. (1980) A 1200 year old tree flourishes in the People's Republic of China. *Fruit Varieties Journal* 34, 23.

Tindall, H.D. (1994) Sapindaceous fruits: botany and horticulture. *Horticultural Reviews* 16, 143–196.

van Welzen, P.C., Lamb, A. and Wong, W.W.W. (1988) Edible Sapindaceae in Sabah. *Nature Malaysiana* 13, 10–25.

Waite, G.K. and Hwang, J.S. (2002) Pests of litchi and longan. In: Pena, J.E., Sharp, J.L. and Wysoki, M. (eds) *Tropical Fruit Pests and Pollinators: Biology, Economic Importance, Natural Enemies and Control.* CAB International, Wallingford, UK, pp. 331–359.

Watson, B.J. (1984) Longan. In: Page, P.E. (ed.) *Tropical Tree Fruits for Australia.* Queensland Department of Primary Industries, Brisbane, Australia, pp. 192–197.

Wong, K.C. (2000) *Longan Production in Asia.* Food and Agricultural Organization of the United Nations, Bangkok, Thailand, 44 pp.

Wongsiri, S. and Chen, P. (1995) The effects of agricultural development on honey bees in Thailand. *Bee World* 76, 3–5.

Yeap, C.K. (1987) Sapindaceous fruits and nuts. *Yearbook of the Western Australia Nut and Tree Crops Association* 12, 16–33.

Zhang, Z.W. (1997) China, the native home of litchi. In: Zhang, Z.W., Yuan, P.Y., Wang, B.Q., Qiu, Y.P. and Li, J.S. (eds) *Litchi: Pictorial Narration of Cultivation.* Pomology Research Institute, Guangdong Academy of Agricultural Science, China, pp. 12–17.

3 Propagation

Sisir K. Mitra[1] and Pravat K. Ray[2]
[1]*Faculty of Horticulture, Bidhan Chandra Krishi Viswavidyalaya, Mohanpur 741235, Nadia, West Bengal, India;* [2]*Horticulture Department, Rajendra Agricultural University, Pusa 848125, Samastipur, India*

Introduction

Litchis are propagated asexually by air-layers (marcots), cuttings or grafts. New plants can also be raised *in vitro* by somatic embryogenesis and shoot-tip culture, although these methods have not been commercialized (Kantharajah *et al.*, 1997; Lai *et al.*, 1997). Seedlings are highly variable, but are the source of new cultivars. Longans can be propagated sexually by seed or vegetatively by cuttings, air-layers or grafts. Only seedlings were used in China before the 10th century, with air-layering developed by the 14th century and grafts by the 16th century (Anonymous, 1978).

Seed

Seedlings are not used for raising new litchi plantations. They may resemble the parent tree, but most have a long juvenile phase, which lasts 10 years or more before they fruit (Pandey and Sharma, 1989; Thakur *et al.*, 2001). Seeds also have a short shelf-life, and are thus not used to maintain or exchange germplasm. The water content of mature seeds is 45% (Kuhn, 1962; Xia *et al.*, 1992, 1993), with germination failing when it falls to 27% (Xia *et al.*, 1992). Chen and Fu (1989) found that no seeds germinated at room temperature after 4–6 days, due to water loss, whereas none germinated outdoors at 27–46°C after 2 days (Ray and Sharma, 1987). Tiny pores in the seed coat result in rapid drying of fresh seed (Lu *et al.*, 1999), whereas dried seeds have a hard coat that is almost impermeable to water.

Fresh litchi seeds germinate within 4–10 days when provided with adequate soil water and aeration (Bolt and Joubert, 1968; Ray and Sharma, 1987; Hartman *et al.*, 1990), and an optimum temperature of 30°C (Xia *et al.*, 1992). Menzel and Paxton (1985) noted that seedlings grown at day/night temperatures of 30/25°C had about four times the dry weight of those grown at 15/10°C (Fig. 3.1). More seeds that were stored in water germinated than those stored in air or vermiculite. Ray and Sharma (1987) reported 75% germination after 7 days in water, and 30% germination after 15 days. Kadman and Slor (1974) also obtained better germination with 2-day-old water-stored seeds than with those stored in air. Damp peat, wet sphagnum or moist sawdust in polythene bags can extend storage life (Cull and Paxton, 1982; Ray and Sharma, 1987). Seeds can be stored for up to 56 days at 8°C (Menzel, 1985). Seeds stored at higher temperatures germinate; however, temperatures of 3°C are detrimental (McClelland, 1944). Fu *et al.* (1990) reported that seeds are viable for up to 60 days when stored at 5°C.

Fig. 3.1. Growth of litchi seedlings at different temperatures (photograph courtesy of Christopher Menzel).

Raising seedlings

For raising plants, seeds should be sown 1–2 cm deep and 8–10 cm apart in a well-drained sandy loam or similar soil. There must be intimate contact between the seed and the soil for water absorption by the embryo and the endosperm. Compost (50:50), vermiculite, peat or sawdust can be added to the medium. Inoculation with mycorrhiza can sometimes improve germination (Kadman and Slor, 1974). However, if the soil is infested with nematodes or other pathogens, it should be sterilized. Immediately following germination, partial shade is desirable, especially during hot weather. Watering should be carried out by a sprinkling can or with a fine spray head on a hose at low pressure. When the seedlings are 10–15 cm tall they are normally transferred to polythene bags, and must be handled carefully, since they die if the roots are damaged. Plants are sometimes grown in bulk trays, dug up and transplanted directly into the field, where they require more frequent watering than those raised in polythene bags.

Seedling rootstocks are usually raised in polythene bags of 0.1 mm gauge and 25 cm × 30 cm, filled with a modified John Innes Compost containing seven parts by volume of loam, three parts peat, and two parts of coarse sand, or similar. After sowing, the plants should be grown in partial shade and watered daily for about 3 months. They are then gradually hardened under full sun before being grafted. The young plants should be fertilized monthly with 10 g of a 10:10:10 NPK mixture and 60 g of dried chicken manure, or equivalent. The seedlings are normally ready for grafting or budding when they are 12–14 months old.

Longans can be propagated by seed; however, most seedlings do not bear for at least 8 years and many have small fruit, poor flesh recovery, poor eating quality, and low market returns. Many of the old orchards in Thailand are based on seedlings, although fruit from these trees receives a low price in the market-place. Seeds stored in intact fruit with 0.05% benomyl gave 85% germination after 30 days (Wong, 1992).

Stem cuttings

Although cuttings are inexpensive and simple to prepare, they can be unreliable. Success depends on good temperature and humidity control for rooting, the type of wood, the stage of growth and the time of year.

Hardwood cuttings from an active flush rooted more readily than those from dormant plants (Paxton *et al.*, 1978; Hartman and Kester, 1986). Litchis also produced roots successfully

Fig. 3.2. Litchi cuttings growing at Nambour in southern Queensland, Australia (photograph courtesy of Brian Paxton).

from softwood (Ochse, 1952; Hayes, 1957; Ahmed, 1961) and semi-hardwood (Galloway, 1922; Albert, 1959; Paxton *et al.*, 1978) (Fig. 3.2). Thus, there are different options on the material to use. In contrast, there seems to be no agreement on the time of year in which the cuttings should be taken. In India, cuttings are best taken in April following the spring flush, or in August and September when it is warm and humid (Bose *et al.*, 1985). In southern Queensland, semi-hardwood and hardwood cuttings collected from 'Bengal' prior to floral initiation in May gave 70% and 80% rooting, respectively (Paxton *et al.*, 1978). Cuttings taken from older wood behind the soft shoot tips failed. In contrast, Ochse (1952) obtained 80% success with 'Brewster' terminal cuttings taken from flushes in April and May in Florida. Dormant wood obtained in December to March callused, but did not root. In Brazil, 72% of cuttings from terminal shoots in July survived (Leonel *et al.*, 1994).

Cuttings are usually 15–20 cm long and 0.8–1.5 cm wide. Thicker cuttings root less readily and have fragile roots. Green woody cuttings rooted better than dark grey ones. Lenka and Das (1981) found that shoots in spring made better cuttings than those in winter. Subterminals were better than terminals in South Africa (Oosthuizen, 1995). The cuttings should have at least three nodes, with one node in the rooting medium. The basal cut should be at an angle, making it easier to cut, easier to place in the medium, and less likely to damage the xylem. Furthermore, making the cut immediately below a node provides a greater concentration of roots. Cuttings are buried 3–4 cm into the medium to reduce drying out and to give support, and take 3–4 months to root. The medium must be kept moist, as the roots die in dry soil. Rooted cuttings should be transplanted into polythene bags filled with good soil and watered regularly. When the roots are firmly established and the top has grown adequately, the plants are set in the field. This usually takes about 15 months after the cuttings have been struck.

Girdling, growth regulators, misting and bottom-heat improve the success rate for cuttings. Girdling the base of the shoot 8 weeks before taking the cuttings increased rooting of 'Brewster' in Hawaii compared with ungirdled controls (Abutiate and Nakasone, 1972). Auxins increase the proportion of cuttings that root, hasten root initiation and increase the number and quality of roots (Bhandary and Shivashankar, 1970; Lenka and Das, 1981; Bose *et al.*, 1985; Mitra and Bose, 1991; Leonel *et al.*, 1994; Oosthuizen, 1995; Brahmachari *et al.*, 1998). Indole butyric acid (IBA) is more effective than naphthalene acetic acid (NAA) and indole acetic acid (IAA). Indole acetic acid is unstable in plants and decomposes in unsterilized solutions, although it remains active in sterile solution for several months.

Indole butyric acid produces a fibrous root system at lower and higher concentrations, whereas the phenoxy compounds 2,4,5-trichlorophenoxyacetic acid (2,4,5-T), 2-(2,4,5-trichlorphenoxy) propionic acid (2,4,5-TP), 2,4,5-trichlorophenoxy-butyric acid (2,4,5-TB) and 4-(2,4-dichlorophenoxy)-butyric acid (2, 4-DB) promote rooting at very low concentrations. The best response was obtained at 100–200 mg/l (soaking for 24 h) or 500–10,000 mg/l (dip method), with higher concentration or longer treatments retarding root and shoot development and reducing rates of survival. Indole butyric acid gave a better response than NAA, whereas mixtures of these hormones were not superior to individual applications (Bhandary and Shivashankar, 1970; Abutiate and Nakasone, 1972).

Mitra and Bose (1991) achieved an 85% success rate with ethephon at 200 mg/l, and IBA at 2000 mg/l. The shoots were treated twice with ethephon at 10-day intervals before the cuttings were taken. Bose *et al.* (1985) observed increased rooting when phenolics such as *p*-hydrobenzoic acid, *p*-coumaric acid and ferulic acid (0.2%) were added to the solutions compared with IBA alone. The best response was obtained with 100–200 mg/l after soaking the cuttings for 24 h, or with 3000–10,000 mg/l by dipping for 5 s. Cuttings were best treated in a bundle rather than singly. The best results (90% success) are usually achieved with a constant or intermittent mist applied over the propagation bed (Bhandary and Shivashankar, 1970; Lenka and Das, 1981; Bose and Mandal, 1972; Paxton *et al.*, 1978; Galán Saúco and Menini, 1989). A temperature of about 30°C provided by electric cables or hot water pipes to the base of the cuttings (Hartman *et al.*, 1981) also assists rooting (Ochse, 1952; Menzel, 1985; Galán Saúco and Menini, 1989).

Air-layering

Air-layering or marcotting has a long history in China and is the most widely used method for propagating litchis. It involves the production of a plant *in situ* from aerial branches. The branch is girdled to remove all the bark near the cambium. This exposed tissue is enclosed in a ball of moist sphagnum or similar, which is wrapped in polythene film to reduce water loss (Fig. 3.3). When roots appear on the stem, the air-layer is cut from the mother plant and planted on its own roots under shade (Fig. 3.4). This method is generally more successful than cuttings, with 80–100% of air-layers surviving (Menzel, 1985; Sharma and Grewal, 1989; Sharma *et al.*, 1990a,b; Kaundal *et al.*, 1993; Cull and Lindsay, 1995). Several factors influence success and establishment in the field, including maturity and thickness of the stem, time of layering, position of the shoot in the crown, wrapping materials, rooting media, application of growth regulators, care in the nursery, and pruning of the shoot after separation from the mother plant.

The best time for air-layering is at the beginning of the monsoon, i.e. June and July in India (Singh *et al.*, 1965; Nijjar, 1981; Sharma and Grewal, 1989; Kanwar and Kahlon, 1986). About 90% of air-layers initiated in mid-July in India survived (Sharma and Grewal, 1989). Best rooting and survival were obtained in spring and early summer compared with autumn and winter in South Africa (Joubert, 1970), Queensland (Campbell, 1981; Cull and Paxton, 1983; Cull and Lindsay, 1995) and New South Wales (Loebel, 1976). In Florida, the recommended time is June, when it is warm and humid (Storey, 1973). Air-layering can be done at any time of the year in Hawaii, although trees are usually air-layered in spring on non-flowering branches, before active growth commences (Yee, 1972). Because air-layers take up to 3 months to develop, those initiated in the hotter and drier parts of the year require added care to prevent the developing roots from drying out.

Healthy, mature (at least 1 year old), upright shoots are preferred for air-layering (Kanwar and Kahlon, 1986). The stem should be 1–2 cm wide at the girdling point, and 40–60 cm long (Joubert, 1970; Campbell and Malo, 1975; Ireta-Ojeda 1975; Kanwar and Kahlon, 1986; Galán Saúco and Menini, 1989; Pandey and Sharma, 1989; Cull and Lindsay, 1995). Roots developed faster on branches with mature vegetative growth compared with wood that has recently flushed (Ireta-Ojeda, 1975; Campbell, 1981; Cull and Lindsay, 1995).

Thin branches gave rise to air-layers with a poor root system, which produced smaller trees

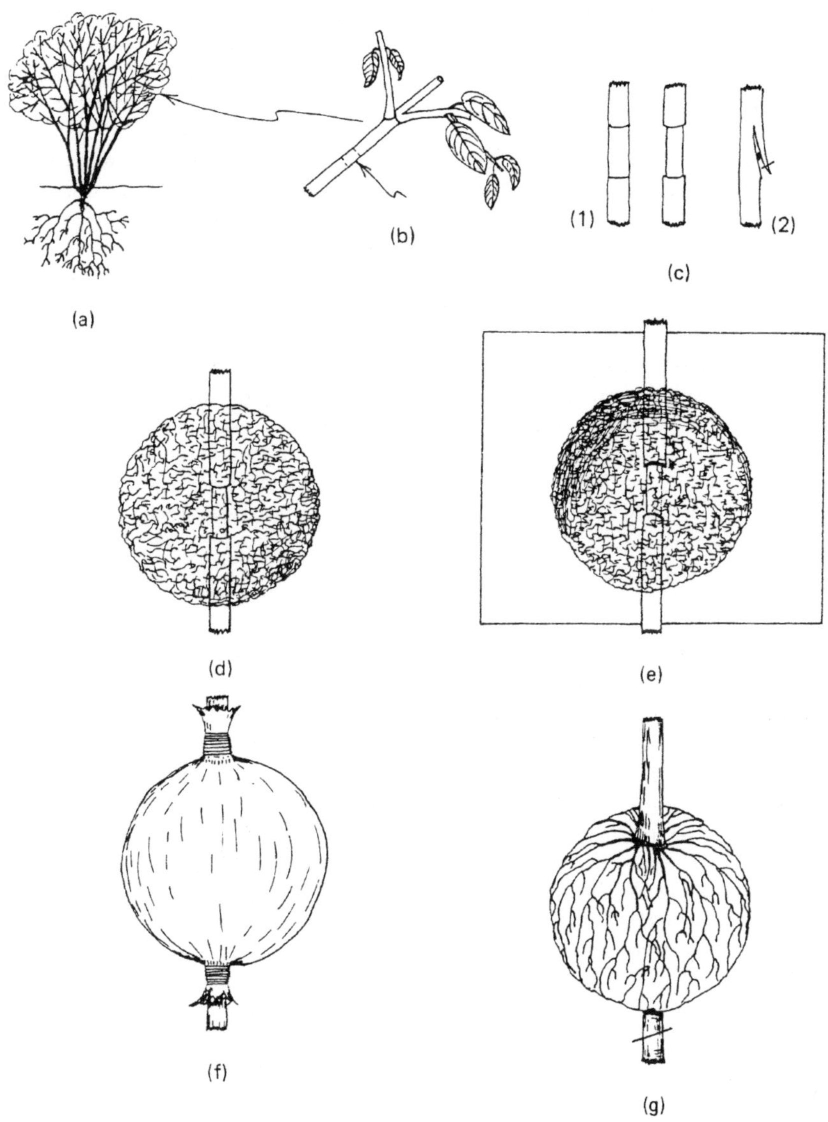

Fig. 3.3. Details of air-layering in litchi (from Soule, 1985, from an original drawing from Ochse *et al.*, 1961, with permission).

that took longer to bear compared with branches with mature growth (Singh and Singh, 1954). Stems more than 2 years old (2 cm diameter) can be used in some cases, but they produce fewer roots compared with younger stems. The larger plants that are produced are also more difficult to handle, and few such air-layers can be taken from an individual tree (Cull and Lindsay, 1995). Shoots in full sun rooted better than those in the shade (Ireta-Ojeda, 1975). Girdling the branch 4 weeks before air-layering promoted rooting compared with ungirdled shoots in Pakistan (Malik and Maqbool, 1977) and Israel (Kadman and Slor, 1974), presumably because starch accumulated above the cut (Jones and Beaumont, 1937; Ray, 2001).

After selecting an appropriate branch, a strip of bark 2–3 cm wide is removed from around the stem by cutting the bark down to the central hardwood with a sharp knife. The soft

Fig. 3.4. Potted litchi air-layer or marcot (from Menzel *et al.*, 2002, with permission).

tissue below is scraped off, ensuring complete removal of the phloem and cambium so there is no callusing. Sometimes the cut is left exposed for a day to ensure that the remaining cambium is killed (Cull and Lindsay, 1995).

Procedures that improve the rooting of cuttings also apply to air-layers. Auxins are beneficial, although the method of application is different (Sadhu *et al.*, 1972; Sharfuddin and Hussain, 1973; Singh and Jawanda, 1981; Sharma *et al.*, 1990b; Dutta *et al.*, 1995). The auxin can be applied to the upper cut, proximal to the apex, as a powder in lanolin, or as a solution in 50% alcohol. Singh *et al.* (1965) showed that, in India, the proportion of 'Early Large Red' air-layers which struck after treatment with auxins at 100–250 mg/l was 89% with IBA, 82% with 2,4,5-T, 72% with NAA, 69% with naphthalene acetamide (NAAm), and 37% in the controls. This improved success was due to superior initiation and subsequent development of roots. The layering period was also reduced from 106 to 60 days with 250 mg IBA/l. The use of peat with IBA resulted in 80% rooting compared with 20% in woodchips (Kadman and Slor, 1974). However, some reports showed that auxins have little impact on rooting (Ireta-Ojeda, 1975; Kadman, 1985).

The preferred medium for air-layers is moistened peat or sphagnum moss (Kadman and Slor, 1974; Campbell, 1981; Menzel, 1985; Sharma and Grewal, 1989; Cull and Lindsay, 1995). These media have virtually replaced other rooting materials such as soil and sawdust. They should be soaked in water for 24 h beforehand, for best results. A quantity of about two handfuls of the selected medium, with the excess water squeezed out, is placed around the stem to enclose the cut surface. This is then wrapped with a piece of polythene film to cover the moss completely. The two ends should be twisted to make sure that no water can seep inside and that the medium does not dry out, and tied off with a piece of string. The use of tightly secured polythene eliminates the need for frequent hand watering.

The air-layer is removed when a good root system has developed and at least six or eight roots have turned from white to creamy brown (Menzel, 1985; Galán Saúco and Menini, 1989; Cull and Lindsay, 1995). Roots normally appear after 6 weeks (Kadman and Slor, 1974; Galán Saúco and Menini, 1989), with the layer ready for removal after 3–4 months (Malik and Maqbool, 1977; Sharma *et al.*, 1990a). Once sufficient roots have formed, the branch is cut below the root ball. Extreme care should be taken during this stage, since the roots are very weak and brittle. The plastic wrapping should be cut from the ball carefully. The new plants are generally planted in 15 × 15 × 18 cm polythene bags. The potting mix should be pushed tightly against the roots, covering them to at least 7–10 cm, and watered. Care should be taken not to allow the pots to dry out, as they may be difficult to re-wet. The air-layers should be kept in shade, watered regularly and protected from strong winds. High humidity or intermittent misting will improve survival (Bose and Mandal, 1972; Menzel, 1985; Galán Saúco and Menini, 1989).

The shoots should be pruned to reduce water loss and to balance the root:shoot ratio. Upper branches are trimmed to form a balanced top and 50–75% of the leaves removed (Hartman *et al.*, 1981; Sharfuddin, 1983; Menzel, 1985; Galán Saúco and Menini, 1989). Sharma *et al.* (1990a) and Kaundal *et al.* (1993) removed 25, 50, 75 or 85% of the leaves at planting and obtained the best survival (75–100% in different cultivars) with 75% defoliation. Kanwar and Kahlon (1986) recommended retaining at least 20 leaflets at transplanting. Bare or defoliated air-layers are difficult to establish (Johnstone, 1956), with the plants best held for two to three flushes before being planted out (Campbell, 1981; Cull and Paxton, 1982; Kadman, 1985; Galán Saúco and Menini, 1989).

Air-layering is very popular amongst Thai and Chinese longan growers, as the plants root easily. Air-layers take 3–5 years to bear, which is much quicker than seedlings (Subhadrabandhu, 1990). However, they are more susceptible to wind damage than grafted trees (Tindall, 1994). Air-layering is usually initiated in the wet season, on healthy branches about 3 cm in diameter with mature leaves. The branch is first girdled or a strip 2.0–2.5 cm wide, extending down to the cambium removed. The exposed surface is scraped to remove the phloem and cambium. Auxins are applied to the upper cut surface of the girdled area (Wong, 2000) and a root ball of wet peat, sphagnum moss or soil enclosed in polythene film is wrapped around the cut. The air-layer is removed from the parent tree after 2–3 months, and planted in a bag for further growth before it is transplanted.

Micro air-layering has been trialled in China (Xu and Zheng, 1999). Twigs that are only 5–14 mm thick are girdled, with the standard rooting medium replaced by wetted cotton wool treated with 400 mg IBA/l. The plantlets are grown in a nursery until they are about 80 cm tall and then transplanted in the field.

Grafting in litchi

Grafting has been practised in litchi for several hundred years. It is most common in China, and other areas, such as Mauritius, that are subject to hurricanes or cyclones (Fig. 3.5). However, the success rate is variable and often low, because of stock–scion incompatibility, poor cambial contact, grafting at the wrong physiological stage, and poor management after grafting (Menzel, 1985).

Venning (1949) suggested that grafting of litchi is difficult because only 30% of the cambium is active at any one time. It is purely by chance that two active areas can be brought together, since the external stem surface offers no clue as to which portion of the cambium is active. Grafting of very young branches or seedlings less than 4 mm in diameter is difficult because the irregularity of the young vascular cylinders makes it almost impossible to secure adequate alignment. He also observed that there was very little persistent unspecialized parenchyma in stems greater than 1 cm in diameter. Pandey and Sharma (1989) mentioned that the success rate with budding and grafting was usually less than 20%; however, in Hawaii, up to 80% of unions were successful when the scion wood (17 mm wide) was ringed (3 mm) 3–4 weeks before grafting. Normally, carbohydrates do not accumulate in litchi wood in Hawaii, whereas girdled branches contained 11% starch (Jones and Beaumont, 1937).

Fig. 3.5. Grafted litchis in Mauritius (photograph courtesy of Christopher Menzel).

Seedlings 1–3 years old are generally used as rootstocks. These may be the same or a different cultivar. However, seedling rootstocks can lead to considerable variability in tree growth and performance. Within litchi, some seed sources produce better rootstocks than others (Ou *et al.*, 1993; Lu and Li, 1998). Abutiate and Nakasone (1972) found that 'Brewster' was better on 'Kwai Mi' than on 'Haak Yip', because the former was more vigorous and had a greater rate of cambial growth. Likewise, 'Wai Chee' was a better rootstock for 'Salathiel' than for 'Tai So' (Menzel, 1985). In China, 'Hexiachuan', 'Luyuexue' and 'Baila' were suitable rootstocks for 'Baitangying' (Lu and Li, 1998). Ye (1987) found up to 99% compatibility with 'Feizixiao' on 'Dazao', whereas 'Nuomici' on 'Dazao' gave only 10% success. 'Huaizhi' and 'Maquechun' were good stocks for 'Nuomici', while 'Heiye' and 'Shanzi' were good stocks for 'Xuehuaizi' (Ou *et al.*, 1992). Ruan (1988) stated that incompatibility usually occurred when early cultivars were combined with late cultivars. Both 'Tianyan' and 'Nuomici' are late cultivars with similar growth rates. 'Nuomici' was compatible with 'Tianyan' and grew and fruited well on 'Huaizhi'.

Little attention has been paid to the use of clonal stocks. Some related forms, such as *Litchi chinensis* ssp. *philippinensis*, *Euphoria cenerea*, commercial longan, *Pseudonephelium fumatum* and *Nephelium mutabile* have been used as rootstocks, but no successful grafts have been reported (Groff, 1921; Pope and Storey, 1933; Eureinoff, 1950; Hayes, 1957; Singh, 1963). In the foothills of the Nilgiris in South India, up to 30% of litchi/longan inarches survived, but they fruited irregularly (Singh, 1963).

Side grafting

In side grafting, the scion is inserted into the side of the stock, which is larger than the scion. Litchis have been side or veneer grafted in Hawaii (Abutiate and Nakasone, 1972), Florida (Cobin, 1948; Nelson, 1954a,b, 1957), Israel (Kadman and Slor, 1974, 1982) and Brazil (Pinheiro *et al.*, 1984). Kadman and Slor (1974) obtained 80–90% success in some years and complete failure in others after grafting semi-hardwood and hardwood scions, 3–5 cm long with three or four buds on to 2-year-old seedlings. There was no difference between grafting in spring or autumn. Seedlings 1 or 2 years old, 1–2 cm in diameter are recommended. The scion should possess three or four buds and be 10–15 cm long and 0.8–1.0 cm wide. Green-brown wood from upright terminals is preferred.

An oblique cut 2.0–2.5 cm long is made in the stock branch with a knife at 20–30° and a matching cut made at the bottom of the scion. The cuts on both sides of the scion should be smooth, each made by one single cut with a sharp knife. The scion is then inserted into the stock and wrapped with tape. The entire union must be thoroughly covered with grafting wax, sealing all openings. It takes 3–4 months for the grafts to establish, with a success rate of 10–50%.

The scions can be placed inside the stock by various methods and they adapt to stocks that are larger than the scion. Thus, side grafting is one of the most rapid methods of changing the cultivar on a tree without altering its framework and delaying production. Nelson (1954a, 1957) successfully top-worked 'Groff' on to the limbs of larger 'Brewster' trees in Florida between April and August, using flushing wood with some green and prominent axillary buds. All existing twigs or secondary branches were removed, with the scions having three or four buds inserted at desired intervals, directions and angles. If only one graft per stem was desired, the rest of the stem could be removed immediately above the insertion. Only secondary vascular tissues form callus in the graft union. The callus is contributed by both scion and stock, with a greater amount coming from the stock in side grafts (Abutiate and Nakasone, 1972).

Whip and tongue grafting

Whip and tongue grafting is best used for stocks that are 1 or 2 years old, 1.0–1.5 cm wide. Scions and stocks that are of equal size should be used. A long slanting cut is made on the lower end of the scion and upper end of the stock. This slanting cut should be about three times as long as the diameter of the scion or stock. In general, the cut surface is 2.5–5 cm long. A whip or tongue is then made by inserting the knife blade at about one-third the length of the cut surface as measured from the pointed end. The blade is drawn slowly into the wood and directed towards the central axis until it is 1–2 cm deep. Cutting against the grain reduces splitting. After both the stock and scion have been prepared, they are inserted into each other, with the tongues interlocking. It is extremely important that the two cambiums are in contact, and this is greatly increased because of the interlocking whips or tongues. The stock and scion are also held more firmly in place (see Fig. 3.6).

Splice (whip) grafting

This method is the same as the whip and tongue graft, except that only one 'tongue' cut is made in either the stock or the scion. A simple slanting cut of the same length and angle is made in the matching stock or scion. These are brought together and wrapped or tied. In South Africa, Bolt and Joubert (1968) obtained good success after splice grafting 'Tai So' seedlings of 3–10 mm diameter. They used scion wood 10 cm long, with at least two swollen buds. Non-terminal wood gave 70% success compared with 35% success with younger wood. Grafts grew more quickly if the leaves were retained on the stock, although the side shoots were usually removed. It was recommended that the scion should be girdled 3 weeks before grafting. Pinheiro *et al.* (1984) obtained 90% success with 'Brewster' using this technique (see Fig. 3.6).

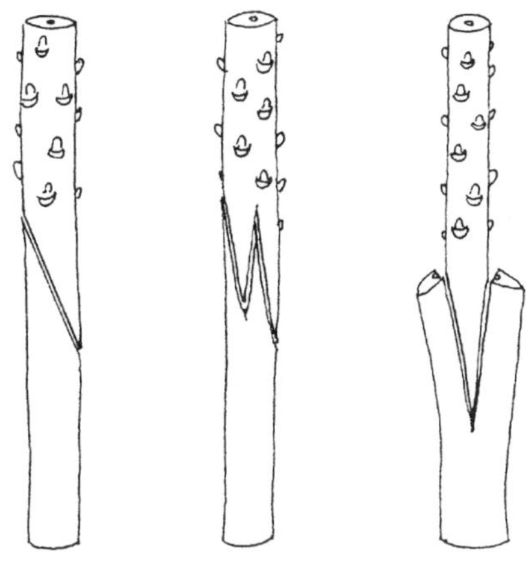

Fig. 3.6. Methods of grafting litchis and longans: (a) splice graft; (b) whip (tongue) graft; and (c) cleft graft (from Soule, 1985, with permission).

Approach grafting

In this technique, the scion continues to grow on its own roots until union with the stock is complete. A length of bark 3–5 cm long is removed from both the scion and the stock and the cut surfaces are held together by a string or tape. After 3–4 months, the top of the stock is removed above the graft and the base of the scion removed below. Thus, the root system of the stock is retained along with the new scion. It is better to sever these parts gradually rather than all at once. Grafting should be done when growth is active and rapid healing of the graft can take place. In India, this is best done during the monsoons from July to October. Although only 10–30% of grafts generally survive, this method has been practised in China (Groff, 1921; Ou *et al.*, 1993; Lu and Li, 1998), India (Singh and Singh, 1954), Hawaii (Yee, 1972), Australia (Loebel, 1976; Chalker, 1981) and Florida (Cobin, 1948, 1954). Galloway (1922), however, considered that approach grafting produced misshapen, unhealthy plants which developed slowly. This technique has also been used for top-grafting trees in the field (Menzel, 1985). Success can be improved by girdling 90% of the circumference of the stock above the union.

Budding

Budding consists of grafting a bud on to a seedling or air-layer, usually as a shield or 'T' bud. It is done when both the rootstock and the scion are actively growing, when the buds can be easily removed. Plump axillary or side buds are cut from the middle of firm, young shoots. A T-shaped cut is made in the bark of the stock, the bud inserted, and rubber bands, soft string or tape used to hold the bud and stock together. The top of the rootstock above the graft is removed when the bud grows and the plants are transplanted after completion of at least one flush.

Other budding techniques that have been tried for litchis are inverted-'T' budding, chip budding and modified Forkert budding (Menzel, 1985). Chip budding usually takes longer to unite and is more complex than 'T' budding, but is practised at times when the bark does not slip. Budwood should be obtained from trees 3–4 months after harvest when the bark parts easily from the wood, and have dormant buds. The most satisfactory buds are those taken from 1-year-old wood about 1 cm in diameter and where the leaves have abscised. However, since litchi normally retains its old leaves, it is usually necessary to cut the leaflet blades on the selected branches.

In chip budding, a downward 45° angle is made below the bud on a scion. A second cut is made about 12–15 mm above the bud, inward and downward in a transverse manner, meeting the first cut. The bud chip is removed and placed near the base of the stock between two nodes where an identical cut has been made. It is essential that the cambium of the stock and scion match properly. Forkert or patch budding consists of removing a square or rectangular piece of bark from the stock and replacing it with a similar patch of bark with the desired bud. The graft is usually wrapped in waxed cloth or budding tape.

Budding has rarely been used for propagation, since the rate of success is much lower than with other techniques such as whip, splice or approach grafting (Bolt and Joubert, 1968; Pinheiro *et al.*, 1984). Bolt and Joubert (1968) reported that shield-budding of 'Tai So' seedlings was unsuccessful in South Africa, because furrows in the wood and cambium prevented complete union between the bud and stock. Pinheiro *et al.* (1984) observed that inverted-'T' and chip buds failed on 1-year- old 'Brewster' seedlings. Nelson (1954a,b), however, had success with shield and chip-buds on small 'Brewster' seedlings in Florida. Buds grew rapidly, with the trees planted out after 18 months. Proper selection of budwood was more critical than the method of budding or the condition of the stock. They recommended the use of flushing wood still green with axillary buds, with the trees being irrigated and fertilized regularly.

Grafting in longan

Longans have been grafted in autumn and spring under shade in Australia. Scions from semi-hard terminals, 0.5 cm in diameter and with leaves attached, were used. The scions were cleft grafted into a node subtended by a leaf, and enclosed in a polythene bag or frame to prevent desiccation (see Fig. 3.7). Seedlings about 12–18 months old were usually used as stocks. In China, whip grafting, side grafting and budding are commonly used. Whip grafting has a relatively large joining surface, heals rapidly and is generally very successful. Side grafting and budding are easier to perform than whip grafting. Tip grafting is commonly used for top-working large trees. In China, seeds for rootstocks are obtained mainly from canning factories (Xu and Zheng, 1999). Once the seeds are taken out of the fruit, the germination rate drops sharply, so it is preferable to use fresh seeds. If the seeds are to be kept for a few days, they should be placed in wet sand before sowing. Seeds are sown in rows at distances of 20–25 cm in the field, and covered with a thin layer of soil. The bed is irrigated and covered with rice straw to assist water conservation.

Fig. 3.7. Grafted longan at Nambour in southern Queensland, Australia (photograph courtesy of Brian Paxton).

Wong (1992) tested different propagation methods and obtained the best success (99%) with air-layers. The air-layers took 2 months to root. Rooting of semi-hardwood cuttings and patch budding gave 59% and 73% success, respectively. Budding was better on 6-month-old (0.4 cm diameter) than on 18-month-old stocks (2.5 cm diameter). Work from Thailand and Queensland suggests that some scion–stock combinations are incompatible. If the scion is thicker than the stock or vice versa, the trees grow slowly. 'Shixia' and 'Dawayan' were not suitable rootstocks for 'Chuliang' in China (Zhang *et al.*, 1998). Incompatibility increased when the grafts were taken from successively higher branches in the canopy (Xu and Zheng, 1999).

Conclusions

Several techniques have been employed for the propagation of litchis and longans, with variable results. Seeds must be protected from desiccation in order for them to germinate, and live for only a few days if left in the open. Some cultivars, especially of litchi, produce a high proportion of fruit with 'chicken tongue' or shrivelled seeds, which is a desirable fruit characteristic. Since these seeds fail to germinate, *in vitro* culture can be used to propagate such cultivars (Yang and Chen, 1987; Lai *et al.*, 1997).

Although air-layering is the most widely practised method of propagation, the rate of success is low in some countries. The effects of tree age, and the position and size of the branches used, need to be considered. Discrepancies in the response to different media, pH and fertilizers require explanation (Menzel, 1985). A better understanding of the influence of weather is also required.

Among the various methods of clonal propagation, grafting has some advantages over the others. Rootstocks have the potential to alter tree size, canopy growth, productivity, fruit quality, and resistance to biotic and abiotic stresses. Unfortunately, no serious effort has been made to realize these advantages. Grafting has been used primarily for top-working, and is exploited for new orchards only in China and Thailand, with 85% success. Culture of rootstocks and preparation of scion wood (girdling), maturity of scion wood and the post-graft environment contribute to success in grafting. Graft incompatibility has often been reported in litchi (Ou *et al.*, 1993; Murray and Murray, 1994), but no work has been done on the mechanism involved. The reasons for failure need to be explored critically. Studies are also required to confirm the continuity of cambium activity along the stem. Molecular techniques may help to identify and characterize genes responsible for certain characters. The potential of other members of the Sapindaceae as rootstocks has scarcely been explored. Longan and rambutan have been reported to be graft-compatible, but only to a limited extent. Other forms need to be tested, including *Litchi chinensis* ssp. *philippinensis*.

One- to two-year-old seedlings of either the same or different cultivars have been used as rootstocks for most grafts. There is a need to standardize the size of the stock for different techniques. Seedlings are not uniform and can affect the growth and performance of the scion. Although air-layers have been used as rootstocks, attention should be focused on the use of specific cultivars.

Longan is propagated commercially by air-layering, with some grafting and budding. Very little information is available on the suitability of different rootstocks, the use of potential clonal material and the mechanism of graft incompatibility.

References

Abutiate, W.S. and Nakasone, N.Y. (1972) Studies on vegetative propagation of the lychee (*Litchi chinensis* Sonn.) with special references to graftage. *Ghana Journal of Agricultural Science* 5, 201–211.

Ahmed, S. (1961) The litchi. *Agriculture in Pakistan* 12, 769–775.

Albert, Y. (1959) The adjustment of some methods for litchi in Israel. MSc (Agric.) thesis, Hebrew University of Jerusalem, Rehovot, Israel.

Anonymous (1978) Studies on tropical fruit trees. *Fruits* 33, 647–665.

Bhandary, K.R. and Shivashankar, Y.T. (1970) Propagation of litchi under mist. *South Indian Horticulture* 18, 74–76.

Bolt, L.C. and Joubert, A.J. (1968) Litchi grafting trials. *Farming in South Africa* 44, 11–13.

Bose, T.K. and Mandal, D.P. (1972) Mist propagation of tropical plants. *Indian Horticulture* 17, 25–26.

Bose, T.K., Dhua, R.S., Mitra, S.K. and Sen, S.K. (1985) Synergism of phenolic substances with IBA in regeneration of root from litchi cuttings. *Indian Journal of Horticulture* 42, 153–155.

Brahmachari, V.S., Singh, M.P. and Biswas, R.C. (1998) Effect of pre-conditioning treatments of shoots and application of auxins on rooting of stem cuttings in litchi. *Orissa Journal of Horticulture* 26, 53–57.

Campbell, C.W. and Malo, S.E. (1975) *The Lychee.* Florida Cooperative Extension Service Fruit Crop Fact Sheet, 2 pp.

Campbell, T.P. (1981) Marcotting litchis. *Queensland Fruit and Vegetable News* 52, 156–157.

Chalker, F. (1981) Propagation of certain tropical fruits. *Combined Proceedings of the International Plant Propagators' Society* 31, 168–176.

Chen, G.Y. and Fu, J.R. (1989) Deterioration of some recalcitrant seeds. *Plant Physiology Communications* 3, 11–14.

Cobin, M. (1948) Notes on the grafting of *Litchi chinensis* Sonn. *Proceedings of the Florida State Horticultural Society* 61, 265–268.

Cobin, M. (1954) Lychee in Florida. *Bulletin of the Florida Agricultural Experiment Station* 546, 1–35.

Cull, B.W. and Lindsay, P. (1995) *Fruit Growing in Warm Climates for Commercial Growers and Home Gardens.* Reed Books, Chestwood, Australia, 192 pp.

Cull, B.W. and Paxton, B.F. (1982) Litchi. *Yearbook of the West Australia Nut and Tree Crop Association* 7, 13–21.

Cull, B.W. and Paxton, B.F. (1983) Growing lychee in Queensland. *Queensland Agricultural Journal* 109, 53–60.

Dutta, P., Biswas, B. and Mitra, S.K. (1995) Regeneration of litchi air-layers by pretreatment of stock plant with cycocel and ethrel. *Current Agricultural Research* 8, 54–57.

Eureinoff, V.A. (1950) The litchi. *Fruits* 5, 325–333.

Fu, J.R., Zhang, B.Z., Wang, X.P., Qiao, Y.Z. and Huang, X.L. (1990) Physiological studies on desiccation, wet storage and cryopreservation of recalcitrant seeds of three fruit species and their excised embryonic axes. *Seed Science and Technology* 18, 743–754.

Galán Saúco, V. and Menini, U.G. (1989) *Litchi Cultivation.* Plant Production and Protection Paper Number 83, Food and Agriculture Organization of the United Nations, Rome, Italy, 203 pp.

Galloway, B.T. (1922) Improved method of propagating the litchi. *Journal of Heredity* 13, 201–206.

Groff, G.W. (1921) *The Lychee and Lungan.* Orange Judd, New York, 188 pp.

Hartman, H.T. and Kester, D.E. (1986) *Plant Propagation: Principles and Practices.* Prentice-Hall of India, New Delhi, India, 727 pp.

Hartman, H.T., Flocker, W.J. and Kofranek, A.M. (1981) *Plant Science: Growth, Development and Utilization of Cultivated Plants.* Prentice Hall, New Jersey, pp. 80–115.

Hartman, H.T., Kester, D.E. and Davies, F.T. (1990) Theoretical aspects of grafting and budding. In: Hartman, H.T. and Kester, D.E. (eds) *Plant Propagation: Principles and Practices.* Prentice Hall, New Jersey, pp. 305–348.

Hayes, W.B. (1957) *Fruit Growing in India.* Kitabistan, Allahabad, India, 180 pp.

Ireta-Ojeda, A. (1975) Studies on litchi propagation by air-layering in Valle de Culiacau. *Agricultura Tecnica en Mexico* 3, 418–423.

Johnstone, H.G. (1956) Direct field planting of lychee layers: a progress report. *Proceedings of the Florida Lychee Growers' Association* 3, 49–53.

Jones, W.M. and Beaumont, J.H. (1937) Carbohydrate accumulation in relation to vegetative propagation of the litchi. *Science* 86, 313.

Joubert, A.J. (1970) The litchi. *Bulletin of the South African Department of Agriculture Technical Service* 389, 1–22.

Kadman, A. (1985) Improvements in the air-layering propagation method for lychee and macadamia trees. *Acta Horticulturae* 158, 143–149.

Kadman, A. and Slor, E. (1974) Experiments with propagation of the litchi (*Litchi chinensis*) in Israel. *Indian Journal of Horticulture* 31, 28–33.

Kadman, A. and Slor, E. (1982) Litchi growing in Israel. *Alon Hanotea* 36, 673–688.

Kantharajah, A.S., E-Bert, G. and Lüdders, P. (1997) *In vitro* techniques for crop improvement in lychee: a review. *Erwerbsobstbau* 39, 116–118.

Kanwar, J.S. and Kahlon, G.S. (1986) Propagation studies in litchi (*Litchi chinensis* Sonn.). *Journal of Research, Punjab Agricultural University* 23, 33–39.

Kaundal, G.S., Grewal, G.P.S. and Deol, I.S. (1993) Studies on propagation and post transplanting success of litchi (*Litchi chinensis* Sonn.) layers. *Scientific Horticulture* 3, 31–36.

Kuhn, G.D. (1962) Dehydration studies of litchi fruits. *Proceedings of the Florida State Horticultural Society* 75, 273–277.

Lai, Z.X., Chen, Z.G., Pan, L.Z. and He, B.Z. (1997) The current status of biotechnology research and prospect for fruit crops of Sapindaceae. *South China Fruits* 2 (6), 27–39.

Lenka, P.C. and Das, R.C. (1981) A note on the effect of IBA and rooting media on rooting of litchi stem cutting under mist. *Orissa Journal of Horticulture* 9, 40–41.

Leonel, S., Rodrigues, J.D. and Cereda, E. (1994) Effect of growth regulators and boric acid on lychee cuttings. *Científica* 22, 105–110.

Loebel, R. (1976) The litchi. *Tropical Fruit Bulletin* 6, 1–12.

Lu, M.Y., Pan, X.C., Pan, J.C. and Yang, M.C. (1999) Observation on the structures of litchi seeds under electronic scanning microscope. *Journal of Guangxi Agriculture and Biological Sciences* 18, 291–293.

Lu, Q.F. and Li, H.P.O (1998) Preliminary report of graft compatibility of Baitangying litchi variety with different root-stocks. *South China Fruits* 27 (3), 32.

Malik, M.N. and Maqbool, M. (1977) Studies on the effect of various physical treatments on the successful establishment of litchi plants through layerage. *Journal of Agricultural Research (Lahore)* 15, 321–327.

McClelland, T.B. (1944) Brief viability of tropical seeds. *Proceedings of the Florida State Horticultural Society* 57, 161–163.

Menzel, C.M. (1985) Propagation of lychee: a review. *Scientia Horticulturae* 25, 31–48.

Menzel, C.M. and Paxton, B.F. (1985) The effect of temperature on growth and dry matter production of lychee seedlings. *Scientia Horticulturae* 26, 17–23.

Menzel, C.M., Bagshaw, J., Campbell, T., Greer, N., Noller, J., Olesen, T. and Waite, G.K. (2002) *Lychee Information Kit.* Queensland Department of Primary Industries, Nambour, Australia.

Mitra, S.K. and Bose, T.K. (1991) Metabolic changes during adventitious root formation in ethrel and IBA treated cuttings of litchi. *Indian Journal of Horticulture* 48, 105–107.

Murray, V.G. and Murray, R.D. (1994) Confronting plant diversity when propagating *Litchi chinensis. Proceedings of the Florida State Horticultural Society* 107, 354–358.

Nelson, R.O. (1954a) Notes on lychee grafting. *Proceedings of the Florida State Horticultural Society* 67, 231–233.

Nelson, R.O. (1954b) Comments on lychee propagation by air-layering and graftage. *Proceedings of the Florida Lychee Growers' Association* 1, 8–11.

Nelson, R.O. (1957) Suggested methods for top-working lychee trees. *Proceedings of the Florida State Horticultural Society* 70, 27–31.

Nijjar, G.S. (1981) *Litchi Cultivation.* Punjab Agricultural University, Ludhiana, India, 11 pp.

Ochse, J.J. (1952) Observations on rooting of lychee cuttings. *Proceedings of the Florida State Horticultural Society* 65, 186–187.

Ochse, J.J., Soule, M.J., Dijkam, M.J., Jr, and Wehlburg, C. (1961) *Tropical and Subtropical Agriculture.* Macmillan, New York, 2 volumes.

Oosthuizen, W.G. (1995) Clonal propagation of litchis by cuttings. *Yearbook of the South African Litchi Growers' Association* 7, 41–43.

Ou, L.X., Zhong, Y.W. and Li, J.S. (1992) A preliminary report on stock–scion incompatibility of litchi. *Litchi Technical Communications* 2, 19–21.

Ou, L.X., Zhong, Y.W. and Li, J.S. (1993) Preliminary report of a study on the compatibility of litchi scion–rootstock combinations. *China Fruits* 2, 30–31.

Pandey, R.M. and Sharma, H.C. (1989) *The Litchi.* ICAR, New Delhi, India, 80 pp.

Paxton, B., Saranah, J. and Chapman, K.R. (1978) The propagation of litchis by cuttings. *Biennial Report of the Maroochy Horticultural Research Station* 1, 34.

Pinheiro, R.V.R., Silva, F.C.C. and Andersen, O. (1984) Propagation of litchi by grafting. *Revista Ceres* 31, 299–304.

Pope, W.T. and Storey, W.B. (1933) Grafting tropical fruit trees in Hawaii. *Circular of the Hawaii Agricultural Experimental Station* 6, 1–24.

Ray, P.K. (2001) Pruning and cincturing. In: Chauhan, K.S. (ed.) *Litchi: Botany, Production and Utilization.* Kalyani Publishers, Ludhiana, India, pp. 125–139.

Ray, P.K. and Sharma, S.B. (1987) Growth, maturing, germination and storage of litchi seeds. *Scientia Horticulturae* 33, 213–221.

Ruan, S.T. (1988) A discussion on litchi grafting. *Litchi Technical Communications* 4, 4.

Sadhu, M.K., Naskar, B.B. and Basu, R.N. (1972) Auxin synergists in the rooting of air-layers of tropical fruit trees. *Indian Agriculturist* 16, 251–257.

Sharfuddin, A.F.M. (1983) Effect of pruning on the survival of detached rooted air-layers in litchi. *Journal of Bangladesh Agricultural University* 11, 39–40.

Sharfuddin, A.F.M. and Hussain, A. (1973) Effect of growth regulators, rooting and wrapping materials on the success of air-layering in litchi. *Journal of Bangladesh Agricultural University* 1, 45–51.

Sharma, R.C. and Grewal, G.P.S. (1989) A note on propagation studies in litchi (*Litchi chinensis* Sonn.). *Haryana Journal of Horticultural Science* 18, 74–76.

Sharma, S.B., Ray, P.K. and Singh, B.K. (1990a) A note on the effect of defoliation on survival of litchi layers in a nursery. *Haryana Journal of Horticultural Science* 19, 129–130.

Sharma, S.B., Ray, P.K. and Singh, B.K. (1990b) Note on rooting and survival of litchi layers. *Indian Journal of Horticulture* 47, 174–176.

Singh, L.B. and Singh, U.P. (1954) *The Lychee.* Superintendent of Printing and Stationery, Lucknow, Uttar Pradesh, India, 87 pp.

Singh, R. (1963) The litchi. In: Singh, S., Krishnamurthi, S. and Katyal, S.L. (eds) *Fruit Culture in India.* ICAR, New Delhi, India, pp. 152–160.

Singh, S. and Jawanda, J.S. (1981) Propagation studies in litchi. *Punjab Horticulture Journal* 21, 184–187.

Singh, S.D., Chugh, C.V. and Singh, K.K. (1965) Seasonal and interaction effects in marcotting of litchi (*Litchi chinensis* Sonn.) with some plant regulators. *Indian Journal of Agricultural Sciences* 35, 101–113.

Soule, J. (1985) *Glossary For Horticultural Crops.* John Wiley & Sons, New York, 898 pp.

Storey, W.B. (1973) The lychee. *Yearbook of the California Avocado Society* 56, 75–86.

Subhadrabandhu, S. (1990) *Lychee and Longan Cultivation in Thailand.* Rumthai Publishers, Bangkok, Thailand, 40 pp.

Thakur, S., Sharma, R.N. and Sharma, R.K. (2001) Breeding. In: Chauhan, K.S. (ed.) *Litchi: Botany, Production and Utilization.* Kalyani Publishers, Ludhiana, India, pp. 28–36.

Tindall, H.D. (1994) Sapindaceous fruits: botany and horticulture. *Horticultural Reviews* 16, 143–196.

Venning, F.D. (1949) Anatomy and secondary growth in the axis of *Litchi chinensis* Sonn. *Quarterly Journal of the Florida Academy of Science* 12, 51–60.

Wong, K.C. (1992) Studies on the propagation, seed storage and fruit storage of *Dimocarpus longan* subspecies *malesianus* variety *malesianus. Acta Horticulturae* 292, 69–71.

Wong, K.C. (2000) *Longan Production in Asia.* Food and Agriculture Organization of the United Nations, Bangkok, Thailand, 44 pp.

Xia, Q.H., Chen, R.Z. and Fu, J.R. (1992) Effects of desiccation, temperature and other factors on the germination of litchi (*Litchi chinensis* Sonn.) and longan (*Euphoria longan* Steud.) seeds. *Seed Science and Technology* 20, 119–127.

Xia, Q.H., Chen, R.Z. and Fu, J.R. (1993) Studies on physiology of lychee (*Litchi chinensis* Sonn.) seeds at different developmental stages. *Acta Scientiarum Naturalium Universitatis Sunyatseni* 32, 80–86.

Xu, X.D. and Zheng, S.Q. (1999) *Photographic Illustration of High Yielding Longan Cultivation.* China Agricultural Press, Beijing, China, pp. 5–29.

Yang, Y.Q. and Chen, Z.F. (1987) Study of inherited character wrinkled seed and embryo culture in longan. *Acta Horticulturae Sinica* 14, 217–222.

Ye, Q.H. (1987) A fast propagation method in litchi trees with grafting. *Litchi Technical Communications* 4, 1–5.

Yee, W. (1972) *The Lychee in Hawaii.* College of Tropical Agriculture, University of Hawaii, Honolulu Circular 366, 1–4.

Zhang, M.C., Hua, Y.Q., Luo, W.L. and Hu, J.S. (1998) Preliminary report on graft incompatibility of longan varieties. *South China Fruits* 27 (2), 35.

4 Biotechnology

Richard E. Litz, Simon Raharjo, Kazumitsu Matsumoto and Witjaksono
Tropical Research and Education Center, University of Florida, 18905 SW 280 Street, Homestead, FL 33031-3314, USA

Introduction

The improvement of litchi and longan by conventional breeding methods has produced few new cultivars. This has been due to the long juvenile period of the species, the apparent lack of genetic variability in the existing germplasm (Menzel, 1992), and the great expenditure that is required in terms of land, time and money. Genetic engineering, including genetic transformation and induced mutations *in vitro*, followed by selection, can potentially be used to alter one or more horticulturally important traits. This has important implications for the improvement of vegetatively propagated fruit trees such as litchi and longan, because serious problems affecting production or quality can be resolved without altering the overall phenotype of the cultivar.

Preliminary attempts to manipulate longan and litchi by somatic cell genetics have only recently been reported. Other biotechnology tools, such as gene cloning and identification of DNA markers that can be applied to taxonomy and systematics, are being developed. However, marker-assisted selection, which could revolutionize conventional breeding, has not been used. The application of some of the tools of biotechnology to longan and litchi has previously been reviewed by Zeng *et al.* (2001) for longan, and by Sarin and Prasad (2003) for litchi. The primary breeding objectives using biotechnology have been defined by Zeng *et al.* (2001) and Lai *et al.* (2000, 2001).

Among fruit crops in the Sapindaceae, cell culture and somatic cell genetics have been applied only to litchi and longan. Zamora *et al.* (1988) described the initiation of multiple shoots of rambutan, *Nephelium lappaceum*, from tip and nodal cultures derived from *in vitro*-germinated seedlings. Induction of roots from individual shoots from these cultures was possible. Desai *et al.* (1986) described the regeneration of soapnut, *Sapindus trifoliatus*, by somatic embryogenesis from leaves of elite trees. The response of soapnut embryogenic cultures to salinity was measured by Unnikrishnan *et al.* (1991), who determined that they could tolerate up to 200 mg of sodium chloride per litre, and that lower concentrations were stimulatory.

The current status of biotechnology in longan and litchi, together with the possible directions of research, are discussed in this review.

Breeding and genetics

Only limited breeding has been conducted for the improvement of any of the fruit species within the Sapindaceae. Nearly all of the cultivars have originated as chance seedlings from open pollinations. Consequently, there is little genetic information available on the important fruit and tree characteristics. According to

 Litchi and Longan: Botany, Production and Uses (eds C.M. Menzel and G.K. Waite)

Menzel (1992), the genetic diversity within litchi collections is limited, since most of the cultivars have been selected in southern China. This conclusion has been confirmed by Ding *et al.* (2000), who observed limited DNA polymorphism amongst cultivated litchis. There have been no efforts to collect litchi material from the original habitats. The same limitations apply to longan (Choo and Ketsa, 1992). Consequently, there should be a concerted effort to collect wild litchis and longans in their centres of origin and diversity in South-east Asia.

The chromosome number of longan is 2n = 2x = 30 (Choo and Ketsa, 1992), and 2n = 2x = 28, 30 or 32 for litchi (Menzel, 1992). Trees are very heterozygous, and are normally propagated vegetatively by air-layering (marcotting), or grafting on to seedling rootstocks. There are no reported root-related problems that limit commercial production of these crops (but see Coates *et al.*, Chapter 13, this volume).

Major breeding objectives and accomplishments

Litchi and longan have become major crops only fairly recently. This is because they have a relatively brief harvest period, and quality and appearance decline rapidly after harvest if fruit are not treated correctly. The fruit are non-climacteric, and unripe fruit do not ripen after harvest. Litchi fruit, in particular, lose their bright red colour within 2–3 days of picking, if not protected from warm dry air. Extending the harvest period and shelf-life would improve production of these crops.

Some important litchi cultivars such as 'Tai So' ('Dazao', 'Mauritius') are very susceptible to anthracnose, *Colletotrichum gloeosporioides*. Control measures in the humid subtropics currently include the application of fungicides from flowering until shortly before harvest. Sustainable production must be based on existing or improved cultivars that require minimal or no chemical sprays.

Litchis and longans are irregular bearers, and selections from a specific growing region often yield erratically elsewhere. The short harvest period and irregular bearing make marketing of the crop very difficult, with prices sometimes falling during the main harvest period. High yields and regular bearing are essential in order to fully exploit their potential.

The most valued litchi and longan cultivars have small aborted seeds, referred to as 'chicken tongues'. These fruit have a high flesh-to-seed ratio. This trait is very common in some cultivars such as 'No Mai Chee' ('Nuomici') and 'Salathiel'. Some longan cultivars such as 'Biew Kiew' also have small seeds, but these are not 'chicken tongues'. Larger fruit, particularly in longan, are desirable.

In the absence of large-scale breeding programmes, the vegetatively propagated selections of longan and litchi have nearly all been derived from open-pollinated seedlings. McConchie *et al.* (1994) reported that longan and litchi are partially sexually compatible, enabling intergeneric hybrids to be recovered, but only if litchi is the maternal parent. Among their progeny, some of the plants produced seedless fruit, which could have interesting implications for selections of superior, seedless, interspecific fruit.

Molecular genetics

Molecular markers

The use of isozyme and RAPD (random amplified polymorphic DNA) markers has been utilized for the identification of litchi cultivars (Aradhya *et al.*, 1995; Degani *et al.*, 1995a; Ding *et al.*, 2000), for determining the role of the pollen parent in fruitlet abscission (Degani *et al.*, 1995b), and for measuring the effects of the pollen parent on outcrossing, yield and fruit characteristics (Stern *et al.*, 1993). Aradhya *et al.* (1995) demonstrated that a number of accessions and cultivars have been misnamed, based on analyses of differences or similarities in enzyme polymorphisms. Although Degani *et al.* (1995a) considered that isozyme polymorphism in litchi is fairly wide and could be used for taxonomy and systematics, Ding *et al.* (2000), working with RAPDs, considered that DNA polymorphism was less than might be expected for a crop with such a long history.

Gene cloning

Ding *et al.* (2001) reported the cloning of ACC (1-amino-cyclopropane-1-carboxylic acid) oxidase from ripening 'Wuye' litchi. Litchi ACC oxidase was shown to have a high level of homology with the enzyme from other plant species, including tomato, peach, carnation, apple, banana, pear and apricot. Ding *et al.* (2001) outlined a genetic transformation procedure to control fruit ripening, using the antisense strategy (Hamilton *et al.*, 1990). However, litchi and longan are non-climacteric, so blocking ethylene biosynthesis would probably not affect fruit ripening.

Micro-propagation

Few attempts have been made to stimulate axillary bud proliferation in explanted shoot tip and nodal cultures of elite longans (Chen *et al.*, 1991, 1997; Chen and Chen, 1996; Wang, 2001). According to Wang (2001), 'Dongi', 'Wulongling', 'Bianxiajian' and 'Honghezi' were regenerated from 3-cm-long shoots that had been explanted on to semi-solid Murashige and Skoog (MS) medium (Murashige and Skoog, 1962), supplemented with 2 μM benzyl adenine (BA) and 1 μM indole acetic acid (IAA). Seasonal conditions influenced establishment of the cultures, with greater success achieved during the summer (Chen *et al.*, 1991; Wang, 2001). Only limited axillary bud proliferation occurred on establishment medium. In order to stimulate axillary bud proliferation, 0.5-cm-long shoot tips were removed, and subcultured on semi-solid proliferation medium consisting of MS supplemented with 1 μM BA and 1 μM IAA. Shoots were rooted on semi-solid, half strength MS containing 0.9 μM BA with 3 μM indole butyric acid (IBA). Rooted plants were established in potting mixture under high relative humidity and ambient greenhouse temperatures.

Wang (2001) claimed that micro-propagation could be modified in order to eliminate witches' broom from affected longan trees. Following explanting on to establishment medium, the protocol involved exposure of the tissue to alternating 40°C and 22°C for 16 h and 8 h, respectively. Although Wang (2001) indicated that witches' broom was caused by a virus, Choo and Ketsa (1992) suggested that the causal agent was a mycoplasma (for further details, refer to Chapters 12 and 13, this volume).

The initiation of multiple shoots of litchi from shoot apices of *in vitro*-germinating seeds (non-clonal) was reported by Kantharajah *et al.* (1992) and Das *et al.* (1999). Multiple shoot formation occurred from the cotyledonary nodes following exposure to 90 μM BA in liquid MS medium with sterile filter paper bridges, and from 4- to 5-week-old *in vitro*-grown seedlings that were exposed to 100 μM BA on alternate days (Das *et al.*, 1999). Rooting was induced by pulsing the shoots with 103 μM IBA for 15 min, followed by root development in potting mixture.

Somatic cell genetics

A prerequisite for applying biotechnology to perennial tree species is the availability of a reliable procedure for regenerating elite trees or cultivars from cell culture. Although longan has been regenerated from cell cultures of mature specimens, this has not been achieved in litchi.

Regeneration

Somatic embryogenesis

Several studies have focused on the induction of embryogenic cultures from non-elite seedling longan tissues in China (Lai and Chen, 1998; Lai *et al.*, 1995, 1997a,b, 1998, 2000; Wei and Yang, 1981). Of greater horticultural significance, embryogenic cultures have been induced from 30-year-old 'Kohala' and 'Selection 12' trees in Florida (Litz, 1988; S. Raharjo and R.E. Litz, unpublished data). The explanted tissue consisted of entire surface-sterilized compound leaves in recent vegetative flushes. Since the original report, the induction medium has been modified slightly, and consists of B5 major salts (without $(NH_4)_2SO_4$) (see Gamborg *et al.*, 1968), MS minor salts and organic components, 400 mg/l glutamine, 60 g/l sucrose, 2–9 μM kinetin, 2–5 μM 2,4-dichlorophenoxyacetic acid (2,4-D), and 2 g/l gellan gum.

Tissue cultures are incubated in the dark at room temperature (25°C). Embryogenic cultures first appear 3–4 weeks after explanting, and are characteristically white and friable. Embryogenic cultures consist almost entirely of pro-embryonal cells and masses (PEMs). On induction medium, globular and early-stage cotyledonary somatic embryos develop from the embryogenic cultures distal to the surface of the medium (Fig. 4.1). For induction, maintenance on semi-solid medium, maturation and germination, cultures are routinely maintained in standard Petri dishes.

For efficient maintenance of embryogenic cultures, they should be inoculated into liquid medium (Lai *et al.*, 1995). Usually, 300 mg embryogenic culture is inoculated into 80 ml maintenance medium in a 250 ml Erlenmeyer flask. Maintenance medium consists of B5 major salts, MS minor salts and organic components, 400 mg/l glutamine, 60 g/l sucrose, and 5 μM 2,4-D. Suspension cultures are maintained at approximately 100 rpm on a rotary shaker under low light at room temperature, and subcultured into fresh medium of the same formulation, every 2–3 weeks (Fig. 4.2). Proliferation of embryogenic cultures is by the induction of embryogenic cells and secondary embryos from the protoderm of PEMs.

Development and maturation can be initiated from embryogenic cultures by subculture on to semi-solid maturation medium (Fig. 4.3), which consists of B5 major salts, MS minor salts and organic components, 400 mg/l glutamine, 20 g/l sucrose, 10% (v/v) filter-sterilized coconut water, and 2.0 g/l gellan gum. Somatic embryos can develop to maturity on this medium (Fig. 4.4). Following germination, cultures are routinely transferred to light (60 μmol quanta/m^2/s), with a 16 h photoperiod supplied by cool white fluorescent bulbs. Germination occurs approximately 10–14 days following transfer to the light, and precedes shoot formation by a few days (Fig. 4.5). Plantlets are transferred to potting soil and hardened in a greenhouse under intermittent mist, with minimal losses (Fig. 4.6).

Embryogenic litchi cultures have been induced only from non-elite, zygotic embryos of 'Xiafanzhi' (Zhou *et al.*, 1996; Yu and Chen, 1997; Yu *et al.*, 2000) and 'Brewster' ('Chenzi') (A. Celo and R.E. Litz, 1996, personal communication). Immature fruit, 3–4 weeks after flowering, are surface-sterilized and dissected under aseptic conditions to remove the zygotic embryo from each fruit. In Florida, the induction medium described for longan has been used effectively for litchi. Cultures are maintained in standard Petri dishes in the dark, at room temperature. Embryogenic cultures generally appear after 5–6 weeks, and consist of PEMs and various developmental stages of somatic embryos, ranging from globular to cotyledonary embryos (Witjaksono and R.E. Litz, Florida, personal communication). On the other hand, Yu and Chen (1997) described the induction of friable litchi embryogenic cultures. These differences in the appearance of induced embryogenic cultures are probably due to differences between cultivars. Similar genotype-specific responses have been reported in avocado (Witjaksono and Litz, 1999a,b).

It is possible to maintain embryogenic litchi cultures either on semi-solid medium, or as suspensions (Yu and Chen, 1998). Maintenance medium consists of induction formulation. Embryogenic cultures proliferate by the formation of secondary somatic embryos, many of which continue to develop in the presence of 2,4-D, to advanced cotyledonary stages and some friable PEMs. Many of the somatic embryos are hyperhydric, a physiological disorder that results in tissue necrosis. Friable cultures can be maintained by subculture on semi-solid medium.

Yu and Chen (1997) reported that maintenance of embryogenic cultures on semi-solid medium resulted in a mixture of PEMs and developing somatic embryos. In order to inhibit somatic embryo development, they incorporated 29 μM silver thiosulphate into the medium, and obtained uniformly friable cultures consisting of PEMs. Approximately 1.5 g of friable PEMs is inoculated into 30 ml of liquid induction medium in 125 ml Erlenmeyer flasks, and diluted by one-third every week. Maintenance of PEMs in suspension for up to a year is accomplished by alternating liquid with semi-solid (containing silver thiosulphate) media. Cultures are maintained on a rotary

Fig. 4.1. Induction of embryogenic longan culture from leaves of mature 'Kohala' longans.

Fig. 4.2. Maintenance of embryogenic 'Kohala' longan cultures in suspension culture.

Fig. 4.3. Maintenance of embryogenic 'Kohala' longan culture on semi-solid medium.

Fig. 4.4. Somatic embryo development in 'Wong Special' longan.

Fig. 4.5. Plantlet recovery from somatic embryos of 'Kohala' longan.

Fig. 4.6. 'Kohala' longan derived from somatic embryo.

shaker at 100 rpm, in the dark or semi-dark, at room temperature.

Maturation of embryogenic cultures can be initiated by subculturing them on to maturation medium, and incubation in the dark; however, germination and conversion rates are poor. Yu *et al.* (2000) observed improved rates of germination of somatic embryos from embryogenic protoplasts (see the section 'Protoplast isolation and culture').

Haploid recovery

Induction of embryogenic cultures from the microspores of longan anthers has been reported in 'Dong Bi', 'Hong He Zi', 'Oolong Mountain' and 'You Tan Ben' (Yang and Wei, 1984; Wei, 1990). Unopened staminate and hermaphroditic flowers were collected from the inflorescences, and closed flower buds containing mid- to late-uninucleate stage microspores, were pre-treated at 4°C for 24 h. Following surface-sterilization and washing, the buds were dissected under aseptic conditions, and the anthers removed and cultured on semi-solid induction medium. Induction medium consisted of MS, supplemented with 5 μM kinetin, 9 μM 2,4-D, 5 g/l activated charcoal and 50 g/l sucrose. Anther cultures were maintained in the dark at 25–27°C.

Following induction, the haploid embryogenic cultures were transferred on to semi-solid maturation medium containing 2–4 μM BA, 0.5 μM NAA, and 16 g/l sucrose. Germination occurred on semi-solid Anderson's rhododendron basal medium (Anderson, 1975) supplemented with 1 μM BA, 0.6 μM IAA, and 16 g/l sucrose. Conditions for haploid embryo development and germination included a 12 h photoperiod at 80 μmol quanta/m^2/s. Plantlets derived from microspores are slow growing. The chromosome number of cells in the root tips of regenerated plants was n = x = 15.

Unopened staminate and hermaphrodite 'Chenzhi' and 'Gushan Jiaohe' litchi flowers containing 3–4-mm-long anthers with microspores at the optimum late-uninucleate stage of development were utilized (Fu and Tang, 1983; Fu, 1990). The flower buds were surface-sterilized, washed with sterile distilled water, and dissected under aseptic conditions. The anthers were removed and plated on induction medium consisting of MS salts and organic components, 5–9 μM 2,4-D, 4.6–9 μM kinetin, 3–5 μM NAA, and 16 g/l sucrose. The cultures were incubated at 25°C, with a 10 h photoperiod (60 μmol quanta/m^2/s). Embryogenic cultures were apparent approximately 10 weeks after explanting, and could be maintained on induction medium.

In order to initiate haploid embryo development and maturation, embryogenic cultures have been subcultured on MS supplemented with 2 μM kinetin, 0.5 μM NAA, 500 mg/l casein hydrolysate, 400 mg/l royal jelly, and 16 g/l sucrose. Early heart-stage embryos were transferred to MS with 2 μM kinetin, 3 μM gibberellic acid (GA_3), 500 mg/l casein hydrolysate, 400 mg/l royal jelly, 1.7 g/l glutamine, and 16 g/l sucrose. The haploid embryos matured and germinated. Examination of the chromosomes in the root tips of regenerated plants indicated that most of the cells were haploid (n = x = 15), although aneuploids and triploids were also present.

Protoplast isolation and culture

Lai (1997) and Lai and Chen (1996) isolated and cultured protoplasts from embryogenic longan suspension cultures that had been derived from zygotic embryos. Embryogenic suspension cultures were incubated for 14 h in an enzyme mixture consisting of 1% Onozuka R-10 cellulase and 1% pectinase with 13% (w/v) mannitol. The protoplasts were washed and suspended in calcium alginate beads. Heart-stage somatic embryos that developed from protoplasts were transferred to maturation medium.

Yu *et al.* (2000) described the isolation and culture of protoplasts from embryogenic litchi suspensions. Four-day-old suspensions were collected by low-speed centrifugation, and plasmolysed for 1 h in CPW (see Frearson *et al.* (1973), as modified by Grosser and Gmitter (1990)), supplemented with 13% w/v mannitol. The cultures were resuspended in filter-sterilized enzyme solution containing 0.8% (w/v) cellulase 'Onozuka' RS, 0.4% (w/v) macerozyme R-10, 11% (w/v) mannitol, and CPW salts. Following overnight incubation in the dark at 26°C and 45 rpm, the cultures were passed through a 37 μm sterile nylon fabric, and centrifuged at

100 *g* for 5 min. The protoplasts were washed twice with CPW salts supplemented with 11% (w/v) mannitol, washed once with CPW without calcium, and resuspended in 10% (v/v) mannitol containing 1.8% (w/v) sodium alginate. The protoplasts in suspension were added drop-wise to sterile 1% (w/v) $CaCl_2.H_2O$ in CPW solution. After 1 h, the solution was replaced with MS medium supplemented with 2 µM zeatin, 2% (v/v) coconut water, 0.5 M glucose and 0.1 M sucrose.

Enhanced survival resulted when a nurse culture, consisting of embryogenic cultures embedded in calcium alginate, was co-cultured with the encapsulated protoplasts. Pro-embryonic masses and globular somatic embryos were released, and somatic embryo development was initiated on MS medium supplemented with B5 vitamins, 5 µM kinetin, 0.5 µM NAA, 500 mg/l glutamine, 8% (w/v) sucrose and 15 g/l agar, in the dark.

In order to stimulate maturation and germination, litchi somatic embryos were transferred to a maturation medium consisting of MS major and minor salts, B5 vitamins, 500 mg/l glutamine, 5% (v/v) coconut water, 50 g/l sucrose, and 9 g/l agar. For germination, mature somatic embryos were subcultured on to medium consisting of MS major and minor salts, B5 vitamins, 500 mg/l glutamine, 5% (v/v) coconut water, 30 g/l sucrose, 5 µM kinetin, 14 µM GA_3 and 7 g/l agar. Cultures were maintained under a 16 h photoperiod (80 µmol quanta/m^2/s). A few of these plants survived transfer to soil.

Genetic manipulation

Genetic transformation

Zeng *et al.* (2001) reported the genetic transformation of longan somatic embryos from embryogenic cultures of 'Sieryuelongyan' zygotic embryos. Somatic embryos were infected with different wild strains of *Agrobacterium rhizogenes*, of which strain R1601 was the most effective. The hairy roots were excised, and cultured on semi-solid MS medium containing 22 µM BA and 23 µM kinetin, in order to stimulate secondary somatic embryo development. Plants that had hairy roots were recovered from the transformed somatic embryos, with transformation confirmed by Southern hybridization and PCR (polymerase chain reaction).

Litchi has been transformed with two genes that are associated with resistance to pathogens (Witjaksono and R.E. Litz, Florida, 1999, personal communication). Embryogenic cultures derived from zygotic embryos of open-pollinated 'Brewster' were incubated for 3 days in liquid maintenance medium at 100 rpm, with acetosyringone-activated *Agrobacterium tumefaciens* strain LBA4404. *Agrobacterium tumefaciens* was electroporated with three different gene constructs: pBI121 containing the selectable marker, neophosphate transferase (*nptII*), and the scorable marker, β-glucuronidase (*gus* or *uidA*); pGPTV-BAR, in which chitinase and β-1,6-glucanase genes were inserted in tandem; and the antifungal protein gene. The *nptII* and *uidA* genes in pBI121 were driven by the 35S promoter. The chitinase, β-1,6-glucanase, and antifungal protein genes and *uidA* in pGPTV-BAR were driven by the double 35S promoter. *Agrobacterium tumefaciens* was then eliminated by incubating the cultures for 2 weeks in maintenance medium supplemented with 200 mg/l cefotaxime.

Following the removal of *A. tumefaciens*, the cultures were transferred on to semi-solid maintenance medium supplemented with 2 g/l glufocinate ammonium to select for pGPTV-BAR, and 100 mg/l kanamycin to select for pBI121. Cefotaxime at 200 mg/l was incorporated into the medium. Expression of the *uidA* gene by the X-Gluc histochemical reaction (Jefferson, 1987), 2 weeks after the end of selection, showed differential transformation efficiency among the constructs. Transformation with pBI121 was most efficient, whereas pGPTV-BAR with chitinase and β-1,6-glucanase were least efficient.

Somatic hybridization

Preliminary studies on the fusion of protoplasts of litchi and longan have been reported (Lai and Chen, 1997). Although limited division of somatic hybrid cells was described, the recovery of somatic embryos was not reported. The utility of litchi + longan somatic hybrids is uncertain, due to their polyploidy and the consequent problems of introgression of useful genes into either parent. McConchie *et al.* (1994) reported that litchi and longan can be sexually hybridized, with hybrids recovered if litchi is used as the female parent.

Cryopreservation

The ability to store embryogenic longan and litchi cultures for extended periods is critical for many cell manipulation studies, because embryogenic cultures lose their morphogenic competence over time. Preliminary studies have demonstrated that embryogenic longan cultures can be cryopreserved, by step-wise or slow cooling at −1°C/min down to −196°C (Matsumoto *et al.*, 2003). The optimum cryoprotectant was 0.5% (v/v) glycerol with 5% (v/v) DMSO (dimethyl-sulphoxide). Embryogenic cultures (approx. 100 mg) were suspended in 2 ml of sterile cryoprotectant in cryogenic vials, and plunged into liquid nitrogen in a cryogenic storage tank. The cryogenic vials were removed from the liquid nitrogen after 48 h, and rapidly thawed in a water bath at 40°C. Embryogenic cultures were transferred on to maintenance medium, and growth resumed. The standardization of this protocol should have an important impact on the storage of experimental material, and for use as a back-up for germplasm collections.

Conclusions

The basic *in vitro* protocol for manipulating elite longan selections is available, and could be utilized to address various plant-breeding objectives. Comparable studies are necessary in order to define a *de novo* regeneration pathway from elite litchi selections. Genetic transformation of litchi, and certainly of longan, is feasible. Longan, because it can be regenerated from embryogenic cultures of elite selections with a high rate of conversion, could be transformed with genes that encode for resistance to fungal diseases such as anthracnose. It should also be possible to apply *in vitro* mutation induction and selection procedures to address certain fungal diseases that affect specific longan cultivars. Genetic transformation could possibly be utilized to develop the preferred 'chicken tongue' seeds, using the pistillate gene from *Arabidopsis thaliana*, which mediates seedlessness. Protoplast technology could be harnessed to produce somatic hybrids between haploids and diploids, as a way of producing seedless triploids. The resolution of problems associated with rapid deterioration of fruit after harvest are more difficult to address at this time, since litchi and longan are non-climacteric.

Some of the new biotechnologies described in this review also have potential for application to other species in the Sapindaceae. *De novo* regeneration, which is a prerequisite for applying somatic cell genetics to elite selections of litchi and rambutan, has not been reported. Following the successful regeneration of mature longan and soapnut from embryogenic cultures, comparable studies should be undertaken with litchi and related species.

Acknowledgements

This research was supported by the Florida Agricultural Experiment Station and approved for publication as Journal Series No. R-08358.

References

Anderson, W.C. (1975) Propagation of rhododendrons by tissue culture. 1. Development of culture medium for multiplication of shoots. *Proceedings of the International Plant Propagators Society* 25, 129–135.

Aradhya, M.K., Zee, F.T. and Manshardt, R.M. (1995) Isozyme variation in lychee (*Litchi chinensis* Sonn.). *Scientia Horticulturae* 63, 21–35.

Chen, Q.Y. and Chen, J.Y. (1996) Shoot tip culture and elimination of virus of longan. *Plant Physiology Communications* 32, 126–127.

Chen, Q.Y., Chen, J.Y., Ke, C. and Song, R.L. (1991) A study of shoot tip culture in longan. *Fujian Fruits* 1, 4.

Chen, Q.Y., Chen, J.Y. and Chen, B. (1997) Studies on improving survival rate of longan in vitro culture. *Journal of Fujian Academy of Agricultural Science* 12, 32–35.

Choo, W.K. and Ketsa, S. (1992) *Dimocarpus longan* Lour. In: Verheij, E.W.M. and Coronel, R.E. (eds) *Plant Resources of South-East Asia*

No.2: Edible Fruits and Nuts. Pudoc, Wageningen, The Netherlands, pp. 146–151.

Das, D.K., Prakash, N.S. and Bhalla-Sarin, N. (1999) Multiple shoot induction and plant regeneration in litchi (*Litchi chinensis* Sonn.). *Plant Cell Reports* 18, 691–695.

Degani, C., Beiles, A., El Batsri, R., Goren, M. and Gazit, S. (1995a) Identifying lychee cultivars by isozyme analysis. *Journal of the American Society for Horticultural Science* 120, 307–312.

Degani, C., Stern, R.A., El Batsri, R. and Gazit, S. (1995b) Pollen parent effect on the selective abscission of 'Mauritius' and 'Floridian' lychee fruitlets. *Journal of the American Society for Horticultural Science* 120, 523–526.

Desai, H.V., Bhatt, P.N. and Mehta, A.R. (1986) Plant regeneration of *Sapindus trifoliatus* L. (soapnut) through somatic embryogenesis. *Plant Cell Reports* 3, 190–191.

Ding, X.D., Lu, L.X., Chen, X.J. and Guan, X. (2000) Identifying litchi cultivars and evaluating their genetic relationships by RAPD markers. *Journal of Tropical and Subtropical Botany* 8, 49–54.

Ding, X.D., Lu, L.X. and Xue, Y.B. (2001) Cloning of ACC oxidase gene in fruit of *Litchi chinensis* Sonn. and its expression in *E. coli*. *Acta Horticulturae* 558, 161–165.

Frearson, E.M., Power, J.B. and Cocking, E.C. (1973) The isolation, culture and regeneration of *Petunia* leaf protoplasts. *Developmental Biology* 33, 130–137.

Fu, L. (1990) Litchi: anther culture. In: Chen, Z., Evans, D.A., Sharp, W.R., Ammirato, P.V. and Sondahl, M.R. (eds) *Handbook of Plant Cell Culture*, Vol. 6: *Perennial Crops*. McGraw Hill, New York, pp. 376–384.

Fu, L. and Tang, D. (1983) Induction of pollen plants of litchi tree (*Litchi chinensis* Sonn.). *Acta Genetica Sinica* 10, 369–374.

Gamborg, O.L., Miller, R.A. and Ojima, K. (1968) Plant cell cultures. I. Nutrient requirements of suspension cultures of soybean root cells. *Experimental Cell Research* 50, 151–158.

Grosser, J.W. and Gmitter, F.G., Jr (1990) Protoplast fusion and citrus improvement. *Plant Breeding Reviews* 8, 339–374.

Hamilton, A.J., Lycett, G.W. and Grierson, D. (1990) Antisense gene that inhibits synthesis of the hormone ethylene in transgenic plants. *Nature* 346, 284–287.

Jefferson, R.A. (1987) Assaying chimeric genes in plants: the gus fusion system. *Plant Molecular Biology Reporter* 5, 387–405.

Kantharajah, A.S., McConchie, C.A. and Dodd, W.A. (1992) *In vitro* embryo culture and induction of multiple shoots in lychee (*Litchi chinensis* Sonn.). *Annals of Botany* 70, 153–156.

Lai, Z.X. (1997) Plant regeneration system of protoplast culture in longan. PhD dissertation, Fujian Agricultural University, Fouzhou, China.

Lai, Z.X. and Chen, Z.G. (1996) A study on the high effective regeneration from protoplasts of longan. *Journal of Fujian Agricultural University* 25, 249–250.

Lai, Z.X. and Chen, Z.G. (1997) A study on protoplast fusion between longan and litchi and the primary culture. In: *Proceedings of the Fifth Congress of the Fujian Society of Genetics*. Fouzhou, China, p. 38 (Abstract).

Lai, Z.X. and Chen, Z.G. (1998) Plant regeneration at high frequency from somatic embryos of longan. *Journal of Fujian Agricultural University* 27, 31–36.

Lai, Z.X., Pan, L.Z. and Chen, Z.G. (1995) Establishment of cell suspension cultures in longan. In: *Proceedings of the Sixth National Conference on Biochemical Engineering*. Chemical Engineering Press, Beijing, China, pp. 434–436.

Lai, Z.X., Pan, L.Z. and Chen, Z.G. (1997a) Study on establishment and maintenance of embryogenic cell lines in longan. *Journal of Fujian Agricultural University* 26, 160–167.

Lai, Z.X., Pan, L.Z. and Chen, Z.G. (1997b) Embryogenesis of high frequency from embryogenic callus in longan. *Journal of Fujian Agricultural University* 26, 271–276.

Lai, Z.X., Chen, Z.G. and He, B.Z. (1998) Somatic embryogenesis in longan. In: Larkin, P.J. (ed.) *Proceedings of the Fourth Asia-Pacific Conference on Agricultural Biotechnology*. UTC Publishing, Canberra, Australia, pp. 369–371.

Lai, Z.X., Chen, C.L., Zheng, L.H. and Chen, Z.G. (2000) Somatic embryogenesis in longan (*Dimocarpus longan* Lour.). In: Jain, S.M., Gupta, P.K. and Newton, R.J. (eds) *Somatic Embryogenesis in Woody Plants*, Vol. 6. Kluwer Academic Publishers, Dordrecht, The Netherlands, pp. 415–431.

Lai, Z.X., Chen, C.L. and Chen, Z.G. (2001) Progress in biotechnology research in longan. *Acta Horticulturae* 558, 137–141.

Litz, R.E. (1988) Somatic embryogenesis from cultured leaf explants of the tropical tree *Euphoria longan* Stend. *Journal of Plant Physiology* 132, 190–193.

Matsumoto, K., Raharjo, S., Dhekney, S., Moon, P.A. and Litz, R.E. (2003). Criopreservacao de longan (*Dimocarpus longan* Lour.). In:

Proceedings of the First Brazilian Congress of Plant Tissue Culture (Abstract).

McConchie, C.A., Vithanage, V. and Batten, D.J. (1994) Intergeneric hybridization between litchi (*Litchi chinensis* Sonn.) and longan (*Dimocarpus longan* Lour.). *Annals of Botany* 74, 111–118.

Menzel, C.M. (1992) *Litchi chinensis* Sonn. In: Verheij, E.W.M. and Coronel, R.E. (eds) *Plant Resources of South-East Asia No.2: Edible Fruits and Nuts*. Pudoc, Wageningen, The Netherlands, pp. 191–195.

Murashige, T. and Skoog, F. (1962) A revised medium for rapid growth and bioassays with tobacco tissue cultures. *Physiologia Plantarum* 15, 473–497.

Sarin, N.R. and Prasad, U.S. (2003) In vitro regeneration and transformation of *Litchi chinensis* Sonn. In: Jaiwal, P.K. and Singh, R.P. (eds) *Plant Genetic Engineering*, Vol. 6: *Improvement of Fruit Crops*. Sci-Tech Publishing, Houston, Texas, pp. 211–222.

Stern, R.A., Gazit, S., El Batsri, R. and Degani, C. (1993) Pollen parent effect on outcrossing rate, yield, and fruit characteristics of Floridian and Mauritius lychee. *Journal of the American Society for Horticultural Science* 118, 109–114.

Unnikrishnan, S.K., Prakash, L., Josekutty, P.C., Bhatt, P.N. and Mehta, A.R. (1991) Effect of NaCl salinity on somatic embryo development in *Sapindus trifoliatus* L. *Journal of Experimental Botany* 42, 401–406.

Wang, J.F. (2001) Shoot tip culture and production of virus-free longan plantlets. *Acta Horticulturae* 558, 155–160.

Wei, W. (1990) Longan: anther and embryo culture. In: Chen, Z., Evans, D.A., Sharp, W.R., Ammirato, P.V. and Sondahl, M.R. (eds) *Handbook of Plant Cell Culture*, Vol. 6: *Perennial Crops*. McGraw Hill, New York, pp. 385–396.

Wei, W.X. and Yang, Y.Q. (1981) The induction of embryoids from longan cotyledons and the culture of test tube plant. *Journal of Fujian Agricultural University* 2, 102–106.

Witjaksono and Litz, R.E. (1999a) Induction and growth characteristics of embryogenic avocado cultures. *Plant Cell, Tissue and Organ Culture* 58, 19–29.

Witjaksono and Litz, R.E. (1999b) Maturation of avocado somatic embryos and plant recovery. *Plant Cell, Tissue and Organ Culture* 58, 141–148.

Yang, Y.Q. and Wei, W.X. (1984) Induction of anther-derived plantlet in longan. *Acta Genetica Sinica* 11, 288–293.

Yu, C.H. and Chen, Z. (1997) Induction of litchi embryogenic calli by immature embryos and anthers culture *in vitro*. *Journal of Fujian Agricultural University* 26, 168–172.

Yu, C.H. and Chen, Z. (1998) Embryogenic suspension culture and protoplast isolation in lichee. *Chinese Journal of Tropical Crops* 19, 16–21.

Yu, C., Chen, Z., Lu, L. and Lin, J. (2000) Somatic embryogenesis and plant regeneration from litchi protoplasts isolated from embryogenic suspensions. *Plant Cell Tissue and Organ Culture* 61, 51–58.

Zamora, A.B., Damasco, O.P. and Barba, R.C. (1988) Shoot culture requirement of rambutan (*Nephelium lappaceum*). In: Umaly, R.C., Umboh, I., Haos, S. and Noor, N.M. (eds) *The Application of Tissue Culture Techniques in Economically Important Tropical Trees*. SEAMEO-BIOTROP, Bogor, Indonesia, pp. 181–185.

Zeng, L.H., Chen, Z.G., Lu, L.X., Zhou, P. and Zheng, X.Q. (2001) *In vitro* transformation mediated by Ri plasmid of *Agrobacterium rhizogenes* and transgenic plant regeneration of longan. *Acta Horticulturae* 558, 149–154.

Zhou, L.N., Kuang, Z.S., Ma, X.J., Chen, J.Q. and Cao, J. (1996) The study of factors affecting somatic embryogenesis in young embryo culture of *Litchi chinensis* L. *Journal of Agricultural Biotechnology* 4, 161–165.

5 Cultivars and Plant Improvement

Christopher M. Menzel[1], Xuming Huang[2] and Chengming Liu[2]
[1]Department of Primary Industries and Fisheries, PO Box 5083, SCMC, Nambour, Queensland 4560, Australia; [2]College of Horticulture, South China Agricultural University, Guangzhou 510642, China

Introduction

Litchi and longan originated in the area encompassing southern China, northern Viet Nam and the Malay Peninsula (Groff, 1921; Menzel and Simpson, 1990, 1991; Menzel *et al.*, 1990). This region is one of the three main cradles of domestication and is the origin of several other important horticultural crops, including tea (*Camellia sinensis*), members of the citrus group (*Citrus sinensis* and others) and kiwifruit (*Actinidia chinensis*). China has the richest germplasm resource of litchi and longan, with a long history of cultivation, selection, breeding and preservation. According to latest registers there are over 300 litchi and longan cultivars and clones, with many ancient varieties still cultivated (Qiu and Zhang, 1996; Wu, 1998).

Numerous wild litchis can be found in the moist forests of Hainan Island from low elevations up to 1000 m, and below 500 m in hilly areas of the Leizhou Peninsula, western Guangdong and eastern Guangxi (Hsu *et al.*, 1964). Litchi is a dominant species in some of these forests and may account for up to 50% of the trees. Wild trees can also be found in Viet Nam, north of Hanoi (Fig. 5.1), although there are fewer pockets of natural rainforest than in China (Papademetriou and Dent, 2002). Wild specimens are similar in general appearance to commercial varieties; however, the flesh or aril is relatively thin and sour.

Wild longan trees and communities have been found in southern China, especially in Yunnan and Guangxi (Li, 1985; Ke *et al.*, 1994; Liu and Ma, 2001). In one study, more than 70,000 trees were observed growing in various areas of south-west Yunnan. Native longans also occur in northern Thailand (Subhadrabandhu and Yapwattanaphun, 2001). There are various tropical subspecies and varieties of longan throughout South-east Asia that are similar to commercial longan, but they have thinner arils and warty skin (Wong, 2001) (Fig. 5.2).

These crops have been taken to most of the tropical and subtropical world in the last 400 years and are very popular in China and South-east Asia, but are less well known in Africa, the Middle East and America. Commercialization around the world has been slow due to the poor cropping of trees in many areas, as well as the short life of the seeds. Litchi production is greatest in China, Viet Nam, Thailand and India, and less important in Bangladesh and Nepal, the Philippines, Indonesia, South Africa, Madagascar, Israel, Australia and the USA. South-east Asia accounts for more than 95% of world production at 2,000,000 t. Longan is significant only in China, Viet Nam and Thailand, with smaller industries in Cambodia, Laos, Myanmar, Indonesia, Malaysia, Australia and the USA. Total production is half that of litchi.

Plants cultivated for a long time or over a wide geographical range are often very diverse, reflecting selection under different environments

Fig. 5.1. Wild litchis in a market near Hanoi, northern Viet Nam (photograph courtesy of Christopher Menzel).

Fig. 5.2. Related non-commercial varieties of longan found throughout Malaysia and Indonesia (photograph courtesy of Wong Kai Choo).

and cultures. However, in litchi and longan, selection for particular fruit traits has narrowed the variation within each crop. Improvements in fruit quality have often been at the expense of productivity and other desirable characteristics. Even though hundreds of varieties are available, in most countries commercial production is based on one or two main cultivars (Tables 5.1 and 5.2). This translates into a relatively short production season in many areas.

Nearly all the cultivars grown throughout the world originated in China, although Viet Nam, India and sections of the industries in Thailand have developed their own varieties from material sent from China. There has also been limited selection in Israel, South Africa, Australia and the USA. Most of the new cultivars have arisen as chance seedlings or seedling selections from known mother trees, with only limited hybridization initiated. Biotechnology is still in its infancy in these crops and no cultivars are sold under licence. Overall, breeding efforts have been negligible compared with the more important tropical fruit such as banana, citrus,

Table 5.1. Major litchi cultivars.

	Major cultivars
China	Sum Yee Hong, Baitangying, Bah Lup, Souey Tung, Fay Zee Siu, Tai So, Haak Yip, No Mai Chee, Kwai May Red, Wai Chee, Lanzhu, Brewster
Viet Nam	Thieuthauhha
Thailand	Tai So, Chacapat, Wai Chee, Haak Yip, Kom
India	Shahi, China, Calcuttia, Bedana, Late Bedana, Longia
Nepal	Mujafpuri, Raja Saheb, Deharaduni, China, Calcuttia
Bangladesh	Bombai, Muzaffarpuri, Bedana, China 3
Indonesia	Local selections
The Philippines	Sinco, Tai So, ULPB Red
Israel	Mauritius, Floridian
South Africa	Mauritius, McLean's Red
Madagascar, Mauritius and Réunion	Mauritius
Australia	Kwai May Pink, Tai So, Souey Tung, Fay Zee Siu, Salathiel, Wai Chee
Florida, USA	Mauritius, Brewster

Table 5.2. Major longan cultivars.

	Major cultivars
China	Fuyan, Wulongling, Chike, Shixia, Dawuyuan, Guangyan, Chuliang, Gushan No.2, Fengko
Viet Nam	Longnhan, Tieuhue (Mekong Delta), Longhungyen (Hanoi)
Thailand	Daw, Chompoo, Haew, Biew Kiew, also Phetsakon (tropical cultivar)
Australia	Kohala, Chompoo, Haew, Biew Kiew
Florida, USA	Kohala

pineapple, mango and avocado. There have been difficulties in the exchange of genetic material because of restrictions imposed by some countries, and the misnaming of many cultivars.

Cultivar development in China

The crops were first mentioned in records from China in the 2nd century BC, while unofficial records date back 1600 years earlier (Chen and Huang, 2001). Tien (1979) discovered a 1200-year-old litchi tree ('Chenzi', 'Brewster'?) in Fujian that still produced fruit. The tree was more than 6 m high, with a crown 7 m wide. Trees of similar age have also been found in Guangdong (Fig. 5.3) and Sichuan, suggesting a long history of cultivation in the region (Zee *et al.*, 1998).

Fig. 5.3. Ancient litchi tree in Guangdong, southern China (photograph courtesy of Keith Chapman).

Cai Xiang listed 32 cultivars in Fujian in his 'Litchi Register' in 1059, while Chen Ding described 43 cultivars in the 17th century, including 17 in Fujian, 6 in Sichuan, 3 in Guangxi, and 14 in Guangdong. The number in Guangdong increased to 58 in the Register compiled by Wu Yingkui in 1826, including some of the main commercial varieties grown today such as 'Guiwei', 'Heiye', 'Huaizhi', 'Sanyuehong' and 'Feizixiao'. The names of cultivars in China and their equivalent name in English are shown in Table 5.3. New orchards were initially based on seedlings; however, air-layering (marcotting) was introduced in the 4th century, and grafting 1000 years later. Propagation by seed was eliminated in the 16th century.

Longan also has a rich diversity in China, with Qiu and Zhang (1996) recording 128 cultivars. The National Longan Germplasm Repository established in Fujian in 1981 holds a collection of 202 specimens, some of which have been cultivated for hundreds of years (Liu and Ma, 2001).

General description of the fruit

The internal and external characteristics of litchi fruit are shown in Fig. 5.4. Longans are similar, except that there are no obvious protuberances or suture line on the skin. Litchi fruit are 3–5 cm in diameter and longan 1.5–3 cm. The shapes of the fruit are highly variable, but tend to be round to egg- or heart-shaped in litchi, and globose to round in longan. At maturity, the skin is various shades of red in litchi, and green-yellow brown in longan. Some litchis also have distinctive yellow blotches on the skin. The skin of litchi has protuberances on each segment, which can be smooth, sharp-pointed or hairlike and sharp, whereas they are absent in longan. The flesh or aril is an outgrowth of the seed stalk, and grows as the fruit develops until it completely covers the seed. It is translucent white in litchi and translucent white to off-white in longan. A good litchi is similar in taste to a prime muscat grape; however, many people prefer longan. Litchi is generally more acidic and juicier than longan, but both are bland when over-mature.

Each fruit contains one chestnut-brown to dark-brown, round, ovoid to oblong seed. In litchi, a high proportion of the seeds may be abortive, small or shrivelled ('chicken tongue' seeds). Cultivars with abortive seeds are preferred and often attract a higher price than full-seeded cultivars. This is less common in longan. Fruit weigh 16–35 g in litchi, with a flesh recovery of 50–70%. Longans weigh 10–20 g, but have a slightly higher flesh recovery of 60–75%.

Characteristics used to identify cultivars

Cultivar descriptions are better developed for litchi than for longan (Menzel *et al.*, 1990; Menzel and Simpson, 1990, 1991). Cultivars can be described in terms of general tree shape

Table 5.3. Names for the main litchi cultivars in China and Australia. 'Kwai May Pink' in Australia is not known as a separate cultivar in China. 'Tai So' is similar to 'Mauritius' in many countries.

China	Meaning	Australia	Thailand
Sanyuehong	Third month red	Sum Yee Hong	
Baitangying	White sugar jar		
Baila	White wax litchi	Bah Lup	
Shuidong	East of the waterways	Souey Tung	
Feizixiao	Concubine laughing	Fay Zee Siu	
Dazao	Big crop	Tai So	Hong Huay
Heiye	Black leaf	Haak Yip	O-Hia
Nuomici	Glutinous rice grain	No Mai Chee	
Guiwei	Cinnamon flavour	Kwai May Red	
Huaizhi	Cherished litchi	Wai Chee	Kim Cheng
Chenzi	Chen purple	Brewster	
Lanzhu	Difficult to obtain		

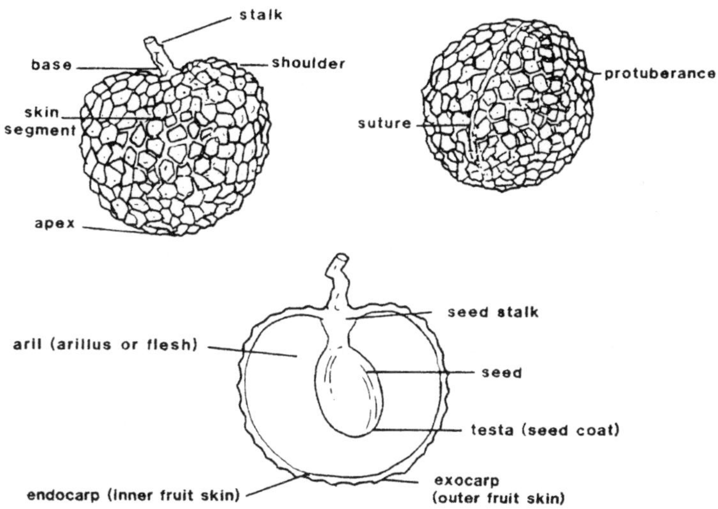

Fig. 5.4. Characteristics of litchi fruit. Longan is similar to litchi, but has smooth skin, with no obvious suture line (from Menzel and Simpson, 1990, adapted from Anonymous, 1978, with permission).

and appearance, along with the shape, size and colour of the leaves, harvest period and several fruit characteristics (Fig. 5.5).

The harvest normally lasts 5–10 weeks for a range of varieties growing in any one location, with cultivars broadly classified as early, mid- or late maturing. However, the order of harvest varies from year to year depending on seasonal conditions, with some variation between different regions, presumably due to differences in environment or culture. Longans normally flower and fruit 1–2 months later than litchis grown in the same location.

Cultivars can be identified by the size and shape of the tree and length and spread of branches, although these can change with weather and culture. For example in litchi, 'Brewster' ('Chenzi') is vigorous and erect with very wide, strong crotch angles; 'Tai So' ('Dazao') is vigorous, with a spreading habit and sharp, weak crotch angles; while 'Wai Chee' ('Huaizhi') is slow-growing, compact and dome-shaped. Litchi trees have long, thin branches and a dense canopy, compared with longan, and are more susceptible to wind damage. In longan, 'Haew' is more brittle than the other Thai cultivars.

Leaves vary in size, shape and colour. For example in litchi, 'Tai So' has large, glossy, dark green leaflets that have an upward curl from the midrib, becoming almost canoe-shaped. 'Bengal' has large leaflets that are mid-green with a distinctive twist along their length. 'Haak Yip' ('Heiye') has dark, glossy green leaflets that are long, narrow-pointed and slightly curled at the tip. 'Wai Chee' leaflets are small, oval, and curve upwards from the midrib and down towards the tip. The new flush of growth is red in 'Wai Chee' and 'Kwai May Pink' and green-bronze in 'Tai So'. In longan, 'Chompoo' has large, narrow leaves compared with the other cultivars in Thailand. A special characteristic of 'Baidum' in Thailand is its small, dark green leaves.

Fruit shape in litchi is very distinctive (Fig. 5.5). The round fruit of 'Kwai May Pink' distinguishes it from the egg shape of 'Tai So' or the heart shape of 'Haak Yip'. The shoulders of the fruit can be smooth or flat as in 'Wai Chee' and 'Kwai May Pink', or uneven as in 'Souey Tung' ('Shuidong') and 'Bengal'. The apex or tip of the fruit can be round as in 'Kwai May Pink' and 'Wai Chee', obtuse or blunt as in 'Souey Tung' and 'Brewster', or pointed as in 'Bengal'.

Typical colours in litchi are bright red ('Bengal'), dull red ('Wai Chee'), purple-red ('Haak Yip'), pink-red ('Brewster') or orange-red ('Kwai May Pink'). The skin can be thick as in 'Wai Chee', 'Bengal' and 'Kwai May Pink', or thin as in 'Haak Yip' and 'Souey Tung'. Skin segments at full maturity can be smooth ('Haak Yip'), swelling ('Wai Chee'), or sharp-pointed

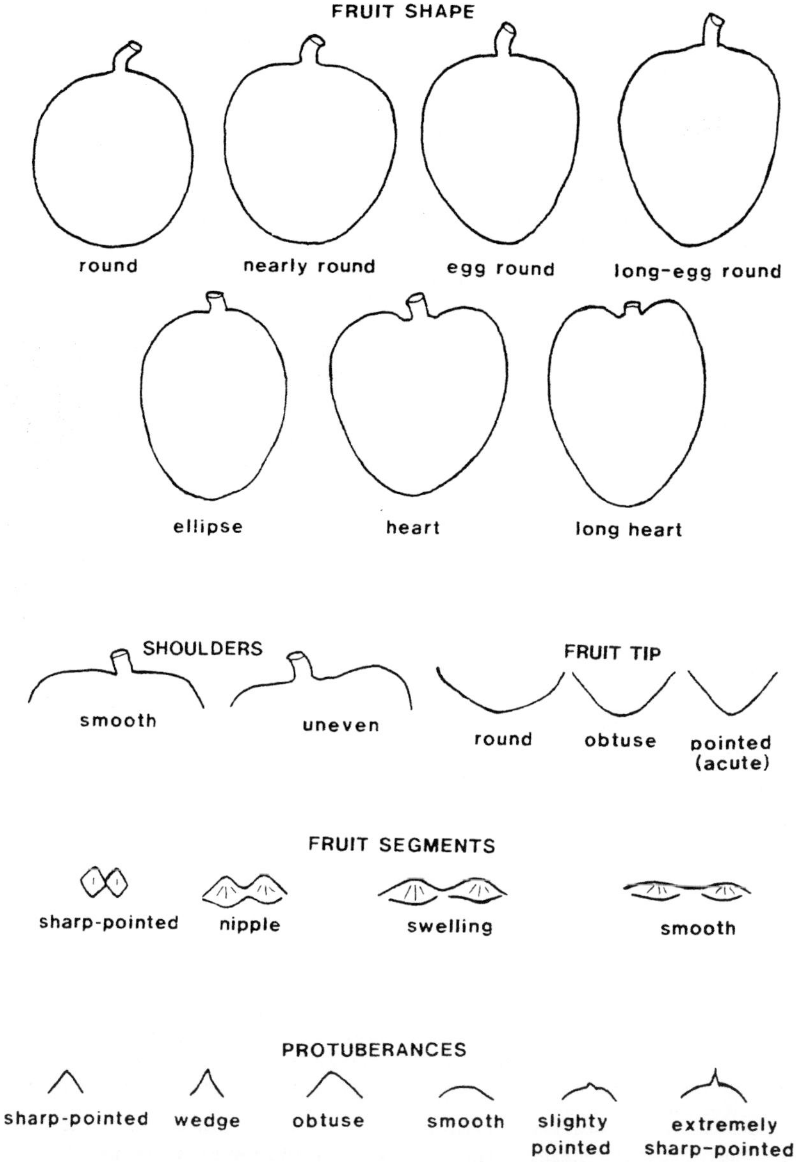

Fig. 5.5. Characteristics used to describe fruit in different litchi cultivars. Longan is similar to litchi, but has smooth skin (from Menzel and Simpson, 1990, adapted from Anonymous, 1978, with permission).

('Kwai May Red'). Similarly, the protuberances on each segment can be smooth as in 'Haak Yip', sharp-pointed as in 'Kwai May Red' and 'Bengal', or hairlike and sharp as in 'Tai So'. The presence or absence of an obvious suture line can distinguish 'Haak Yip' and 'Souey Tung'. The skin of longan is smooth without any obvious suture line. 'Chompoo' has light green-brown skin compared with yellow-brown in the other Thai cultivars. 'Biew Kiew' has a thicker skin than some of the other Thai cultivars and is less susceptible to damage during transport.

The texture, juiciness, taste and aroma of the flesh can aid description, although experience is needed to make clear distinctions.

For example in litchi, 'Wai Chee' is 'watery', 'Kwai May Red' is 'firm', 'Kwai May Pink' is 'spicy', and 'Bengal' is 'very sweet'. The proportion of small seeds is important, but varies with season and orchard. In Australia, most 'Salathiel' fruit have small seeds, while 'Bengal', 'Souey Tung', 'Haak Yip' and 'Wai Chee' have few small seeds. This characteristic varies in 'Tai So' and 'Kwai May Pink'. 'Biew Kiew' longan has a sweet crisp aril, whereas 'Daw' is tougher and less sweet. It has smaller seeds than 'Haew'.

Litchi in China and Taiwan

Several cultivars are important in Guangdong, whereas in Fujian and Guangxi, one or two dominate (Table 5.4). 'Nuomici' ('No Mai Chee') and 'Guiwei' ('Kwai May Red') have excellent eating quality and a high proportion of chicken tongues or aborted seeds. 'Feizixiao' ('Fay Zee Siu') is also popular because of its excellent eating quality and size (28–32 g). Some cultivars are best eaten fresh while others are more suitable for canning or drying. Cultivars exported include 'Sanyuehong' ('Sum Yee Hong'), 'Feizixiao', 'Heiye' ('Haak Yip'), 'Guiwei', 'Huaizhi' ('Wai Chee') and 'Nuomici'. 'Heiye' is the most popular cultivar in Taiwan and accounts for over 80% of plantings. The remaining cultivars include 'Feizixiao', 'Nuomici', 'Sanyuehong' and 'Sah Keng' (Chung Rey Yen, Taiwan, 2002, personal communication).

Descriptions of the major cultivars are based on Fu *et al.* (1985), Zhang *et al.* (1997), Wu (1998) and the authors' data. The shape of the skin segments and protuberances is more reliable than fruit size, shape or taste for identification (Table 5.5).

- 'Sanyuehong' ('Sum Yee Hong', 'Sun Yueh Hong', 'Third Month Red') has a history of 700 years in China and is widely cultivated in Guangdong, Guangxi, Hainan and Taiwan. Fruit mature in May in Guangdong, the third month of the Chinese lunar calendar (earlier in Hainan and Guangxi). The tree is open, spreading and of medium size, with long, thin, fragile branches. The leaflets are long, narrow, shiny dark green and much thicker than in other cultivars. The fruit weigh 37–42 g and are heart-shaped or uneven heart-shaped, with thick skin and flat, unevenly sized segments. The flesh is waxy white, juicy and sweet-acid (17% total soluble solids or TSS, and 0.3% acid). Seeds are mostly large, giving a flesh recovery of 65%. Although 'Sanyuehong' is of only average eating quality, it fetches a good price in the market because it is one of the first cultivars available.
- 'Baitangying' ('White Sugar Jar') is indigenous to western Guangdong. The tree is open with dark brown branches and oval,

Table 5.4. Importance and characteristics of different litchi cultivars in Guangdong and Fujian. From Chen and Huang (2001). 'Wai Chee' accounts for over 80% of plantings in Guangxi.

Cultivar	Area (ha)	Harvest	Comments
Guangdong			
Baila	27,000	Early June	Mainly for export; alternate bearing
Baitanying	27,000	Early June	Unreliable flowering
Heiye	34,000	June	Suitable for export, canning and drying
Feizixiao	27,000	Mid-June	Regular bearing and high yields
Guiwei	60,000	Late June to early July	Unreliable flowering; many large seeds some years
Nuomici	60,000	Late June to early July	Unreliable flowering and susceptible to cracking; high market prices
Huaizhi	40,000	Early to mid-July	Regular bearing; suitable for canning and drying
Fujian			
Lanzhu	25,000	Early to mid-July	Regular bearing; suitable for fresh market and drying
Wuye	13,000	Early to mid-July	Regular bearing; suitable for fresh market, canning and drying

Table 5.5. A key to the major litchi cultivars in China (from Menzel and Simpson, 1990, adapted from Anonymous, 1978).

A. *The protuberances are protruding and hard*
1. The protuberances are relatively fine, dense and sharp-pointed. The skin segments are small and irregularly arranged. *Group 1.* 'Dazao' type.
2. The protuberances are large, sharp and short-pointed. The skin segments are small and regularly arranged. *Group 2.* 'Guiwei' type.
3. The protuberances are relatively blunt and short. The skin segments are relatively large, and irregular in size and arrangement. There are often small segments among the normal skin segments. *Group 3.* 'Jingfeng' type.

B. *The protuberances are hairlike or sparse, fine and sharp-pointed*
1. The skin segments are irregular in size and arrangement. The fruit shoulder is extremely wide and pronounced. The stalk is thick and strong. *Group 4.* 'Sanyuehong' type.
2. The skin segments are regular in size and arrangement. The fruit shoulder is flat. *Group 5.* 'Heiye' type.

C. *The protuberances are smooth or not evident*
1. The skin segments are obviously protruding, usually long and narrow, and arranged in longitudinal rows. *Group 6.* 'Nuomici' type.
2. The skin segments are smooth or slightly protruding, usually near round in shape and irregular in arrangement. *Group 7.* 'Huaizhi' type.

light green leaflets. Fruit weigh 25 g and are uneven heart-shaped with smooth, bright red skin. The flesh accounts for about 72% of fruit weight and is waxy white, crisp and very sweet (18% TSS and 0.05% acid). Seeds are normally large, giving a flesh recovery of 65%. General fruit quality is rated as 'superior', with fruit harvested in late May in Guangdong.

- 'Baila' ('Bah Lup', 'Dianbaibaila') is grown widely in western Guangdong, southern Hainan and eastern Guangxi. The tree is medium in vigour, dome-shaped and reasonably productive. Leaflets are long, narrow, dark glossy green with short, pointed tips. Fruit are available in late May in Guangdong and are nearly heart-shaped, 25 g, with thin, soft, smooth, brilliant purple-red skin. The flesh is waxy white, juicy and delicately sweet (18% TSS and 0.12% acid). The fruit contain a large oval seed, giving a flesh recovery of 72%. Fruit are rated better than 'Sanyuehong' but not as good as 'Baitangying'. A proportion of the crop is exported each year, with cropping being biennial in many orchards.
- 'Shuidong' ('Souey Tung', 'Yuanzhi') tolerates wet soils and is commonly planted along waterways in Guangdong. Trees have been distributed to Australia, but the cultivar is not widely planted there. Productivity is average in northern Australia and poor in southern districts. The tree is relatively low, with thin, long, open, spreading branches that hang down. Leaflets are large, flat, dark glossy green and pointed. The inflorescences are large compared with many other cultivars. Fruit weigh 25 g and are heart-shaped with distinctive uneven shoulders. The skin is thin, smooth, and dull purple-red. The fruit tip is obtuse or blunt. The flesh is soft, juicy and sweet. Seeds are mostly of medium size, giving a flesh recovery of 72%. The fruit ripen from late May to early June in Guangzhou, and between November and December in northern Queensland.
- 'Feizixiao' ('Fay Zee Siu', 'Smile of the Imperial Concubine') has a long history in China and is reported to have 'delighted the Emperor and his concubines'. The fruit are described as having 'the colour of amber, the size and shape of a goose egg, and the sweetness of honey' (Plate 20). It is widely cultivated in Guangdong, Guangxi, Hainan and Taiwan and has also been distributed to Australia. Productivity is variable in Australia, even in northern districts. The tree is vigorous, with long, sparse, fragile branches. Leaflets are large, narrow

and deep glossy green. Fruit weigh 30 g and are round to oval. The skin is thin with dense sharp-pointed protuberances. Fruit are splotchy green, yellow-red at maturity. The flesh is firm, sweet and slightly fragrant (18% TSS and 0.2% acid). Seed size is variable, giving a flesh recovery of about 80% when there are many small seeds. Fruit mature in early June in Guangzhou and in October in northern Queensland.

- 'Dazao' ('Tai So', 'Big Crop', 'Mauritius', 'Hong Huay') has a history of 500 years in Guangdong, but is also found in most of the litchi-growing areas of the world (Plate 21). A similar cultivar, 'Mauritius', selected from seedlings in Mauritius in the 1870s (Ledin, 1957) forms the basis of industries in southern Africa (Plate 8), Israel and Florida. Practically all the trees in South Africa are descendants of a few trees imported from Mauritius. 'Hong Huay' in Thailand is also very similar. 'Dazao' ('Tai So') was the most important litchi cultivar in Australia, but has now lost favour because it bears erratically in most districts. There are also problems with the trees splitting and the fruit turning brown or cracking before harvest. Trees are vigorous and spreading, with an open crown, and have branches that tend to split from the trunk. Leaflets are large, glossy dark green and almost canoe-shaped. Fruit weigh 24 g and are oblong, with uneven shoulders. The bright red skin is thin, with small, sharp-pointed protuberances. The flesh is milky white, slightly tough, with 15% TSS and 0.30% acid. Flavour is sweet-acid when immature, sweet when fully ripe, and bland when overripe. Seeds are generally large, giving a flesh recovery of 75%. Fruit mature in early June in Guangdong and in early December in northern Queensland.
- 'Heiye' ('Haak Yip', 'Hak Ip', 'Black Leaf','O-Hia') has a history of over 600 years in China and is the most widely grown cultivar in southern districts (Plate 22). It is also popular in Thailand, where it is known as 'O-Hia', but has undergone limited distribution elsewhere. There are small plantings in Australia, although production is erratic and fruit quality rated as 'average'.The trees and fruit resemble 'Shuidong' ('Souey Tung'). They are vigorous and dome-shaped, with sparse foliage and long branches. The leaflets are long, narrow-pointed and slightly curled at the tip. The name of the cultivar refers to the 'very dark glossy green leaves'. The fruit weigh 19 g and are uneven heart-shaped or oval, with tough, smooth, dark red skin. The flesh is milky white, juicy, sweet and slightly fragrant (18% TSS and 0.4% acid). Seeds are of medium size, giving a flesh recovery of 65%. Quality is rated 'above average' in China. Fruit are harvested in late June in Guangdong and in late December in northern Queensland. 'Heiye' is the main cultivar used for canning in South-east Asia.
- 'Nuomici' ('No Mai Chee', 'No Mai Tsz', 'Glutinous Rice Cake') has a history of more than 200 years in Guangdong and is widely grown in the Pearl River Delta, as well as in various districts of Guangxi, Fujian and Taiwan. It is one of the most highly prized cultivars in China and usually sells for three to four times the price received for other cultivars (Plate 27). Fruit are suitable for the fresh market and for drying. The tree is umbrella-shaped, with a dense canopy and slim branches that hang down. The leaflets are small, soft and thin, with wavy edges. The fruit weigh 25 g and nearly all are small-seeded, giving a flesh recovery of 85%. The fruit are heart-shaped and bright red at maturity. The skin is thin and smooth, and the flesh juicy, with a distinctive sweet fragrant flavour. Fruit appear on the market in late June in Guangzhou. Cropping is irregular and the fruit susceptible to cracking; however, the high price received guarantees a good return to most growers. 'Nuomici' ('No Mai Chee') has been introduced to Australia and Israel, but is not widely grown or cropped as heavily as in China.
- 'Guiwei' ('Kwai May Red', 'Kwai Mi') is native to Luogan in Guangzhou, and is widely grown in Guangdong, Guangxi and southern Taiwan (Plate 24). It is another highly prized cultivar. 'Guiwei' was introduced into Australia where it is known as 'Kwai May Red', but it is not widely planted. A similar cultivar, 'Kwai May

Pink', forms the basis of the industry there (Plate 25). In Australia, trees resemble those of 'Kwai May Pink', but are more spreading. Leaflets are slightly larger than 'Kwai May Pink' and are flatter. The tree is dome-shaped, with sparse, slim, hard branches. Leaflets are short, pointed at the tip and curl upward from the midrib. The fruit weigh 18 g and are round, with thin, bright red skin. The segments are swollen and the protuberances sharp-pointed. The flesh is milky white, crisp, sweet and fragrant (19% TSS and 0.18% acid). Seeds are small in most seasons, giving a flesh recovery of up to 80%. The fruit mature in late June in Guangdong, the same season as 'Nuomici', with consumers in China generally rating 'Guiwei' as superior. In Australia, fruit are almost identical to those of 'Kwai May Pink', except that 'Kwai May Red' has red rather than pink-orange skin, firmer flesh, a higher proportion of small seeds (50–60%), higher flesh recovery (70–80%), and a slightly better flavour.

- 'Huaizhi' ('Wai Chee', 'Cherished Litchi', 'Kim Cheng') is one of the most common cultivars in Guangdong and Guangxi and is also popular in Thailand ('Kim Cheng') and Australia (Plate 28). It crops heavily in most areas in China, although the trees are slow to establish after planting. Productivity is better in southern Queensland than in northern districts. Trees are low and dome-shaped, with thick, hard branches and compact foliage. The small leaflets are oval and curve upwards from the midrib and down along their length. The fruit weigh 20 g and are round and dark red. The skin is thick and tough, with large, flat segments. The flesh is waxy white, juicy and sweet (19% TSS and 0.25% acid). The seeds are nearly all large, giving a flesh recovery of 70%. Fruit quality in China is rated as 'above average'. Although fruit of 'Huaizhi' have full flavour, their large seeds and soft flesh reduce eating quality and therefore the price in Asia, compared with 'Heiye' ('Haak Yip'), 'Guiwei' ('Kwai May Red') and 'Nuomici' ('No Mai Chee'). Fruit are available in early July in Guangdong, and in early February in southern Queensland. The mature fruit can be left to hang on the tree for several days, adding flexibility to harvesting.
- 'Chenzi' ('Chen Family Purple', 'Brewster'), is widely cultivated in Fujian and was taken to Florida by Reverend W.M. Brewster in 1903 (Groff, 1951). It was the major cultivar in Florida, although 'Mauritius' is now more important. Production in Florida is very erratic. It was also sent to Australia, but is not popular there. A similar cultivar, 'Floridian', was sent to Israel from California (Goren *et al.*, 2001). The tree is open, with a dome-shaped crown and dense branches that hang down. The leaflets are long, oval, and pointed at the tip. The fruit weigh 15 g and are mostly heart-shaped with dark purple-red skin that has small segments and sharp protuberances. The flesh is waxy white, juicy, fragrant and slightly sour at maturity (19% TSS, 0.22% acid). There are many small seeds, giving a flesh recovery of 75%. Fruit quality in China is rated as 'average'. Fruit mature in mid- to late July in Fujian and in late December in southern Queensland.
- 'Lanzhu' ('Difficult to Obtain') has a history of 300 years in China, and is one of the most popular cultivars in Fujian. The tree is vigorous and semi spherical, with a dense canopy and slim branches that hang down slightly. The leaflets are long and oval, with straight edges and short, pointed tips. The fruit weigh 26 g and are heart-shaped to nearly round, and yellow-purple red. The skin is thin, with large segments and pointed protuberances. The flesh is milky white, juicy and slightly sour (16% TSS). The seeds are generally aborted (flesh recovery of 75%) and fruit quality is rated as 'above average'. Fruit are available from late June to early July in Fujian.
- 'Sah Keng' was developed in Taiwan in the 1970s and is a seedling of 'Haak Yip' (Yen, 1984). It was introduced into Australia but is not grown commercially outside Taiwan. 'Sah Keng' produces large fruit with small seeds, with significant variation amongst trees in a single orchard. Fruit are available mid-season. Yields are heavy but irregular. Trees are of medium height, dome-shaped, with short, fragile branches and

mid-green leaves. Fruit weigh 30–35 g and are heart-shaped with purple-red skin. The skin segments are swollen and protuberances blunt. The flesh is soft and sweet. Seeds are variable, often small, giving a flesh recovery of 75%. Fruit quality is rated as 'good or better'.

Litchi in Viet Nam and Thailand

Northern Viet Nam includes part of the area where litchi originated (Papademetriou and Dent, 2002) (Plate 10). Eight varieties are grown commercially; however, a single cultivar, 'Thieuthauhha', introduced from China about 100–300 years ago, accounts for 80% of plantings. Because the industry is based on a single cultivar, the harvest is unduly short, lasting only 4–6 weeks from late May to early July. Fruit of 'Thieuthauhha' weigh 20 g and are blotchy yellow-red, with the segments swollen at maturity. The flesh is firm, sweet and fragrant (16% TSS and 0.28% acid). The seeds are small or abortive, giving a flesh recovery of 75%. Fruit are suitable for the fresh market but sometimes discolour after canning. Expansion of the industry in Viet Nam has been dependent on this cultivar.

Thailand has both tropical and subtropical litchis (Subhadrabandhu, 1990; Subhadrabandhu and Yapwattanaphun, 2001; Yapwattanaphun and Subhadrabandhu, 2001). The cultivars exploited in the subtropics are mainly of Chinese origin, whereas production in the tropics is based on local selections from material introduced from China (Plate 70). 'Hong Huay' ('Dazao', 'Tai So') accounts for 65% of production and is grown in the north. 'Kim Cheng' ('Huaizhi', 'Wai Chee'), 'O-Hia' ('Baidum', 'Heiye', 'Haak Yip') and 'Chakrapad' ('Chacapat') are also grown around Chiang Mai and Chiang Rai. 'Kom' accounts for 11% of plantings and is the main 'tropical' cultivar. Other cultivars grown around Bangkok include 'Sam Prae Kaew', 'Kaloke Bai Yao' and 'Luk Lai'. The quality of these selections is poorer than those in the north. A description of some of the cultivars is given below (Menzel and Simpson, 1991; Yapwattanaphun and Subhadrabandhu, 2001). The Thai authors provide a description for 'Hong Huay', but since it is the same as 'Dazao' or 'Tai So', no further details are presented here.

- 'Kom' crops under tropical conditions. The trees are vigorous and erect and have long, strong branches and dense foliage. Leaflets are narrow, pointed, medium in size and dark green. They are generally flat, but curve downwards slightly towards the tip. The new flush of growth is red, changing to green with maturity. Fruit weigh 8–20 g and are long heart-shaped to nearly round. The very thick skin is blotchy yellow-purple red at maturity. Shoulders are flat or even and the fruit apex is obtuse. The skin segments are smooth at maturity and variable in size, shape and arrangement. The protuberances are sharp-pointed. Fruit are borne in small, loose clusters. The flesh is tough to fibrous and mild in taste, becoming bland at maturity. Seed and fruit size are in proportion, with small fruit having 'chicken tongues'. Flesh recovery ranges from 60% to 80%. Fruit are available early in the season.
- 'O-Hia' ('Baidum', 'Heiye', 'Haak Yip') resembles 'Haak Yip' but does not match it in all characteristics. Trees are of medium size with dense foliage on long, thin branches (not as long as 'Haak Yip'). Leaflets are large, narrow, dark green and slightly curled upwards from the midrib. The new flush of growth is red-brown. Fruit weigh 21 g and are heart-shaped. The skin changes from blotchy yellow to deep red with maturity. Skin segments are irregular in size, shape and arrangement, with smooth to obtuse protuberances. The flesh is juicy and sweet. Seeds are mostly plump, giving a flesh recovery of 65%. Fruit are available mid-season.
- 'Chakrapad' ('Chacapat') produces very large fruit late in the season. Fruit are sweet and acceptable in Thailand, but are often acidic in Australia. Fruit production in Australia is also average. Trees are moderately vigorous, erect, and have long branches and dense foliage. Leaflets are small, long, narrow, pointed and dark green. They curl upwards from the midribs and downwards along their length towards the tip. The new growth is green. Fruit weigh 30 g and are

round to slightly heart-shaped. The skin is thin and soft, deep red with yellow markings. Shoulders are flat and the fruit tip round. Skin segments are swelling, with obtuse protuberances. The flesh is moderately juicy, remaining slightly acid when fully ripe in Thailand, and very acid in Australia. Seeds are nearly all large, giving a flesh recovery of 65%.

- 'Luk Lai' trees are medium in vigour and dome-shaped, with many short upright branches and dense foliage. Leaflets are mid-green, and broad oval in shape. The leaflets are curled slightly up from the midrib to the margins and curled down along their length towards the tip. Leaflets characteristically hang vertically from the branches. The colour of the new growth is brown-green changing to light green with advancing maturity. Fruit weigh 17 g and are almost round to slightly egg-shaped, with flat shoulders and an obtuse apex. The skin is yellow-red changing to deep red with maturity. The segments are smooth and irregular in size and arrangement. The skin is relatively thick, and the protuberances very sharp and strong. Fruit are normally produced in small, tight clusters with even maturity. The flesh is moderately crisp and separates easily from the seed. Flavour is subacid to sweet, becoming bland once the fruits are fully mature. Most fruit contain a medium to large, dark-brown seed, giving a flesh recovery of 60%. Flavour is acceptable if the fruit are picked at the proper stage of maturity, but the large seeds limit commercial acceptance of this variety.

Litchi in India, Bangladesh and Nepal

Most of the cultivars grown in India are selected seedlings from material sent from China that can crop in the hot and dry conditions (Ghosh *et al.*, 2001). Of the 30 or so varieties grown, only six are commercially important: 'Shahi' ('Muzaffarpur'), 'China', 'Calcuttia', 'Bedana', 'Late Bedana' and 'Longia' (Papademetriou and Dent, 2002). These cultivars generally have large fruit and excellent eating quality.

In the hilly areas of Nepal, commercial production is based on various seedlings, whereas in the plains there are established cultivars that probably came from India ('Majafpuri', 'Raja Saheb', 'Deharaduni', 'China' and 'Calcuttia'). The history of litchi in Bangladesh is unclear, although the species was probably introduced from Myanmar in the 1800s, with direct Chinese and Indian imports soon following. The most important cultivars are 'Bombai', 'Muzaffarpuri', 'Bedana' and 'China Number Three'. 'Bombai' is the oldest cultivar. 'Bedana' has the best quality but is low-yielding. A description of some of these cultivars is provided below (Menzel and Simpson, 1991; Ghosh *et al.*, 2001; Papademetriou and Dent, 2002).

- 'Shahi' ('Muzaffarpur', 'Rose Scented') is the most popular cultivar in Bihar and is also found in other parts of India, as well as in Bangladesh. Fruit weigh 23 g and are oval with crimson-red skin. The flesh is juicy, sweet and fragrant, and the seeds variable in size. Yields are heavy and regular, with large trees carrying 100–150 kg of fruit early in the season, although the fruit can split.
- 'China' ('Purbi', 'Calcuttia', 'Bengalia', 'Bombaiya') is an important cultivar in India, which ripens when most of the other cultivars have been harvested. Its origin has not been determined, although there is a similar cultivar in Bangladesh known as 'China Number Three'. Trees are relatively small, high-yielding, but alternate bearing. Fruit weigh 25 g and are orange-red. The flesh is soft, juicy and very sweet, but is inferior to 'Shahi'. Seeds are normally small.
- 'Early Bedana' ('Early Seedless') is a popular early cultivar from Bihar, Uttar Pradesh, the Punjab and Bangladesh. Trees are medium in height and yield, with regular fruit production. Fruit weigh 17 g and are oval or heart-shaped with rough, deep red skin at maturity. The flesh is white, juicy and sweet. Overall fruit quality is rated as 'good.'
- 'Late Bedana' ('Late Seedless') is a late cultivar from northern India. Trees are vigorous and productive, with average yields of 60–80 kg for 10-year-old specimens.

Fruit are of medium size with good flesh recovery. The flesh is creamy white, juicy and sweet. Seeds are usually small.

- 'Bombai' is an important early cultivar from West Bengal and Bangladesh. Trees are vigorous and productive. Ripe fruit are an attractive deep red, with grey-white, juicy and sweet flesh. It is similar to 'China' grown in other areas.
- 'Dehra Dun' ('Dehra Rose') is an important cultivar from Uttar Pradesh and the Punjab. Trees are of medium height and produce moderate to heavy yields. Fruit are bright pink-red at harvest and very attractive. They have small seeds, but are susceptible to cracking.

Litchi in the Philippines and Indonesia

The Philippines produces many tropical fruit including banana, pineapple, mango and citrus (Papademetriou and Dent, 2002). Litchi is also grown, but on a much smaller scale. 'Mauritius' and a local selection from China, 'Sinco', dominate production in the hilly areas, while 'UPLB Red', a cultivar introduced from Thailand, is planted in the lowlands. The Department of Agriculture is also evaluating two new selections, 'Olan' (from Thailand) and an unnamed selection for the warmer areas. Fruit of 'UPLB Red' weigh 14 g and are ovate to almost round, with rough, dark-red skin. The flesh is sweet (20% TSS) and comprises 60% of fruit weight. Fruit of 'Olan' weigh 26 g and are ovate, with a TSS of 18% and a flesh recovery of 60%. Trees bear regularly.

Litchi is a minor fruit in Indonesia, where plantings are less extensive than longan or rambutan (Papademetriou and Dent, 2002). Commercial production is based on local selections.

Litchi in southern Africa, Israel, Australia and the USA

The litchi was taken to southern Africa 200 years ago and, since that time, commercial industries have developed in South Africa, Zimbabwe, Mozambique, Madagascar, Réunion and Mauritius (Milne, 1999; Jahiel and Abraham, 2001). Production is greatest in Madagascar and South Africa. About 84% of the plantings in South Africa are based on 'Mauritius' ('Dazao', 'Tai So') (Plate 8) and the rest on 'McLean's Red' ('Bengal') (Plate 23). Almost the entire high rainfall area of the eastern seaboard of Madagascar has informal plantings of 'Mauritius'. Réunion and Mauritius lie about 800 km east and have traditionally grown litchi for export to France. Here also, 'Mauritius' is the only cultivar widely grown.

The litchi industry in Israel covers some 300 ha in a Mediterranean climate from 200 m below sea level to 300 m above sea level (Goren *et al.*, 2001). About 80% of the orchards are planted to 'Mauritius' and the remainder to 'Floridian', which may be a seedling of 'Chenzi' ('Brewster') introduced from California.

Litchi was introduced into Australia by Chinese immigrants in the mid-1800s, although commercial production only commenced in the late 1970s (Menzel and McConchie, 1998; Menzel, 2002). 'Kwai May Pink' accounts for more than 50% of plantings, with 'Tai So' ('Dazao'), 'Souey Tung' ('Shuidong'), 'Fay Zee Siu' ('Feizixiao'), 'Salathiel' and 'Wai Chee' ('Huaizhi') being the other main cultivars. A description of some of these varieties is provided (Menzel and Simpson, 1991).

- 'Bengal' ('McLean's Red') is a seedling of the Indian cultivar 'Purbi' sent to Florida in 1929 (Loomis, 1955). At one time, it was the second most important cultivar after 'Tai So' in Australia, but has now lost favour (Plate 23). A similar cultivar, 'McLean's Red', makes up 16% of plantings in South Africa (Milne, 1999). Fruit are attractive and pleasant tasting but have large seeds, poor flesh recovery, and ripen unevenly. Average cropping is disappointing, although trees can have very high yields in an 'on year'. Trees are vigorous and spreading, with thin branches, but are reasonably resistant to wind damage. Leaflets are large, mid-green and have a distinctive twist or curl along their length. The new flush of growth is red-brown. The fruit weigh 23–27 g and are formed in large clusters of up to 50 or more. The skin is thick, very rough and an attractive bright red. The fruit are round-egg-shaped to lop-sided heart-shaped, with uneven shoulders and a distinctive pointed tip. The protuberances are sharp-pointed to wedge-shaped. The flesh is soft, sweet and moderately juicy. The flavour of the fruit deteriorates if they are left to hang. There are very few abortive seeds. Under drought, the aril is undeveloped and does not cover the seed at the pointed end, giving a flesh recovery of 50% or lower. For these reasons 'Bengal' is not considered a good marketing type.

- 'Kwai May Pink' is thought to have originated in China and resembles 'Guiwei' ('Kwai May Red'). It is popular in Australia, with large numbers of trees planted, but is relatively unknown elsewhere. Bearing ability is good in most districts. It has a long harvest period, possibly due to the development of acceptable sweetness and flavour well before the fruit mature. Fruit are available mid-season. Trees are large, very erect, and have long, slim branches that point upwards. They tolerate strong winds to some degree. Leaflets are narrow, long, oval and shiny light green. They curl upwards slightly from the midrib and downwards along the length. The new flush of growth is an attractive red. Fruit weigh 20 g and are round with very rough, thick skin. The skin colour changes from yellow to yellow-pink to orange-pink at maturity, with some green on the shoulders. Flesh is crisp, sweet and aromatic. Fruit are sweet well before they are fully mature. Seeds are variable, with up to 70% of the fruit having chicken tongues, giving a flesh recovery of 70%. Fruit are exported to South-east Asia from Australia.
- The original 'Salathiel' tree was found growing in northern Queensland, but its parentage is unknown. Yields are variable in subtropical districts and light in tropical areas. Fruit are harvested late, just before 'Wai Chee' ('Huaizhi'). Trees are small and compact and sometimes produce long branches with undeveloped leaves. Leaflets are small, broad and curve down slightly at the tip. The tip of the leaflet is round, with a short, distinctive point. The new flush is red, changing to green with maturity. Fruit weigh 20 g and are egg- to ball-shaped and are borne in small, loose clusters (Plate 26). The skin is thick and moderately rough, with prominent markings. It changes from blotchy yellow to deep red at maturity. The fruit tip is obtuse to round. The flesh is thick, crisp, and sweet long before the fruit are fully coloured. Most fruit have chicken-tongue seeds, giving a flesh recovery of 75%. Occasionally, fruit can be almost seedless, although these fruit are very small and unmarketable. Fruit attract a high price in domestic markets and are also exported to Asia.

Litchi production in the USA is confined to Florida, California and Hawaii, with Florida being the most important. The crop was introduced into Florida about 120 years ago, but the industry has been plagued by irregular production, severe frosts in many years and severe hurricanes (Knight, 2001). The total area under crop is 240 ha, based mainly on 'Mauritius' ('Tai So'), with smaller plantings of 'Brewster' ('Chenzi', 'Chen Purple'). Several other cultivars are currently being evaluated (Campbell and Campbell, 2001a). 'Kaimana', developed in Hawaii about 20 years ago from a population of 'Haak Yip' seedlings (Hamilton *et al.*, 1992; Zee *et al.*, 1998), forms the basis of a small industry in Kona and Hilo. A description of 'Kaimana' is given below.

'Kaimana' can bear heavily in Kona, Hawaii, and in some parts of Australia. Fruit mature mid-season. Trees are medium in size and spreading, with long, strong branches. Leaflets are large, elongated and mid-green. The new flush of growth is green. Fruit weigh 25 g and are heart-shaped with purple-red skin. The skin segments are swollen and the protuberances smooth when the fruit are mature. The flesh is crisp, sweet and of excellent quality. Seeds are medium in size.

Longan in China and Taiwan

China is the largest producer of longan, with 608,500 t from 465,600 ha in 2000. The most important areas are Guangdong (346,000 t), Guangxi (150,900 t) and Fujian (104,000 t), followed by Yunnan, Hainan and Sichuan. The main cultivars in Guangxi are 'Dawuyuan' and 'Guangyan' (Liu and Ma, 2001). 'Fuyan', 'Wulongling' and 'Chike' are popular in Fujian, and 'Shixia' and 'Chuliang' are popular in Guangdong. An overview of the Chinese cultivars is based on information collected by the authors.

- 'Fuyan' ('Eye of Fortune', 'Lucky Eye') has a long history in China and has been cultivated in Quanzhou Municipality in Fujian for over 800 years. The crown is open and dome-shaped, with long, oval, mid-green leaflets that are obtuse at the tips. Fruit weigh 14 g and are oblate, with thin, yellow skin at maturity. The flesh is firm, transparent, succulent, sweet (14% TSS) and separates easily from the seed. Seeds are of medium size, giving a flesh recovery of 70%. Fruit are available from late August to early September in Fujian and are best canned. Although the quality of 'Fuyan' is rated 'below average', it crops heavily and is resistant to witches' broom.
- 'Wulongling' ('Wulong Mountain', 'Black Dragon Peak') is a mid-season cultivar from Fujian, where it has been cultivated for more than 130 years. The tree is vigorous with an open canopy and dome-shaped crown. The leaflets are light green and canoe-shaped. The fruit weigh 13 g and are round, with a thick, yellow-brown skin. The flesh separates easily from the seed and is milky white, semi-transparent, soft and sweet (22% TSS). The seeds are medium in size, giving a flesh recovery of 67%. It is one of the most important cultivars in Fujian for drying, but can also be eaten fresh. Cropping is heavy in many orchards. Fruit mature from late August to early September in Fujian.
- 'Chike' ('Red Skin') is mainly cultivated in Fujian. The tree is spreading, with a dome-shaped crown. Leaflets are dark green and long oval in shape, with an obtuse tip and an asymmetric wedge-shaped base. The fruit weigh 15 g and are oblate, with thick, red-brown skin. The flesh separates easily from the seed and is transparent, crisp and sweet (16% TSS). Production is biennial in some districts, with fruit available from late August to early September in Fujian.
- 'Shixia' ('Ten Leaves', 'Stone Gorge') has been known for about 170 years and is mainly cultivated in Guangdong and south-east Guangxi. The tree crown is spherical and open, with the branches typically hanging down. The leaflets are dark glossy green, thick, broad-oval and wavy at the edges. Fruit weigh only 9 g and are round or oblate, with thin, yellow-brown skin. The flesh separates easily from the seed and is creamy to waxy yellow, crisp and sweet (21% TSS). Seeds are small, giving a flesh recovery of 68%. Fruit are harvested in late August in Fujian, with eating quality rated as 'superior'. It is the best fresh-eating cultivar and produces heavy crops on long panicles. Fruit normally sell for twice the price of the other cultivars.
- 'Dawuyuan' ('Large Black Round'), nicknamed 'litchi longan' due to its large fruit, has been cultivated in Guangxi for more than 200 years. The crown is open and dome-shaped. Leaflets are dark green, long oval, slightly pointed at the tip, and asymmetric at the base with wavy edges. The inflorescences are long and the fruit clusters crowded. The fruit weigh 18 g and are round, with moderately thick, yellow-brown skin. The flesh is thick, waxy white, semi-transparent, crisp and sweet (18% TSS). Seeds are large, black and round, giving a flesh recovery of 75%. The fruit mature in late August in Guangxi.
- 'Guangyan' ('Broad Longan') is native to Guangxi and is the most extensively cultivated variety in the Province, where trees more than 400 years old are common. The tree is semi-open and dome-shaped, with slim, dense branches. The trunk is grey-brown with cracked bark. The leaflets are dark green, long oval, blunt at the tip and slightly wavy at the edges. The fruit are round or oblate and weigh 10 g. The skin is green-brown with distinct but smooth segments. The flesh is yellow-white, semi-transparent, crisp and sweet (23% TSS), and has a flesh recovery of 65%. The seeds are round, glossy black and weigh 1.3 g. Fruit quality is rated 'above average' and suitable for fresh consumption and drying.

The crop is harvested in mid-August in Guangxi, with the trees being resistant to witches' broom.

- 'Chuliang', named after the village in Guangdong where it was selected, won the Gold Prize in the First National Exposition of Agricultural Products in 1992. It accounts for more than 30% of the area under longan in China, with about 25,000 ha cultivated in Guangdong up to 1996. The crown is semi-spherical and open. The leaflets are dark, glossy green, canoe-shaped or long-oval, with protruded veins on the lower surface. Fruit weigh 13 g and are oblate, with yellow-brown skin. The flesh separates easily from the seed and is waxy white, crisp, sweet (22% TSS) and fragrant. Flesh recovery is 72%. Fruit mature in late July to early August in Guangdong.
- 'Gushan Number Two' was selected as an elite tree in Jiedong County in Guangdong in 1983 and won the Gold Prize in the Second National Exposition of Agricultural Products in 1995. The fruit weigh 10 g and have thin, yellow-brown skin. The flesh is creamy white, semi-transparent, crisp and sweet (20% TSS) with a recovery of 70%. It produces fruit in early August in Guangdong.

Longan is also a major crop in Taiwan, with 130,050 t from 1200 ha (Liu and Ma, 2001). Productivity is much higher than in mainland China, with 'Fenke' ('Fengko', 'Dusty Skin') accounting for 90% of production (Wong, 2000). The other important cultivars are 'Hongko' and 'Chingko'. Fruit of 'Fenke' weigh 10 g, with a flesh recovery of 65%. The flesh is slightly whitish, crisp and sweet (21% TSS). Fruit do not detach readily from the stalk. Yields are heavy in Taiwan, with harvests from early to mid-August.

Longan in Viet Nam and Thailand

Longan production in Viet Nam is mainly concentrated around the Mekong Delta in the south of the country, with plantings in this area accounting for 70–80% of the crop (Wong, 2000). The rest of the industry is based in the traditional litchi-growing area around Hanoi, with smaller plantings in the south-east. The cultivars 'Longnhan', 'Tieuhue' and 'Xuongcomvang' are planted in the south, and 'Longhungyen' in the north. 'Longnhan' and 'Tieuhue' are true tropical ecotypes and provide two crops per year or three crops in two years. 'Longhungyen' only crops in subtropical climates and produces a single crop each year.

Total production of longan in Thailand in 1998 was 238,000 t from 41,504 ha, of which about 20% was exported (Subhadrabandhu and Yapwattanaphun, 2001). Longan is the main tropical fruit exported from the country. 'Daw' ('E-Daw') is the most popular cultivar and accounts for 73% of plantings. Other cultivars in the north include 'Chompoo' ('Sri Chompoo'), 'Haew' ('E-Haew') and 'Biew Kiew'. 'Phetsakon' is a tropical ecotype grown in the central region of the country. An overview of the most important cultivars is given below (Menzel *et al.*, 1990; Subhadrabandhu, 1990).

- 'Daw' is the most consistent bearer. Fruit are large, with a big seed, thin skin, crisp sweet flesh and generally good flavour. Fruit are available early in the season and do not keep well on the tree. The fruit can be eaten fresh or processed.
- 'Chompoo' has large fruit borne in long clusters, with excellent quality and small seeds. The flesh turns slightly pink after processing and is therefore suitable only for fresh eating. Cropping is not as regular as 'Daw' in Thailand, mainly because of poor flowering. It requires high fertility and good management to yield heavily. Fruit are harvested mid-season. There are also commercial plantings of this cultivar in Australia.
- 'Haew' is a mid- to late-season cultivar that has large fruit with firm flesh and excellent eating quality. Fruit should be tree-ripened for best flavour. The fruit stalk is hard, and hence is difficult to fold for packing. However, the postharvest life of 'Haew' is good because its thick skin reduces water loss. The fruit are suitable for canning. The main shortcoming of 'Haew' is its alternate bearing habit. Plantings of this cultivar have been established in Australia.

- 'Biew Kiew' is highly regarded but is slow to come into production and can crop irregularly in some areas. Trees are also susceptible to witches' broom. Fruit are large with crisp, creamy coloured flesh and excellent quality. Seeds are often small, giving a good flesh recovery. Fruit hang on the tree a few days once they are mature without loss of quality. Fruit are harvested late in the season. 'Biew Kiew' has also been planted in Australia.
- 'Baidum' bears regularly and has an acceptable flavour and long harvest period, but the fruit are small and susceptible to skin cracking, scales and sooty mould. It is not as popular as the other cultivars.
- 'Phetsakon' is grown in the central region of Thailand in Samut Sakhon and Ratchaburi. It flowers readily under tropical conditions, with fruit available earlier than that produced in the more traditional northern areas. However, fruit quality is inferior compared with the main cultivars grown in Chiang Mai.

Longan in Australia and the USA

Commercial longan production in Australia is no more than litchi, with 1200 t produced (Menzel and McConchie, 1998; Nicholls, 2001). About 80% of the orchards are located near Mareeba on the Atherton Tableland, with smaller plantings along the east coast, south to Byron Bay in northern New South Wales. The main cultivars exploited include 'Chompoo', 'Haew' and 'Biew Kiew' from Thailand, and 'Kohala' from Hawaii. 'Kohala' is the earliest cultivar and has large fruit, but the taste is bland in some districts.

Longan was introduced to Florida about 70 years ago, although a commercial industry developed only in the 1980s with 150 ha planted mainly around Miami and Dade Counties (Campbell and Campbell, 2001b). The orchards are generally based on 'Kohala' (see below) introduced from Hawaii in the 1950s (Whitman and Wirkus, 1958), although other cultivars are being evaluated. There are small plantings of 'Biew Kiew' from Thailand and 'Diamond River' from Malaysia.

- 'Kohala' was sent to Florida from Hawaii in 1954 (Knight *et al.*, 1968) and was subsequently distributed to Australia, Israel and elsewhere. It produces large, sweet fruit with a high proportion of aril. Thinning of flowers and fruit is required in heavy production seasons to reduce the proportion of small, unmarketable fruit. Fruit quality is inferior compared with the better cultivars from Thailand.

Evaluation of litchi and longan cultivars

There is little published information on the yields and quality of different cultivars in both species. Many authors provide information on quality, but these are often based on general descriptions or data collected from single trees. In replicated experiments, data may still be based on only three or four replicate trees for each cultivar. Cultivars for evaluation are also often planted at different times and may not bear commercial crops for 4–5 years. Surprisingly, few data sets are available from China, the home of both crops. There is a need to systematically evaluate the productivity of the major commercial cultivars.

Ghosh *et al.* (2001) examined the performance of litchi cultivars in West Bengal, India over 10 years. Data were collected on yield, fruit cracking, fruit weight and concentrations of sugar and acid in the fruit. 'Bombai', 'Muzaffarpur' and 'Rose Scented' had the highest yields, whereas 'Bedana' and 'China' had the largest fruit and highest flesh recovery, associated with a high proportion of fruit that had small or shrivelled seeds (Table 5.6). Total soluble solids was greater than 16% in all the cultivars. Many cultivars in India are susceptible to fruit cracking under the hot dry conditions experienced there; thus the incidence of this problem was also of interest to the authors. Cultivars such as 'Early Large Red', 'Rose Scented', 'Purbi' and 'Delhi', with thick skin and many tubercles per unit area, crack more than cultivars with thin skin and few tubercles. Overall, 'Bombai', the most important commercial variety, along with 'Muzaffarpur' and 'Rose Scented', had superior production, even though 'Bedana' had better quality.

Table 5.6. Characteristics of litchi in West Bengal. Trees planted in 1977, with data collected from 1991 to 2000. From Ghosh *et al.* (2001). TSS is total soluble solids.

Cultivar	Fruit wt (g)	Seed wt (g)	TSS (%)	Yield (kg/tree)
Bedana	24.2	2.8	19.2	21.1
Bombai	22.0	3.7	18.0	38.1
China	23.7	3.1	18.2	22.1
Early Large Red	20.8	4.1	17.5	28.3
Elaichi	17.6	2.4	17.8	14.3
Muzaffarpur	21.4	3.7	17.9	37.3
Purbi	19.3	4.2	18.3	26.9
Rose Scented	20.1	3.6	18.4	35.6

Table 5.7. Characteristics of litchi in the Canary Islands, Spain. Trees were planted in 1995. From Galán Saúco *et al.* (2001).

Cultivar	Full bloom	Fruit wt (g)	Yield (kg/tree)
Brewster	22 March	18.6	1.0
Calcuttia Late	29 March	24.7	30.1
Early Large Red	12 March	20.9	21.1
Fay Zee Siu	11 March	27.1	13.5
Haak Yip	15 March	21.9	17.2
Kwai May Pink	5 April	20.8	11.2
Sum Yee Hong	22 March	30.9	9.8
Tai So	12 March	27.9	7.6
Wai Chee	22 March	16.8	6.4

Galán Saúco *et al.* (2001) established a litchi cultivar evaluation block at the Cuevae del Polvo Experimental Station on Tenerife in the Canary Islands in 1995 to record information on flowering, yield and fruit quality. Trees flowered between February and April, with full bloom lasting less than a month. Fruit were harvested from mid-July to late August. Many of the trees produced good yields a few years after planting, suggesting that litchi was well adapted to conditions on Tenerife. High-yielding cultivars included 'Brewster' ('Chenzi') from Florida, 'Calcuttia Late' and 'Early Large Red' from India, and 'Haak Yip' ('Heiye') from Taiwan. Fruit of these cultivars weighed 17–31 g, with 'Sum Yee Hong' ('Sanyuehong'), 'Tai So' ('Dazao') and 'Fay Zee Siu' ('Feizixiao') producing fruit heavier than 27 g. Total soluble solids ranged from 17% to 21%. Ten of the cultivars had a flesh recovery greater than 70%. A summary of the data recorded is provided in Table 5.7. Galán Saúco *et al.* (2001) concluded that litchi had good potential for commercial production in the Canary Islands.

Winston and O'Farrell (1993) evaluated 38 litchi cultivars on the Atherton Tableland in northern Queensland over 20 years. All cultivars failed to crop when minimum temperatures were 1.1–2.3°C warmer than the long-term averages for June, July and August. Panicles emerged in June, while 50% of flowers opened during August and September. Fruit were harvested between November and January. 'Bengal, 'Haak Yip', 'Souey Tung' ('Shuidong') and 'Tai So' were the most productive cultivars, with maximum yields of 58 kg/tree. Average yields in these cultivars were about 40 kg/tree.

Mean fruit weight in their trial ranged from 11 g in 'Groff' from Hawaii, to 34 g in 'Chacapat' ('Chakrapad') from Thailand. 'Fay Zee Siu' ('Feizixiao'), 'Tai So' ('Dazao'), 'Brewster' ('Chenzi'), 'Bengal', 'Sah Keng' (Taiwan) and 'Kiamana' (Hawaii) produced fruit weighing more than 20 g. 'Salathiel' had more than 90% of fruit with chicken tongues and a flesh recovery of more than 75%. 'Souey Tung' ('Shuidong'), 'Fay Zee Siu', 'Tai So' (early season), 'Haak Yip' ('Heiye'), 'Kwai May Pink' (mid-season), and 'Wai Chee' ('Huaizhi'), 'Luk Lai' and 'Salathiel' (late season) were recommended for commercial production.

Winston *et al.* (1993) recorded similar information on 18 longan cultivars in the same area over 10 years. Cropping was irregular, with the best yields produced in years with extended periods of daily minima below 12°C. Mean yields for trees 3–13 years old ranged from 3 to 79 kg/tree, and from 7 to 35 kg/tree for trees 3–9 years old. 'Chompoo' and 'Biew Kiew' produced the heaviest yields amongst the older trees (greater than 45 kg/tree), while 'Duan Yu' ('Dawuyuan') and 'Fuhko Number Two' had the highest yields amongst the younger trees (greater than 30 kg/tree). Trees flowered and cropped 1–2 months later than the litchis.

Average fruit weight in the longan experiments in northern Queensland ranged from 5 to 11 g, with only 'Kohala', 'Kay Sweeney', 'Homestead' (Hawaii and Florida) and 'Biew Kiew' (Thailand) having fruit heavier than 10 g. Total soluble solids generally ranged from 16%

to 21%, with most cultivars rated as 'crisp and sweet'. Mean flesh recovery ranged from 57% to 70%. Only 'Duan Yu' ('Dawuyuan'), 'Kohala', 'Haew' and 'Biew Kiew' had more than 80% of their fruit exceeding the minimum commercial standard with a diameter greater than 22 mm. Cultivars recommended for commercial plantings included 'Kohala' (early season), 'Fuhko Number Two' (mid-season), and 'Haew' and 'Biew Kiew' (late season).

Collections of genetic material

The litchi and longan industries in most producing regions outside China are based on introduced cultivars. Seeds are too short-lived for germplasm collection; hence most cultivars are maintained as trees. Plant material can be distributed as air-layers (marcots), cuttings or graft wood.

Numerous collections of these species can be found in Fujian, Guangdong and Guangxi, with some of these centres having hundreds of accessions (Chen and Huang, 2001; Liu and Ma, 2001). There has also been some attempt to preserve wild specimens found growing in forests. Wild longan and related species have been collected from South-east Asia and evaluated at the University of Agriculture, Malaysia, in Bintulu, Sarawak (Wong, 2001). This institution maintains the largest collection of *D. longan* subspecies *malesianus*. There are large collections of litchi and longan at research institutes in Taiwan, especially at the Chia-yi Agricultural Experimental Station (Yen *et al.*, 1984).

Thailand has collections of these crops in Chiang Rai, Chiang Mai, Lamphun and at Kasetsart University in Bangkok (Yapwattanaphun and Subhadrabandhu, 2001). There are collections in Viet Nam, with 33 litchi accessions maintained at the Research Institute of Fruits and Vegetables (RIFAV) in Hanoi (Papademetriou and Dent, 2002). The Thai, Vietnamese and Chinese cultivars differ in tree and fruit characters, but overall the differences are small. India has collections of local selections of litchi in Bihar, West Bengal and the Punjab (Ghosh *et al.*, 2001).

There are collections of litchi at various research centres and universities in Israel. The Volcani Centre and The Hebrew University of Jerusalem are evaluating about 80 accessions introduced from China, Thailand, Viet Nam, India, South Africa and Australia (Goren *et al.*, 2001). Potential new cultivars for the industry include 'No Mai Chee' ('Nuomici') and 'Wai Chee' ('Huaizhi'), along with two local selections.

South Africa has a collection of about 40 litchi cultivars at Nelspruit. Froneman (1999) detailed the release of two new cultivars, 'Fay Zee Siu' (early maturing with average fruit weight of 24–34 g) and 'Wai Chee' (late maturing with heavy yields) imported from Australia in the late 1980s. CIRAD in Réunion has a collection of about 50 litchi cultivars at various locations on the island (Normand and Bouffin, 1994).

Florida has a collection of 27 litchi cultivars and a smaller number of longans, which are being evaluated at the State University of Florida at Homestead and at the Fairchild Tropical Garden in Miami (Campbell and Campbell, 2001a,b). There are collections of litchi and longan in Hilo, Hawaii (50 cultivars of litchi), and at various centres in northern Australia (Aradhya *et al.*, 1995; Menzel, 2002). New material is still being taken to different locations, such as Central America, for commercial evaluation (Fig. 5.6)

Fig. 5.6. Litchi cultivars such as 'Tai So' have been introduced to Nicaragua and other countries in Central America (photograph courtesy of Christopher Menzel).

Genetic analysis

The chromosome number for litchi has been reported as 2n = 28, 30, 32 or 34 (Tindall, 1994). Similar data for longan show that 2n = 20. There is little information available on the inheritance of morphological or physiological characters, although Yang and Chen (1987) indicated that shrivelled seed was inherited in longan. Dwivedi (1996) and Dwivedi and Mitra (1996) found that sugar-to-acid ratio and aril weight were linked in 11 litchi cultivars in West Bengal. The cultivars were grouped into four clusters on the basis of six fruit characteristics. The largest cluster contained five cultivars, and the other clusters, two cultivars each. It was suggested that the best approach to breeding new cultivars would be to cross members from the different groups.

Different cultivars can have the same names, and the same cultivars different names in different places. Identification has traditionally been based on fruit characteristics but these can change with environment and culture (Wu, 1998; Qiu and Zhang, 1996). More sophisticated and reliable techniques are based on polymorphism of isozymes or on DNA fingerprinting with RAPD (random amplified polymorphic DNA) or AFLP (amplified fragment length polymorphism).

Aradhya *et al.* (1995) used eight enzymes to study the diversity amongst 49 litchi accessions in Hawaii. The 40 different cultivars were classified into three main groups. The largest group had 32 accessions and was divided into three subgroups, representing 'Tai So' ('Dazao') types (Groups 1a and 1b) and 'No Mai Chee' ('Nuomici') types (Group 1c). There were many instances of mislabelling of cultivars ('Mauritius' = 'Kwai Mi'). The cultivar 'Groff', reportedly a seedling of 'Haak Yip' ('Heiye'), resembled 'Brewster' ('Chenzi') rather than 'Haak Yip'. Degani *et al.* (1995) conducted a similar analysis of 30 cultivars using seven enzymes in Israel. The 18 different cultivars were clustered into two large groups and three pairs of similar cultivars. 'Haak Yip', 'Kaimana', 'Souey Tung' ('Shuidong') and 'Mauritius' ('Dazao', 'Tai So') were in the first major group, and 'Kwai May Pink', 'McLean's Red', 'Bengal' and 'Salathiel' in the second group. 'Groff' and 'Sah Keng' were similar, as were 'No Mai Chee' and 'Chacapat' ('Chakrapad'). Isozyme analyses have also been conducted in South Africa and Thailand, with virtually identical results (Cilliers and Visser, 1995; Yapwattanaphun and Ramingwong, 1999).

Liu (2001) was able to construct the phylogenetic tree of 62 litchi and longan accessions using RAPD analysis. 'Dazao' and 'Zaohong', 'Feizixiao' and 'Zimali', 'Ziniangxi' and 'Zhengchengdaguoli', and 'Baiye' and 'Guahong' were the same cultivar. Ding *et al.* (2000) analysed the DNA of 34 litchi cultivars in China using RAPD markers and divided them into four main groups. These authors suggested that litchi cultivars have a relatively narrow genetic base.

DNA analysis can assist the development of new cultivars by identifying potential parents with known traits. The analysis can be used to compare the DNA fingerprints of mutants that produce or eliminate the specific trait with those of related genotypes. This approach was investigated by Liu (2001) in litchi. He obtained two RAPD loci linked to seed abortion, and 18 RAPD loci linked to earliness. These markers have great potential for identifying potential cultivars possessing the characters of seed abortion or early maturity. Based on the markers, Liu was able to construct a genetic map within a population of 76 F_1 plants produced from a cross between 'Maguili' (large-seeded and very late) and 'Sanyuehong' ('Sum Yee Hong', aborted-seed and very early). This map covered 1982 centimorgans of the litchi genome and involved 107 RAPD loci and 25 linkage groups.

Ding *et al.* (2001) studied the genetics of litchi seedlings in Fujian using a population of 68 plants developed from a cross between 'Wuye' and 'Luhebao'. Twenty-nine of the 75 primers isolated resulted in amplification of 294 polymorphic DNA fragments, which were useful for genetic mapping. About 14% of the markers were present only in 'Wuye', and about 10% present only in 'Luhebao'. The authors concluded that differences in the genetic map between the two cultivars were small.

Less information is available concerning longan genetics. Chen and Liu (2001) studied the relationship between five cultivars in China using RAPD markers with 36 primers. Their data showed that 'Reyin 17' was a new cultivar related to 'Shixia', and quite separate from 'Dawuyuan', 'Guangyan' and 'Zhuliang'. Wong and Gan (1992) studied the diversity in tropical longans in Sarawak using profiles of peroxidase. Four distinct varieties were identified: 'isau', 'sau', 'kakus' and 'mata kuching', with the first three having greater horticultural value.

Crop improvement

The abundant genetic resources of litchi and longan in South-east Asia have provided good material for developing new cultivars, especially in China, which has a long history of crop improvement (Menzel, 1991; Papademetriou and Dent, 2002). New cultivars have been developed mainly through the selection of seedlings after open-pollination, or from variants (mutations) in existing cultivars. Deliberate cross-pollination is rare. Most of the modern cultivars have been developed in China, with new varieties still being released in Guangdong, Fujian and Guangxi (Table 5.8) (Plate 29). 'Salathiel' (Plate 26) was a chance litchi selection from an unknown parent discovered in northern Queensland in the 1980s (Menzel and Simpson, 1991). 'Sah Keng' and 'Kiamana' were developed from small populations of litchi seedlings in Taiwan and Hawaii (Yen, 1984; Hamilton *et al.*, 1992). The main longan industries in Viet Nam and central Thailand, and the litchi industry in India, are based on seedlings of cultivars imported from China. There are small selection programmes in Israel and South Africa that have yet to deliver substantial numbers of new cultivars.

Most of the present gene pool has been selected on the basis of fruit quality, with a preference for large fruit having bright red or yellow skin, a small seed or seed abortion, and sweet, crisp flesh (Menzel and Simpson, 1990). Selection for these characteristics has generally been at the expense of productivity (Menzel, 2001). Other important attributes for commercial cultivars include sound tree structure, resistance to pests and diseases, and good shelf-life. Seedlings generally resemble the parent tree, but few bear fruit regularly.

Table 5.8. New litchi and longan cultivars in China.

Cultivar	Selection criteria
Litchi	
Maguili (Litchi of Magui)	Extremely late, maturing from early to mid-August in Guangzhou; fruit weigh 40 g (Huang, 2002)
Wuheli (Seedless Litchi)	Up to 99% of fruit seedless; fruit weigh 22 g (Wu, 1998)
Jianjianghongnuo (Jianjiang Red Glutinous)	High rate of chicken tongue; fruit weigh 25–38 g (Wu, 1998)
Qingzhouhongli (Red Litchi of Qingzhou)	Fruit weigh 45 g and have a flesh recovery of 80% (Peng *et al.*, 2001a)
Liqiuli (Autumn Litchi)	Very late, maturing from early August in Guangxi (Xie, 1995)
Dongliu No.1	Very late, maturing from late July to early August in Fujian; wide adaptation (Su, 1995)
Longan	
Chuliang	Fruit maturing from late July to early August; flesh recovery of 72%; high yielding (Xu and Zheng, 1999)
Lidongben (Beginning of Winter)	Extremely late, maturing in early November (Xu and Zheng, 1999)
Qingkebaoyuan (Green Skin Treasure)	Fruit weigh 17 g; late, maturing in late September; good quality (Xu and Zheng, 1999)
Gushan No.2	Maturing in early August; good quality and high yielding (Xu and Zheng, 1999)
Shuangmamu (Twin Fruit Tree)	Early, maturing from late July to early August; flesh recovery of 72%; crops on poor soils (Xu and Zheng, 1999)
Mingjiao No.4	High proportion of fruit with small seeds (Huang *et al.*, 2001)

Several new litchis have been developed recently in China from selections within existing cultivars. Seven clones of 'Sanyuehong' ('Sum Yee Hong') have been selected with seed abortion rates varying from 63% to 97% (Zhou, 1992). Clone '90–9' was the most promising of the group, with large fruit (38 g), 81% seed abortion, 72% flesh recovery, 18% TSS, and fruit maturing 1 week earlier than the standard 'Sanyuehong'. A selected line of 'Huaizhi' ('Wai Chee') produced slightly smaller fruit (15 g) than the standard 'Huaizhi', but had 90–100% small seeds (Liu and Ma, 2001). Fruit matured a week later than the standard. Liu (2001) compared the DNA fingerprints of these cultivars using RAPD. It was shown that the groups shared more than 99.6% of their genes, suggesting that only small changes in genetics were involved. Li *et al.* (1999) outlined the selection of eight improved clones of 'Dahongpao' litchi in Ibin Prefecture, with three clones having larger fruit (24 g) and greater productivity than the original cultivar.

Yen *et al.* (1984) discussed a litchi breeding and selection programme initiated in Taiwan. Twenty cultivars were evaluated, along with a similar number of local selections. Important selection criteria were early or late maturity to extend the production season, high yields, large fruit and high quality. Other important attributes included resistance to downy blight, fruit borers and bark beetles, and suitability for processing and storage. Results of controlled hybridization indicated that 'Kang Wei', 'Hwai Li', 'Sah Keng' and 'Haak Yip' ('Heiye') were the best mother plants. Cultivars developed from this programme are still being evaluated.

The litchi industry in India is based on selections from material sent from China. Many of these cultivars resemble 'Bengal' grown in Australia and Florida. There are about 35 varieties classified into 15 groups. 'Saharanpur' was selected about 20 years ago in the city of that name because of its sweet flavour and low incidence of fruit cracking (2%) compared with seven other cultivars (Lal and Nirwan, 1980). 'Ches-2' is a late cultivar selected from a population of 'China' (Papademetriou and Dent, 2002). The fruit weigh 21 g, have a deep red skin and are less prone to sunburn compared with some of the other cultivars.

A litchi evaluation programme was initiated in South Africa in 1992 to determine the potential of local and overseas cultivars. In the following year, a supplementary breeding programme was commenced, consisting of 3000 seedlings from open-, self- and cross-pollination (Froneman and Oosthuizen, 1995; Froneman, 1999; Rheedar, 1999). Various selections of 'Mauritius' were included in the programme along with 'Souey Tung' ('Shuidong'), 'Haak Yip' ('Heiye') and 'Kwai May Pink'. About 1000 seedlings were to be planted each year. Of the several cultivars imported by the South African Litchi Growers' Association in the 1980s, only 'Wai Chee' ('Huaizhi') and 'Fay Zee Siu' ('Feizixiao') have been released (Froneman, 1999). Several promising selections and hybrids are currently under evaluation, but it is too early to predict likely yields in the different growing areas at this stage.

Israeli researchers have imported and evaluated more than 80 litchi cultivars in the past 25 years. 'No Mai Chee' and 'Wai Chee', along with two local selections, have been suggested for semi-commercial plantings (Goren *et al.*, 2001). 'BD 5–27' was selected in Bet-Dagan as a seedling of 'Floridian' that was pollinated by 'Mauritius' (Goren *et al.*, 1999). The fruit are larger than 'Mauritius' and ripen several days earlier. 'BD 17–70', selected in Bet-Dagan, is a hybrid between 'Kiamana' (♀) and 'No Mai Chee' (♂). It produces fruit that are larger than 'No Mai Chee'. Fruit quality is rated as 'good'. Goren *et al.* (2001) indicated that it could take 10 years to fully evaluate a cultivar introduced from overseas with a history of commercial production, and possibly 20 years for a new line.

Storey *et al.* (1953) selected 'Groff' from a population of 500 'Tai So', 'Brewster' and 'Haak Yip' seedlings in Hawaii, although these are not regarded as premium cultivars. In any case, 'Groff' has never been grown commercially outside Hawaii. 'Kaimana' was developed in Hawaii about 20 years ago from a population of 'Haak Yip' seedlings (Hamilton *et al.*, 1992). Knight (2001) described a small breeding programme initiated in Florida in the early 1960s in which 1200 seedlings arising from open-pollination of 'Tianyan' ('Sweet Cliff') mother trees were planted out in Miami. The majority of the trees grew well but never fruited. A few selections that fruited were evaluated in Florida and Israel ('9–34', and '11–57' or 'Garnet'). Campbell and Campbell (2001a) indicated that

'Garnet' produced fruit every 2–3 years in Florida and had moderate tolerance to anthracnose, *Colletotrichum gloeosporioides*. These authors considered that 'Garnet' had good commercial potential if productivity could be improved.

The National Longan Germplasm Repository established at the Fujian Academy of Agricultural Sciences Fruit Research Institute in Fuzhou has a collection of more than 200 cultivars and strains. The improvement of longan in China has mainly focused on the selection of improved types or seedlings from existing cultivars (Liu and Ma, 2001). Recent selections made in Fujian, Guangdong, Guangxi and Sichuan include 'Shuinan One', 'Songfenben', and the extremely late 'Lidongben'. 'Songfengben' is a late-maturing seedling discovered in Fujian, which was superior to 19 other cultivars compared in a trial from 1985 to 1994. Fruit weigh 14.5 g and are very sweet (23% TSS). They mature in mid-October, 35 days later than the leading local cultivar, 'Wulongling' (Yang, 1996).

Huang and Xu (1980) described a longan selection programme that was initiated in Fuzhou in the 1960s to improve yield and fruit quality in China. A major objective of the programme was to develop cultivars with small seeds. Five promising strains with this character have been produced, including one cultivar with superior fruit (Huang *et al.*, 2001). 'Minjiao Number Four' was selected from a 100-year-old tree growing in Putian. Average yields were about 35 kg/tree, with 64–95% of fruit having small seeds. Average fruit weight was 9 g, with a flesh recovery of 73%. 'Linglong' was a chance seedling of 'Daguoguangyan' discovered in Guangxi in 1990 (Peng *et al.*, 2001b). Trials over 10 years showed that young trees were productive, with 65% of 3-year-old specimens bearing 3–5 kg/tree. Fruit weighed 13 g and had a sweet, pleasant flesh (21% TSS). Harvest was in late August to early September in Guangxi.

The longan industries in Viet Nam and Thailand are based on selections from material originally sent from China (Wong, 2000). Improved cultivars have been selected from thousands of trees planted in the different growing areas. Both countries have developed ecotypes that can produce more than one crop a year under tropical conditions.

'Kohala' was selected in Hawaii and sent to Florida in the 1950s and now forms the basis of commercial plantings around Miami (Knight *et al.*, 1968). It generally performs better than cultivars from China and Thailand (Campbell and Campbell, 2001b). This cultivar has also been distributed to Australia.

The development of better cultivars is very slow because it takes several years for most seedlings to bear fruit. When they do fruit, less than 1% of the seedlings are found to be worthy of selection. Zheng *et al.* (2001) calculated that it took about 40 years to develop a new cultivar in these crops using traditional breeding methods. Future efforts in plant breeding need to concentrate on the cross-pollination of selected cultivars with desirable traits. Seedlings for evaluation can be planted close together at a density of 2000–2500 trees/ha compared with standard orchard densities of 70–280 trees/ha.

Use of related species

Wild plants have been sought as sources of disease resistance or dwarfing in other subtropical trees, and this approach has potential in litchi and longan. The Sapindaceae family contains many genera and species in the tropics and warm subtropics, although there have been few attempts to assess the horticultural value of this material.

The *Litchi* genus contains two other forms, which have not been commercialized (Menzel, 1991). Subspecies *philippinensis* is found in the Philippines (Luzon, Sibuyan, Samar and Mindanao) and Papua New Guinea at high elevations, while subspecies *javensis* is recorded in the Malay Peninsula and Indonesia. Philippines litchi has long, oval fruit with long, thorn-like protuberances. Fruit split in the middle when ripe, displaying an inedible aril that only partly covers the seed. Subspecies *javensis* is a rare specimen found in Chinese gardens in West Java and Indo-China and has fruit similar to cultivated litchi, but with a thinner aril. It flowers and fruits regularly in the tropics. Many of the Malayan specimens belong to the subspecies *chinensis*. There have been no efforts to assess the potential for hybrids between these different races.

Froneman, S. (1999) Two new litchi cultivars released: new litchi cultivars lengthen the South African production season. *Neltropika Bulletin* 305, 3–6.

Froneman, S. and Oosthuizen, J.H. (1995) ITSG litchi breeding programme. *Yearbook of the South African Litchi Growers' Association* 7, 35–37.

Fu, L.Y., Wu, S. X. and Zhou, Q.M. (1985) *An Album of Guangdong Litchi Varieties in Full Colour.* Guangdong Province Scientific and Technology Commission, Guangzhou, China, 78 pp.

Galán Saúco, V., Fernández Galván, D. and Hernández Delgado, P.M. (2001) Behavior of different litchi cultivars in Canary Islands. *Acta Horticulturae* 558, 83–88.

Ghosh, B., Mitra, S.K. and Sanyal, D. (2001) Litchi cultivars of West Bengal, India. *Acta Horticulturae* 558, 107–113.

Goren, M., Degani, O. and Gazit, S. (1999) New improved litchi cultivars for Israel: evaluation and recommendations for 1999. *Alon Hanotea* 53, 222–225.

Goren, M., Tomer, E. and Gazit, S. (2001) Litchi cultivars in Israel. *Acta Horticulturae* 558, 129–133.

Groff, G.W. (1921) *The Litchi and Lungan.* Orange Judd, New York, 188 pp.

Groff, G.W. (1951) Additional notes upon the history of the 'Brewster' lychee. *Proceedings of the Florida State Horticultural Society* 64, 285–289.

Hamilton, R.A., Ito, P.J. and Rapoza, H. (1992) 'Kaimana', an improved new lychee cultivar for Hawaii. *Commodity Fact Sheet Lych-3 (B) Fruit.* Hawaii Agricultural Co-operative Extension Service, University of Hawaii, Hawaii.

Hsu, H.H., Peng, C., Hsuch, J.H. and Tsai, Y.F. (1964) The discovery of wild lychee trees. *Acta Horticulturae Sinica* 3, 415.

Huang, J. and Xu, X. (1980) A preliminary report on the selection program for aborted-seeded longan cultivars. *Fujian Fruits* No. 1, 10–13.

Huang, J.S., Xu, X.D., Zheng, S.Q. and Xu, J.H. (2001) Selection for aborted-seeded longan cultivars. *Acta Horticulturae* 558, 115–118.

Huang, X.M. (2002) Lychee production in China. In: Papademetriou, M.K. and Dent, F.J. (eds) *Lychee Production in the Asia-Pacific Region.* Food and Agricultural Organization of the United Nations, Bangkok, Thailand, pp. 41–54.

Jahiel, M. and Abraham, D. (2001) The growth of the Malagasy litchi field face to the hardening of European control in the matters of sulphured residue. *Acta Horticulturae* 558, 63–71.

Ke, G.W., Wang, C.C. and Tang, Z.F. (1994) Palynological studies on the origin of longan cultivation. *Acta Horticulturae Sinica* 21, 323–328.

Knight, R.J., Jr (2001) The lychee's history and current status in Florida. *Acta Horticulturae* 558, 41–44.

Knight, R.J., Jr, Manis, W.E., Kosel, G.W. and White, C.A. (1968) Evaluation of longan and lychee introductions. *Proceedings of the Florida State Horticultural Society* 81, 314–318.

Lai, Z.X., Chen, C.L. and Chen Z.G. (2001) Progress in biotechnology research in longan. *Acta Horticulturae* 558, 137–141.

Lal, H. and Nirwan, R.P.S. (1980) Saharanpur selection: a new litchi cultivar. *Progressive Horticulture* 12, 75–77.

Ledin, R.B. (1957) A note on the fruiting of the Mauritius variety of lychee. *Yearbook and Proceedings of the Florida Lychee Growers' Association* 4, 57.

Leenhouts, P.W. (1971) A revision of *Dimocarpus* (Sapindaceae). *Blumea* 19, 113–129.

Leenhouts, P.W. (1978) Systematic notes on the Sapindaceae-Nephelieae. *Blumea* 24, 395–403.

Li, Y.Q. (1985) An investigation on wild longan (*Dimocarpus longana* Lour.). *Acta Horticulturae Sinica* 12, 223–227.

Li, Z.X., Liang, C.W. and Wang, E.P. (1999) Study on the selection of Dahongpao litchi variety. *South China Fruits* 28 (5), 26.

Litz, R.E. (1988) Somatic embryogenesis from cultured leaf explants of the tropical tree *Euphoria longan* Stend. *Journal of Plant Physiology* 132, 190–193.

Liu, C.M. (2001) Construction of molecular genetic linkage map and germplasm resources analysis of litchi (*Litchi chinensis* Sonn.). PhD thesis, South China Agricultural University, Guangzhou, China.

Liu, X.H. and Ma, C.L. (2001) Production and research of longan in China. *Acta Horticulturae* 558, 73–92.

Loomis, H.F. (1955) Bengal, a promising large-clustered Indian lychee. *Proceedings of the Florida Lychee Growers' Association* 2, 9–12.

McConchie, C.A., Vithanage, V. and Batten, D.J. (1994) Intergeneric hybridisation between litchi (*Litchi chinensis* Sonn.) and longan (*Dimocarpus longan* Lour.). *Annals of Botany* 74, 111–118.

Menzel, C.M. (1991) Litchi. In: Verheij, E.W.M and Coronel, R.E. (eds) *Plant Resources of South-East Asia Vol. 2: Edible Fruit and Nuts.* Pudoc, Wageningen, The Netherlands, pp. 191–195.

Menzel, C.M. (2001) The physiology of growth and cropping in lychee. *Acta Horticulturae* 558, 175–184.

Menzel, C.M. (2002) The development of the lychee industry in Australia. *Acta Horticulturae* 575, 391–396.

Menzel, C.M. and McConchie, C.A. (1998) Lychee and longan. In: Hyde, K.W. (ed.) *The New Rural Industries*. Rural Industries Research and Development Corporation, Canberra, Australia, pp. 288–295.

Menzel, C.M. and Simpson, D.R. (1990) Performance and improvement of litchi cultivars: a review. *Fruit Varieties Journal* 44, 197–215.

Menzel, C.M. and Simpson, D.R. (1991) A description of litchi cultivars. *Fruit Varieties Journal* 45, 45–56.

Menzel, C.M., Watson, B.J. and Simpson, D.R. (1990) Longan. In: Bose, T.K. and Mitra, S.K. (eds) *Fruits: Tropical and Subtropical*. Naya Prokash, Calcutta, India, pp. 522–546.

Milne, D.L. (1999) Lychee production and research in South Africa. In: Menzel, C.M., Waite, G.K. and Bronson, R. (eds) *Proceedings of the Fifth National Lychee Conference*. Australian Lychee Growers' Association, Twin Waters, Queensland, pp. 25–31.

Nicholls, B.G. (2001) A perspective of the Australian longan industry. *Acta Horticulturae* 558, 45–47.

Normand, F. and Bouffin, J. (1994) Management of litchi genetic resources in Réunion. *Fruits* 49, 376–382.

Papademetriou, M.K. and Dent, F.J. (2002) *Lychee Production in the Asia-Pacific Region*. Food and Agricultural Organization of the United Nations, Bangkok, Thailand, 128 pp.

Peng, H.X., Li, Y.C. and Huang, D.J. (2001a) 'Qinzhou red litchi': a new big fruit form variety of litchi. *Acta Horticulturae Sinica* 28, 276.

Peng, H.X., Wu, C.D. and Zhang, C.T. (2001b) Breeding report on late longan variety Linglong. *China Fruits* No. 5, 5–6.

Qiu, W.L. and Zhang, H.Z. (1996) *Encyclopaedia of China Fruits: Longan*. China Forestry Press, Beijing, China, 85 pp.

Rheedar, S. (1999) Suitable litchi cultivars and selections. *Neltropika Bulletin* 306, 13.

Storey, W.B., Hamilton, R.A. and Nakasone, H.Y. (1953) Groff: a new variety of lychee. *Circular of the Hawaii Agricultural Experimental Station* 39, 1–8.

Su, H.F. (1995) 'Dongliu No. 1': a new litchi cultivar. *China Agricultural Extension* 6, 21.

Subhadrabandhu, S. (1990) *Lychee and Longan Cultivation in Thailand*. Rumthai Publishers, Bangkok, Thailand, 40 pp.

Subhadrabandhu, S. and Yapwattanaphun, C. (2001) Lychee and longan production in Thailand. *Acta Horticulturae* 558, 49–57.

Tien, S. (1979) A 1,200-year-old litchi tree flourishes in China. *Progressive Horticulture* 10, 37–39.

Tindall, H.D. (1994) Sapindaceous fruits: botany and horticulture. *Horticultural Reviews* 16, 143–196.

Wang, J.F. (2001) Shoot tip culture and production of virus-free longan plantlets. *Acta Horticulturae* 558, 155–160.

Whitman, W.F. and Wirkus, L.V. (1958) Rare Fruits Council Activities 1957–58. *Proceedings of the Florida State Horticultural Society* 71, 278–287.

Winston, E.C. and O'Farrell, P.J. (1993) Evaluation of lychee (*Litchi chinensis* Sonn.) cultivars on the Atherton Tableland of North Queensland. *Fruit Varieties Journal* 47, 83–91.

Winston, E.C., O'Farrell, P.J. and Young, K.E. (1993) Yield and fruit quality of longan (*Dimocarpus longan* Lour.) cultivars on the Atherton Tableland of tropical North Australia. *Fruit Varieties Journal* 47, 153–160.

Wong, K.C. (2000) *Longan Production in Asia*. Food and Agricultural Organization of the United Nations, Bangkok, Thailand, 44 pp.

Wong, K.C. (2001) Variations of local longan in Malaysia. *Acta Horticulturae* 558, 103–106.

Wong, K.C. and Gan, Y.Y. (1992) The diversity of *Dimocarpus longan* subspecies *malesianus* variety *malesianus* in Sarawak. *Acta Horticulturae* 292, 29–39.

Wu, S.X. (1998) *Encyclopaedia of China Fruits: Litchi*. China Forestry Press, Beijing, China, 221 pp.

Xie, C.P. (1995) 'Liqiuli': a late maturing litchi cultivar. *China Crop Industry* 3, 55–56.

Xu, X.D. and Zheng, S.Q. (1999) *Photographic Illustration of High-Yielding Longan Cultivation*. China Agricultural Press, Beijing, China, 116 pp.

Yang, W.H. (1996) Songfengben, an uncommon and promising late longan variety. *China Fruits* 1, 52.

Yang, Y.Q. and Chen, Z.F. (1987) Study of the inherited character wrinkled seed and embryo culture in longan. *Acta Horticulturae Sinica* 14, 217–222.

Yapwattanaphun, C. and Ramingwong, K. (1999) Identification of lychee cultivars by electrophoresis of isozymes. *Thai Journal of Agricultural Science* 32, 321–326.

Yapwattanaphun, C. and Subhadrabandhu, S. (2001) Litchi cultivars in Thailand. *Acta Horticulturae* 558, 93–101.

Yen, C.R. (1984) Seeded and seedless fruit growth of 'Sah Keng' litchi. *Journal of Agricultural Research of China* 33, 257–264.

Yen, C.R., Laio, Y.W. and Tien, Y.J. (1984) The cultivars of litchi (*Litchi chinensis* Sonn.) and their improvement in Taiwan. *Journal of the Chinese Society for Horticultural Science* 30, 213–222.

Zee, F.T.P., Chan, H.T., Jr and Yen, C.R. (1998) Lychee, longan, rambutan and pulasan. In: Shaw, P.E., Chan, H.T., Jr and Nagy, S. (eds) *Tropical and Subtropical Fruits.* AgScience, Auburndale, Florida, pp. 290–335.

Zhang, Z.W., Yuan, P.Y., Wang, B.Q. and Qui, Y.P. (1997) *Litchi Pictorial Narration of Cultivation.* Pomology Research Institute, Guangdong Academy of Agricultural Sciences, Guangzhou, China, 189 pp.

Zheng, Q., Hu, G. and Chen, D. (2001) Biotechnology in longan and litchi breeding: applications and prospects. *Acta Horticulturae* 558, 143–148.

Zhou, B.P. (1992) Selection and exploitation of elite 'Sanyuehong' in Zhongshan. *Guangdong Fruits* No. 1, 31–32.

6 Flowering

Tom L. Davenport[1] and Raphael A. Stern[2]
[1]*University of Florida, Tropical Research and Education Center, 18905 SW 280 Street, Homestead, FL 33031–3314, USA;* [2]*MIGAL, Galilee Technology Center, PO Box 831, Kiryat Shmona 11016, Israel*

Introduction

This chapter deals with the factors governing shoot initiation, floral induction, and flower development in litchi and longan. The centre of origin for both crops is southern China, and the limits of commercial production are generally between 12° and 35° latitude. Commercial plantings in China occur primarily in the southern provinces, ranging from 18° to 25°N latitude, in Guangdong (about 68% of total production), Guangxi, Fujian, Hainan, Yunnan, and in Taiwan (Zhang, 1997a; Chen and Huang, 2001; Ghosh, 2001; Liu and Ma, 2001). Details of litchi production in China are discussed by Zhang (1997a) and Chen and Huang (2001). Liu and Ma (2001) reviewed similar information for longan.

Indian production is centred primarily in the eastern states of Bihar, West Bengal and Uttar Pradesh, with less in Tripura, Orissa, Punjab, Himachal Pradesh, Assam and the Nilgris hills (20–27°N latitude) (Ghosh, 2001). The main litchi-growing area in Viet Nam is near Hanoi, whereas longan is more important in the south, near the Mekong Delta. Most of the Thai production is centred around Chiang Mai (19°N latitude), but commercial orchards growing tropical cultivars can be found as far south as Samut Songkhram (13°N latitude) (Subhadrabandhu, 1990; Subhadrabandhu and Yapwattanaphun, 2001a).

Litchi-growing in South Africa is limited to the north-east of the country (25°S latitude). Production in Madagascar (13–25°S latitude) is primarily by small landholders on the coastal plains along the eastern shore (Ghosh, 2001; Jahiel and Abraham, 2001), with some production at higher elevations. The small islands of Mauritius and Réunion (20–22°S latitude) in the Indian Ocean have irregular but substantial production of litchis (Ghosh, 2001). Litchi and longan plantings in Australia span the eastern coast from Cairns (17°S latitude) to northern New South Wales (32°S latitude) (Menzel, 2001).

Those areas producing less than 1000 t of litchi and longan per year include Indonesia (3–10°S latitude), Spain (39–41°N latitude) (Menzel, 2001), southern Florida (25–27°N latitude) (Whitman and Wirkus, 1958; Campbell, 1970; Crane *et al.*, 1998; Campbell and Campbell, 2001), northern Israel (32–34°N latitude) (Goren *et al.*, 2001), Brazil (20–24°S latitude) (Yamanishi *et al.*, 2001), Honduras (15–16°N latitude) (Rindermann and Gomez-Cruz, 1997), Nayarit (19°N latitude) and Sinaloa (25°N latitude) in Mexico (Rindermann and Gomez-Cruz, 1997), and experimental plantings in the Canary Islands (28°N latitude) (Galán Saúco, 1990; Galán Saúco *et al.*, 2001).

Annual temperatures vary markedly across this range of latitudes. Polar air normally reaches the higher latitudes during winter, but does not always have the power to reach the more tropical lower latitudes where it might reduce temperatures, except at higher elevations. It is well established that the lower the latitude in which litchis or longans are grown, the less dependable is flowering (Menzel, 2001). This is because cool temperatures are necessary for floral induction (Groff, 1943; Young and Harkness, 1961; Menzel and Simpson, 1994). In contrast, the upper limit of production is governed by the occurrence of frosts around the time of anthesis (Groff, 1943; Cobin, 1950; Menzel and Simpson, 1994). Best yields occur when temperatures periodically approach freezing, followed by warm days and nights that allow flower development and pollination (Menzel, 1983, 1991; Batten and McConchie, 1995).

Various reviews have considered the impact of weather and tree physiology on flowering (Menzel *et al.*, 2000; Liu and Ma, 2001; Subhadrabandhu and Yapwattanaphun, 2001a; Stern and Gazit, 2003). Using this background information, we propose a model for vegetative and reproductive growth of litchi and longan (Fig. 6.1) that is similar to those of mango (Davenport and Núñez-Eliséa, 1997) and citrus (Davenport, 1990).

Phenology and shoot development

In litchi and longan, periods of stem elongation are separated by periods of stasis, when the terminal bud is at rest (quiescent), or a terminal inflorescence has formed. The interval between successive episodes of stem elongation depends on pruning, available water and nutrients, and the weather. For the purposes of our discussion, we refer to that part of stem developing during the current episode of stem elongation as a 'shoot' (whether vegetative or floral), and define this in recognition of the loose way in which the term is used in the literature. Similarly, we define a 'flush' as a collection of synchronous 'shoots'.

The pattern of shoot and flush development is fundamental to flowering and fruit production in litchi and longan (Table 6.1). Batten and McConchie (1995) showed that flowering of litchi could be induced after a new shoot had been initiated, provided that the new bud was no more than a few millimetres long. This research, along with work on other tropical trees (Davenport, 1990, 2000, 2003c; Núñez-Eliséa *et al.*, 1993, 1996; Davenport and Núñez-Eliséa,

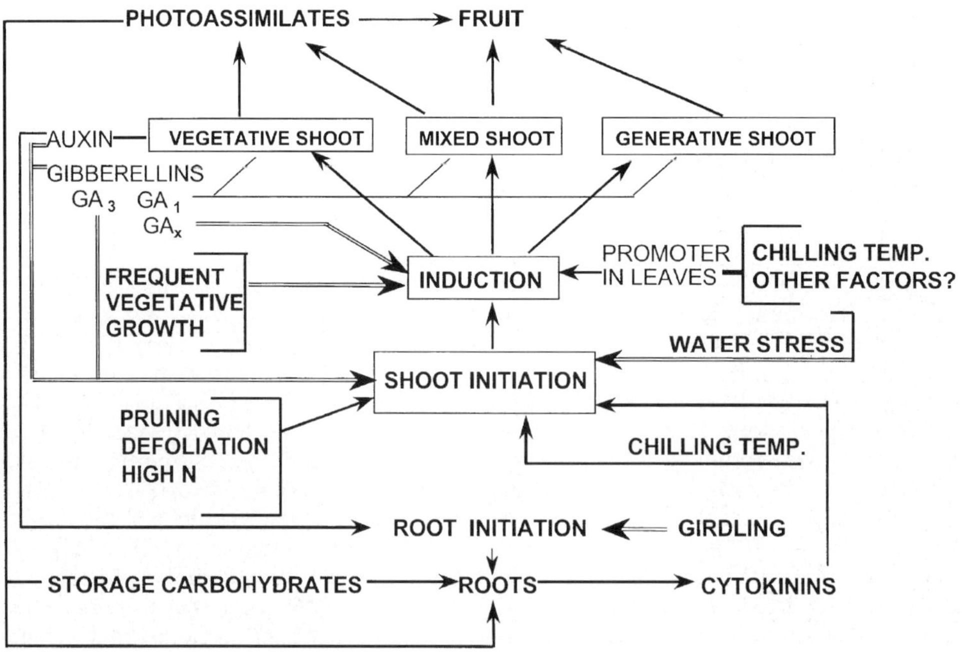

Fig. 6.1. Conceptual model of litchi/longan flowering. Single lines indicate promoters, and double lines inhibitors.

1997), is consistent with the idea that an inductive signal acts on actively growing buds. Flushes tend to occur on groups of stems that are connected by a common scaffold within the canopy (Davenport *et al.*, 2000; Zheng *et al.*, 2001). Spatial variation in the timing of flush development over a canopy can lead to variations in flowering.

Types of shoots

Under normal circumstances, one of three shoots initiates growth from apical or axillary buds in resting stems (Das and Choudhury, 1958; Robbertse *et al.*, 1995). Vegetative shoots produce only leaves at each node; generative shoots give rise to inflorescences; and mixed shoots bear both leaves and inflorescences at each node (Fig. 6.2). A fourth type of shoot, referred to as a 'leafy inflorescence', is distinct from mixed shoots. It commences producing only leaves, but makes a transition midway to form lateral inflorescences in the distal portion of the shoot. These are defined as vegetative to floral or V/F transition shoots (Fig. 6.2).

Stems and shoots have an apical meristem that is never fully committed to form only leaves or flowers (Batten and McConchie, 1995;

Table 6.1. Effect of the timing of autumn vegetative flushes on the flowering of litchi in Florida (Mustard and Lynch, 1959).

Month of vegetative flush	No. of tagged terminals	Percentage of shoots next April		
		Dormant	Generative	Vegetative
September	63	17.5	76.2	6.3
October	76	48.7	42.1	9.2
November	74	29.7	36.5	33.8

Fig. 6.2. Vegetative (A), mixed (B), generative (C), and vegetative/floral (V/F) transition shoots (D) in litchi.

Núñez-Eliséa *et al.*, 1996). Surrounding the resting meristem in both litchi (Shukla and Bajpai, 1974a; Scholefield, 1982; Naphrom *et al.*, 2001) and longan (Liang and Chen, 1965; Ke *et al.*, 1988; Wang *et al.*, 1988; Wang and Ke, 1992; Qiu *et al.*, 2001a,b) is a spiralling whorl of nodes, each composed of a primordial compound leaf and a lateral meristem. The primordial leaves of the outermost whorl form protective bracts that surround the meristem. The bracts are variable in size, similar to miniature leaves, but with a brownish cast. At initiation, the pre-existing primordial leaves and/or lateral meristems within each node begin to develop, depending on the type of shoot, i.e. vegetative, generative or mixed. This early development of pre-existing lateral meristems constitutes the first phase of shoot differentiation. The apical meristem then forms more nodes, providing further leaf primordia and lateral meristems, before returning to rest in the case of a vegetative shoot, or differentiating in the case of an inflorescence.

Vegetative shoots

In warm weather, the pre-existing primordial leaves develop to form true compound leaves. If favourable conditions persist throughout the initiation and differentiation of new nodes, the new primordial leaves develop into compound leaves at the distal part of the shoot (Batten and McConchie, 1995). The lateral meristems at the base of each expanding leaf are repressed, and form a spiral of a few lateral nodes, each forming a primordial leaf and lateral or axillary meristem. The primordial leaves of the outermost spiral of nodes form the minute bracts that protect the new lateral buds.

Vegetative shoot development takes 6 weeks or longer, depending on the weather (Olesen *et al.*, 2002). During this period, the apex produces about eight leaves before returning to rest (Fig. 6.2A). Expanding leaves are initially light pink to red, soft and limp, turning light green and then dark green and stiff, 1–2 months after full expansion (Shukla and Bajpai, 1974a; Menzel *et al.*, 1987). The limits of these episodes of growth are recorded on each branch as segments consisting of short, compressed internodes alternating with long internodes. Davenport and Núñez-Eliséa (1997) defined the regions of compressed internodes as intercalations, and the entire segment of long internodes terminating in an intercalation as an intercalary unit, for citrus and mango. Litchi intercalations are less compressed than those of citrus or mango, but are easily recognized.

Generative shoots

Generative shoots typically form in the apical and/or lateral buds when shoots emerge during cool weather (Batten and McConchie, 1995). In contrast to vegetative shoots in which true compound leaves arise from the primordial leaf tissues, the leaf primordia remain relatively quiescent, and the lateral meristems at each node are instead evoked (Shukla and Bajpai, 1974b; Scholefield, 1982; Qiu *et al.*, 2001a,b). The lateral meristems organize themselves to form short-lived apical meristems, which give rise to primordial leaves, and secondary lateral meristems, which may develop into tertiary or higher levels of branching, with leaf development repressed at each level. If inductive conditions continue during the formation of new nodes at the terminal apical meristem, then the same pattern of lateral meristem development prevails, thus extending the length of the thyrsoid or panicle. Elongation of the articulated thyrsoidal structures at each node terminates in dichasial floral structures as described by Mustard (1960) and Robbertse *et al.* (1995) (Fig. 6.2B).

Mixed and transition shoots

Initiation of mixed shoots occurs in mild weather, at temperatures lower than those required for purely vegetative shoots, and higher than those for purely floral shoots. In this case, both leaf primordia and lateral meristems develop, resulting in shoots with a leaf and an inflorescence at each node (Fig. 6.2C).

Transition shoots (Fig. 6.2D) commonly appear in litchi when shoots emerge during rapidly changing temperatures. These initiate one type of shoot, either vegetative or generative, which finish as the other. Batten and McConchie (1995) described such shoots as 'leafy inflorescences', produced when plants were transferred from warm to cool conditions soon after bud break. Vegetative to flowering (V/F) transition shoots occur in south Florida when cold weather arrives soon after the initiation of vegetative shoots. Bud determination in mango is also progressive and sensitive to fluctuations in vegetative/floral signals. Transferring mango plants from warm to cool conditions soon after the initiation of vegetative shoots resulted in a high proportion of V/F shoots, while transferring plants initiating floral shoots from cool to warm conditions resulted in a high proportion of F/V shoots (Núñez-Eliséa *et al.*, 1996).

Role of growth substances in shoot initiation

Initiation is the onset of shoot development in resting buds, regardless of the type of shoot. It commences with cell division and elongation of dormant cells in the leaf primordia (vegetative shoots), lateral meristems (generative shoots) or both (mixed shoots), followed by cell division in the apical meristem, with subsequent new lateral leaf primordia and meristems in the distal portion of the shoot. Shoot initiation is cyclic, occurring two or more times per year, but may also be stimulated by pruning, defoliation and irrigation (Fig. 6.1). Davenport (1990) and Davenport and Núñez-Eliséa (1997) proposed that shoot initiation in citrus and mango is regulated by the ratio of a shoot initiation promoter and a shoot initiation inhibitor. It was further proposed that cyclic changes in these components also regulate shoot initiation in litchi and longan (Davenport, 2000). A high promoter:inhibitor ratio favours shoot initiation, and a low ratio favours continued rest (Fig. 6.3).

Evidence elicited from citrus and mango indicates that the promoter is a cytokinin (CK) produced in growing root tips and transported in xylem to the buds. This concept is supported by the observation of higher concentrations of endogenous dihydrozeatin (DHZ) in xylem sap of vigorous rootstocks with frequent flushes on 'Alphonso' mango, compared with low levels in non-vigorous, dwarfing rootstocks producing fewer flushes (Murti *et al.*, 2000).

The inhibitor may be an auxin that is produced in, and transported at relatively high concentrations from, expanding leaves, tapering off as they age (Davenport *et al.*, 1980). It is proposed that auxin synthesized in young leaves inhibits growth of resting buds, and is transported to roots that are connected to stems that have recently flushed (Lomax *et al.*, 1995). As those stems age, the growing roots synthesize CKs that are transported in xylem, and accumulate in these same stem buds. This is despite the fact that most of the water in the xylem flows out of the leaves.

Although a distinct alternation in root and shoot growth has been demonstrated in large avocado trees (Ploetz *et al.*, 1993), such alternation was not apparent in young litchis (Marler and Willis, 1996). These relationships may explain why flushes occur frequently in young trees, due to the close proximity of stems and roots. Consistent with this hypothesis is the observation that pruning or defoliation stimulates shoot initiation (Campbell, 1994; Goren and Gazit, 1996; Olesen *et al.*, 1999). Such treatments not only remove the source of auxin production, but also increase CK concentrations in xylem sap (Bangerth, 1994),

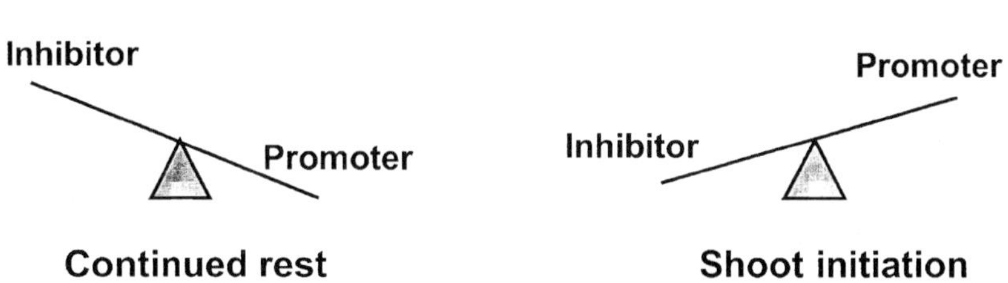

Fig. 6.3. Proposed relationship between shoot initiation, and promoters and inhibitors (Davenport, 2000).

which would quickly shift the CK:auxin ratios to favour bud break.

Cytokinins (*CKs*)

Cytokinins are substituted adenine compounds that induce cell division in the presence of auxin in tobacco pith or similar tissue culture systems (Horgan, 1984). Their effects on plants, however, extend far beyond this limited definition. An overview of CK biochemistry and action can be found in McGaw and Burch (1995), while their involvement in flowering is discussed by Bernier (1988). The most common CKs are zeatin (Z), its riboside (ZR) or glucoside; dihydrozeatin (DHZ), its riboside; and N6-(δ^2-isopentenyl) adenine (2iP) and its riboside (iPA). These substances are intimately involved in shoot initiation (Davenport and Núñez-Eliséa, 1997).

Synthetic CKs, N6-benzyladenine (BA), its riboside (BAR), and kinetin, mimic the action of the native hormones in some plants. Roots are the major source of CKs in higher plants (Torrey, 1976), although other tissues with active cell division also produce or metabolize CKs (Carmi and van Staden, 1983). The current view is that the majority of CKs are synthesized in the root tips near the cap (Itai and Birnbaum, 1996) during active root growth, and distributed to stems through transpiration. However, the root cap is hydraulically isolated from the remainder of the root, raising the question of how these substances access the transpiration stream.

Shoot initiation in litchi and longan is intricately associated with CKs. Chen and Ku (1988) reported that foliar application of kinetin at 200 mg/l to 5-month-old 'Haak Yip' ('Heiye') litchi stems in the autumn, about 2 weeks after ethephon (200 mg/l) treatment, stimulated bud break 1 month earlier than in control stems. This treatment resulted in 60% of the stems flowering, and in combination with ethephon raised the flowering rate to 80% compared with no flowering in the controls. It was not revealed whether the absence of flowering in the controls was due to a lack of bud break or to the initiation of leaves. Chen (1991), however, found that 100 μg kinetin in a 5 μl drop of 1 M acetic acid applied weekly to buds from 6 weeks before initiation through generative shoot development stimulated bud break 1 week early, and increased the proportion of generative shoots over non-treated controls. Das *et al.* (1999) reported generation of multiple lateral shoots from litchi stems cultured with BA. Exogenous application of CKs, presumably during cool weather, resulted in greater bud break and more branched inflorescences than in controls (Puiman *et al.*, 1985). Núñez-Eliséa *et al.* (1990) reported similar results for mango when a synthetic CK, thidiazuron, was applied during cool weather. The same product applied during warm weather resulted in many lateral vegetative shoots, supporting the concept that CKs are involved in shoot initiation, but not in floral induction.

Further evidence of the role of CKs comes from measurements by bioassay, of CK activity, which increased in litchi buds (Liang *et al.*, 1983) and xylem sap (Chen, 1990; Naphrom *et al.*, 2001) before floral initiation. In longan, glucosylated conjugates of Z and ZR accumulated in resting buds, but they were hydrolysed to free CKs during shoot initiation (Chen *et al.*, 1997). The apices of initiated shoots had concentrations of free-CKs several-fold higher than those of resting buds.

The evidence for the role of CKs in floral induction from measurements of endogenous hormone levels is scant. Buds of litchi (Puiman *et al.*, 1985) and longan (Huang and Huang, 1996) had a higher concentration of CKs in 'on' flowering years than 'off' flowering years, and generative longan shoots had higher CK concentrations than vegetative shoots (Chen *et al.*, 1997; Qiu *et al.*, 2001a,b). In general, high CK activity promotes shoot initiation, but not flowering.

Auxins

Auxins are primarily synthesized in young leaves and seeds (Davies, 1995), and transported to root tips. Unlike CKs, they typically inhibit shoot initiation. Indole-3-acetic acid (IAA) is the natural auxin in angiosperms (Davies, 1995). Shigeura (1948) was the first to report that foliar application of the sodium salt of naphthalene acetic acid (NAA) in November and December (late autumn and early winter) inhibited initiation of vegetative litchi shoots in autumn, in Hawaii. This delay in flushing promoted flowering in winter, whereas stems that flushed during autumn did not flower. The most effective application depended on the concentration used. For example, four bi-weekly sprays were necessary with 50 mg/l, commencing in late September. Higher concentrations, e.g. 150–400 mg/l, required fewer applications, commencing in October or November.

Applications of 100–150 mg/l of the sodium salt of NAA to litchis in Florida in early autumn discouraged vegetative flushes and increased flowering and yields over controls in most years (Ledin, 1953, 1954, 1955; Young, 1957a). No increases in yield were observed when the weather discouraged vegetative flushing before flowering (Ledin, 1954, 1955; Young, 1957a). Responses in other experiments were inconsistent when treatments were applied during vegetative flushes in late summer or early autumn (Nakata, 1955; Mustard *et al.*, 1956), or when high rainfall and high leaf nitrogen concentrations promoted frequent leaf growth (Mustard *et al.*, 1956).

Endogenous concentrations of auxin in terminal shoots decreased as they aged (Veen and Jacobs, 1969). Liang *et al.* (1987) and Chen (1990) observed higher concentrations of IAA in young litchi leaves than in mature leaves. Liang *et al.* (1987) also noted initiation of generative shoots only when concentrations of IAA were low.

Auxin is also the primary hormone regulating root initiation and development (Torrey, 1976), and the high concentrations that suppress shoots may promote root growth. New roots, in turn, are sources of CKs that promote shoot initiation. Such an interaction between stems and roots might account for the observed rise in CK concentrations as auxin concentrations fall.

Ethylene

Ethylene is a phytohormone involved in abscission, ripening and senescence (Abeles *et al.*, 1992). It was thought to have a role in the floral induction of mango on the basis of the effectiveness of smudge smoke (ethylene is one of its constituents) and ethephon (an ethylene precursor) on flowering (Barba, 1974). However, Davenport and Núñez-Eliséa (1991) found either no response or a vegetative response with ethephon, and suggested that any effect of ethylene on flowering was due to a change in the timing of shoot initiation, not a direct effect on induction.

For litchi, there are several reports of ethephon enhancing the number of flowering shoots (Sittichaikasem, 1974; Huang and Weng, 1978; Subhadrabandhu, 1986; Subhadrabandhu and Koo-Duang, 1987), and one where vegetative shoots were enhanced (Subhadrabandhu and Koo-Duang, 1987). As argued above, the effect of ethylene on litchi may be on shoot initiation, not floral induction. The argument is strengthened by research on longan, which showed that ethephon increased CK concentrations and bud break (Qiu *et al.*, 2001a,b).

Ethephon has been tested for its ability to defoliate young vegetative flushes occurring in late autumn, when it is too warm for flowering. Subsequent shoots do not appear before spring, when temperatures are too warm for flowering (Olesen *et al.*, 1999). Defoliation of the autumn shoots stimulates the initiation of lateral shoots a few weeks after treatment. If cool weather coincides with this, the shoots will flower.

Gibberellins (GAs)

More than 100 GAs have been described, but only a few have a specific physiological role (Sponsel, 1995). GA_1 is considered to be the primary GA involved in shoot elongation (Rademacher, 2000a), while GA_3 inhibits flowering in citrus and mango (Davenport, 2003c). For mango, the effect is through the inhibition of shoot initiation (Kachru *et al.*, 1971; Davenport and Núñez-Eliséa, 1997).

Few reports are available describing the effects of exogenous GAs on shoot initiation or flowering of litchi or longan. Zong and Wu (1983) reported that application of GA_3 at 200 mg/l to litchis in mid-October and late November did not affect flower initiation in January compared with unsprayed controls. The gibberellin GA_3 at 50–1000 mg/l did not affect growth or fruit set when applied five times to 'Brewster' litchi trees, from before flowering to fruit set (Young *et al.*, 1960).

Chen (1990, 1994) found that the dominant GAs in litchi were 1, 4, 17 and 20, and in longan were 1, 8, 9 and 32. In general, the concentrations of GAs in the two species are lowest during rest (Naphrom *et al.*, 2001). Chen (1994) found that the highest concentrations of free GA_1-like and GA_8-like substances in 'Ken Fur' longan were present during leaf flushing. In contrast, the concentrations of GA_9-like and GA_{32}-like substances increased substantially at the start of bud dormancy and flower initiation, respectively. Chen proposed that GA_{32}-like substances were involved in floral induction in this species. Other experiments showed that injection of GA_4 or GA_7 into the centre of the shoot apex induced flowering.

Some researchers have investigated the role of GAs on shoot initiation and flowering through application of plant growth retardants such as paclobutrazol, often combined with other chemicals that inhibit specific enzymes involved in GA biosynthesis (Rademacher, 2000a,b). Applied as foliar sprays or root drenches, these retardants affect the phenology, morphology and hormonal fluxes of litchi and longan (Li and Chang, 1987; Chaitrakulsub *et al.*, 1989; Menzel and Simpson, 1990; Liang and Yu, 1991; Huang and Huang, 1996). The effects on flowering are highly variable, with no change recorded in some experiments, and enhancement in others. These findings are somewhat at odds with the situation in citrus and mango, where growth retardants promote flowering (Davenport and Núñez-Eliséa, 1997).

Abscisic acid (ABA)

Abscisic acid is the hormone responsible for the opening and closing of stomata (Mansfield and McAinsh, 1995). A few measurements of concentrations of ABA have been made in litchi (Hou *et al.*, 1987; Chen, 1990) and longan (Qiu *et al.*, 2001a,b), but the role of this hormone in flowering has hardly been addressed.

Potassium chlorate

Potassium chlorate is a chemical unrelated to any plant growth regulator. It stimulates out-of-season flowering in longan, but not in litchi, and it sometimes damages the leaves (Jungyoosuk, 1999; Subhadrabandhu and Yapwattanaphun, 2000, 2001b). The effect appears to be similar to that of potassium nitrate in mango (Davenport and Núñez-Eliséa, 1997). 'See Chompoo', 'E-Daw' and 'Haew' are the most responsive cultivars in Thailand (Manochai, 2000). 'E-Daw' flowered 3 weeks after a November (autumn) soil drench at 4–8 g/m^2 (Manochai *et al.*, 1999), while 'See Chompoo' flowered at doses of 1–4 g/m^2 (Khao-Sumeru *et al.*, 1999). For flowering to occur, it was necessary to have mature leaves on resting stems (Manochai *et al.*, 1999). Foliar sprays (Sritontip *et al.*, 1999) and trunk injection (Viriyalongkorn *et al.*, 1999, cited in Subhadrabandhu and Yapwattanaphun, 2001b) were also effective. Protocols for use of the chemical were summarized by Khao-Sumeru *et al.* (1999) and Subhadrabandhu and Yapwattanaphun (2001b). Application of the chemical to out-of-season, easy-to-flower, 'Tawai' has resulted in the production of longans nearly all year round in Thailand (Subhadrabandhu and Yapwattanaphun, 2001b).

The results of experiments in Taiwan, using gunpowder and potassium chlorate, were somewhat different. Trees flowered after a single application nearly any time of the year (Yen *et al.*, 2001), as in Thailand, but rates of 13–26 g/m^2 were required. Potassium chlorate, applied at rates shown to be effective for particular growing areas, has the potential to increase longan production in the tropics.

Concerns about the health risks associated with the use of potassium chlorate have been prompted by the knowledge that it is an ingredient of some herbicides, matches, cleaning agents, fireworks and explosives. Potassium chlorate caused an explosion in a longan drying plant in 1999, which claimed many lives in the San Pa Tong district of Chiang Mai. Recent research in Thailand has recorded anaemia, thrombocytopaenia, high serum creatinine, haematuria, pyuria and methaemoglobinaemia, all symptoms of blood, kidney, and urinary tract damage, in a significant proportion of longan growers (Wewatanadate *et al.*, 2001). Most growers did not use any personal protection, such as gloves or respirators, when applying the material. The same care should be used in application of potassium chlorate as with any other hazardous agricultural product.

Effects of environment on shoot induction

A simple model of shoot induction has been developed for litchi, citrus and mango based on the balance between putative flowering and vegetative promoters, neither of which is known, and the time of shoot initiation (Fig. 6.4; see Davenport, 2000). Some of the characteristics of the flowering promoter are well understood. Mature leaves are thought to be a major source of the promoter in mango, which is often effective with only a fraction of a leaf per stem (Davenport and Ying, 2003). This promoter is graft-transmissible (Kulkarni, 1988) and can be transported over long distances (Davenport and Ying, 2003), but is short-lived, effectively being lost within days of leaves being removed (Davenport *et al.*, 2001b). Citrus is somewhat different in that the promoter is thought to be synthesized in the stem tip (Davenport, 2003a,b). Little is known of the flowering promoter in litchi other than that mature leaves are required (T.L. Davenport and Z. Ying, unpublished results, 2003).

Temperature

It has long been recognized that, in common with a range of other tropical tree species, cool temperatures promote flowering in litchi (Shukla and Bajpai, 1974a) and longan (Groff, 1921, Young and Harkness, 1961; Menzel *et al.*, 1990; Ke *et al.*, 1998). Nakata and Watanabe (1966) determined that cool nights of 14°C induced flowering of litchi under short (6 h) and long (18 h) photoperiods. Thus, litchi was deemed a non-photoperiodic flowering plant. These experiments also demonstrated that cool temperatures stimulate shoot initiation.

Cool nights of 10°C or less stimulate shoot initiation and floral induction (Young, 1956b, 1957b, 1970; Young and Harkness, 1961), if temperatures remain low through early shoot development (Batten and McConchie, 1995). Under such conditions, generative shoots appear within 1 month (Shukla and Bajpai, 1974a).

Menzel and Simpson (1988) conducted experiments on several cultivars of litchi grown in pots and exposed to 15°/10°, 20°/15°, 25°/20° or 30°/25°C until shoot initiation occurred (within 8 weeks in all regimes). There were differences between cultivars, but overall, plants exposed to 30°/25° or 25°/20°C initiated vegetative shoots; those exposed to 20°/15°C, a combination of generative, mixed and vegetative shoots; and those exposed to 15°/10°C, generative and mixed shoots, with a

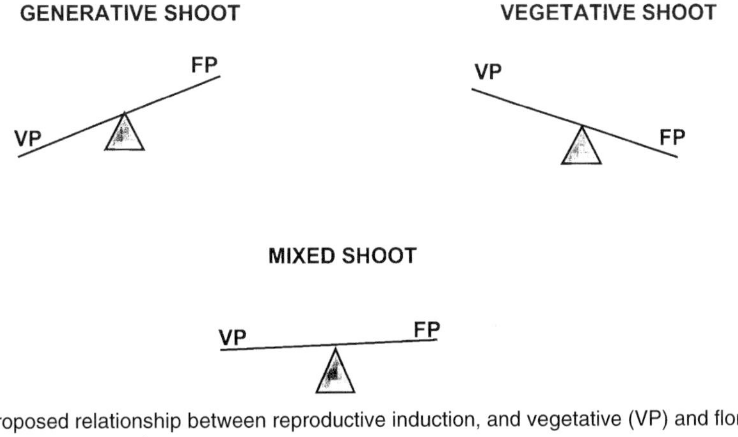

Fig. 6.4. Proposed relationship between reproductive induction, and vegetative (VP) and floral (FP) promoters (Davenport, 2000).

higher ratio of generative to mixed shoots than at 20°/15°C.

Menzel and Simpson (1995) observed that maximum/minimum temperatures of 25°/20°C or above stimulated vegetative growth in 'Kwai May Pink' and 'Casino' litchis. Floral induction occurred when plants were exposed to temperatures of 15°/5°C for 10 weeks. Although there were some differences between the two cultivars, intermediate temperatures resulted in mixed shoots. Stored starch reserves did not correlate with vegetative or floral development. The longer the period that 'Wai Chee' plants were kept at 15°C (4–10 weeks) before they were transferred to 30°/25°C, the greater was the flowering (Menzel and Simpson, 1995). A minimum of 4 weeks' exposure was required for floral induction. The greatest proportion of mixed shoots occurred when plants were transferred after 6 or 8 weeks, whereas 10 weeks of cool temperatures gave the maximum proportion of generative shoots. Increasing periods from 1 h to 24 h per day of 20°C interrupting a constant 15°C, reduced flowering (Menzel and Simpson, 1995).

Batten and McConchie (1995) took a different approach by examining the responses of buds at different stages of development. Buds that had not elongated when the plants were transferred from warm to cool conditions, or which were actively growing and no more than a few millimetres long, generally flowered. Actively growing buds that were slightly longer, up to several millimetres in length, generally produced transition inflorescences or no flowers, while buds that were longer still, rarely flowered. The strict conclusion was that temperatures experienced by the whole plant during early shoot development affected bud determination, but the results are also consistent with the hypothesis that buds need to be actively growing to flower. N. Jarassamrit (unpublished results, 1996) collected similar data for longan.

In order to make field estimates of the critical temperature for flowering, Olesen *et al.* (2002) assumed that early shoot development was the time of bud determination in litchi. Floral bud determination occurred at mean daily temperatures below 17–19°C, a result similar to that reported by Menzel and Simpson (1995) for plants grown in controlled environments.

Water relations

It is a common misconception that drought or a water deficit induces flowering in tropical fruit trees (Menzel and Simpson, 1994). Citrus is one of the few species in which water stress has a direct effect on floral induction (Davenport, 2003c). Litchi and mango, however, do not respond in the same way as citrus (Menzel *et al.*, 1989; Chaikiattiyos *et al.*, 1994, Núñez-Eliséa and Davenport, 1994). A moderate or severe water deficit with dawn leaf water potentials (Ψ_{leaf}) of –2.0 and –3.5 MPa did not induce flowering in litchis grown in a greenhouse (Chaikiattiyos *et al.*, 1994). Droughted plants only flowered when they were placed in a low-temperature regime when re-watered (Menzel *et al.*, 1989). Several authors have shown that drought can increase flowering and yield in litchi (Nakata and Suehisa, 1969; Menzel *et al.*, 1989); however, these responses are related to the effects of the water deficit on the timing of shoot initiation rather than on floral induction. Drought is only effective if it coincides with, or is followed by, low temperatures. Shoot initiation is inhibited when Ψ_{leaf} is lower than –2.0 MPa (Menzel *et al.*, 1989). Early experiments in Hawaii demonstrated that a mild drought in 'Kwai May', for 6 months prior to flowering, prevented vegetative flushing in autumn and promoted inflorescences, compared with irrigated controls that flushed vegetatively (Nakata and Suehisa, 1969).

Drought has been used to manipulate autumn vegetative flushing and to improve flowering and yields of litchi in Israel. Irrigation is withheld from mid-September until the commencement of cool, wet weather in December, when the trees initiate inflorescences (Stern and Gazit, 1993; Stern *et al.*, 1993, 1998; Goren and Gazit, 1996). Stern *et al.* (2003) reported elevated concentrations of the ribosides of the cytokinins Z and DHZ in 'Mauritius' xylem sap during the drought, which were proportional to the level of water deficit in the plants. The use of drought to reduce the risk of autumn flushes requires different strategies in different environments. In some areas, the control of water levels in the root zone is impractical. Although there is no information on the Ψ_{leaf} needed to achieve the desired response, Menzel *et al.* (1995b) reported that application of 60 mm every 3 weeks is adequate to maintain high levels of plant water in all but the driest orchards of Australia.

Nitrogen

Cobin (1950) observed that 'where the nutrition level is high and where abundant moisture and hot temperatures prevail, one vegetative flush will be succeeded by a succession of vegetative flushes without the appearance of a general bloom'. The need for vegetative rest during autumn and early winter prompted Cobin to recommend that fertilizer not be applied during this period to promote flowering in early spring in Florida. However, there is no evidence of a direct link between floral induction and leaf or soil nitrogen (Menzel *et al.*, 1989), even though nitrogen is the most important nutrient affecting vegetative flushing (Menzel *et al.*, 1995a; Zheng *et al.*, 2001).

Management of plant nitrogen concentrations is challenging, making a relationship between flowering and nitrogen difficult to establish. Results of efforts to optimize the rate and time of nitrogen applications for consistent flowering have been inconclusive (Young, 1956b, 1957b; Young and Harkness, 1961; Menzel and Simpson, 1988; Menzel *et al.*, 1995a; Li *et al.*, 2001). Koen and co-workers (Koen *et al.*, 1981a,b,c; Koen and Smart, 1982) and others (Mitra, 1988; Menzel *et al.*, 1992; Liu and Ma, 2001) have concluded that nitrogen concentrations of 1.3–1.5% in leaf dry matter at flowering or fruit set are generally optimal for litchi production in South Africa, Hawaii, India and Australia, with higher concentrations reducing yields in some years. Difficulties in relating fertilizer applications to consistent flowering and yields may be related to the fact that leaf nitrogen concentrations can take months or years to adjust following soil applications, due to storage in soils and trees (Menzel *et al.*, 1995a).

Pruning and girdling

Pruning is necessary in order to manage canopy size and shape, especially in high-density plantings (Goren, 1990). Tip pruning after harvest synchronizes the postharvest vegetative flush throughout the orchard (Goren and Gazit, 1993). Olesen *et al.* (2002) demonstrated the impact of pruning litchis at different times on the subsequent development of vegetative and reproductive flushes. They developed a model for the pruning of trees along the eastern seaboard of Australia, based on the dependency of flush development on temperature and irradiation.

Girdling (ringing or cincturing) involves removing a narrow band of bark by cutting through the cambium layer to the wood. The band can be connected to form a closed ring or an overlapping spiral around the trunk. It is generally applied to scaffolding limbs or at the base of tree trunks.

Girdling stops shoot initiation while the cut remains open (Menzel and Paxton, 1986) and can increase flowering (Morse and Oosthuizen, 1993; Li and Xiao, 2001). Menzel *et al.* (1988) and Zhang (1997b) reported that girdling after the postharvest flush had matured delayed the next flush until winter. Girdling disrupts the downward flow of photo-assimilates and auxin from the leaves (Lomax *et al.*, 1995), which can reduce root growth. The inhibition of root growth may then depress the production of CKs produced in the root tips (Chen *et al.*, 1985).

Early experiences in Hawaii indicated that the best time to girdle was in early autumn (Nakata, 1956). If vegetative flushes occurred during or after girdling, then the trees generally did not flower (Nakata, 1956; Menzel and Paxton, 1986). Girdling experiments have also been conducted on longan in China (Wu *et al.*, 2000), and litchi in Florida (Shigeura, 1948; Young, 1957a, 1959, 1977). The most effective time to girdle in Florida was mid-September (Young, 1956a). Complete girdling of scaffolding branches often weakened or killed trees (Young, 1957a). Girdling only 75% of the circumference was unsuccessful, as was banding branches with butyl rubber (Young, 1959).

The inflorescence and flowers

Shukla and Bajpai (1974a) considered the appearance of enlarged leaf primordia and lateral meristems on the elongating main axis to be the first indications of floral differentiation in litchi. The emerging inflorescence is initially similar to a vegetative flush, and it is only when the lateral meristems develop into secondary inflorescences, or start producing small leaves in the case of mixed shoots, that it is possible to identify the shoot. Each panicle produces tens to hundreds of small flowers (Fig. 6.5): two functional male types (M_1 and M_2), and one functional female (F) type (Joubert, 1985; Costés, 1988).

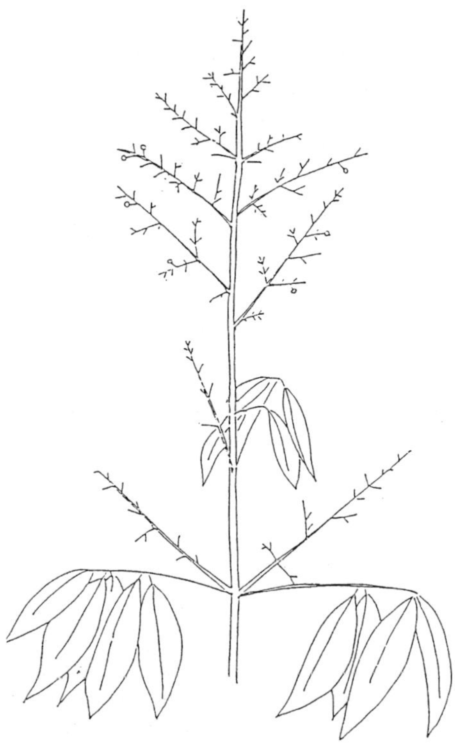

Fig. 6.5. A schematic representation of a litchi inflorescence (Costés, 1988).

Robbertse *et al.* (1995) described the morphology of the flowering axis and the position of the flowers. They identified the inflorescence as a determinate, compound thyrse, because its side branches are cymose (side branches with flat-topped or convex flower clusters). The flowers are borne on partial inflorescences or dichasia (Fig. 6.6). Usually, the first flower to open (number one) is the M_1; the second to open (number two) are two female flowers inserted laterally beneath the number one flower; the third group to open (number three) are four M_2 flowers inserted laterally beneath the number two flowers. The remaining eight (number four) buds are usually insignificant, but in richly branched inflorescences, they may also develop. Many variations on this scheme have been encountered; however, in most of them, the M_1 flowers are absent and the female flowers occupy the first, and sometimes the second, position in the dichasium. Anthesis within the panicle as a whole is from base to tip (acropetal anthesis between dichasia), whereas anthesis within the dichasia occurs from the tip to the base (basipetal anthesis within the dichasia).

Shukla and Bajpai (1974a) described the differentiation of individual flowers, and found that development takes place in an acropetal manner: first the calyx, then the stamens, and finally, the carpels. Litchi flowers are 3–8 mm wide and 7–12 mm long, and are borne on short, 2–6 mm, pedicels (Liu, 1954). They have a cup-shaped calyx, with four or five short, serrated sepals, and no petals. The stamens and pistil are inserted into the nectary located

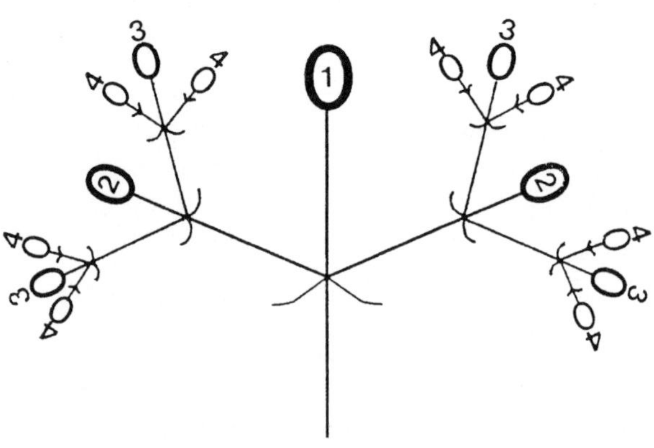

Fig. 6.6. Litchi dichasium in lateral view, with numbers showing the sequence of flower anthesis (Robbertse *et al.*, 1995).

within the calyx. The ovules are anatropous, i.e. they stand upside-down, with the top of the nucellus facing the funiculus (Fahn, 1990), and contain integuments, a nucellus, and an obturator. Each stamen has a two-celled anther, which splits longitudinally (Mustard *et al.*, 1953; Liu, 1954).

The three types of flower (Fig. 6.7) open in succession on the same inflorescence (Liu, 1954; Das and Choudhury, 1958; Joubert, 1985; Stern and Gazit, 1996). As all the flowers have male and female reproductive tissues, they are classified as hermaphrodites. However, each type exhibits distinct degeneration of the male or female tissues. They were named Type I, II and III by Mustard (1960), according to the order in which they opened, and as male, hermaphrodite and pseudo-hermaphrodite by Singh and Singh (1954). The terminology based on their sexual functionality (Male 1, Female and Male 2) described by Costés (1988) is preferred.

The following description (Fig. 6.7) is based on reports by Liu (1954), Mustard *et al.* (1953), Mustard (1960), Scholefield (1982), Costés (1988) and Moncur (1988) for litchi, and by Xu (1991) for longan. The M_1 flower has a rudimentary pistil, which appears as a conspicuous pink, pubescent protuberance. The ovary contains two partially formed ovules with no embryo sac. The pistil is surrounded by six to eight stamens with hairy filaments about 6 mm long. The nectar disc is small.

The F flower has a fully developed pistil, which has a two-lobed superior ovary containing two anatropous ovules, a short style, and a bifurcated stigma (Fig. 6.8). The surface of the ovary is pubescent, with protuberances that persist and give the fruit its rough surface. Usually, only one of the lobes of the ovary develops into a fruit, with the other aborting. Occasionally, however, the two lobes may develop, producing twin fruit. Six to eight stamens, which have very short filaments less than 1.5 mm long, surround the pistil. The anthers contain little viable pollen and do not dehisce, so the flower is functionally female. The nectar disc is large.

The M_2 flower has a prominent pistil (smaller than that of the F flower), with a short style that ends in a two-lobed stigma. About 20% of the ovules contain a mature embryo sac at anthesis. However, the pistil is non-functional, as the lobes of the stigma do not open. It is surrounded by six to eight stamens, which are similar in appearance and function to those in M_1 flowers. The nectar disc is smaller than that of the F flower, but much larger than that of the M_1 flower.

Floral development

Three distinct waves of flowering occur in litchi inflorescences, with each wave consisting of the same flower type. The first wave consists of M_1 flowers, the second of F flowers, and the third of M_2 flowers (Liu, 1954; Costés, 1988; Goren *et al.*, 1998). There is little overlap in the same inflorescence; however, each inflorescence is not synchronized on the same tree, even less so on different trees. Thus, in a single block, each wave lasts for 7–12 days, with an overlap of 1–3 days (Fig. 6.9). Such overlap facilitates cross-pollination, even on individual trees.

Several irregularities have been observed in flowering. The M_1 flowers are sometimes absent, and flowering commences with F flowers. Lack

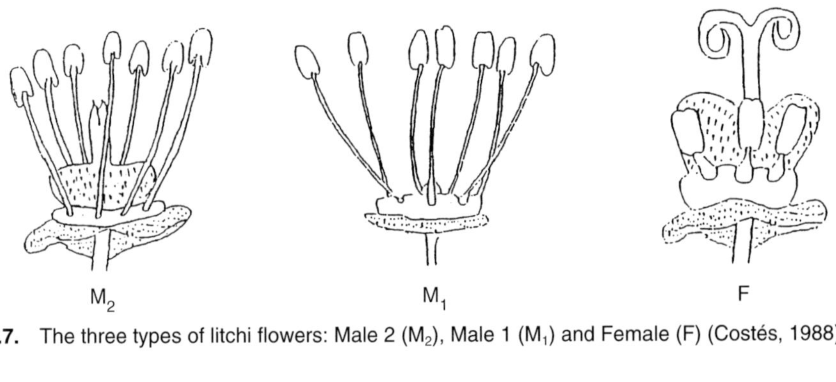

Fig. 6.7. The three types of litchi flowers: Male 2 (M_2), Male 1 (M_1) and Female (F) (Costés, 1988).

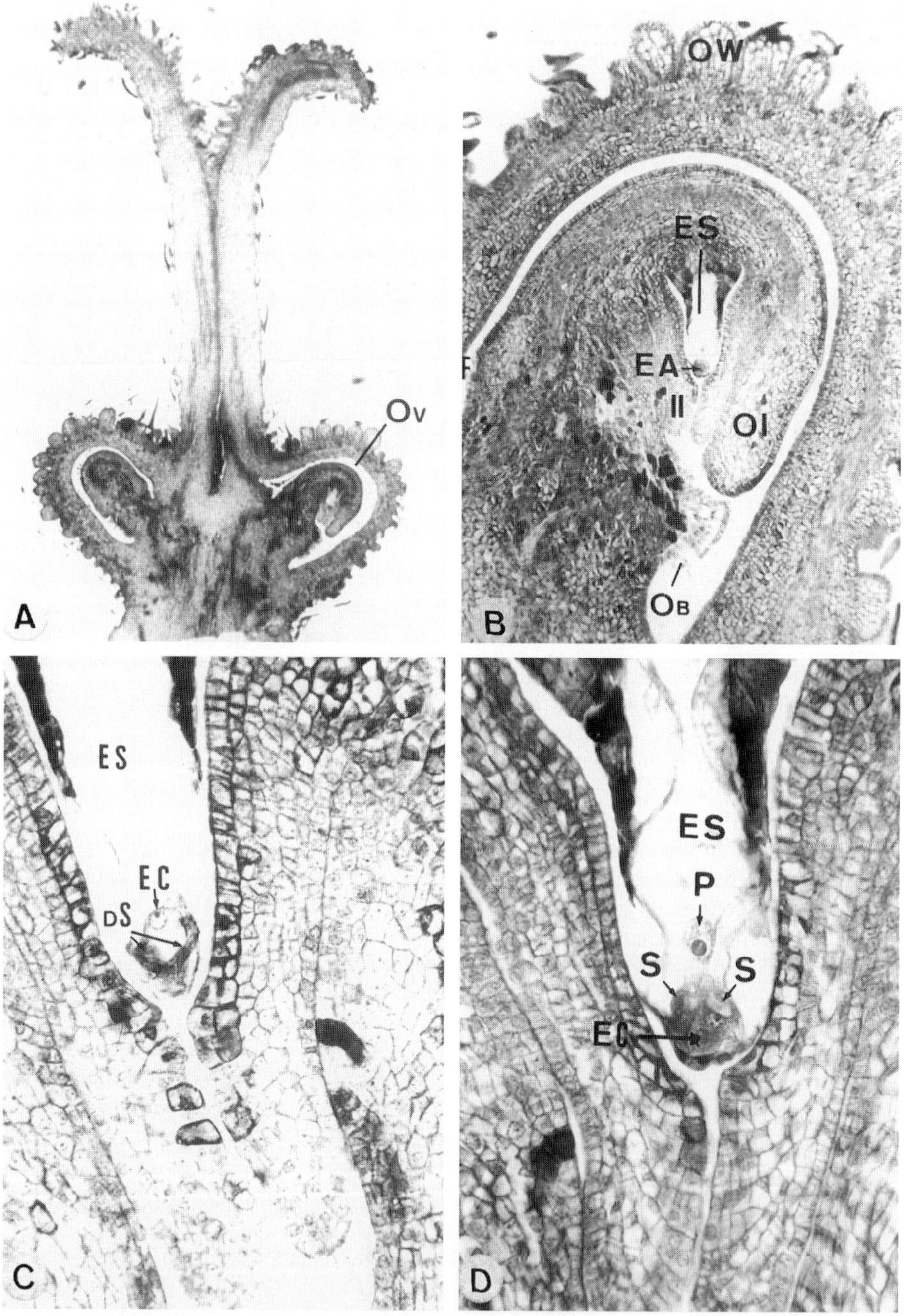

Fig. 6.8. Longitudinal section of a 2-day-old female 'Mauritius' flower. (A) Pistil with a two-lobed, bi-locular ovary (Ov) (×25). (B) Normal ovule, composed of funiculus, obturator, two integuments and an embryo sac containing an egg apparatus and a polar nucleus (×100). (C) Embryo sac, containing a normal egg cell and two degenerate synergids (×400). (D) Normal embryo sac, containing a normal polar nucleus, egg cell and two synergids (×400). ES = embryo sac; EC = egg cell; EA = egg apparatus; S = synergid; DS = degenerate synergid; P = polar nucleus; OI = outer integument; II = inner integument; OB = obturator; OW = ovary wall; and F = funiculus (Stern *et al.*, 1996).

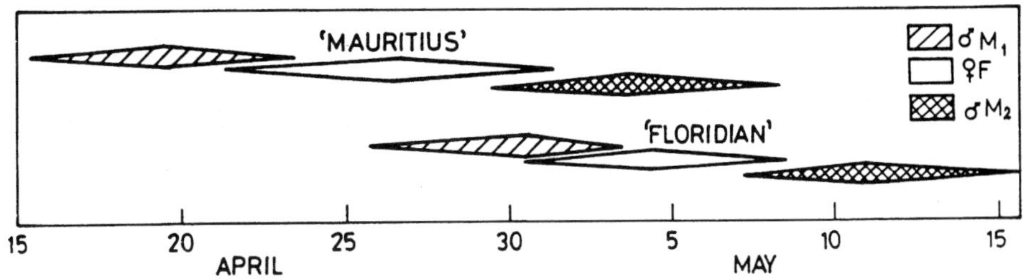

Fig. 6.9. Male flowering (M_1), female flowering (F), and pseudo-hermaphroditic flowering (M_2) of 'Mauritius' and 'Floridian' in Israel. The peak of flowering is represented by the centre of the rhombus, while its edges represent the start and end of flowering (Degani *et al.*, 1995).

of the M_1 stage occurs frequently in young trees, but also occurs in mature trees (Robbertse *et al.*, 1995; Goren *et al.*, 1998). Nakata (1956) reported the failure of the M_1 stage after girdling, or when trees had not flushed for a long period. Some trees produce a second flowering if the first flowering fails. Both F and M_2 flowers may appear in this second bloom, while the M_1 bloom is absent (Mustard *et al.*, 1953). This phenomenon is well known in 'Fay Zee Siu' ('Feizixiao') in China (Wang and Qiu, 1997) and in Israel (Goren *et al.*, 1998), and may result in a heavy fruit set.

Sex ratio

The sex ratio depends on the cultivar and the environment. The proportion of F, M_1 and M_2 flowers varied from 1.00:0.16:0.32 ('Mountain Litchi') to 1.00:2.30:3.90 ('Brewster') amongst eight cultivars in Florida (Mustard *et al.*, 1953), and from 1.00:0.16:0.90 ('Calcuttia Late') to 1.00:1.60:2.40 ('Dheradun') amongst five cultivars in India (Chadha and Rajpoot, 1969). 'Mauritius' had a ratio of 1.00:1.10:1.10 in Israel (Stern *et al.*, 1993) and 1.00:1.20:1.50 in Réunion (Costés, 1988). Chaturvedi (1965) found that 32% of flowers were female in 'Early Large Red' in India, while the proportion of female flowers in five cultivars at four locations in Queensland, Australia, varied from 16% to 43%, with almost as much variation within cultivars as between them (e.g. the proportions of female flowers on 'Bengal' at the four locations were 20, 26, 34 and 43%) (Menzel and Simpson, 1992).

In China, 'Fay Zee Siu' produced long inflorescences with a high proportion of male flowers, whereas short inflorescences had a high proportion of female flowers (Wang and Qiu, 1997). Wu *et al.* (2001) reported that pruning the inflorescence 6 weeks prior to anthesis increased the female:male ratio and initial fruit set.

Temperature is the main environmental factor affecting sex ratio. When 2-year-old plants of five cultivars were kept at 15°/10°, 20°/15°, 25°/20° or 30°/25°C, the lowest temperature regime resulted in the highest proportion of female flowers. The average proportion of female flowers was 72, 49, 27 and 11%, respectively, under the different regimes. 'Bengal' and 'Souey Tung' produced no female flowers at 25°/20° or 30°/25°C, while 'Wai Chee' produced no female flowers at 30°/25°C (Menzel and Simpson, 1991).

Ovule development and pistil receptivity

As litchi flowers do not have petals or well-developed sepals, there is no distinct time of flower opening or anthesis, as in many other species. Hence, anthesis is defined in this crop as the stage when the flowers mature. In female flowers, this is when the two stigmatic lobes spread.

Determination of ovule normality or fertility based on anatomy needs to be corroborated with functional tests. Stern *et al.* (1996) used fruit set after hand pollination as their criterion. A positive correlation was found between initial fruit set and the number of flowers with a normal embryo sac. Similar relationships have been found in other fruit trees (Burgos and Egea, 1993).

The pistil is receptive when the stigma supports pollen germination, the style and ovary enable normal growth of pollen tubes, and the ovule is mature and viable. The stigma is receptive for several days. At anthesis, the surface of the stigma is covered with long, plump papillae and is shiny white. Over time, the stigma curves, turning from white to matt-white, white-brown, then brown after 2–3 days. The stigma remains receptive until it turns white-brown, and most of the papillae have collapsed (Moncur, 1988; McConchie and Batten, 1989; Stern *et al.*, 1997). When pollen germination or fruit set was used as the criterion for receptivity, it was found that receptivity lasted for 5 days at 20°/17°C or 22°/17°C, but less than 2 days at 33°/27°C. Wetting the stigma induces premature browning and inhibits fertilization (McConchie and Batten, 1989; Stern *et al.*, 1997).

Ovule maturity tends to lag behind stigma receptivity in litchi. The highest number of mature ovules occurs several days after anthesis, when stigma receptivity is declining. Thus, hand pollination should not be performed before anthesis (Stern *et al.*, 1997), with the pollen tube taking 48 h to reach the ovule.

Several researchers have studied the anatomy of litchi ovule megasporogenesis and megagametogenesis (Banerji and Chaudhuri, 1944; Liu, 1954; Mustard, 1960; Joubert, 1967; Stern *et al.*, 1996, 1997). Little information is available for longan. The megaspore mother cell or megasporocyte can be identified in the nucellus about 10 days before the embryo sac matures. During meiotic division, it forms a linear tetrad of haploid megaspores, of which the chalazal one (at the base of the ovary where it is attached to the funiculus and the point at which the vascular tissues enter and spread into the ovule) elongates to form the embryo sac. After three mitotic divisions, a mature monosporic, eight-nucleate embryo sac (polygonum type) is formed.

The antipodals degenerate at maturity and disappear after fusion of the two polar nuclei. The nucellus is absorbed during embryo sac development, except for the nucellar epidermis and nucellar cap at the micropylar end. Two integuments envelop the mature embryo sac. The outer integument is poorly developed on the ventral side of the anatropous ovule and appears only as a protuberance when viewed in cross-section. An obturator, consisting of three to four layers of enlarged stigmatoid tissue, is present on the outer integument, forming a ring around the micropylar end of the mature ovule. The anatomy of the pistil, ovule and embryo sac is presented in Fig. 6.8.

Mustard (1960) observed embryo sac retardation and degeneration in 'Brewster' during the first 5 days of anthesis. About 40% of the ovules had a mature embryo sac, although only 12% of the ovules were full. Stern *et al.* (1996) studied the anatomy of 2-day-old F flowers of 'Mauritius' and 'Floridian' in 11 orchards in Israel, and found that close to half of the ovules lacked an embryo sac, whereas the great majority of sacs lacked egg cells and synergids. The proportion of flowers with a normal ovule at different locations varied from 3% to 27% (Stern *et al.*, 1996). To determine whether this behaviour limits productivity, and to determine the optimum age for pollination, Stern *et al.* (1997) studied changes in stigmatic surface, the anatomy of the ovule, receptivity to pollination, and initial set. They found that, at anthesis, most 'Mauritius' ovules were still immature. Two-thirds were sterile and did not have an embryo sac. Both ovules were sterile in 41% of the flowers. A majority of flowers had embryo sacs 2 days after anthesis (76% of the flowers had one or two ovules with embryo sacs). Only embryo sacs containing a polar nucleus, an egg cell, and at least one synergid were considered fertile (Stern *et al.*, 1996). One or more of these elements was lacking or degenerated in most of the sacs. The lack of a normal egg cell, and even more so of a normal synergid, was the main factor limiting fertility. On the first day of anthesis, only 6% of flowers were fertile compared with 21% on the fifth day (Table 6.2).

Pollen viability

The research in litchi is relatively extensive, whereas little information is available for longan. All of the three flower types produce pollen. The stamens of the two male flowers have long hairy filaments, with yellow anthers that contain 1900–5200 pollen grains. Stern (1992) and Costés (1988) found that M_2 anthers contain

Table 6.2. The effects of flower maturity (1, 3 and 5 days after anthesis) and stage of bloom (1, 3 and 5 days from the beginning of female anthesis) on the percentage of flowers with normal structures. Data are the means of about 30 flowers per treatment (Stern *et al.*, 1997).

Age of flowers (day)	Days from start of bloom			
	1	3	5	Mean
1	3.4	3.8	10.7	6.0
3	7.1	16.7	19.2	14.3
5	17.9	23.3	22.2	21.1

more pollen than M_1 anthers: 4200 vs. 2690 for 'Mauritius' in Réunion; 3300 vs. 2000 for 'Mauritius'; and 3000 vs. 1900 for 'Floridian' in Israel. In China, 2500–3000 pollen grains per anther were reported for 'Wai Chee' ('Huaizhi') (Wang and Qiu, 1997).

Anthesis in male flowers is when the first anthers dehisce. Anthers in the same flower mature and dehisce gradually over 2–3 days (Das and Choudhury, 1958; Chaturvedi and Saxena, 1965). Pollen sacs on the same anther may even dehisce at different times during the day (Shalem-Galon, 1980). When viewed under a light microscope, dry litchi pollen grains are oblong, swelling to become triangular or rectangular after soaking in glycerin or water. The average pollen grain in ten Indian cultivars was 22 μm long, and ranged from 18 to 27 μm (Das and Choudhury, 1958; Chaturvedi and Saxena, 1965). Similarly, pollen grains from 'Mauritius' and 'Floridian' were elongate under the scanning electron microscope (SEM), 10 μm wide × 20 μm long, with three elongate germination pores (Stern and Gazit, 1998). There were no noticeable differences between the pollen grains of the two cultivars, or between the two male flowers (Stern, 1992). In contrast, Wang and Qiu (1997) reported that sculpturing patterns differed between cultivars. The released pollen grains are binucleate (Mustard *et al.*, 1953; Liu, 1954).

Liu (1954) identified defective, non-functional, small, and often shrivelled pollen grains in 'Haak Yip' ('Heiye'): 18 μm long compared with 25 μm for functional grains. She also encountered triangular pollen grains without germination pores. A considerable proportion of semi-quadrangular grains were found in pollen sampled from orchards in Israel, especially from M_1 flowers (Stern, 1992; Stern and Gazit, 1998).

About three times more M_2 pollen germinated than M_1 pollen in the hanging-drop test (Mustard *et al.*, 1953; Costés, 1988). Fivaz *et al.* (1994) found a small advantage of M_2 over M_1 pollen in four out of five cultivars in South Africa, using *in vitro* methods. Stern and Gazit (1998) found a consistent advantage of M_2 over M_1 pollen in five cultivars across different temperature regimes, *in vitro*, with only 5–12% germination for M_1 pollen, and 55% for M_2 pollen. Pollen-tube growth was also faster for M_2 pollen (750 μm after 24 h at 25°C versus 350 μm for M_1). In contrast, Singh (1962) found 62% germination for 'Calcuttia' M_1 pollen compared with 48% for M_2 pollen, using aceto-carmine staining. He did not find any differences in pollen-tube growth.

In vitro germination does not fully characterize pollen viability. Hundreds of pollen grains reach each flower in the case of litchi, but only one pollen tube needs to penetrate the embryo sac for fertilization. Success is usually reported as the proportion of hand-pollinated flowers with pollen tubes in the ovary, ovule or embryo sac.

Stern and Gazit (1998) found that the best temperature for *in vivo* germination was 22°/17° or 27°/22°C. Tubes reached the ovule in about 35% of hand-pollinated flowers after 48 h, with no difference between 'Mauritius' and 'Floridian', or between M_1 and M_2 pollen. In contrast, more M_2 pollen reached the embryo sac. M_2 pollen is slightly richer in protein (8.1% vs. 6.7%) and sugars (10.9 vs. 9.9%) than M_1 pollen (Stern, 1992). M_2 flowers also secrete more nectar and sugar than do M_1 flowers (Stern and Gazit, 1996), and may act as stronger sinks, with the pollen benefiting from a better supply of nutrients.

Stern *et al.* (1996) and Stern and Gazit (1998) studied the effects of temperature on litchi ovule and pollen genesis. Two-year-old 'Mauritius' and 'Floridian' plants with emerging inflorescences were kept for 2 months before anthesis at 22°/12°, 27°/17° or 32°/22°C, for ovule studies, and at 22°/17°, 27°/22° or 32°/27°C, for pollen studies. 'Floridian' was more sensitive to temperature than 'Mauritius'.

At high temperatures, 'Floridian' had no functional ovules and sterile pollen, whereas 'Mauritius' had normal ovules and viable pollen. These differences reflected the climate where they are grown in China. 'Mauritius' ('Dazao') is cultivated near Guangzhou (latitude 23°N), whereas 'Floridian' ('Chenzi') is cultivated in cooler Putian (latitude 25°N).

Pollination

The stigmas, anthers and nectaries are exposed in litchi, and are available to insects. A large number of species has been found on flowers in India (Dhaliwal *et al.*, 1977), Florida (Butcher, 1956, 1957), Israel (Pivovaro, 1974), South Africa (Du Toit, 1994) and Australia (King *et al.*, 1989). Nectar production occurs mainly in the morning (Butcher, 1957; Pandey and Yadava, 1970) and insect activity is generally higher in the morning than in the afternoon, with pollination occurring when the insects visit one of the two male flowers followed by a female flower.

Litchi inflorescences are a rich source of nectar, and are highly attractive to insects. Female flowers had greater nectar volumes and sugar concentrations than M_2 flowers, which in turn were higher than M_1 flowers (Stern and Gazit, 1996). A typical 'Mauritius' inflorescence has over 100 female flowers open at the same time, and a typical tree has hundreds of such inflorescences.

Anther dehiscence takes place throughout the day and night, mostly from morning to mid-afternoon (Das and Choudhury, 1958; Chadha and Rajpoot, 1969; Wang and Qiu, 1997). Pollen is not attractive to most of the pollinators. Pandey and Yadava (1970) found that honeybees collect pollen mainly at the beginning and end of flowering.

Bees are the most important pollinators (Butcher, 1956, 1957; McGregor, 1976; Free, 1993). Pollination by honeybees has been reported to increase yields by 2.5–2.9 times compared with orchards without bees (Chen, 1993). Only a few of the several *Apis* species in China are involved in pollination, with the Chinese honeybee, *Apis cerana cerana*, being the main pollinator. This species was domesticated 2000 years ago and is used for honey production in litchi-growing areas. The European honeybee, *A. mellifera*, was introduced to China about 100 years ago (Crane, 1990; Chen, 1993), and also pollinates litchi. The European honeybee is the most important species in Florida, South Africa, Israel and Australia, with hives moved to orchards in bloom. It is a very effective pollinator, with the bees touching the stigmas on almost every visit to the female flowers (King *et al.*, 1989).

A bee can easily fill its stomach with nectar from just a few litchi flowers (Free, 1993), so both the quality and quantity of nectar can have a large bearing on the efficiency of honey production. Stern and Gazit (1996) found a correlation between bee density on the M_1, F and M_2 flowers and sugar in the nectar, but a negative correlation between bee density and nectar volume for M_2 and F flowers. The proportion of glucose (43%), fructose (39%) and sucrose (18%) in the nectar was similar for the three flower types.

The amount of pollen found on bees sampled from F and M_2 flowers was twice as high as that found on bees sampled during M_1 bloom, reflecting the pronounced differences in the quantity and quality of the nectar present in the different flower types (Stern and Gazit, 1996). Bees visiting M_1 blooms will be 'disappointed' with the meagre yield, and will abandon them, whereas bees visiting M_2 flowers will be rewarded, will visit many flowers, and will be dusted with plenty of pollen.

Pollination has been reported to range from 0 to 10%, notwithstanding the large number of bees visiting during the first days of female bloom (Stern and Gazit, 1996). The few bees visiting M_1 flowers resulted in little pollen adhering to the bees (less than 10 pollen grains per bee). When the M_2 bloom commenced, pollination increased to 20% or 30%, reaching 80–90% within a few days. The greater attractiveness of M_2 flowers resulted in hundreds of pollen grains adhering to each bee. Consequently, the number of germinating pollen grains per stigma increased from one or two to about 20 (Stern and Gazit, 1996).

McConchie and Batten (1991) provided information on the relationship between fruit set and pollination. They found that hand pollination of all the F flowers on a 'Bengal' inflorescence (225 flowers) produced only nine fruit, whereas pollination of only 47 or 48 flowers and removal of all other female flowers yielded 13 or 23 fruit for early- and late-opening flowers, respectively. The greater fruit set of the late flowers may reflect a higher fertility rate, or the use of the more potent M_2 pollen (Fig. 6.9).

Longan flowering and pollination

There have been few studies on longan flowering, although its botany is similar to that of litchi (Subhadrabandhu, 1990; Yaacob and Subhadrabandhu, 1995; Zee *et al.*, 1998). The inflorescences are compound dichasia, terminal, usually leafless, erect, with widely branched panicles up to 30 cm long. Flowering in each panicle progresses as staminate flowers (M_1), functionally female hermaphrodite flowers (F), and finally, functionally male hermaphrodite flowers (M_2). Flowers are small and yellow-brown with five petals. The staminate flowers have no more than eight stamens in a single row on a light-brown disc. The stamens are hairy with two-lobed anthers. The F flowers have a bicarpellate ovary on the disc. The exterior of the ovary is covered with short hairs. Usually, only one locule in each carpel grows and develops into a fruit. At the end of the stigma, there are two lobes, which are shorter than those of litchi. The eight stamens on the F flower are reduced, with a semi-sessile filament. The male-to-female bloom ratio is about 5:1 in well-grown trees.

As with litchi, there is an overlap of flower types on individual trees, since not all inflorescences develop at the same time. The overlap in each orchard may be a few weeks, depending on the cultivar and environment. Pollination is carried out by many insects, especially by the honeybee. *Apis mellifera* is the most important pollinator in Thailand (Waite and Hwang, 2002), whereas *Trigona iridipennsis* contributes in Thailand and India (Boonithee *et al.*, 1991). Pollination occurs from early morning to mid-afternoon.

Conclusions

Litchi and longan have phenologies similar to those of citrus and mango (Fig. 6.1). Shoots must first be initiated from resting stem buds, then determined as either leaves or flowers. Shoot initiation is regulated by the ratio between an initiation promoter, perhaps a CK from the roots, and a shoot inhibitor, perhaps auxin from the leaves (Fig. 6.3). It is proposed that low CK:auxin ratios in resting buds promote continued rest, and that high ratios stimulate shoot initiation.

Shoot induction is dependent on the interaction of an age-dependent, vegetative promoter (GA_x in Fig. 6.1) from leaves with a short-lived, floral promoter (unknown substance) synthesized in the leaves during cool weather. The ratio of these substances governs whether the shoots produce leaves or flowers, or both (Fig. 6.4).

The occurrence of defective pollen and especially ovules in litchi and longan is widespread. Although wind can play a role in litchi pollination, insects appear to be essential for commercial production. There is usually high pollinator activity on F blooms due to the strong attraction of bees and other insects to the nectar. Interplanting of cultivars to encourage pollen transfer from M_2 to F flowers may be beneficial.

Acknowledgements

This research was supported by the Florida Agricultural Experiment Station and approved for publication as Journal Series No. R-09784. Trevor Olesen and David Turner kindly read an earlier draft of this chapter.

References

Abeles, F.B., Morgan, P.W. and Saltveit, M.E., Jr (1992) *Ethylene in Plant Biology*, 2nd edn. Academic Press, San Diego, 414 pp.

Banerji, I. and Chaudhuri, K.L. (1944) A contribution to the life history of *Litchi chinensis* Sonn. *Proceedings of the Indian Academy of Science Section B* 19, 19–27.

Bangerth, F. (1994) Response of cytokinin concentration in the xylem exudate of bean (*Phaseolus vulgaris* L.) plants to decapitation and auxin treatment, and relationship to apical dominance. *Planta* 194, 439–442.

Barba, R.C. (1974) Induction of flowering of the mango by chemical spray. *Proceedings of the Crop Science Society of the Philippines* 5, 154–160.

Batten, D.J. and McConchie, C.A. (1995) Floral induction in growing buds of lychee (*Litchi chinensis*) and mango (*Mangifera indica*). *Australian Journal of Plant Physiology* 22, 783–791.

Bernier, G. (1988) The control of floral evocation and morphogenesis. *Annual Review of Plant Physiology and Plant Molecular Biology* 39, 175–219.

Boonithee, A., Juntawong, N., Pechhacker, H. and Huttinger, E. (1991) Floral visits to select crops by four *Apis* species and *Trigona* sp. in Thailand. In: van Heemert, C. and de Ruijter, A. (eds) *Proceedings of the Sixth International Symposium on Pollination*, Tilburg, The Netherlands, pp. 74–80.

Burgos, L. and Egea, J. (1993) Apricot embryo-sac development in relation to fruit set. *Journal of Horticultural Science* 68, 203–208.

Butcher, F.G. (1956) Bees pollinate lychee blooms. *Proceedings of the Florida Lychee Growers' Association* 3, 59–60.

Butcher, F.G. (1957) Pollinating insects on lychee blossoms. *Proceedings of the Florida State Horticultural Society* 70, 326–328.

Campbell, C.W. (1970) Minor tropical fruit cultivars in Florida. *Proceedings of the Tropical Region of the American Society of Horticultural Science* 14, 66–73.

Campbell, R.J. (1994) Fall pruning induces blooming in young lychee trees. *Proceedings of the Florida State Horticultural Society* 107, 348–350.

Campbell, R.J. and Campbell, C.W. (2001) Evaluation and introduction of lychee cultivars in south Florida. *Acta Horticulturae* 558, 119–123.

Carmi, A. and van Staden, J. (1983) Role of roots in regulating the growth rate and cytokinin content of leaves. *Plant Physiology* 73, 76–78.

Chadha, K.L. and Rajpoot, M.S. (1969) Studies on floral biology, fruit set and its retention and quality of some lychee varieties. *Indian Journal of Horticulture* 26, 124–129.

Chaikiattiyos, S., Menzel, C.M. and Rasmussen, T.S. (1994) Floral induction in tropical fruit trees: effects of temperature and water supply. *Journal of Horticultural Science* 69, 397–415.

Chaitrakulsub, T., Sidthiwong, P., Gemma, H. and Oogaki, C. (1989) Effects of SADH and MH on vegetative growth of longan (*Euphoria longana* Lamk.) in different orchard sites. *Japanese Journal of Tropical Agriculture* 33, 73–80.

Chaturvedi, R.B. (1965) Preliminary studies in the sex distribution, pollination and fruit development in litchi (*Litchi chinensis* Sonn.). *Allahabad Farmer* 39, 49–51.

Chaturvedi, R.B. and Saxena, G.K. (1965) Studies on the blossom biology of litchi (*Litchi chinensis* Sonn.). *Allahabad Farmer* 39, 10–13.

Chen, C.M., Ertl, J.R., Leisner, S.M. and Chang, C.C. (1985) Localization of cytokinin biosynthesis sites in pea plants and carrot roots. *Plant Physiology* 78, 510–513.

Chen, H. and Huang, H. (2001) China litchi industry: development, achievements and problems. *Acta Horticulturae* 558, 31–39.

Chen, W.S. (1990) Endogenous growth substances in xylem and shoot-tip diffusate of lychee in relation to flowering. *HortScience* 25, 314–315.

Chen, W.S. (1991) Changes in cytokinins before and during early flower bud differentiation in lychee (*Litchi chinensis* Sonn.). *Plant Physiology* 96, 1203–1206.

Chen, W.S. (1994) Gibberellins in buds of *Euphoria longana* Lam. during different stages of growth. *Botanical Bulletin Academic Sinica* 35, 39–43.

Chen, W.S. and Ku, M.L. (1988) Ethephon and kinetin reduce shoot length, and increase flower bud formation in lychee. *HortScience* 23, 1078.

Chen, W.S., Huang, Q.W. and Ku, W.C. (1997) Cytokinin from terminal bud of *Euphoria longana* during different growth stages. *Physiologia Plantarum* 99, 185–189.

Chen, Y. (1993) *Apiculture in China*. Agricultural Publishing House, Beijing, China, 180 pp.

Cobin, M. (1950) *The Lychee in Florida*. University of Florida Agricultural Experiment Stations Bulletin No. 471, 24 pp.

Costés, E. (1988) Analyze architecturale et modelisation du litchi (*Litchi chinensis* Sonn.). PhD thesis, University of Montpellier, France.

Crane, E. (1990) *Bees and Beekeeping: Science Practice and World Resources*. Heinemann Newnes, Oxford, UK, 614 pp.

Crane, J.H., Balerdi, C.F., Campbell, R.J. and Knight, R.J., Jr (1998) *The Lychee in Florida*. Fact Sheet HS-6, Cooperative Extension Service, IFAS, University of Florida, Gainesville [http://edis.ifas.ufl.edu/MG051].

Das, C.S. and Choudhury, K.R. (1958) Floral biology of litchi (*Litchi chinensis* Sonn.). *South Indian Horticulture* 6, 17–22.

Das, D.K., Prakash, N.S. and Bhall-Sarin, N. (1999) Multiple-shoot induction in *Litchi chinensis*. In: Kishor, P.B.K. (ed.) *Plant Tissue Culture and Biotechnology: Emerging Trends*. University Press, Hyderabad, India, pp. 166–170.

Davenport, T.L. (1990) Citrus flowering. *Horticultural Reviews* 12, 349–408.

Davenport, T.L. (2000) Processes influencing floral initiation and bloom: the role of phytohormones in a conceptual flowering model. *HortTechnology* 10, 733–739.

Davenport, T.L. (2003a) Principles of climatic effects on flowering in citrus. In: *Proceedings of the International Society of Citriculture*, Vol. 1, pp. 289–291.

Davenport, T.L. (2003b) Leaves not necessary for citrus floral induction. In: *Proceedings of the International Society of Citriculture*, Vol. 2, pp. 660–661.

Davenport, T.L. (2003c) Management of flowering in three tropical and subtropical fruit tree species. *HortScience* 38, 1331–1335.

Davenport, T.L. and Núñez-Eliséa, R. (1991) Is endogenous ethylene involved in mango floral induction? *Acta Horticulturae* 291, 85–95.

Davenport, T.L. and Núñez-Eliséa, R. (1997) Reproductive physiology. In: Litz, R.E. (ed.) *The Mango: Botany, Production and Uses*. CAB International, Wallingford, UK, pp. 69–146.

Davenport, T.L. and Ying, Z. (2003) Further characterization of the mango florigenic promoter. In: *Proceedings of the Twenty-Eighth Annual Meeting of the Plant Growth Regulation Society of America*, p. 26.

Davenport, T.L., Morgan, P.W. and Jordan, W.R. (1980) Reduction of auxin transport capacity with age and internal water deficits in cotton petioles. *Plant Physiology* 65, 1023–1025.

Davenport, T.L., Zheng, Q. and Li, Y. (2000) Reliable flowering of lychee (*Litchi chinensis* Sonn.) in Florida. In: *Proceedings of the Twenty-Seventh Annual Meeting of the Plant Growth Regulation Society of America*, pp. 312–315.

Davenport, T.L., Kulkarni, V. and White, T.L. (2001a) Minimum number of leaves necessary for floral induction of mango. In: *Proceedings of the Twenty-Eighth Annual Meeting of the Plant Growth Regulation Society of America*, p. 67.

Davenport, T.L., Kulkarni, V. and White, T.L. (2001b) Longevity of the florigenic promoter in mango. In: *Proceedings of the Twenty-Eighth Annual Meeting of the Plant Growth Regulation Society of America*, p. 53.

Davies, P.J. (1995) *Plant Hormones: Physiology, Biochemistry and Molecular Biology*. Kluwer Academic, Boston, Massachusetts, 833 pp.

Degani, C., Stern, R.A., El-Batsri, R. and Gazit, S. (1995) Pollen parent effect on the selective abscission of Mauritius and Floridian lychee fruits. *Journal of the American Society for Horticultural Science* 120, 523–526.

Dhaliwal, H.S., Srivastava, S. and Adlakha, R.L. (1977) Insect pollination of lychee, *Litchi chinensis* Sonn., in the valley areas of the Indian Himalayas. In: *Proceedings of the Twenty-Sixth International Apiculture Congress*. Apimondia Publishing House, Bucharest, Hungary, p. 396.

Du Toit, A.P. (1994) Pollination of avocados, mangoes and litchis. *Plant Protection Research Institute, South Africa Plant Protection News Bulletin* 35, pp. 4–5.

Fahn, A. (1990) *Plant Anatomy*, 4th edn. Pergamon Press, New York, 611 pp.

Fivaz, J., Robbertse, P.J. and Gazit, S. (1994) Studies on the morphology, viability and storage of pollen grains of litchi (*Litchi chinensis* Sonn.). *Yearbook of the South African Litchi Growers' Association* 6, 9–12.

Free, J.B. (1993) *Insect Pollination of Crops*, 2nd edn. Academic Press, London, 684 pp.

Galán Saúco, V. (1990) *Los Frutales Tropicales en los Subtropicos. I. Aguacate – Mango – Litchi y Longan*. Ediciones Mundi-Prensa, Spain, 133 pp.

Galán Saúco, V., Fernández Galván, D. and Hernández Delgado, D.P.M. (2001) Behavior of different litchi cultivars in Canary Islands. *Acta Horticulturae* 558, 83–88.

Ghosh, S.P. (2001) World trade in litchi: past, present and future. *Acta Horticulturae* 558, 23–30.

Goren, M. (1990) High density litchi orchards by reducing tree height. *Alon Hanotea* 44, 699–704.

Goren, M. and Gazit, S. (1993) Small-statured litchi orchards: a new approach to the growing of litchi. *Acta Horticulturae* 349, 69–74.

Goren, M. and Gazit, S. (1996) Management of lychee orchards in Israel. In: *Proceedings of the Fourth National Lychee Seminar*, Australian Lychee Growers' Association, Yeppoon, Australia, pp. 74–77.

Goren, M., Degani, O. and Gazit, S. (1998) Flowering phenology of 10 lychee cultivars in the Volcani Center, Israel. *Alon Hanotea* 52, 59–63.

Goren, M., Tomer, E. and Gazit, S. (2001) Litchi cultivars in Israel. *Acta Horticulturae* 558, 129–133.

Groff, G.W. (1921) *The Lychee and Lungan*. Orange Judd Company, New York, 188 pp.

Groff, G.W. (1943) Some ecological factors involved in successful lychee culture. *Proceedings of the Florida State Horticultural Society* 56, 134–155.

Horgan, R. (1984) Cytokinins. In: Wilkins, M.B. (ed.) *Advanced Plant Physiology.* Pitman Publishing, London, pp. 53–75.

Hou, X., Liang, L., Ji, Z., Lee, P. and Li, M. (1987) Studies on the assay method for endogenous abscisic acid (ABA) and the fluctuation of its contents in the buds of litchi during floral initiation. *Acta Horticulturae Sinica* 14, 12–16.

Huang, P. and Weng, S. (1978) The effect of delayed shoot treatment of fruit bearing of litchi trees. *Journal of the Chinese Society for Horticultural Science* 24, 121–126.

Huang, Q.W. and Huang, Q.W. (1996) Changes in endogenous hormone contents in relation to flower bud differentiation and on-year or off-year fruiting of longan. *Journal of Tropical and Subtropical Botany* 4, 58–62.

Itai, C. and Birnbaum, H. (1996) Synthesis of plant growth regulators by roots. In: Waisel, Y., Eshel, A. and Kafkafi, U. (eds) *Plant Roots The Hidden Half.* Marcel Dekker, New York, pp. 273–284.

Jahiel, M. and Abraham, D. (2001) The growth of Malagasy litchi field face to the hardening of European control in the matter of sulphured residue. *Acta Horticulturae* 558, 63–71.

Joubert, A.J. (1967) Die bloei, embriosak-, embrio-en verugontwikkeling von *Litchi chinensis* Sonn. cultivar HLH Mauritius. MSc thesis, University of the Witwatersrand, Johannesburg, South Africa.

Joubert, A.J. (1985) *Litchi chinensis.* In: Halevy, A.H. (ed.) *Handbook of Flowering,* Vol. 5. CRC Press, Boca Raton, Florida, pp. 204–210.

Jungyoosuk, P. (1999) Off-season longan: good or bad news for growers? *Kehakarn-Kaset* 23, 73–74.

Kachru, R.B., Singh, R.N. and Chacko, E.K. (1971) Inhibition of flowering in mango (*Mangifera indica* L.) by gibberellic acid. *HortScience* 6, 140–141.

Ke, G.W., Huang, J.H. and Wang, C.C. (1988) Morpho-anatomical observations on the degradation of longan inflorescences. *China Fruits* 1, 22–24.

Ke, G.W., Tang, J.F. and Liu, R.F. (1998) Study on the reasons causing low yield of longan fruit trees and possible improvements. *South China Fruits* 27 (1), 25–26.

Khao-Sumeru, Y., Sritontip, C. and Changenraja, S. (1999) *Summary on the Use of Potassium Chlorate and Sodium Chlorate for Inducing Off-Season Flowers in Longan.* Institute of Agricultural Research and Training, Lampang, Thailand, 5 pp.

King, J., Exley, E.M. and Vithanage, V. (1989) Insect pollination for yield increases in lychee. In: Batten, D.J. (ed.) *Proceedings of the Fourth Australian Conference on Tree and Nut Crops.* Exotic Fruit Growers' Association, Lismore, Australia, pp. 142–145.

Koen, T.J. and Smart, G. (1982) Effect of optimal manuring in the production and fruit quality of litchi trees. *Information Bulletin of the Citrus and Subtropical Fruit Research Institute* 117, 1–2.

Koen, T.J., Langenegger, W. and Smart, G. (1981a) Stikstofbemesting van lietsjiebome. *Subtropica* 2, 9–11.

Koen, T.J., Langenegger, W. and Smart, G. (1981b) Nitrogen fertilization of litchi trees. *Information Bulletin of the Citrus and Subtropical Fruit Research Institute* 107, 9–11.

Koen, T.J., Langenegger, W. and Smart, G. (1981c) Determination of the fertilizer requirements of litchi trees. *Information Bulletin of the Citrus and Subtropical Fruit Research Institute* 103, 9–12.

Kulkarni, V.J. (1988) Further studies on graft-induced off-season flowering and fruiting in mango (*Mangifera indica* L.). *Journal of Horticultural Science* 63, 361–367.

Ledin, R.B. (1953) Lychee. *Annual Report of the Florida Agricultural Experiment Station,* Gainesville, Florida, pp. 314–315.

Ledin, R.B. (1954) Lychee. *Annual Report of the Florida Agricultural Experiment Station,* Gainesville, Florida, p. 285.

Ledin, R.B. (1955) Lychee. *Annual Report of the Florida Agricultural Experiment Station,* Gainesville, Florida, pp. 306–307.

Li, C. and Xiao, Y. (2001) Girdling increases yield of 'Nuomici' litchi. *Acta Horticulturae* 558, 233–235.

Li, M.X. and Chang, B.S. (1987) The effects of plant hormones on flower-forming and fruit-setting and their applied effects in production. *Science and Technology Communications, Litchi* 1, 5.

Li, Y.C., Davenport, T.L., Rao, R. and Zheng, Q. (2001) Nitrogen, flowering and production of lychees in Florida. *Acta Horticulturae* 558, 221–224.

Liang, G.J. and Yu, G.X. (1991) Effects of PP333 on mature, non-flowering and leaflet-on-panicle litchi trees. *Scientia Horticulturae* 48, 319–322.

Liang, L., Ji, Z. and Puiman, L. (1983) Fluctuations of endogenous cytokinin contents in the differentiating flower buds in the litchi (*Litchi chinensis* Sonn.). *Journal of the South China Agricultural College* 4, 37–44.

Liang, T.G. and Chen, L.L. (1965) Preliminary observations on the flower bud differentiation of 'Honghezi' longan in Fuzhou. *Acta Horticulturae Sinica* 4, 13–18.

Liang, W., Liang, L., Ji., Z. and Li, P. (1987) The fluctuation of endogenous gibberellin and indole-3-acetic acid in *Litchi chinensis* shoot-tips during floral initiation. *Acta Horticulturae Sinica* 14, 145–152.

Liu, S.Y. (1954) Studies of *Litchi chinensis* Sonn. PhD thesis, University of Michigan, Ann Arbor, Michigan.

Liu, X.H. and Ma, C.L. (2001) Production and research of longan in China. *Acta Horticulturae* 558, 73–82.

Lomax, T.L., Muday, G.K. and Rubery, P.H. (1995) Auxin transport. In: Davies, P.J. (ed.) *Plant Hormones: Physiology, Biochemistry and Molecular Biology.* Kluwer Academic, Boston, Massachusetts, pp. 509–530.

Manochai, P. (2000) *Longan.* Department of Horticulture, Faculty of Agricultural Production, Mae Jo University, Thailand, 114 pp.

Manochai, P., Suton, V., Viriya-alongkorn, W., Charasamrit, N. and Ussahatanon, S. (1999) Effect of potassium chlorate on flowering of longan cvs. E-Daw and See Chompoo. In: *Proceedings of the National Symposium on Plant Hormones for the Production of Off-Season Fruit.* Research Council of Thailand, Bangkok, Thailand, pp. 1–8.

Mansfield, T.A. and McAinsh, M.R. (1995) Hormones as regulators of water balance. In: Davies, P.J. (ed.) *Plant Hormones: Physiology, Biochemistry and Molecular Biology.* Kluwer Academic, Boston, Massachusetts, pp. 598–616.

Marler, T.E. and Willis, L.E. (1996) Root and stem growth patterns of young 'Mauritius' lychee trees. *HortScience* 31, 815–818.

McConchie, C.A. and Batten, D.J. (1989) Floral biology and fruit set in lychee. *Proceedings of the Second National Lychee Seminar.* Australian Lychee Growers' Association, Cairns, Australia, pp. 71–74.

McConchie, C.A. and Batten, D.J. (1991) Fruit set in lychee (*Litchi chinensis* Sonn.). Variations between flowers, panicles and trees. *Australian Journal of Agricultural Research* 42, 1163–1172.

McGaw, B.A. and Burch, L.R. (1995) Cytokinin biosynthesis and metabolism. In: Davies, P.J. (ed.) *Plant Hormones: Physiology, Biochemistry and Molecular Biology.* Kluwer Academic, Boston, Massachusetts, pp. 98–117.

McGregor, S.E. (1976) *Insect Pollination of Cultivated Crop Plants.* US Department of Agriculture Handbook No. 496, pp. 247–248.

Menzel, C.M. (1983) The control of floral initiation in lychee: a review. *Scientia Horticulturae* 21, 201–215.

Menzel, C.M. (1991) Litchi. In: Verheij, E.W.M. and Coronel, R.E. (eds) *Plant Resources of South-East Asia* Vol. 2: *Edible Fruits and Nuts.* Pudoc, Wageningen, The Netherlands, pp. 191–195.

Menzel, C.M. (2001) The physiology of growth and cropping of lychee. *Acta Horticulturae* 558, 175–184.

Menzel, C.M. and Paxton, B.F. (1986) Effect of cincturing on growth and flowering of lychee: preliminary observations in subtropical Queensland. *Australian Journal of Experimental Agriculture* 26, 255–259.

Menzel, C.M. and Simpson, D.R. (1988) Effect of temperature on growth and flowering of litchi (*Litchi chinensis* Sonn.) cultivars. *Journal of Horticultural Science* 63, 349–360.

Menzel, C.M. and Simpson, D.R. (1990) Effect of paclobutrazol on growth and flowering of lychee (*Litchi chinensis*). *Australian Journal of Experimental Agriculture* 30, 131–137.

Menzel, C.M. and Simpson, D.R. (1991) Effect of temperature and leaf water stress on panicle and flower development of litchi (*Litchi chinensis* Sonn.). *Journal of Horticultural Science* 66, 335–344.

Menzel, C.M. and Simpson, D.R. (1992) Flowering and fruit set in lychee (*Litchi chinensis* Sonn.) in subtropical Queensland. *Australian Journal of Experimental Agriculture* 32, 105–111.

Menzel, C.M. and Simpson, D.R. (1994) Lychee. In: Schaffer, B. and Anderson, P.C. (eds) *Handbook of Environmental Physiology of Fruit Crops*, Vol. II: *Sub-Tropical and Tropical Crops.* CRC Press, Boca Raton, Florida, pp. 123–145.

Menzel, C.M. and Simpson, D.R. (1995) Temperatures above 20°C reduce flowering in lychee (*Litchi chinensis* Sonn.). *Journal of Horticultural Science* 70, 981–987.

Menzel, C.M., Carseldine, M.L. and Simpson, D.R. (1987) The effect of leaf age on nutrient composition of non-fruiting litchi (*Litchi chinensis* Sonn.). *Journal of Horticultural Science* 62, 273–279.

Menzel, C.M., Watson, B.J. and Simpson, D.R. (1988) The lychee in Australia. *Queensland Agricultural Journal* 114, 19–27.

Menzel, C.M., Rasmussen, T.S. and Simpson, D.R. (1989) Effects of temperature and leaf water stress on growth and flowering of litchi (*Litchi chinensis* Sonn.). *Journal of Horticultural Science* 64, 739–752.

Menzel, C.M., Watson, B.J. and Simpson, D.R. (1990) Longan. In: Bose, T.K. and Mitra, S.K. (eds) *Fruits: Tropical and Subtropical.* Naya Prokash, Calcutta, India, pp. 522–546.

Menzel, C.M., Carseldine, M.L., Haydon, G.F. and Simpson, D.R. (1992) A review of existing and proposed new leaf nutrient standards for lychee. *Scientia Horticulturae* 49, 33–53.

Menzel, C.M., Haydon, G.F. and Simpson, D.R. (1995a) Growth of lychee (*Litchi chinensis* Sonn.) under variable nitrogen supply in sand culture. *Journal of Horticultural Science* 70, 757–767.

Menzel, C.M., Oosthuizen, J.H., Roe, D.J. and Doogan, V.J. (1995b) Water deficits at anthesis reduce CO_2 assimilation and yield of lychee (*Litchi chinensis* Sonn.) trees. *Tree Physiology* 15, 611–617.

Menzel, C.M., Olesen, T., McConchie, C., Wiltshire, N., Diczbalis, Y. and Wicks, C. (2000) *Lychee, Longan and Rambutan: Optimizing Canopy Management.* Rural Industries Research and Development Corporation No. 00/29, Canberra, Australia.

Mitra, S.K. (1988) Litchi. In: Bose, T.K., Mitra, S.K. and Sadhu, M.K. (eds) *Mineral Nutrition of Fruit Crops.* Naya Prokask, Calcutta, India, pp. 366–371.

Moncur, M.W. (1988) *Floral Development of Tropical and Subtropical Fruit and Nut Species.* CSIRO, Division of Water and Land Resources, Melbourne, Australia, 181 pp.

Morse, R.L. and Oosthuizen, J.H. (1993) Litchi flower manipulation by girdling. *Yearbook of the South African Litchi Growers' Association* 7, 5–10.

Murti, G.S.R., Upreti, K.K., Reddy, Y.T.N. and Kurien, R.M. (2000) Cytokinin and abscisic acid in xylem sap of mango cv. Alphonso grafted on rootstocks of contrasting vigour. *Indian Journal of Horticulture* 57, 221–223.

Mustard, M.J. (1960) Megagametophytes of the lychee (*Litchi chinensis* Sonn.). *Proceedings of the American Society for Horticultural Science* 75, 292–304.

Mustard, M.J. and Lynch, S.J. (1959) Notes on lychee panicle development. *Proceedings of the Florida State Horticultural Society* 72, 324–327.

Mustard, M.J., Liu, S. and Nelson, R.O. (1953) Observations of floral biology and fruit setting in lychee varieties. *Proceedings of the Florida State Horticultural Society* 66, 212–220.

Mustard, M.J., Nelson, R.O. and Goldweber, S. (1956) Exploratory study dealing with the effect of growth regulators and other factors on the fruit production of the lychee. *Proceedings of the Florida Lychee Growers' Association* 3, 33–38.

Nakata, S. (1955) Floral initiation and fruit set in lychee, with special reference to the effects of sodium naphthaleneacetate. *Botanical Gazette* 117, 126–134.

Nakata, S. (1956) Lychee flowering and girdling. *Hawaii Farm Science* 4, 4–5.

Nakata, S. and Suehisa, R. (1969) Growth and development of *Litchi chinensis* as affected by soil-moisture stress. *American Journal of Botany* 56, 1121–1126.

Nakata, S. and Watanabe, Y. (1966) Effects of photoperiod, and night temperature on flowering of *Litchi chinensis. Botanical Gazette* 127, 146–152.

Naphrom, D., Potjanapimon, C. and Subhadrabandhu, S. (2001) Changes in cytokinin and gibberellin-like substances in stem apex of lychee cv. Hong Huay prior to flowering. *Acta Horticulturae* 558, 199–203.

Núñez-Eliséa, R. and Davenport, T.L. (1994) Flowering of mango trees in containers as influenced by seasonal temperature and water stress. *Scientia Horticulturae* 58, 57–66.

Núñez-Eliséa, R., Caldeira, M.L. and Davenport, T.L. (1990) Thidiazuron effects on growth initiation and expression in mango (*Mangifera indica* L.). *HortScience* 25, 1167.

Núñez-Eliséa, R., Davenport, T.L. and Caldeira, M.L. (1993) Bud initiation and morphogenesis in 'Tommy Atkins' mango as affected by temperature and triazole growth retardants. *Acta Horticulturae* 341, 192–198.

Núñez-Eliséa, R., Davenport, T.L. and Caldeira, M.L. (1996) Control of bud morphogenesis in mango (*Mangifera indica* L.) by girdling, defoliation and temperature modification. *Journal of Horticultural Science* 71, 25–40.

Olesen, T., Menzel, C.M., Wiltshire, N. and McConchie, C.A. (1999) Manipulating flushing cycles and flowering in lychee. In: Menzel, C.M, Waite, G.K. and Bronson, R. (eds) *Proceedings of the Fifth National Lychee Conference.* Australian Lychee Growers' Association, Twin Waters, Queensland, Australia, pp. 47–52.

Olesen, T., Menzel, C.M., Wiltshire, N. and McConchie, C.A. (2002) Flowering and shoot elongation of lychee in eastern Australia. *Australian Journal of Agricultural Research* 53, 977–983.

Pandey, R.S. and Yadava, R.P.S. (1970) Pollination of litchi (*Litchi chinensis*) by insects, with special reference to honeybees. *Journal of Apiculture Research* 9, 103–105.

Pivovaro, S.Z. (1974) Studies of the floral biology and the influence of growth regulators on fruit set and drop of *Litchi chinensis* Sonn. MSc thesis, Hebrew University of Jerusalem, Rehovot, Israel.

Ploetz, R.C., Ramos, J.L. and Parrado, J.L. (1993) Periodicity of shoot and root growth in grafted avocado. *Tropical Agriculture* 70, 248–251.

Puiman, L., Ji, Z. and Liang, L. (1985) The relationship between cytokinins and the differentiation of flower buds in 'on' year and vegetative buds in 'off' year trees of litchi (*Litchi chinensis* Sonn. cv. No Mi Chi). *Journal of the South China Agricultural University* 6, 1–8.

Qiu, J., Luo, X. and Wu, D. (2001a) Regulation of flower bud differentiation in longan. *Acta Horticulturae* 558, 225–228.

Qiu, J.D., Wu, D.Y. and Zhang, H.L. (2001b) A study on flower differentiation of 'Shixia' longan (*Dimocarpus longana* Lour. cv. Shixia). *Journal of the South China Agricultural University* 22, 27–30.

Rademacher, W. (2000a) Growth retardants in agriculture and horticulture. In: *Proceedings of the Twenty Seventh Annual Meeting of the Plant Growth Regulation Society of America*, pp. 87–92.

Rademacher, W. (2000b) Growth retardants: effects on gibberellin biosynthesis and other metabolic pathways. *Annual Review of Plant Physiology and Plant Molecular Biology* 51, 501–531.

Rindermann, R.S. and Gomez-Cruz, M.A. (1997) *El Litchi: la Fruita Mas Fina del Mundo: Mercado Mundial y Perspectivas para México.* Centro de Investigaciones Economicas, Sociales y Tecnologicas de la Agroindustria y la Agricultura Mundial (CIESTAAM) de la Universidad Autonoma Chapingo, Chapingo, Mexico, 121 pp.

Robbertse, H., Fivaz, J. and Menzel, C.M. (1995) A reevaluation of tree model, inflorescence morphology, and sex ratio in lychee (*Litchi chinensis* Sonn.). *Journal of the American Society for Horticultural Science* 120, 914–920.

Scholefield, P.B. (1982) A scanning electron microscope study of flowers of avocado, litchi, macadamia and mango. *Scientia Horticulturae* 16, 263–272.

Shalem-Galon, M. (1980) Lychee: fertilization, fruit set and storage. MSc thesis, Hebrew University of Jerusalem, Rehovot, Israel.

Shigeura, G. (1948) Blossom-bud formation and fruit setting in the litchi. *Hawaii Agricultural Experiment Station Report* 1946–48, pp. 138–140.

Shukla, R.K. and Bajpai, P.N. (1974a) Blossom bud differentiation and ontogeny in litchi (*Litchi chinensis* Sonn.). *Indian Journal of Horticulture* 31, 224–228.

Shukla, R.K. and Bajpai, P.N. (1974b) Studies on vegetative growth in litchi (*Litchi chinensis* Sonn.). *Indian Journal of Horticulture* 31, 148–153.

Singh, L.B. and Singh, U.P. (1954) *The Litchi.* Superintendent of Printing and Stationery, Lucknow, UP, India, 87 pp.

Singh, S.N. (1962) Studies on the morphology and viability of the pollen grains of litchi. *Horticulture Advance* 6, 28–52.

Sittichaikasem, S. (1974) The effects of some environmental factors and certain growth regulator treatments on flowering of *Litchi chinensis* Sonn. var. Hong Huay. MSc thesis, Kasetsart University, Bangkok.

Sponsel, V.M. (1995) Gibberellin biosynthesis and metabolism. In: Davies, P.J. (ed.) *Plant Hormones: Physiology, Biochemistry and Molecular Biology.* Kluwer Academic, Boston, Massachusetts, pp. 66–97.

Sritontip, C., Khao-Sumeru, Y. and Changenraja, S. (1999) Effect of potassium chlorate on flowering in longan cv E-Daw. In: *Proceedings of the National Symposium on Plant Hormones for the Production of Off-season Fruit.* Research Council of Thailand, Thailand, pp. 30–37.

Stern, R.A. (1992) Elucidation of the factors affecting litchi productivity in Israel, and the development of methods to improve its yield. PhD thesis, Faculty of Agriculture, Hebrew University, Rehovot, Israel.

Stern R.A. and Gazit, S. (1993) Autumnal water stress checks vegetative growth and increases flowering and yield in litchi (*Litchi chinensis* Sonn.). *Acta Horticulturae* 349, 209–212.

Stern, R.A. and Gazit, S. (1996) Lychee pollination by the honeybee. *Journal of the American Society for Horticultural Science* 121, 152–157.

Stern, R.A. and Gazit, S. (1998) Pollen viability in lychee. *Journal of the American Society for Horticultural Science* 123, 41–46.

Stern, R.A. and Gazit, S. (2003) The reproductive biology of the lychee. *Horticultural Reviews* 28, 393–453.

Stern, R.A., Adato, I., Goren, M., Eisenstein, D. and Gazit, S. (1993) Effects of autumnal water stress on litchi flowering and yield in Israel. *Scientia Horticulturae* 54, 295–302.

Stern, R.A., Eisenstein, D., Voet, H. and Gazit, S. (1996) Anatomical structure of two-day-old litchi ovules in relation to fruit set and yield. *Journal of Horticultural Science* 71, 661–671.

Stern, R.A., Eisenstein, D., Voet, H. and Gazit, S. (1997) Female 'Mauritius' litchi flowers are not fully mature at anthesis. *Journal of Horticultural Science* 72, 19–25.

Stern, R.A., Meron, M., Naor, A., Wallach, R., Bravdo, B. and Gazit, S. (1998) Effect of fall irrigation level in 'Mauritius' and 'Floridian' lychee on soil and plant water status, flowering intensity, and yield. *Journal of the American Society for Horticultural Science* 123, 150–155.

Stern, R.A., Naor, A., Bar, N., Gazit, S. and Bravdo, B.A. (2003) Xylem-sap zeatin-riboside and dihydrozeatin-riboside levels in relation to plant and soil water status and flowering in 'Mauritius' lychee. *Scientia Horticulturae* 98, 285–291.

Subhadrabandhu, S. (1986) Studies of plant growth regulator effects on tropical and subtropical tree fruits of Thailand. *Acta Horticulturae* 175, 291–297.

Subhadrabandhu, S. (1990) *Lychee and Longan Cultivation in Thailand.* Rumthai Publication, Bangkok, Thailand, 40 pp.

Subhadrabandhu, S. and Koo-Duang, A. (1987) Effect of ethephon on flowering of two lychee (*Litchi chinensis* Sonn.) cultivars. *Acta Horticulturae* 201, 181–186.

Subhadrabandhu. S. and Yapwattanaphun, C. (2000) Regulation of flowering time for longan (*Dimorcarpus longan*) production in Thailand. *Journal of Applied Horticulture* 2, 102–105.

Subhadrabandhu, S. and Yapwattanaphun, C. (2001a) Lychee and longan production in Thailand. *Acta Horticulturae* 558, 49–57.

Subhadrabandhu, S. and Yapwattanaphun, C. (2001b) Regulation off-season flowering of longan in Thailand. *Acta Horticulturae* 558, 193–198.

Torrey, J.G. (1976) Root hormones and plant growth. *Annual Review of Plant Physiology* 27, 435–459.

Veen, H. and Jacobs, W.P. (1969) Transport and metabolism of indole-3-acetic acid in coleus petiole segments of increasing age. *Plant Physiology* 44, 1157–1162.

Waite, G.K. and Hwang, J.S. (2002) Pests of litchi and longan. In: Pena, J.E., Sharp, J.L. and Wysoki, M. (eds) *Tropical Fruit Pests and Pollinators: Biology, Economic Importance, Natural Enemies and Control.* CAB International, Wallingford, UK, pp. 331–359.

Wang, B. and Qiu, Y. (1997) Growth and fruiting. In: Zhang, Z.W., Yuan, P.Y., Wang, B.Q. and Qui, Y.P. (eds) *Litchi Pictorial Narration of Cultivation.* Pomology Research Institute, Guangdong Academy of Agricultural Sciences, Guangzhou, China, pp. 66–88.

Wang, C.C. and Ke, G.W. (1992) Studies on the morphological differentiation of flower buds of longan. *Journal of the Fujian Academy of Agricultural Science* 7, 55–58.

Wang, C.C., Huang, J.H. and Ke, G.W. (1988) An observation on the inflorescence development and the sequence of flower differentiation of longan. *Journal of the Fujian Academy of Agricultural Science* 3, 68–71.

Wewatanadate, P., Voravong, R., Mahawana, T. and Saraprug, D. (2001) Health effects in potassium chlorate-exposed longan growers. *Internal Medicine Journal of Thailand* 17, 94–97.

Whitman, W.F. and Wirkus, L.V. (1958) Rare Fruit Council activities 1957–58. *Proceedings of the Florida State Horticultural Society* 71, 278–287.

Wu, D., Lin, X., Ye, Q. and Wang, W. (2001) Improvement of fruit set in secondary panicles of 'Feizixiao' litchi by removal of the primary panicles. In: Huang, H.B and Huang, X.M. (eds) *First International Symposium on Litchi and Longan.* Programme and Abstracts. South China Agricultural University, Guangzhou, China, p. 42.

Wu, D.Y., Qiu, J.D., Zhang, H.L. and Lou, X.Z. (2000) A study on flowering promotion by ringing in longan (*Dimocarpus longana* Lour.). *Scientia Agriculturae Sinica* 33, 40–43.

Xu, S.X. (1991) Scanning electron microscopic studies on the organ development of longan flower *Euphoria longan* Lour. Steud. *Acta Botanica Sinica* 33, 938–942.

Yaacob, O. and Subhadrabandhu, S. (1995) Longan (*Dimocarpus longan* Lour.). In: *The Production of Economic Fruits in South-East Asia.* Oxford University Press, Oxford, UK, pp. 255–263.

Yamanishi, O.K., Machado Filho, J.A. and Kavati, R. (2001) Overview of litchi production in São Paulo State, Brazil. *Acta Horticulturae* 558, 59–62.

Yen, C.R., Chau, C.N., Chang, J.W. and Tzeng, J.C. (2001) Effects of chemicals on flowering in longan. *Journal of the Chinese Society for Horticultural Science* 47, 195–200.

Young, T.W. (1956a) Branch girdling and root pruning trials on lychees in Florida. *Proceedings of the Florida Lychee Growers' Association* 3, 16–18.

Young, T.W. (1956b) Two-year performance of 529 Brewster lychee trees. *Proceedings of the Florida Lychee Growers' Association* 3, 42–47.

Young, T.W. (1957a) Lychee: branch girdling. *Annual Report of the Florida Agricultural Experiment Station*, Gainesville, Florida, pp. 353–354.

Young, T.W. (1957b) Lychee fruitfulness. *Proceedings of the Florida Lychee Growers' Association* 4, 9–13.

Young, T.W. (1959) Lychee: branch banding. *Annual Report of the Florida Agricultural Experiment Station*, Gainesville, Florida, p. 356.

Young, T.W. (1970) Some climatic effects on flowering and fruiting of 'Brewster' lychees. *Proceedings of the Florida State Horticultural Society* 83, 362–367.

Young, T.W. (1977) Effect of branch girdling on yield of severely pruned 'Brewster' lychee trees. *Proceedings of the Florida State Horticultural Society* 90, 251–253.

Young, T.W. and Harkness, R.W. (1961) Flowering and fruiting behavior of Brewster lychees in Florida. *Proceedings of the Florida State Horticultural Society* 74, 358–363.

Young, T.W., Popenoe, J. and Campbell, C.W. (1960) Lychee culture. *Annual Report of the Florida Agricultural Experiment Station*, Gainesville, Florida, p. 316.

Zee, F.T.P., Chan, H.T., Jr and Yen, C.R. (1998) Lychee, longan, rambutan and pulasan. In: Shaw, P.E., Chan, H.T., Jr and Nagy, S. (eds) *Tropical and Subtropical Fruits*. Agscience, Auburndale, Florida, pp. 290–335.

Zhang, Z. (1997a) Local cultivars and selections with potential. In: Zhang, Z.W., Yuan, P.Y., Wang, B.Q. and Qui, Y.P. (eds) *Litchi Pictorial Narration of Cultivation*. Pomology Research Institute, Guangdong Academy of Agricultural Sciences, Guangzhou, China, pp. 48–57.

Zhang, Z. (1997b) Cultivation practices for high yield. In: Zhang, Z.W., Yuan, P.Y., Wang, B.Q. and Qui, Y.P. (eds) *Litchi Pictorial Narration of Cultivation*. Pomology Research Institute, Guangdong Academy of Agricultural Sciences, Guangzhou, China, pp. 106–139.

Zheng, Q., Davenport, T.L. and Li, Y.C. (2001) Stem age, winter temperature and flowering of lychee in south Florida. *Acta Horticulturae* 558, 237–240.

Zong, Y.W. and Wu, S.Y. (1983) Effects of growth regulators on the induction of flowers of lychee. *Guangdong Agricultural Science* No. 5, 17–19.

7 Fruit Set, Development and Maturation

A. Litchi

Huibai Huang
College of Horticulture, South China Agricultural University, Guangzhou 510642, China

Introduction

Litchi is one of the most valuable members of the Sapindaceae, which also contains the related longan, *Dimocarpus longan* Lour., and the tropical rambutan, *Nephelium lappaceum*, and pulasan, *Nephelium mutabile*. All these fruit contain an edible aril enclosing a single seed.

Litchis can be round, ovoid or heart-shaped and are 2.5–4.0 cm or more in diameter, depending on the cultivar. The fruit have a thin or thick, leathery, tuberculate skin or pericarp that is yellow-red to red. The juicy aril is white and translucent and envelops a dark-brown seed. Some cultivars produce a proportion of aborted seeds and thus have a higher flesh recovery than others, while a few rare strains produce small seedless fruit. According to data cited by Deng *et al.* (1999), the composition per 100 g of flesh is: moisture, 83.6 g; protein, 0.7 g; fat, 0.1 g; carbohydrate, 15.0 g; Ca, 4.0 mg; P, 32.0 mg; Fe, 0.7 mg; carotene, trace; thiamine, 0.02 mg; riboflavin, 0.07 mg; niacin, 1.1 mg; and ascorbic acid, 15.0 mg.

Embryology, histology and organogenesis

Double fertilization takes place 2–3 days after pollination, followed by division of the nucleus of the primary endosperm (Lü *et al.*, 1985). The zygote then begins to divide after 3–7 days (Lü *et al.*, 1985; Joubert, 1986). In normal-seeded 'Haok Yiy' ('Heiye'), the nucellus degenerated 20–30 days after anthesis (DAA), when the liquid endosperm in the embryo sac was copious (Lü *et al.*, 1985). The embryo reached the globular stage after 20 days, the heart-stage with a rudimentary cotyledon after 30 days, and the torpedo stage after 40 days. The liquid endosperm was absorbed by the developing cotyledon after 50 days. Development of the embryo was restricted by day 10 in aborted-seeded 'Lühebao', with one of the two ovules having atrophied by day 15.

In the aborted-seeded 'No Mai Chee' ('Nuomici'), embryo development slowed after 30 days and aborted after 40–50 days, at the torpedo stage (Qiu *et al.*, 1994; Xiang *et al.*, 2001). The volume of the liquid endosperm began to decrease after 25 days and it disappeared after 40 days. Embryos aborted after the formation of the globular embryo on day 25, and before the heart-stage on day 40. A microscopic study showed that cell division in 'Wai Chee' ('Huaizhi') was very active in the ovary wall before anthesis, but relatively quiescent during bloom (Li, 2001). A second wave of cell division occurred in the pericarp 14 days after anthesis. Cell division ceased in various parts of the pericarp at different times: at day 19 for the inner mesocarp, day 32 for the outer mesocarp, and day 47 for the endocarp and the epicarp. A longer duration of cell division leads to a larger pericarp and a larger fruit, as shown by a comparison of red and white strains of 'Nuomici' (Wang *et al.*, 2000). Difference in fruit size amongst cultivars was related to variations in the number of cells rather than to their final size (Li *et al.*, 2002).

Environmental factors around bloom may also affect cell division, as they do in other fruit species. Among these factors, temperature is most important. On 'Fay Zee Jin' ('Feizixiao') trees growing in Guangdong, China, early female flowers may open 3 weeks before the last. Fruit set on 25 March was 1.5 times higher than fruit set on 15 April (Li, 2001). Average pericarp weight was 2.5 g and 1.4 g, respectively, at the end of Stage I of fruit ontogeny. The later-set fruit were exposed to higher temperatures during the growth of the ovary wall and pericarp, and had a shortened Stage I phase compared with the early fruit.

There is much divergence of opinion about the origin of the litchi aril. Banerji and Chaudhuri (1944) suggested that the primordium arose from the obturator. Joubert (1969, 1986) supported this, stating 'the obturator in the litchi can be regarded as a *proaril* since it gives rise to the aril after fertilization'. Huang *et al.* (1983) obtained valid microscopic evidence in 'Kwai May' ('Guiwei'), which showed that the primordium was not derived from the obturator, but from a site immediately above it on the funicle, on day 35 (Fig. 7.1). The first cells of the primordium appeared 9 days after anthesis, on the side opposite the micropyle. The obturator *per se* began to break down on day 16 and it eventually vanished. Later, Steyn and Robbertse (1992) claimed that the aril did not originate from the obturator. Ye *et al.* (1992) suggested that the primordium originated from the outer integument rather than from the funicle.

The stigmas usually open into two or bifurcate lobes, occasionally into three or trifurcate, rarely into four or even more. According to observations in China, multilobes form when low temperatures occur during ovary formation. This needs to be verified through experiments conducted under controlled conditions. The ovary generally has two loculi, one that grows and one that atrophies (Fig. 7.2). Sometimes, normal well-developed twin fruit set on one pedicel.

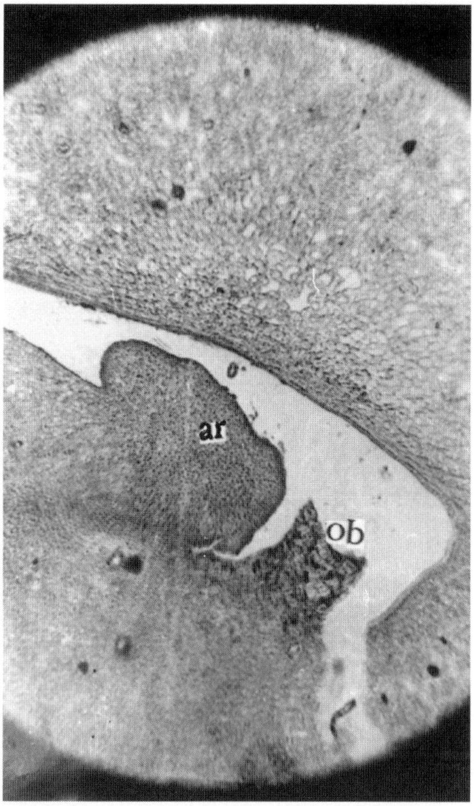

Fig. 7.1. Fruit development 35 days after anthesis in 'Kwai May' ('Guiwei') litchi. The aril primordium (ar) is developing just above the obturator (ob), which is disintegrating (Huang *et al.*, 1983).

Types of fruit

There are four types of fruit: normal, aborted or seedless fruit, and 'hollow' fruit. Aborted-seeded fruit are preferred by consumers, since they have a high flesh recovery. Large seedless fruit are very rare. Normal-seeded fruit have a dark-brown seed containing a viable embryo when mature, and a fully developed aril.

Aborted-seeded fruit have a well-developed aril that fills the whole space provided by the pre-formed pericarp, regardless of whether the pericarp is large or small. The seed is small and shrivelled, with an empty cavity and a dead or rudimentary stunted embryo. Shrivelled seeds in litchi are called 'chicken tongues'; however, botanically this phenomenon is called 'stenospermocarpy', which means 'fruit

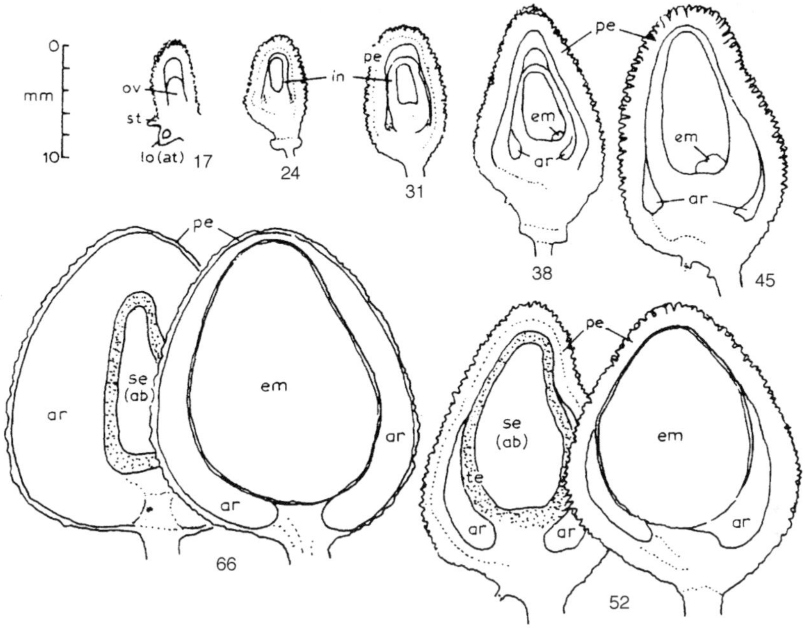

Fig. 7.2. Longitudinal sections of 'Kwai May' ('Guiwei') litchi after anthesis (days indicated). st = style; lo (at) = atrophied locule; ov = ovule; pe = pericarp or fruit skin; ar = aril; em = embryo; te = testa (seed coat); se (ab) = aborted seed; and in = integument. The embryo grew rapidly 38 days after anthesis. Sections on the left-hand side at the bottom are fruit with aborted seeds. The expanding aril squeezed the hollow seeds and stretched the skin (Huang and Xu, 1983).

with slim or narrow seeds'. This term was first used for seed abortion in grapes, but is now used for many other species.

Stenospermocarpy can arise due to the occurrence of abnormal ovules or incomplete fertilization. Lü *et al.* (1989) suggested that serial genes determined the type of embryo that develops. Stenospermocarpy can also occur after unfavourable weather (Stern *et al.*, 1996). This helps to explain why the proportion of aborted seeds in some cultivars, such as 'Guiwei', varies from year to year in the same orchard. Stenospermocarpy may also depend on the type of pollination, with seeds from self-pollinated flowers being more likely to abort than seeds from cross-pollination (Stern *et al.*, 1993). Xiang *et al.* (2001) showed that the rate of stenospermocarpy in a particular cultivar varied with the pollen source. However, some cultivars, such as 'Nuomici', have up to 90% stenospermocarpy even when they are cross-pollinated. 'Guiwei' was more erratic, with 5–95% aborted seeds when cross-pollinated. These data could be used in the proper arrangement of different cultivars in single orchards.

The development of normal embryos is accompanied by increasing concentrations of polyamines such as putrescine, spermidine and spermine in the ovule (Chen, 2000). The ovules of normal-seeded 'Heiye' had a higher concentration of abscisic acid (ABA), and lower concentrations of indole acetic acid (IAA), gibberellins (GAs) and cytokinins (CTKs) compared with aborted-seeded 'Lühebao'. The concentration of the phenolic inhibitor, *p*-hydroxybenzoic acid (*p*-HBA), was ten times higher in aborting ovules than in normal ovules 22 days after anthesis.

Seedless fruit, botanically termed 'parthenocarpic', do not have any seeds because the ovary is not fertilized. These types of fruit are rare in commercial varieties. One of the fruit on a twin-fruit stalk is sometimes parthenocarpic, although the companion fruit is usually normal or at least aborted-seeded. It is conceivable that plant growth substances are translocated from the actively growing companion fruit to its neighbour to enable it to grow. 'Hollow fruit' usually remain small even when mature, with an underdeveloped pericarp enveloping a hollow space and sheltering a rudimentary, stunted aril and a very small seed (Huang and Qiu, 1987).

'Tai So' ('Dazao'), 'Shuili' and 'H-1224' produce more of these malformed fruit than other cultivars. Hollow fruit often follow a period of temperatures below 16°C shortly after bloom (Chen *et al.*, 1997). Such temperatures possibly injure the ovules, inhibiting cell division. The production of hollow fruit can reduce the productivity of some commercial orchards. This anomaly is undesirable for growers, but it may assist researchers in studying the mechanisms of fruit development.

Artificial induction of shrivelled seeds was achieved in 'Huaizhi' by Liang and Qiu (1998) when maleic hydrazide or MH was applied at 1000 mg/l about 2 weeks after bloom, when the liquid endosperm was full. The fruit weighed about the same as non-sprayed controls, but had 10% more aril. No chemical residues were detected in the fruit. Sprays applied earlier or later were less effective. Xiang *et al.* (2002) established that the best treatment for 'Huaizhi' and 'Xuehuaizi' was a double spray of 800 mg/l, applied 15 and 25 days after female bloom. Addition of β-napthoxyacetic acid or β-NOA at 30 mg/l assured better set and fruit size than MH alone. Maleic hydrazide inhibited cell division and the development of the embryo, but did not appreciably affect the pericarp and the aril if it was sprayed at the correct time. Attempts to induce shrivelled seeds in other cultivars have met with less success, indicating genetic variation in stenospermocarpy. Stern *et al.* (1997) reported that shrivelled seeds could be induced in 'Floridian' by spraying trichlorophenoxy propionic acid or 2,4,5-TP at 100 mg/l when the fruit weighed 1 g, whereas the rate of seed abortion induced was less than 30%. In contrast, 'Mauritius' produced many unmarketable, small fruit when subjected to this treatment (Stern *et al.*, 1995).

There have been some attempts to produce parthenocarpic fruit in order to reduce losses following rain or overcast weather during bloom. 'Hexiachuan' produces seedless, aborted-seeded and normal-seeded fruit, but the seedless fruit are small (Zhang *et al.*, 1990; Qiu *et al.*, 1998). Treatments that can be used to increase the size of these parthenocarpic fruit need to be investigated. Large-fruited strains named as 'Wuheli' produced varying proportions of seedless fruit on Hainan Island. One of these strains had a higher concentration of auxin in the pericarp compared with normal-seeded 'Huaizhi' and a lower concentration of abscisic acid (ABA) (Zhou *et al.*, 1998).

Attempts to induce the production of seedless parthenocarpic fruit from normal-seeded cultivars using growth regulators have met with little success, although Wang *et al.* (1997) induced seedless fruit in 'Feizixiao' with 2,4-dichloro-phenoxy acetic acid or 2,4-D at 20 mg/l. This report implies that success is restricted by the ability of the cultivar to set fruit parthenocarpically.

There are different categories of cultivars, which roughly correspond to the types of fruit produced. Normal-seeded cultivars, such as 'Heiye', 'Dazao' and 'Huaizhi', have a low incidence of aborted seeds, while aborted-seeded cultivars, such as 'Nuomici' and 'Lühebao', have a very low incidence of normal seeds. Other cultivars, such as 'Guiwei', 'Lanzhu' and 'Feizixiao', have a variable proportion of aborted seeds, depending on the weather at flowering and pollination. Cultivars such as 'Hexiachuan', which yield a high proportion of seedless fruit, are rare.

The 'ball skin versus bladder effect' concept

Within a given cultivar, the weight of the seed coat or testa varies in aborted-seeded fruit, where the embryo is absent or negligible (Huang and Xu, 1983). The litchi pericarp grows before the aril. The weights of the aril and whole fruit are correlated with the weight of the pericarp, irrespective of the fruit having a normal or aborted seed (Huang and Xu, 1983; Huang and Qiu, 1987). Pericarp weight was correlated with seed coat weight in both aborted- and normal-seeded fruit when the seed coat of normal seeds was carefully separated from the cotyledon (Table 7.1). In contrast, the weights of the aril and whole fruit are negatively correlated with the weight of the cotyledon. These results have been verified by different correlation tests.

These data prompted the development of the 'ball skin versus bladder effect' concept in fruit growth (Huang and Xu, 1983; Huang and Qiu, 1987). The negative correlations between the aril and embryo (mainly cotyledon) indicate that fruit expansion growth, mainly aril enlargement during Stage II, does not rely on the embryo. In fact, the embryo, which consists mainly of the cotyledon, invades the space provided by the pre-formed pericarp for the aril, and competes with the aril for assimilates. This competition has been substantiated by several observations. Total soluble solids (TSS) accumulation and aril growth are faster in aborted-seeded fruit than in normal-seeded fruit (Fig. 7.3).

The peak rate of solute influx into the embryo occurs 1–2 weeks before that in the aril (Fig. 7.4). The vascular system, which enters the fruit through the pedicel, leads directly to and spreads all over the seed and the peel (Huang and Qiu, 1987). The aril is derived from and attached to a confined region of the integument, and is partially isolated from the pericarp and the seed coat. The aril consists exclusively of parenchyma and contains no vascular elements, which are common in the pulp of other fruit, and thus nutrient transport within the aril is less efficient than in the embryo. Although a normal embryo does not favour aril growth, a fully developed embryo implies a long lifespan for the liquid endosperm compared with fruit where the embryo aborts too early.

The direction of fruit development is linked closely with the development of both the embryo and the aril. The embryo may abort soon after fertilization or degenerate later. Figure 7.5 illustrates that a long lifespan of the liquid endosperm supports the growth of the aril and embryo and gives rise to a normal-seeded fruit. If the liquid endosperm degenerates halfway through fruit development then the embryo dies,

Table 7.1. Correlations between growth of the various components in the fruit of 'Kwai May' ('Guiwei') litchi. Partial correlations (*r* values) are based on a sample of 120 fruit (Huang and Qiu (1987). $*P < 0.05$, $**P < 0.01$, $***P < 0.001$.

Variables	Fruit with normal seed	Fruit with aborted seed
Aril versus pericarp	0.45***	0.50***
Aril versus seed coat	0.31***	0.24**
Aril versus embryo (mainly cotyledon)	–0.34***	–
Pericarp vs. seed coat	0.22*	0.25*
Pericarp vs. embryo (mainly cotyledon)	0.32***	–
Embryo (mainly cotyledon) versus seed coat	0.32***	–

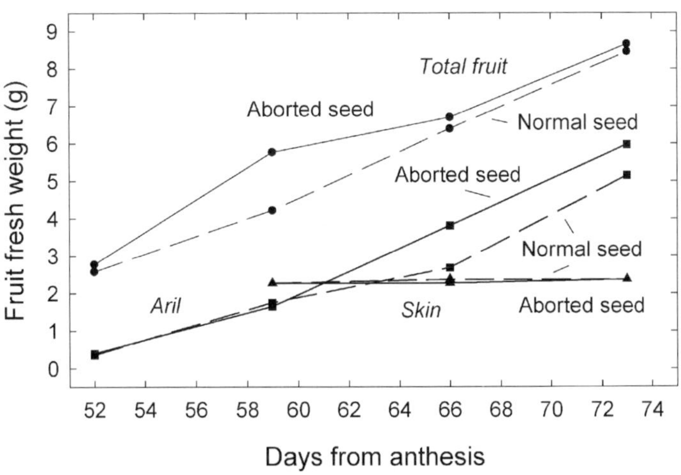

Fig. 7.3. Changes in fresh weight in normal- or aborted-seeded 'Kwai May' ('Guiwei') litchi during development (Huang and Xu, 1983). Total fruit weight and aril weight are significantly different ($P = 0.05$) in the two groups on days 59 and 66, respectively. Aborted-seeded fruit also had higher TSS on days 66 and 73.

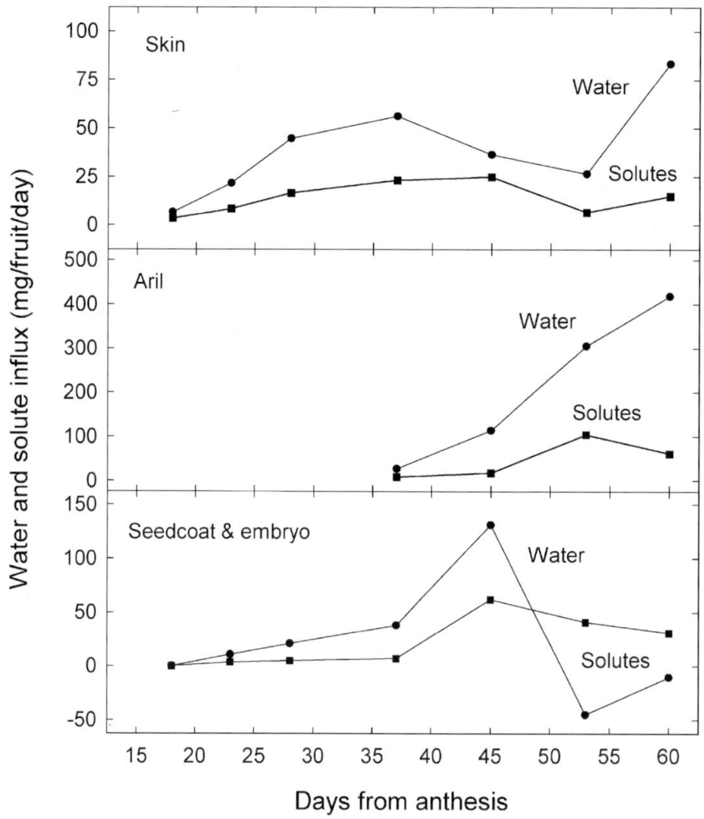

Fig. 7.4. Influx of water and solutes into 'Haak Yip' ('Heiye') litchis (Huang and Qiu, 1987).

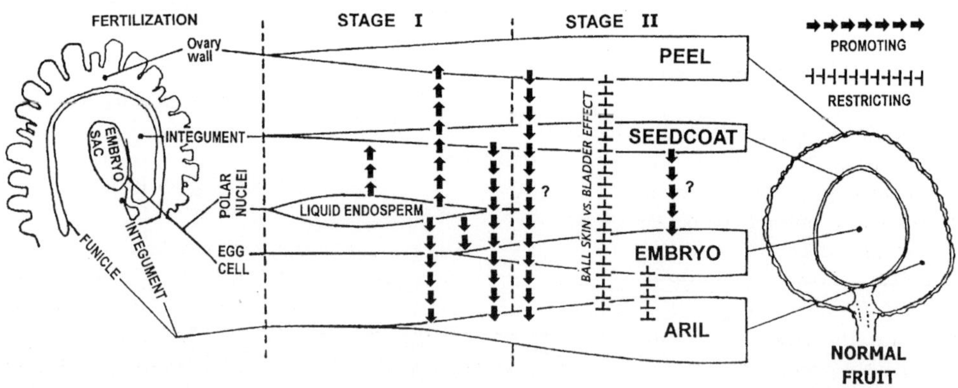

Fig. 7.5. Relationship amongst different tissues of litchis.

producing an aborted-seeded fruit with a full aril. The aril but not the embryo can continue to grow, leading to an aborted-seeded fruit with a fully developed aril (Fig. 7.6A). Normal and aborted-seeded fruit on the same tree potentially have comparable weights. A study in 'Guiwei' showed that a population of fruit with aborted seeds had a similar distribution of pericarp weight to a population with normal seeds (n = 120). This suggests that a fruit with a seed that aborts late can still produce a significant pericarp (Huang and Qiu, 1987). If the liquid

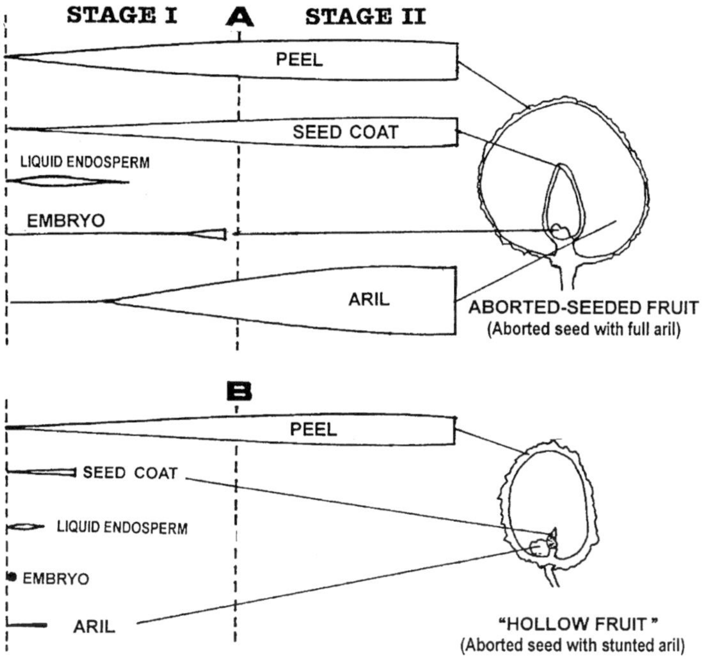

Fig. 7.6. The formation of different tissues in aborted-seeded litchis with (A) fully developed and (B) stunted arils.

endosperm dies soon after the main period of fruit drop, then a 'hollow fruit' with a rudimentary aril and a tiny seed will develop (Fig. 7.6B).

Fruit growth stages

Fruit growth in litchi is sigmoidal (Fig. 7.7) and ranges from 50 days in 'San Yee Hong' ('Sanyuehong'), to 90 days or more in 'Huaizhi' in southern China. These variations are mainly due to differences in the duration of Stage I, when the pericarp is growing (Figs 7.8 and 7.9). Not all parts of the fruit grow at the same time. Stage I is dominated by growth of the pericarp and the seed coat, while Stage II is dominated by growth of the aril and the embryo, mainly the cotyledon (Fig. 7.9). Growth is initiated in the embryo of 'Guiwei' about 5 weeks after anthesis (Fig. 7.8), whereas the pericarp and the seed coat grow soon after fertilization. This implies that a large pericarp is required for a large fruit. The 'ball skin versus bladder effect' notion signifies the restraints that the preformed pericarp exerts on the expanding aril, while providing a growing space and shelter for the latter (Huang and Xu, 1983; Huang and Qiu, 1987). Stage II can be further divided into two substages, mainly cotyledon growth (Stage IIa), and mainly aril growth (Stage IIb) for normal-seeded fruit, but not for aborted-seeded ones (Fig. 7.9).

Fruit abscission

Normal-seeded cultivars have two main periods of fruit abscission (Mustard *et al.*, 1954; Chang, 1961; Chadha and Rajpoot, 1969; Joubert, 1986), but this notion does not apply to aborted-seeded varieties. Yuan and Huang (1988) reported three to four waves of fruit abscission in different cultivars based on relative abscission rate (RAR), with an obvious preharvest drop in cultivars with aborted seeds (Fig. 7.10). The first wave of abscission occurred 1 week after full bloom (AFB) and was associated with a lack of fertilization. Wave II occurred around 3 weeks AFB, before the liquid endosperm was full. A third wave occurred 6–7

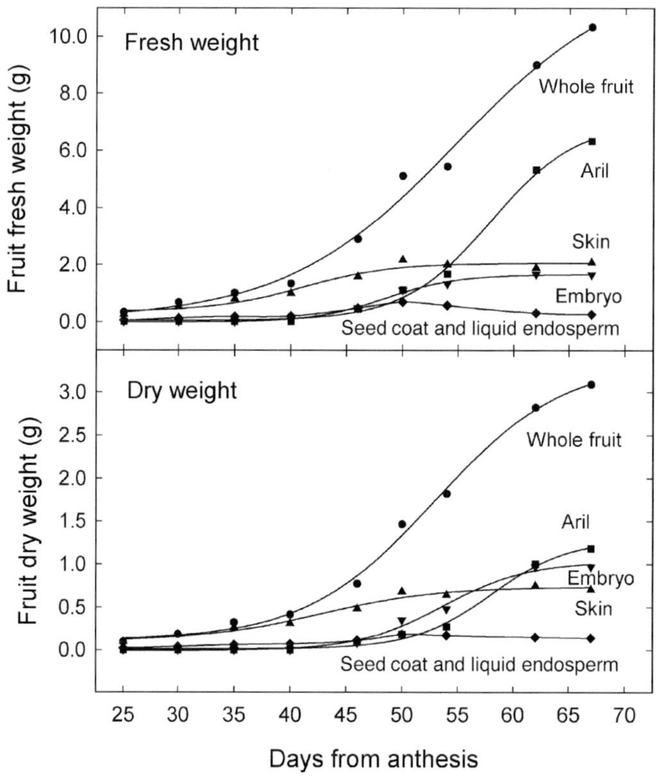

Fig. 7.7. Changes in fresh and dry weight of 'Wai Chee' ('Huaizhi') litchi fruit (Huang and Qiu, 1987).

weeks AFB, when the embryo grew rapidly. Wave IV was specific to cultivars with aborted seeds and occurred 2–3 weeks before harvest. Qiu *et al.* (1998) confirmed this pattern.

Litchis are self-compatible, since orchards based on single cultivars can yield heavily. However, Stern *et al.* (1993) and Degani *et al.* (1995) found that outcrossed fruit showed higher survival rates and produced heavier seeds than selfed fruit of 'Floridian' (but not of 'Mauritius'). The initial fruit drop occurs after pollination, with the loss of defective female flowers, unfertilized flowers and some of the fertilized ovaries. The proportion of 2-day-old fertile flowers abscising ranged from 3% to 27% in 11 orchards in Israel (Stern *et al.*, 1996). In extreme cases, all the fruit may be lost, leading to bare panicles after prolonged rain or overcast weather.

A panicle usually blooms in the order of male flowers (M_1), functional female flowers (F), and functional male flowers (M_2). M_1 pollen was found to be less fertile than M_2 pollen and led to unsatisfactory fertilization (Stern and Gazit, 1996). Competition for nutrients between flowers and fruit within panicles also caused excessive drop. It has been noted that M_2 flowers are usually plentiful and might deplete assimilates needed for the fertilized ovaries. In an emasculation trial, initial set was enhanced by removing M_2 flowers a few days after female bloom. The proportion of fruit set recorded 11 days after flower removal was 63% in the control and 73% in the emasculation treatment for 'Nuomici', and 51% vs. 61%, respectively, for 'Feizixiao' (Li and Wang, 1999). Excessive mid-term drop can substantially reduce final yield and is associated with poor vitality of the liquid endosperm and the failure of the embryo, which serves as the source of growth substances for the fruit.

Summer vegetative flush can sometimes be detrimental, causing fruit to drop before they enter Stage II. An additional early peak of root growth in young trees of 'Nuomici', which is a shy bearer, coinciding with an early summer leaf flush, caused a heavy fruit drop (Huang, 2002).

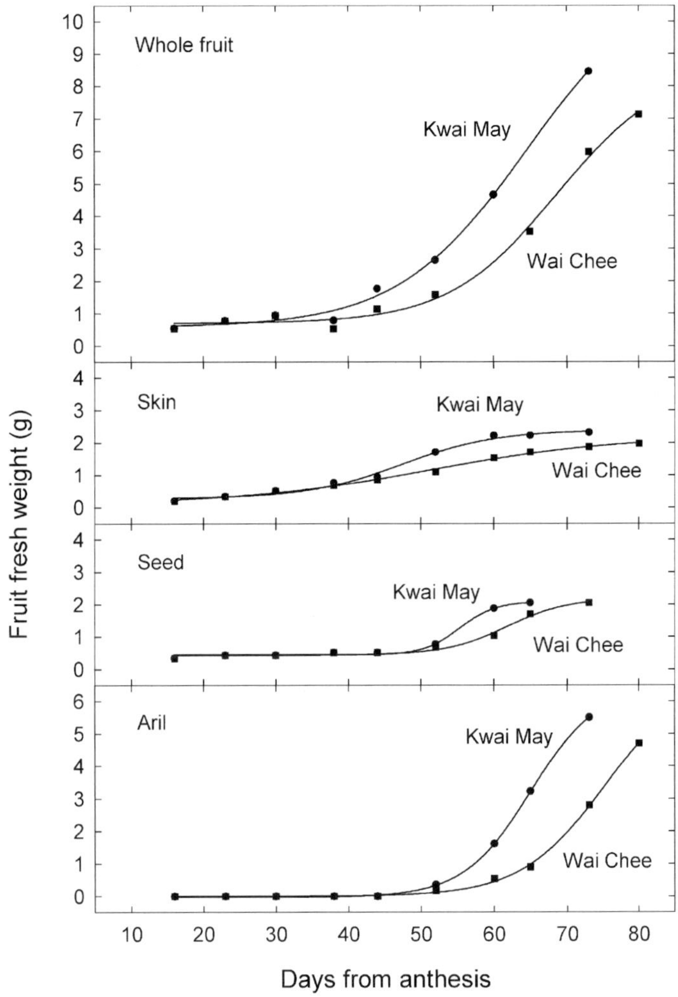

Fig. 7.8. The developmental patterns of 'Kwai May' ('Guiwei') and 'Wai Chee' ('Huaizhi') litchi (Huang and Xu, 1983).

A trial with aborted-seeded 'Nuomici' in China showed that trees with summer flushes on 50% of terminal shoots lost 59% of their fruit during the period between 22 May and 5 June (female flowers in late April) compared with trees with summer flushes on 7% of terminal shoots, which lost 43% of their fruit (J.G. Li, unpublished results, 2000). Therefore, the extent of mid-term drop can be mediated by competition amongst sinks. Fruit drop can thus be intensified by summer flushes, although such an effect usually occurs only within individual twigs or shoots (Huang, 1981). Growers of citrus in China remove summer flushes to prevent excessive fruit drop.

Fruit drop is usually excessive in some elite litchi cultivars such as 'Nuomici', and hand removal of summer flushes to improve fruit retention is practised, although it is less common than in citrus since litchi trees are much larger. Generally, flushing in summer is uncommon on litchi trees, or individual branches, that carry a good crop. Hieke *et al.* (2002), working with 'Tai So' and several other cultivars, found that pruning of one side of a tree to induce summer flushes did not affect fruit growth on the unpruned side. They concluded that 'the production of young leaves does not usually increase fruit abscission in adjacent branches'.

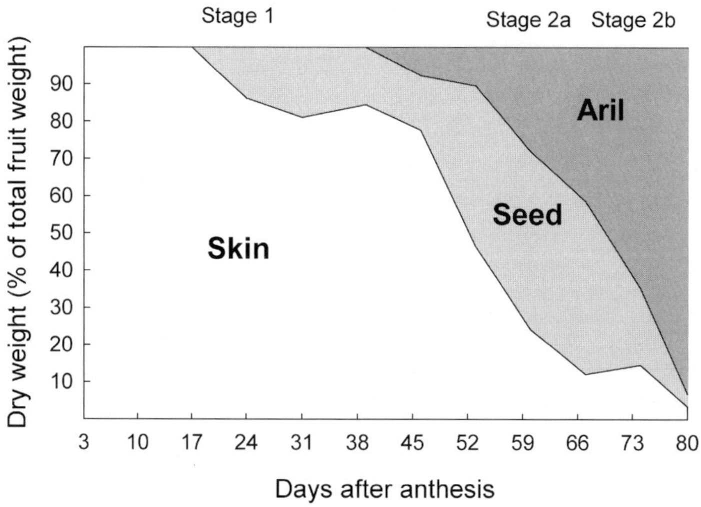

Fig. 7.9. Changes in dry weight in 'Wai Chee' ('Huaizhi') litchi. The skin and seed coat dominated growth in the first 53 days after bloom (Li, 2001).

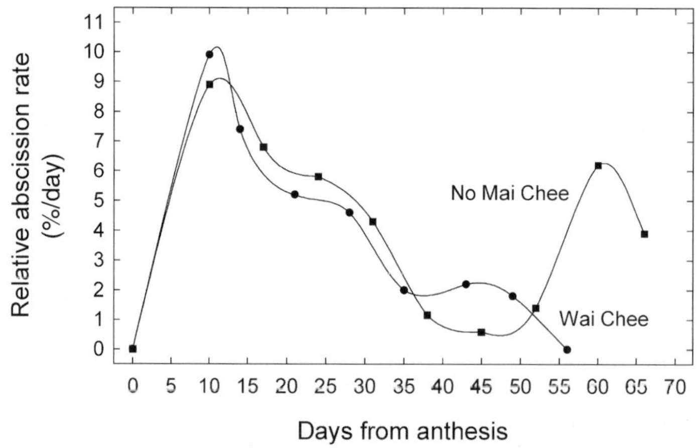

Fig. 7.10. Fruit drop patterns, with separation of Waves I, II, III and IV in normal-seeded 'Wai Chee' ('Huaizhi') and aborted-seeded 'No Mai Chee' ('Nuomici') litchi. 'Wai Chee' was harvested 75–80 days after bloom, and 'No Mai Chee' 70–75 days after bloom (Yuan and Huang, 1988).

Results of the previously mentioned experiment imply a negative impact of sink (young) leaves and a positive impact of source (mature) leaves on fruit growth in individual twigs. However, fruit can acquire assimilates from more distant sources within the same tree. Fruit may acquire photosynthates from beyond the bearing twig if the twig is totally deprived of photosynthetic capability by defoliation or by spraying with photosynthesis inhibitors. Yuan and Huang (1988) showed that fruit set on panicles depended on current photosynthesis of the foliage on a twig for its development. This was demonstrated by an experiment in which an application of DCMU (Diuron, a weedkiller and photosynthesis inhibitor) at full bloom greatly reduced fruit set, while application of a nucleotide enhanced photosynthesis and fruit set.

In an experiment conducted on 'H-1224', where fruiting twigs were girdled at their bases, the treatment with 50 leaves left above the girdle resulted in about half the number of fruit retained compared with that with 100 leaves left.

Application of a nucleotide increased fruit set in overcast weather, and was even more effective when combined with naphthalene acetic acid or NAA (Yuan and Huang, 1988, 1991). Normal fruit set and growth was achieved through the support of assimilates from outside the defoliated, non-girdled twig (Yuan and Huang, 1992).

Yuan and Huang (1993), working with non-bearing 4- to 5-year-old trees of shy-cropping 'Nuomici', found a peculiar peak of root growth soon after full bloom from late April to late May, which was followed by a leaf flush and the complete loss of fruit. This pattern of root growth was not evident in mature trees of 'Nuomici' or in young trees of regular-cropping 'Huaizhi' (Zhou *et al.*, 1996). Girdling prior to or during bloom enhanced fruit set by inhibiting root growth and thus eliminated root–fruit competition. These observations were substantiated by field and laboratory experiments, which showed that girdling prevented the transport of ^{14}C-photosynthates to the roots, and that concentrations of carbohydrates in fruit were much higher on the girdled trees than on non-girdled trees (Huang, 2002).

Preharvest fruit drop is peculiar to aborted-seeded and seedless strains. Fruit loss is accentuated by heavy rains and summer typhoons in coastal southern China. The economic losses caused by preharvest drop are distressing to growers, especially to those who grow elite and high-value cultivars such as 'Nuomici'. Yuan and Huang (1988) suggested that a well-developed embryo might help prevent preharvest abscission. Growers in southern China usually feel safe with respect to preharvest drop in normal-seeded cultivars such as 'Huaizhi', and are satisfied with partially aborted-seeded cultivars such as 'Guiwei'. Stern *et al.* (1997) indicated that cultivars such as 'Floridian' and 'Mauritius', which have normal seeds, exhibited no preharvest drop in Israel.

Growth substances and fruit development

Auxins

Ji *et al.* (1992) found that the concentration of auxin in young fruitlets was low, but rose sharply after 10 days in the seeds of 'Huaizhi', reaching a peak on day 25, when the liquid endosperm was full. Another peak was observed on day 60, when the cotyledons were fully grown. Concentration declined sharply from 10 days after bloom in aborted-seeded fruit of 'Nuomici', with a small increase 35 days after bloom (DAB), when the pericarp grew rapidly. Low concentrations of auxins during fruit maturation were related to the lack of an embryo, and dilution caused by the rapid influx of water into the expanding aril (Xiang *et al.*, 1994).

Xu (1962) reported that one or two sprays of 2,4,5-T at 25–100 mg/l soon after bloom improved fruit retention in a range of cultivars, although higher concentrations or repeated sprayings reduced average fruit weight. Spraying with 2,4-D gave the best response in terms of fruit set, but reduced fruit size at rates above 5 mg/l. The best time to apply 2,4-D was about 2 weeks after bloom (J.G. Li and H.B. Huang, unpublished results). Prasad and Jauhari (1963) reported that fruit retention and yield were increased with 2,4,5-T and naphthalene acetic acid (NAA) at 35–100 mg/l. Hoda *et al.* (1973) found that applications of NAA at 10 mg/l and 2,4-D at 15 mg/l increased fruit retention, especially when they were preceded by a 10 g/l spray of zinc sulphate.

Yuan and Huang (1991) reported that two consecutive sprays of NAA at 10 mg/l applied 7 and 17 days after full bloom gave a sixfold increase in final set in 'H-1224' compared with unsprayed controls. Naphthalene acetic acid at 10 mg/l also reduced preharvest drop in some aborted-seeded cultivars in China (J.G. Li and H.B. Huang, unpublished results). Earlier studies in Israel did not report better fruit retention after application of NAA or 2,4,5-T (Pivovaro, 1974; Stern *et al.*, 1995), although Stern *et al.* (1997, 2001) found that fruit set in 'Floridian' was increased by 2,4,5-TP at 100 mg/l applied 4 weeks after female bloom. Both fruit set and fruit weight were increased by the application of 2,4,5-TP at 67 mg/l at the same stage, to 'Feizixiao' and 'Heiye' in Guangxi, whereas earlier sprays were ineffective. The application of 3,5,6-trichloro-2-pyridyl-oxy acetic acid (3,5,6-TPA) at 20 or 25 mg/l increased fruit retention when applied to 2 g 'Mauritius' (Stern and Gazit, 1997) and 'Heiye' (Stern *et al.*, 2001) fruit.

Gibberellins

Ji *et al.* (1992) and Xiang *et al.* (1994) showed that the concentration of gibberellins (GAs) in fruit peaked shortly after anthesis and at the start of rapid growth of the aril. An upsurge of GAs on a per fruit basis linked to the development of the cotyledons, was detected during Stage II. Different types of GAs were dominant in different stages of fruit ontogeny: GA_3 and GA_5 in Stage I; GA_4, GA_7 and GA_9 in Stage IIa; and GA_3 in Stage IIb (Xiang *et al.*, 1994). A spray of GA_3 at 50 mg/l applied 5 weeks after bloom reduced the third wave of fruit drop compared with unsprayed controls in 'Huaizhi' (Ji *et al.*, 1992). Increases in fruit retention and size with the GAs were also reported in India by Srivastava and Singh (1969), Singh and Lal (1980) and Thakur *et al.* (1990). On young 'Nuomici' trees, mid-term and preharvest abscission was greatly reduced by trunk girdling carried out around bloom. These responses coincided with an increase in concentrations of GAs and a decrease in ABA in the fruit, suggesting that GA plays a role in fruit set (Zhou *et al.*, 1999).

Cytokinins

Concentrations of cytokinins (CTKs) were high in normal seeds of 'Huaizhi' about 10 DAB, when the liquid endosperm was full and starting to decline (Ji *et al.*, 1992). The concentrations of CTKs in aborted seeds of 'Nuomici' on a per fruit basis were highest shortly after bloom, declined 50 DAB, and rose sharply during fruit maturation (Xiang *et al.*, 1994). These results suggest that both the endosperm and the aril are rich sources of CTKs. Several forms of cytokinins have been isolated in litchi. Dihydrozeatin (diHZ) was dominant during bloom in the ovary, with lower concentrations of zeatin (ZT), dihydrozeatin riboside (diHZR) and zeatin riboside (ZR). Zeatin riboside was dominant during rapid embryo growth but it declined if the embryo aborted (Xiang *et al.*, 1994). A dip of 6-benzylaminopurine (6-BA) at 50 mg/l, especially with GA_3 at the same concentration 6 weeks after bloom, increased fruit weight by increasing the size of the pericarp and aril in aborted-seeded 'H-1224'. In contrast, these chemicals were ineffective on trees carrying normal fruit (Huibai Huang, 1983, unpublished results).

Abscisic acid

Yuan and Huang (1988) showed that the multiwaves of fruit abscission in litchi were associated with increases in the concentration of abscisic acid (ABA) in the seed, and decreases in indole acetic acid (IAA). Seed ABA during fruit ontogeny decreased at three different stages: after fertilization; in the early stage of liquid endosperm development; and at the beginning of rapid embryo growth. Fruit drop was greater when seed ABA was high. The difference in concentration of ABA between persisting and abscising fruit also supports the proposition that there is a close link between abscission and ABA. It is thus suggested that preharvest drop in aborted-seeded cultivars such as 'Nuomici' is associated with lower auxin and higher ABA concentrations compared with those in normal-seeded cultivars such as 'Huaizhi'. Shading and lack of pollination, which accentuated abscission, also increased the concentration of ABA in fruitlets (Yuan and Huang, 1988). The concentrations of CTKs and GAs before harvest were lower, and the concentrations of ABA higher, in the aborted-seeded cultivar 'Nuomici', which was prone to preharvest fruit drop. In contrast, CTK, GA and auxin concentrations were higher, and ABA lower, in the normal-seeded cultivar 'Huaizhi' (Xiang *et al.*, 1994; Qiu *et al.*, 1998). It appears that the effects of ABA are antagonized by IAA, GAs and CTKs produced in a developing embryo.

Ethylene

Akamine and Goo (1973) reported that ethylene production by the ovules of 'Tai So' (referred to as 'Kwai Mi') was high shortly after flowering, but it subsequently declined and was undetectable 40 DAB. Different patterns of ethylene production from 1 month after full bloom (12 May) onwards were demonstrated in

'Nuomici' and 'Feizixiao' by Wang and Huang (2002) in China (Fig. 7.11). In 'Nuomici', ethylene declined to very low concentrations 1 month after flowering, with a small rise around 56 DAB, when the pericarp turned red. In contrast, concentrations of ethylene were uniformly low in 'Feizixiao'. Jiang *et al.* (1986) found that the pericarp of 'Nuomici' produced 80 times more ethylene than the aril, although there are no data supporting a direct role for ethylene in litchi fruit growth. The actively proliferating parenchyma cells in the aril are initially elongated and slim, but become irregular in shape during the later phase of isodiametric growth. The involvement of ethylene in the growth of these cells seems probable and needs to be investigated, since this substance is active in the growth of other fruit.

Fruit maturation

Litchi fruit maturation occurs during the later part of Stage II of fruit development, along with the proliferation and enlargement of the aril cells. Akamine and Goo (1973) classified the fruit as non-climacteric, based on observations that ripening in 'Tai So' was not accompanied by increased respiration and ethylene production. Ethylene production in fertilized ovaries was high shortly after flowering, but was undetectable 40 DAB. Jiang *et al.* (1986) substantiated these views by monitoring the release of CO_2 and ethylene, and the absorption of O_2 by 'Huaizhi' and 'Nuomici'. Lin *et al.* (1990) showed that dipping the fruit of 'Guiwei' in 50 g/l of ethephon increased respiration in the pericarp but not in the aril, and did not induce cyanide-insensitive respiration, one of the features of non-climacteric fruit. Gaur and Bajpai (1978) found that the peak of respiration in 'Calcuttia' coincided with the start of aril development at day 34. There was another increase between days 46 and 49 termed 'on tree rudimentary climacteric'. Wang and Huang (2002) found that ethylene production declined from high to low values and recorded a small peak when the pericarp of 'Nuomici' turned red about 3 weeks before harvest, whereas ethylene was low and no peak was discerned in 'Feizixiao' (Fig. 7.11).

Sugars

Sucrose, fructose and glucose are the main sugars present in the aril. Sucrose was the main sugar in 'Groff' during initial aril development, whereas fructose and glucose dominated later, indicating the transformation of sucrose into other sugars (Fig. 7.12). Different data on the relative ratios of sucrose and monosaccharides were collected in China: sucrose was dominant in 'Nuomici' whereas this finding was reversed in 'Feizixiao' (Wang *et al.*, 2002b). Fructose and glucose were transformed into sucrose in

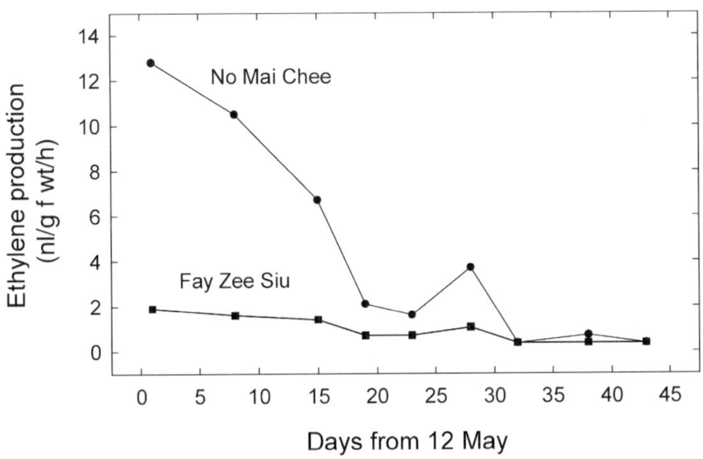

Fig. 7.11. Ethylene production in 'No Mai Chee' ('Nuomici') and 'Fay Zee Siu' ('Feizixiao') litchi during fruit development (Wang, 2001).

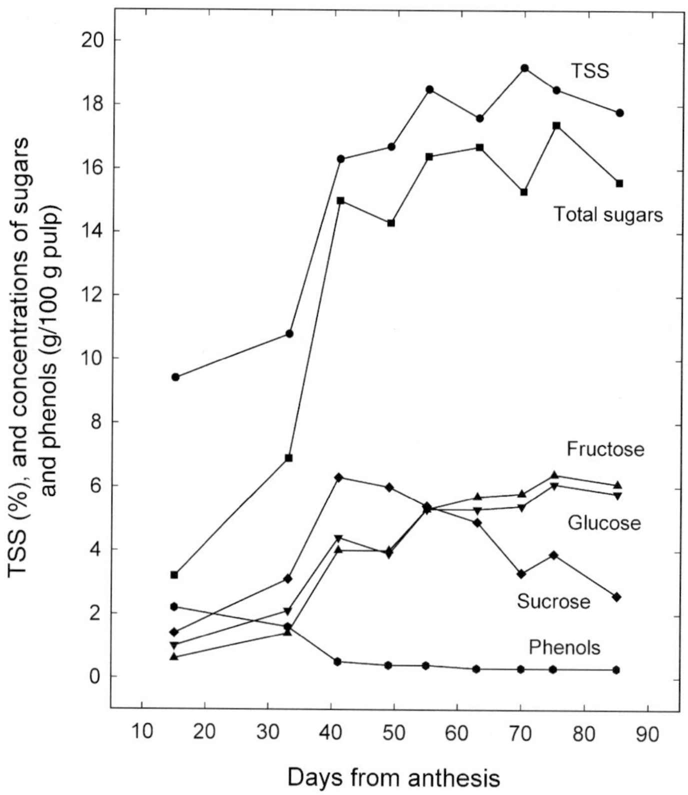

Fig. 7.12. Total soluble solids, total sugars, total phenols, sucrose, glucose and fructose in the aril of 'Groff' litchi from anthesis (Paull *et al.*, 1984).

'Guiwei' and 'H-1224', with more sucrose present than the two monosaccharides up until harvest. These differences were associated with differences in pH of the aril. The pH of 'Groff' increased from 2.0 to 3.5 at harvest, whereas it rose from 3.1 to 5.5 and from 3.2 to 4.0 in 'Guiwei' and 'H-1224', respectively (Huang *et al.*, 1986). There were two peaks of invertase activity at a pH of 5.5 and a pH of 7.5 for 'Feizixiao' (Wang, 2001). The ratio of sucrose (S), glucose (G) and fructose (F) in ripe fruit varied with cultivar. The ratio was 2.7:1.6:1.0 for 'Brewster' (Chan *et al.*, 1975), 1.0:2.1:2.1 for 'Groff' (Paull *et al.*, 1984) and 1.7:1.0:1.1 for 'Guiwei' (Huang *et al.*, 1986).

The absence of invertase and higher activities of sucrose synthase (SS) and sucrose phosphate synthase (SPS) resulted in sucrose accumulation in 'Nuomici', whereas the higher activity of acid invertase and relatively lower activities of SS and SPS resulted in the accumulation of reducing sugars in 'Feizixiao' (Wang *et al.*, 2002b).

The concentration of sugars often declines when the fruit approaches full maturation in 'Feizixiao' and some other cultivars. This 'de-sugaring' results in the fruit becoming bland in taste. As pointed out by Huang and Qiu (1987), water influx into a growing aril increased in Stage IIa as the fruit matured, while solute influx steadily declined (Fig. 7.4). This dilution, coupled with the loss of sugars through respiration, can present a practical problem for determining harvest date in some cultivars such as 'Feizixiao', where pericarp coloration lags behind sugar accumulation. Growth regulators have variable effects on sugar accumulation in litchi. Transfusion of 6-BA (6-benzylaminopurine) at 400 mg/l through a cotton thread inserted into the panicle stem 3 weeks before harvest lowered the concentrations of reducing sugars during maturation, but did not affect sucrose in 'Feizixiao' (Wang, 2001). In contrast, transfusion of ABA at 800 mg/l increased the concentration of reducing sugars. The effects of 6-BA were associated with the decreased activity of acid and neutral invertases, whereas ABA increased the activity of acid invertase.

Acids

The concentration of acids increases in the initial phase of aril growth, but then declines (Fig. 7.13). Several organic acids have been identified, with succinic acid dominant (Paull *et al.*, 1984). Since the respiratory quotient is low in litchi, at around 0.5 during the fruitlet period and 0.9 during maturation, succinic acid is logically one of the main intermediate products accumulated in the tricarboxylic acid cycle. Malic acid is also found, whereas citric acid is routinely low (Fig. 7.13). Gaur and Bajpai (1978) reported that the concentration of ascorbic acid increased rapidly along with total sugars, since the synthesis of the vitamin depends on an adequate supply of the precursor hexose sugars. The concentration of this acid declined slightly in mature fruit when sugar accumulation slowed. Chan and Kwok (1974) also found levulinic, phosphoric, glutaric, malonic and lactic acids in litchi.

Uptake of nutrients

The uptake of potassium by fruit is much higher than that of phosphorus, calcium and magnesium (Fig. 7.14). Potassium is translocated

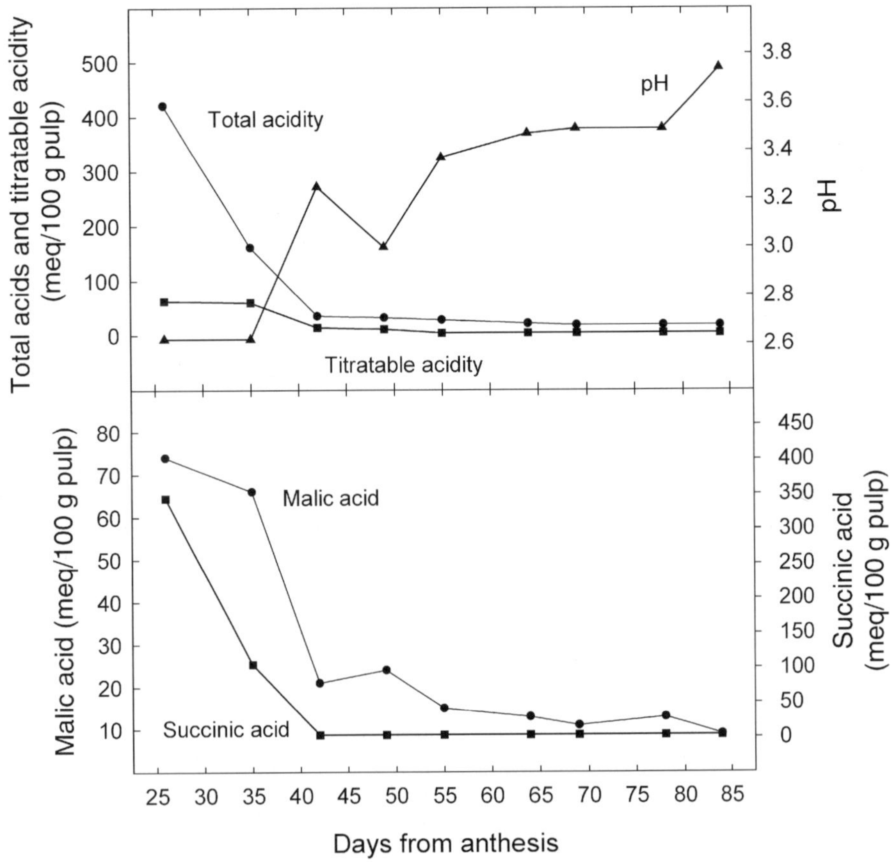

Fig. 7.13. Concentrations of total acidity and titratable acidity, pH, and concentrations of succinic and malic acid in the aril of 'Groff' litchi from anthesis (Paull *et al.*, 1984).

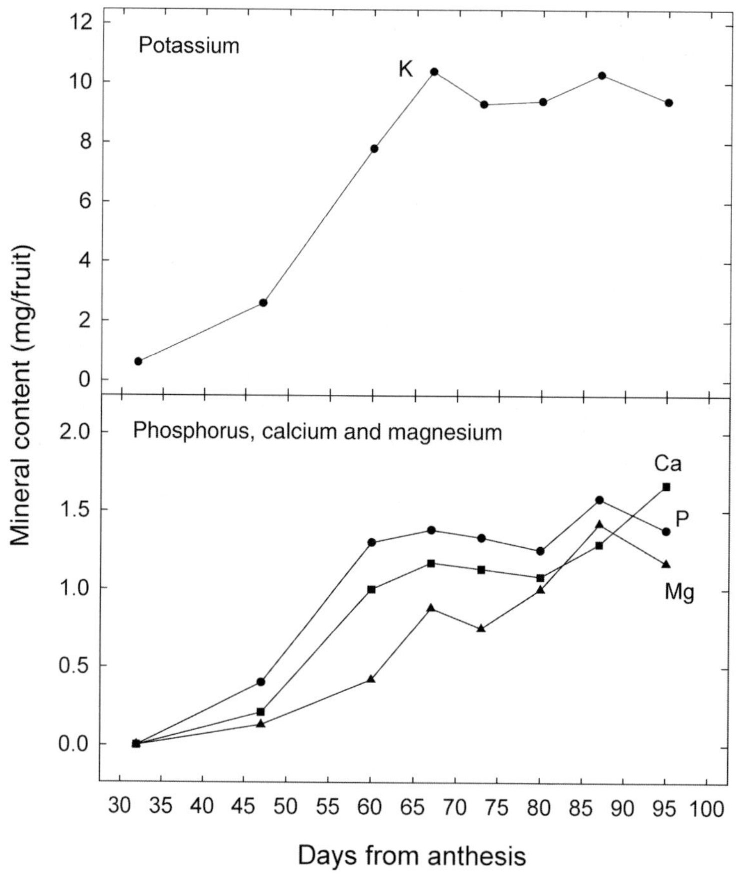

Fig. 7.14. Changes in the amount of nutrients in 'Groff' litchi (Paull *et al.*, 1984).

along with sugars, contributing to water uptake through lower osmotic potential (Li and Huang, 1995). The pericarp continues to take up calcium throughout fruit growth, but this is reduced under drought (Li, 1991; Li and Huang, 1994). The uptake of calcium parallels aril growth on a per fruit basis, whereas the uptake of phosphorus and magnesium increases during early aril development and subsequently decreases (Paull *et al.*, 1984).

Colour

The colour of the pericarp is produced by a combination of chlorophyll, carotenoids, flavones and anthocyanins. Chloroplasts are distributed in the outer mesocarp, particularly between the protuberances (Underhill and Critchley, 1992). Anthocyanins are responsible for the red colour and are found in the outer mesocarp and exocarp (Underhill and Critchley, 1994). Lee and Wicker (1991), using high performance liquid chromatography (HPLC), identified three anthocyanins in the pericarp of 'Brewster': cyanidin-3-rutinoside, malvidin-3-acetylglucoside and cyanidin-3-glucoside. These results were verified in 'Huaizhi' by Pang (2001), using HPLC coupled with mass spectrometry. Zhang *et al.* (2000), however, claimed that the major anthocyanin in 'Huaizhi' was malvidin-3-glucoside. This discrepancy might be due to the use of different extraction methods.

Red pigmentation occurs in conjunction with the degradation of chlorophyll (Paull *et al.*, 1984; Thakur *et al.*, 1990) and carotenoids. The poor colour of some cultivars when they are mature and harvestable, especially in 'Feizixiao', occurs because pericarp pigmentation lags behind sugar accumulation in the aril. The difference in peak sugar content and colour development may be as great as 10 days in some cultivars (Fig. 7.15). Slow-degrading chlorophylls present as a green background mask the appearance of the red surface and possibly impede anthocyanin formation (Wang *et al.*, 2002a). If harvest is deferred, the fruit of 'Feizixiao' turn almost fully red, but are less sweet due to 'de-sugaring'.

In 'Feizixiao', the rate of chlorophyll degradation was increased by dips of ABA at 100 mg/l and jasmonic acid at 50 mg/l. This was accompanied by enhanced chlorophyllase activity in the pericarp compared with untreated controls. In contrast, these chemicals increased the formation of anthocyanins. Fruit dipped in 6-BA at 100 mg/l had higher and sustained chlorophyll concentrations, and lower chlorophyllase activity and red colouring compared with controls (Wang, 2001). Sharma *et al.* (1986) showed that ethylene at 400 mg/l advanced coloration in 'Shahi' by 8 days compared with untreated controls.

Urea at 10–20 g/l delayed fruit maturation in 'China' by 5–12 days. Spraying with GA_3 at 25–50 mg/l, chlormequat at 2000 mg/l and daminozide at 1000–2000 mg/l also delayed ripening (Ray and Sharma, 1986). Wang *et al.* (2001) found that the activities of ACO and ACC oxidase (1-aminocyclopropane-1-carboxy oxidase), which regulate ethylene biosynthesis, were enhanced by ABA and jasmonic acid, although a dip of ethephon at 400 mg/l did not decrease chlorophylls or increase anthocyanins

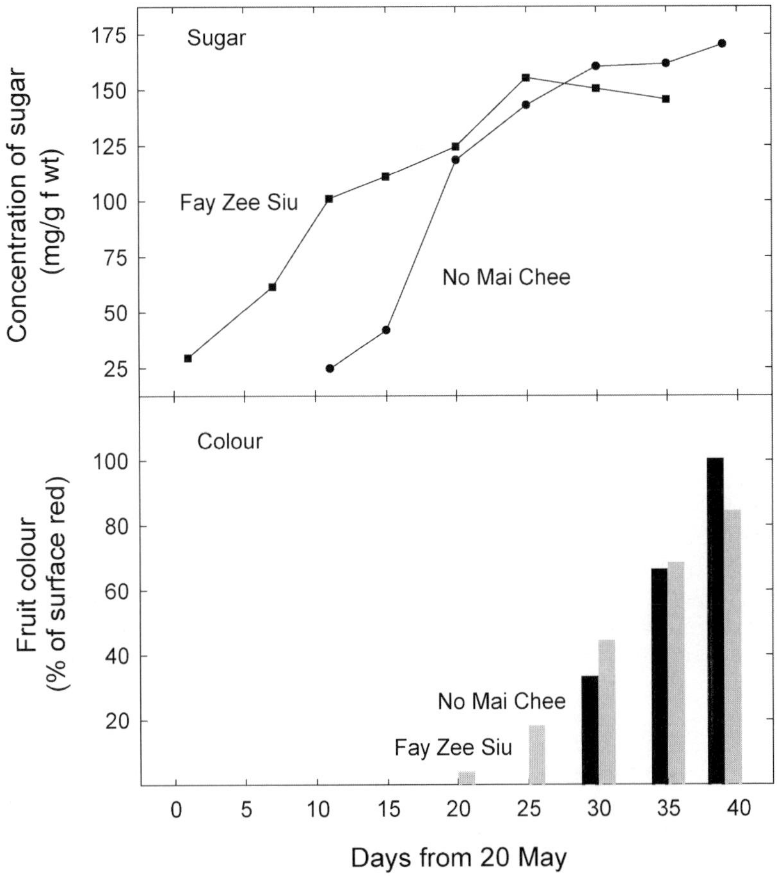

Fig. 7.15. Colour development and sugar accumulation in 'No Mai Chee' ('Nuomici') and 'Fay Zee Siu' ('Feizixiao') litchi (Wang, 2001).

in the pericarp of 'Feizixiao'. Repeated sprays of the ethylene inhibitor, silver thiosulphate (STS), from 60 DAB until harvest (85 DAB), delayed coloration of 'Nuomici' compared with untreated controls (Yin *et al.*, 1999).

Colour development can be promoted by diffuse light, as shown in bagging experiments with 'Feizixiao' (Hu *et al.*, 2001). Bags were made of adhesive-bonded fabric, with about 58% of the total sunlight being transmitted through the bag. These bags reduced potential damage to anthocyanins caused by direct sunlight. Under direct sunlight, the shaded part of the fruit received almost no light and the synthesis of anthocyanins was inhibited, resulting in uneven colouring on the fruit. Bags are also used on litchis growing in the lower canopy in orchards in South Africa, which typically have better quality than exposed fruit (Fig. 7.16). Experiments have shown that UDP glucose-flavonoid 3-*O*-glucosyl transferase (UFGT) rather than phenylalanine ammonia lyase (PAL), chalcone isomerase (CHI) or dihydroflavonol 4-reductase (DFR), is the main enzyme that regulates anthocyanin biosynthesis in the pericarp (Wang, 2001).

Mechanisms of fruit maturation

Some features are common amongst the Sapindaceae, including litchi, longan (Ke *et al.*, 1989) and lanzone, *Lansium domesticum* (Paull *et al.*, 1987). These species have an abrupt 'switch-over' from slow to rapid aril growth, accelerated sugar and anthocyanin accumulation, and degradation of acids and chlorophyll. Hirai (1980) stated that the drastic changes in loquat fruit that occur during maturation were somewhat similar to those that take place in grapes, *Vitis vinifera*, at *veraison* (a special viticultural term denoting the commencement of ripening in grapes). Huang (1983) proposed that a similar phenomenon occurs in litchi.

Sugar accumulation in the aril of litchi far exceeds the multiplication rate and enlargement of the cells. Being a strong sink, the cells accumulate increasingly greater concentrations of sugars once the fruit starts to mature, indicating that the sink activity of the aril is activated by internal metabolic changes. The abrupt increase in membrane-bound adenosine triphosphatase (ATPase) in early June in Guangdong was associated with the rapid enlargement of aril cells and the influx of sugars (Wang, 2001). The presence of enzymes related to the hydrolysis of sucrose, such as invertase and sucrose synthase, also leads to greater sink activity.

Concentrations of ABA in the seeds of 'Huaizhi' with viable embryos increased quickly when the aril and embryo started growing rapidly, and declined as the fruit matured (Yuan and Huang, 1988). Qiu *et al.* (1998) reported that concentrations of ABA increased during

Fig. 7.16. Bagging of litchi clusters in South Africa (photograph courtesy of Christopher Menzel).

maturation in three cultivars. Wang and Huang (2002) also showed that there was a steady rise in the concentration of ABA in the aril and pericarp during fruit maturation. This was more noticeable in 'Nuomici' than in 'Feizixiao' (Fig. 7.17). Applications of ABA accelerated colour development and sugar accumulation, whereas CTKs retarded these processes. Panicle micro-girdling experiments showed that ABA was translocated via the phloem and increased during fruit maturation, whereas CTKs were translocated via the xylem, retarding fruit maturation and senescence. Applications of 6-BA before colour break, however, did not affect maturation (Huang *et al.*, 2001; Yuan *et al.*, 2002). It is believed that litchi fruit maturation is closely related to ABA, whereas ethylene is involved in the degradation of pericarp chlorophylls (Wang and Huang, 2002). The interaction between ABA and ethylene is still open to investigation.

Conclusions

Litchi is an arillate fruit, having a pericarp or fruit skin, an aril or flesh, and a single normal or aborted seed. The aril originates from the outer integument. The ovary normally has two loculi, although generally only one develops. The pericarp grows first and the aril later, with the pre-formed pericarp providing a growing space and shelter for the aril. The extent of aril growth is thus largely determined by the extent of pericarp growth, while the size of the pericarp is determined by the number of cells initiated. This forms the basis for the so-called 'ball skin versus bladder effect'. The growth of the aril does not rely on the embryo, whereas the embryo uses the space that could be taken up by the aril.

There are four types of fruit: normal-seeded, aborted-seeded (stenospermocarpy), seedless fruit (parthenocarpy) and 'hollow fruit'.

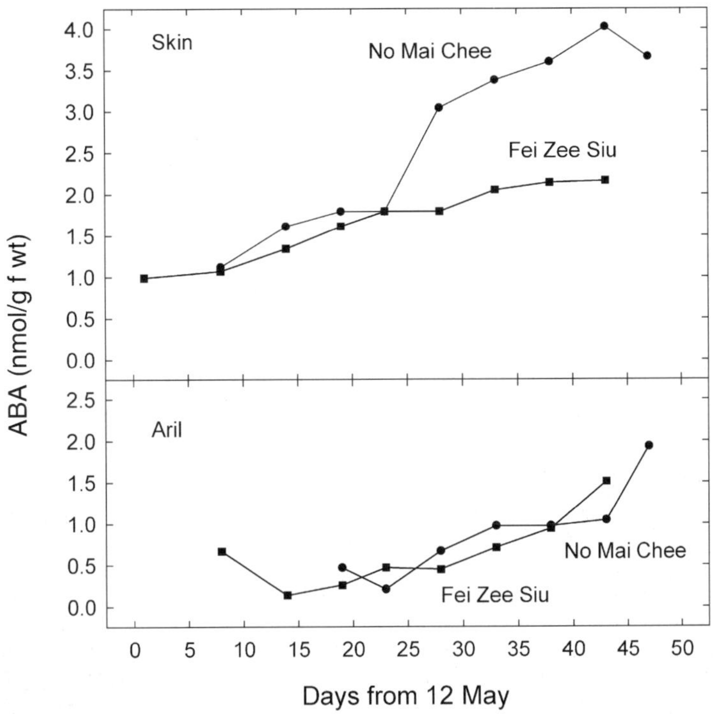

Fig. 7.17. Concentrations of abscisic acid (ABA) in the pericarp and aril of 'Fay Zee Siu' ('Feizixiao') and 'No Mai Chee' ('Nuomici') litchi during fruit development. Full bloom was on 8 April for 'Fay Zee Siu', and on 12 April for 'No Mai Chee'. Colour burst was on 9 June for 'Fay Zee Siu', and on 19 June for 'No Mai Chee'. Rapid sugar accumulation commenced on 29 May for 'Fay Zee Siu', and on 9 June for 'No Mai Chee' (Wang and Huang, 2002).

Commercial cultivars are categorized as normal-, aborted-, partially aborted-seeded, or seedless. Fruit growth is sigmoidal. Stage I is dominated by the growth of the pericarp and the seed coat, while Stage II is dominated by growth of the aril and the cotyledons. Stage II can further be divided into mainly growth of the cotyledons (Stage IIa) and mainly aril growth (Stage IIb) for normal-seeded fruit (this does not apply to aborted-seeded fruit). There are several waves of fruit abscission, with preharvest drop being peculiar to aborted-seeded cultivars.

Litchi fruit are non-climacteric. Maturation occurs during Stage II, along with the proliferation and enlargement of aril cells. Sucrose, fructose and glucose are the major sugars. There are monosaccharide-prevalent cultivars and disaccharide-prevalent cultivars, associated with different activities of various enzymes. Succinic acid is the dominant acid, whereas malic acid and citric acid are less important. The uptake of potassium by the fruit is much higher than that of phosphorus, calcium and magnesium. Red pigmentation in the pericarp occurs in conjunction with the degradation of chlorophyll and carotenoids. UDP glucose-flavonoid 3-*O*-glucosyl transferase (UFGT) regulates anthocyanin biosynthesis in the pericarp. Fruit maturation has been linked with changes in the concentrations of ABA.

References

Akamine, E.K. and Goo, T. (1973) Respiration and ethylene production during ontogeny of fruit. *Journal of the American Society for Horticultural Science* 98, 381–383.

Banerji, I. and Chaudhuri, K.L. (1944) A contribution to the life history of *Litchi chinensis* Sonn. *Proceedings of the Indian Academy of Science* Section B 19, 19–27.

Chadha, K.L. and Rajpoot, M.S. (1969) Studies on floral biology, fruit set, and its retention and quality of some litchi varieties. *Indian Journal of Horticulture* 26, 124–129.

Chan, H.T. and Kwok, S.C.M., Jr (1974) Nonvolatile acids in lychee. *Journal of Food Science* 39, 792–793.

Chan, H.T., Kwok, S.C.M., Jr and Lee, C.W.O. (1975) Sugar composition and invertase activity in lychee. *Journal of Food Science* 40, 772–774.

Chang, C.C. (1961) Studies on the fruit shedding of lychee. *Journal of Agricultural Association of China* 35, 43–50.

Chen, G.P., Zhou, X.Z. and Feng, L.H. (1997) Low temperature induced 'hollow fruit' in litchi. *Journal of Fruit Science* 14, 71–72.

Chen, W. (2000) Studies on the physiology and biochemistry of embryo development in *Litchi chinensis* Sonn. PhD dissertation, Fujian Agricultural and Forestry University, Fuzhou, Fujian, China.

Degani, C., Stern, R.A., El-Batsri, R. and Gazit, S. (1995) Pollen parent effect on the selective abscission of 'Mauritius' and 'Floridian' lychee fruitlets. *Journal of the American Society for Horticultural Science* 120, 523–526.

Deng, X.P., Han, Z.H. and Li, S.H. (1999) *Fruit Tree Biology*. Higher Education Press, Beijing, China, Appendix 1.

Gaur, G.S. and Bajpai, P.N. (1978) Some aspects of developmental physiology of the litchi fruit. *Indian Journal of Horticulture* 35, 173–177.

Hieke, S., Menzel, C.M., Doogan, V.J. and Lüdders (2002) The relationship between fruit and leaf growth in lychee (*Litchi chinensis* Sonn.). *Journal of Horticultural Science & Biotechnology* 77, 320–325.

Hirai, M. (1980) Sugar accumulation and development of loquat fruit. *Journal of the Japanese Society for Horticultural Science* 49, 347–353.

Hoda, M.N., Syamal, N.B. and Chhonkar, V.S. (1973) Effect of growth substances and zinc on litchi fruit drop. *Indian Journal of Horticulture* 30, 456–460.

Hu, G.B., Chen, D.C., Li, P., Ouyang, R., Gao, F.F., Wang, W.H. and Dong, J. (2001) Effects of bagging on fruit coloration and phenylalamine ammonia lyase and polyphenol oxidase in 'Feizixiao' litchi. *Acta Horticulturae* 558, 273–278.

Huang, H.B. (1981) Studies on the effects of summer pruning on fruiting, grape quality and vine capacity of Muscat Hamburg grapevines. *Acta Agriculturae Universitatas Pekeinensis* 1, 45–48.

Huang, H.B. (1983) On the research on grape veraison and the possible veraison-like phenomenon in litchi. *Science and Technology in Viticulture* 3, 1–3.

Huang, H.B. (2002) Unfruitfulness of young litchi trees in relation to their peculiar root growth behavior: an overview. *Acta Horticulturae* 575, 737–743.

Huang, H.B. and Qiu, Y.X. (1987) Growth correlations and assimilate partitioning in the arillate fruit of *Litchi chinensis* Sonn. *Australian Journal of Plant Physiology* 14, 181–188.

Huang, H.B. and Xu, J.K. (1983) The developmental patterns of fruit tissues and their correlative relationships in *Litchi chinensis* Sonn. *Scientia Horticulturae* 19, 335–342.

Huang, H.B., Jiang, S.Y. and Xie, C. (1983) The initiation of aril and ontogeny of fruit in *Litchi chinensis* Sonn. *Journal of South China Agricultural College* 4, 78–83.

Huang, H.B., Cheng, G.W. and Gao, F.F. (1986) Studies on fruit development in *Litchi chinensis* Sonn. II. Some physiological and biochemical features during maturation. *Acta Horticulturae Sinica* 13, 9–15.

Huang, X.M., Huang, H.B., Li, J.G., Wang, H.C. and Gao, F.F. (2001) Why do detached litchi fruit senesce more rapidly? *Acta Horticulturae* 558, 217–220.

Ji, Z.L., Liang, L.F., Liu, J.L. and Wang, G.T. (1992) Studies on the changes of endogenous hormone contents in litchi (*Litchi chinensis* Sonn.) fruits during development. *Journal of South China Agricultural University* 13, 93–98.

Jiang, J.P., Su, M.X. and Lee, P.W. (1986) The production and physiological effects of ethylene during ontogeny and after harvest of litchi fruits. *Acta Phytophysiologica Sinica* 12, 95–103.

Joubert, A.J. (1969) 'n Studie van die arillus by die lietsjie (*Litchi chinensis* Sonn.). *Tydskrif vir Natuuurwetenskappe* 9, 242–250.

Joubert, A.J. (1986) Litchi. In: Monselise, S.P. (ed.) *Handbook of Fruit Set and Development.* CRC Press, Boca Raton, Florida, pp. 233–246.

Ke, G.W., Wang, C.C. and Huang, J.H. (1989) The aril initiation and ontogeny of longan fruit. In: Lin, J.B. and Gong, J.Z. (eds) *Proceedings of the Fujian Symposium on Longan*, Fuzhou, Fujian, China, pp. 54–56.

Lee, H.S. and Wicker, L. (1991) Anthocyanin pigments in the skin of lychee fruit. *Journal of Food Science* 56, 466–468, 483.

Li, J.G. (1991) Studies on the mechanisms of fruit cracking in litchi. MSc thesis, South China Agricultural University, Guangzhou, China.

Li, J.G. (2001) Ontogenetic studies of fruit development and minimizing cracking in *Litchi chinensis* Sonn. PhD thesis, South China Agricultural University, Guangzhou, China.

Li, J.G. and Huang, H.B. (1994) Analyzing the cause of fruit cracking increased by rain after a long period of drought. In: Cheng, K.S. and Zhang, S.L. (eds) *Advances in Horticulture.* Agricultural Publisher, Beijing, China, pp. 361–364.

Li, J.G. and Huang, H.B. (1995) Physico-chemical properties and peel morphology in relation to fruit-cracking susceptibility in litchi fruit. *Journal of South China Agricultural University* 16, 84–89.

Li, J.G. and Wang, Z.H. (1999) The influence of second male blossoming on initial set and the countermeasures. *South China Fruits* 28, 27–28.

Li, J.G., Huang, X.M., Huang, H.B. and Zhou, B.Y. (2002) A cytological and physiological study of large-fruited and small-fruited litchi cultivars. *Journal of Fruit Science* 19, 158–162.

Liang, Y.G. and Qiu, S.D. (1998) A study on the artificially inducing techniques from large-seeded fruit to small-seeded and high quality fruit in lychee (*Litchi chinensis* Sonn.). In: *Proceedings of the XXV International Horticultural Congress*, Brussels, Belgium, p. 355.

Lin, Z.F., Sun, G.T., Chen, F., Lin, G.Z. and Liu, S.X. (1990) The cyanide-insensitive respiration of litchi fruit. *Acta Phytophysiologica Sinica* 16, 367–372.

Lü, L.X., Chen, J.L. and Chen, X.J. (1985) An observation on the process of embryo development in litchi. *Subtropical Plant Research Communications* 1, 1–5.

Lü, L.X., Yu, X.L., Ye, M.Z. and Xiao, Z.T. (1989) Studies on the mechanism of embryonic development in litchi (*Litchi chinensis* Sonn.). *Journal of Fujian Agricultural College* 18, 149–155.

Mustard, M.J., Liu, S.Y. and Nelson, R.O. (1954) Observations of floral biology and fruit setting in lychee varieties. *Proceedings of the Florida State Horticultural Society* 66, 212–220.

Pang, X.Q. (2001) Study on the stability of anthocyanins from lychee pericarp. MSc thesis, South China Agricultural University, Guangzhou, China.

Paull, R.E., Chen, N.J., Deputy, J., Huang, H.B., Cheng, G.W. and Gao, F.F. (1984) Litchi growth and compositional changes during fruit development. *Journal of the American Society for Horticultural Science* 109, 817–821.

Paull, R.E., Goo, T. and Chen, N.J. (1987) Growth and compositional changes during development of lanzone fruit. *HortScience* 22, 1252–1253.

Pivovaro, S.Z. (1974) Studies on the floral biology and the influence of growth regulators on fruit set, size and drop of *Litchi chinensis* Sonn. MSc (Agr) thesis, Hebrew University of Rehovot, Israel.

Prasad, A. and Jauhari, O.S. (1963) Effect of 2,4,5-trichlorophenoxyacetic and alpha naphthaleneacetic acids on 'drop stop' and size of litchi fruits. *Madras Agricultural Journal* 50, 28–29.

Qiu, Y.P., Zhang, Z.W. and Qiu, R.X. (1994) Study on embryo and endosperm development of *Litchi chinensis* Sonn. *Chinese Bulletin of Botany* 11, 45–47.

Qiu, Y.P., Xiang, X., Wang, B.Q., Zhang, Z.W. and Yuan, P.Y. (1998) Endogenous hormone balance in three types of litchi fruit and their fruit set mechanism. *Journal of Fruit Science* 15, 39–43.

Ray, P.K. and Sharma, S.B. (1986) Delaying litchi harvest by growth regulators or urea spray. *Scientia Horticulturae* 28, 93–96.

Sharma, S.B., Ray, P.K. and Rai, R. (1986) The use of growth regulators for early ripening of litchi (*Litchi chinensis* Sonn.). *Journal of Horticultural Science* 64, 533–534.

Singh, U.S. and Lal, R.K. (1980) Influence of growth regulators on setting, retention and weight of fruits in two cultivars of litchi. *Scientia Horticulturae* 12, 321–326.

Srivastava, R.P. and Singh, L. (1969) Effect of growth substances on the quality of litchi. *Horticultural Science* 1, 1–6.

Stern, R.A. and Gazit, S. (1996) Lychee pollination by the honeybee. *Journal of the American Society for Horticultural Science* 121, 152–157.

Stern, R.A. and Gazit, S. (1997) Effect of 3,5,6-trichloro-2-pyridyl-oxyacetic acid on fruitlet abscission and yield of 'Mauritius' litchi (*Litchi chinensis* Sonn.). *Journal of Horticultural Science* 72, 659–663.

Stern, R.A., Gazit, S., El-Batsri, R. and Degani, C. (1993) Pollen parent effect on outcrossing rate, yield, and fruit characteristics of 'Floridian' and 'Mauritius' lychee. *Journal of the American Society for Horticultural Science* 118, 109–114.

Stern, R.A., Kigel, J., Tomer, E. and Gazit, S. (1995) 'Mauritius' lychee fruit development and reduced abscission after treatment with the auxin 2,4,5-TP. *Journal of the American Society for Horticultural Science* 120, 65–70.

Stern, R.A., Eisenstein, D., Voet, H. and Gazit, S. (1996) Anatomical structure of two-day-old litchi ovules in relation to fruit set and yield. *Journal of Horticultural Science* 71, 661–671.

Stern, R.A., Nadler, M. and Gazit, S. (1997) 'Floridian' litchi yield is increased by 2,4,5-TP spray. *Journal of Horticultural Science* 72, 609–615.

Stern, R.A., Stern, D., Miller, H., Xu, H. and Gazit, S. (2001) The effect of the synthetic auxins 2,4,5-TP and 3,5,6-TPA on yield and fruit size of young 'Fei Zi Xiao' and 'Hei Ye' litchi trees in Guangxi Province, China. *Acta Horticulturae* 558, 285–288.

Steyn, E.M.A. and Robbertse, P.J. (1992) Is aril development in litchi triggered by pollen tube growth? *South African Journal of Botany* 59, 258–262.

Thakur, S., Kumar, R., Brabmachari, V.S. and Sharma, R.K. (1990) Effect of different growth regulators on fruit-set, retention and size of litchi. *Indian Journal of Horticulture* 47, 305–306.

Underhill, S.J.R. and Critchley, C. (1992) The physiology and anatomy of lychee (*Litchi chinensis* Sonn.) pericarp during fruit development. *Journal of Horticultural Science* 67, 437–444.

Underhill, S.J.R. and Critchley, C. (1994) Anthocyanin decolorisation and its role in lychee pericarp browning. *Australian Journal of Experimental Agriculture* 34, 115–122.

Wang, B.Q., Qiu, Y.P., Xiang, X., Yuan, P.Y. and Zhang, Z.W. (1997) Study on the fluctuation of endogenous hormones during litchi (*Litchi chinensis* Sonn.) fruit setting and its induced parthenocarpy. *Acta Horticulturae Sinica* 24, 19–24.

Wang, H.C. (2001) Studies on the pigmentation, sugar metabolism and hormonal regulation of litchi fruit during maturation. PhD dissertation, South China Agricultural University, Guangzhou, China.

Wang, H.C. and Huang, H.B. (2002) A preliminary study of the key hormone in litchi fruit maturation. In: Lei, J.J., Chen, R.Y. and Chen, H.B. (eds) *Advances in Horticulture*. Guangzhou Press, Guangdong, China, pp. 183–189.

Wang, H.C., Wei, B.W., Gao, F.F. and Huang, H.B. (2000) Litchi fruit skin studies: skin structure and cell division in relation to fruit cracking. *Journal of South China Agricultural University* 21, 10–13.

Wang, H.C., Huang, X.M. and Huang, H.B. (2001) Litchi fruit maturation studies: changes in abscisic acid (ABA) and 1-aminocyclopropane-1-carboxylic acid (ACC) and the effect of cytokinin and ethylene on coloration in cv. Feizixiao. *Acta Horticulturae* 558, 267–272.

Wang, H.C., Huang, X.M. and Huang, H.B. (2002a) A study of the causative factors retarding pigmentation in the fruit of 'Feizixiao' litchi. *Acta Horticulturae Sinica* 29, 408–412.

Wang, H.C., Huang, H.B. and Huang, X.M. (2002b) A study of sugar accumulation and related enzyme activities in the litchi fruit. *Acta Horticulturae Sinica* 30, 1–5.

Xiang, X., Zhang, Z.W., Qiu, Y.P., Yuan, P.Y. and Wang, B.Q. (1994) A study of fruit set physiology in relation to endogenous hormones in litchi (*Litchi chinensis* Sonn.) fruits. *Acta Horticulturae Sinica* 21, 1–6.

Xiang, X., Ou, L.X., Qiu, Y.P., Yuan, P.Y. and Chen, J.Z. (2001) Embryo abortion and pollen parent effects in 'Nuomici' and 'Guiwei' litchi. *Acta Horticulturae* 558, 257–260.

Xiang, X., Ou, L.X., Qiu, Y.P., Chen J.Z. and Li, Z.Q. (2002) Studies on the induction of stenospermocarpy in litchi normal-seeded cultivars. *Journal of Fruit Science* 19, 111–114.

Xu, E. (1962) Studies on growth regulator sprays to reduce fruit drop in litchi. Postgraduate thesis, South China Agricultural College, Guangzhou, China.

Ye, X.L., Wang, F.X. and Qian, N.F. (1992) Embryological studies of *Litchi chinensis. Acta Botanica Yunnanica* 14, 59–65.

Yin, J.H., Gao, F.F., Ye, Z.X. and Chen, D.C. (1999) Effect of ethylene on maturation of litchi fruit. *Journal of Fruit Science* 16, 272–275.

Yuan, R.C. and Huang, H.B. (1988) Litchi fruit abscission: its patterns, effect of shading and relation to endogenous abscisic acid. *Scientia Horticulturae* 36, 281–292.

Yuan, R.C. and Huang, H.B. (1991) Effect of NAA, NAA plus nucleotides on fruit set of lychee. *Yearbook of the Australian Lychee Growers' Association* 1, 46–50.

Yuan, R.C. and Huang, H.B. (1992) Improvement of fruit-set in *Litchi chinensis* Sonn. through the regulation of source–sink relationships. *Journal of South China Agricultural University* 13, 136–141.

Yuan, R.C. and Huang, H.B. (1993) Regulation of root and shoot growth and fruit-drop of young litchi trees by trunk girdling in view of source–sink relationships. *Journal of Fruit Science* 10, 195–198.

Yuan, W.Q., Huang, J.S., Xian, Q.W., Dai, X.Y., Lin, Z.Y. and Huang, X.M. (2002) Effects of micro-girdling at panicle stalk on the maturation of litchi fruit. In: Lei, J.J., Chen, R.Y. and Chen, H.B. (eds) *Advances in Horticulture*. Guangzhou Press, Guangdong, China, pp. 215–219.

Zhang, D.L., Quantick, P.C. and Grigor, J.M. (2000) Changes in phenolic compounds in litchi (*Litchi chinensis* Sonn.) fruit during postharvest storage. *Postharvest Biology and Technology* 19, 165–172.

Zhang, Z.W., Qiu, Y.P. and Xiang, X. (1990) A preliminary study of parthenocarpy in litchi. *Journal of Fruit Science* 7, 234–235.

Zhou, B.Y., Ji, Z.L., Ye, Y.C., Zhao, X.D. and Ye, Y.X. (1998) Changes of endogenous hormones in litchi fruits during fruit development. *Acta Horticulturae Sinica* 25, 236–240.

Zhou, X.J., Wu, D.Y., Huang, H.B., Lai, J. and Huang, D.G. (1996) Manipulating growth and fruiting of young litchi trees by spiral girdling. *Acta Horticulturae Sinica* 23, 13–18.

Zhou, X.J., Huang, D.Y., Huang, H.B. and Wu, D.Y. (1999) Carbohydrate and endohormone status in relation to fruit set as influenced by trunk spiral girdling of young litchi tree. *Acta Horticulturae Sinica* 26, 77–80.

B. Longan

Raphael A. Stern
MIGAL, Galilee Technology Center, PO Box 831, Kiryat Shmona 11016, Israel

Introduction

Longan trees grown from air-layers usually come into bearing in their third or fourth year in South-east Asia. Fruit development is generally similar to that in litchi, although the trees are more prolific, with a lower rate of fruit abscission than in litchi. The non-commercial types of longan, found throughout much of Malaysia, Indonesia and Borneo, usually have thinner arils than the commercial varieties, and have a warty or rougher skin. Many of these 'wild' forms remain green at maturity.

Fruit morphology

According to Morton (1987), Subhadrabandhu (1990), Tindall (1994) and Yaacob and Subhadrabandhu (1995), longan fruit are arillate, globose, and sometimes have pronounced shoulders. The fruit are dull green while developing, changing to dull yellow-brown during ripening. The inflorescences may carry up to 80 individual fruit, which vary in weight from 5 to 20 g and from 1.2 to 3.0 cm in diameter. Premium commercial fruit typically weigh 12–18 g (Watson, 1984). The pericarp or skin is thin but tough and leathery, and the protuberances on the skin are relatively superficial compared with those on litchis.

In high-quality cultivars, the edible part or aril accounts for up to 60–75% of total fruit fresh weight at harvest. It is white, translucent, sweet (total soluble solids of 16–25% in fresh fruit, and 75% in dried fruit), and musky compared with litchi (Knight *et al.*, 1968; Morton, 1987). When the fruit are dried, the aril becomes shrivelled and brown, and shrinks away from the shell. The fruit are known as 'longan nuts' in this form, and represent an increasing share of the longan market in China, Viet Nam and Thailand.

The fruit have one small, shiny black seed, which is ovoid with a white scar on the top, known in Thailand as 'dragon eye'. The seeds range from 1.0 to 1.5 cm in diameter, with only a few cultivars having shrivelled, shrunken or aborted seeds. In China, cultivars such as 'Min Jiao 64–1', 'Min Jiao 64–2' and 'Baiha' have been selected for this characteristic (Zheng *et al.*, 1994). Fruit with such seeds are usually smaller than those with normal seeds (Zheng *et al.*, 1996).

Fruit development

The time from anthesis to harvest in longan (4–5 months) is much longer than in litchi (2–3 months), with some variation amongst different cultivars. In Thailand, the traditional crop flowers from December to February, with harvesting from June to August. In southern districts, tropical cultivars flower and crop after treatment with potassium chlorate (Wangsin and Pankasemsuk, 2003). This chemical is also used in Viet Nam to produce more than one crop per year in tropical cultivars. In northern Queensland, Australia, flowering commences in August or September, with the main harvest from January to March.

The pattern of fruit development in longan is similar to that in litchi, although there have been fewer studies dealing with this subject in longan (Li, 1984; Ke *et al.*, 1992; Chen *et al.*, 1995; Xu *et al.*, 1997; Zhou *et al.*, 1999). The longan ovary is bi-lobed, with one ovule in each lobe. The two lobes stand at right angles to the pedicle and style. Usually, only one ovary develops into a fruit, while the second atrophies, but remains attached. The developing ovule gradually moves into alignment with the pedicel.

Ke *et al.* (1992), Zheng *et al.* (1994), Chen *et al.* (1995) and Xu *et al.* (1997) studied the development of longan cultivars in China, with two distinct phases of fruit growth distinguished (Fig. 7.18). The first phase lasted for about 50 days, after which fruit weighed 1 g, and were composed mainly of peel or pericarp (65%) and seed (35%). At this time, the aril forming at the base of the seed just above the obturator was hardly recognizable, but started growing rapidly a week or so later. The tiny embryo ('heart' stage) protruded from the seed coat, but could only just be seen with the naked eye. The remnants of the endosperm were still present in the seed cavity at this stage.

The second phase lasted for 60 days or more, depending on the cultivar. During this period, the embryo grew rapidly, while the endosperm disappeared. About 80 days after anthesis, the embryo filled the seed cavity with two well-developed cotyledons, while the seed coat was hard and had finished growing along with the pericarp. The aril continued to grow for another 30–60 days, while the pericarp stretched and thinned from 300 μm at day 50, to 100 μm at harvest (Ke *et al.*, 1992).

Several studies have investigated the relative growth rate of the fruit and its components in southern China (Ke *et al.*, 1992; Zheng *et al.*, 1994; Chen *et al.*, 1995; Xu *et al.*, 1997). The growth of the whole fruit, pericarp, aril and seed followed a sigmoidal pattern (Fig. 7.18). The pericarp and the seed reached their maximum fresh weight at about 80 days, whereas the aril continued to grow until the fruit were mature.

Fruit abscission and fruit size

The proportion of fruit that is initially set and which is carried to harvest is much higher than that in litchi. At harvest, the longan inflorescence may carry 60–80 fruit (about 20% fruit set) compared with only 15–25 fruit in litchi (5% fruit set). Unlike litchi, which has a heavy fruit drop over three or four abscission cycles, longan has only two peaks of abscission during fruit development. The first cycle occurs 3–20 days after fertilization, and the second 30–60 days before harvest (Zee *et al.*, 1998). Zhou *et al.* (1999) found a reduction in the concentrations of auxins and gibberellins in 'Chuliang' fruit at these two times, and suggested a link between abscission and the growth substances. Gibberellins are sometimes applied to longan trees during an 'off-year' in China in an attempt to reduce fruit drop (Liu and Ma, 2001).

The heavy set in longan is often associated with small fruit at harvest, with Knight *et al.* (1968) reporting a strong relationship between individual fruit weight and tree yield. Removing parts of the inflorescence at flowering increased

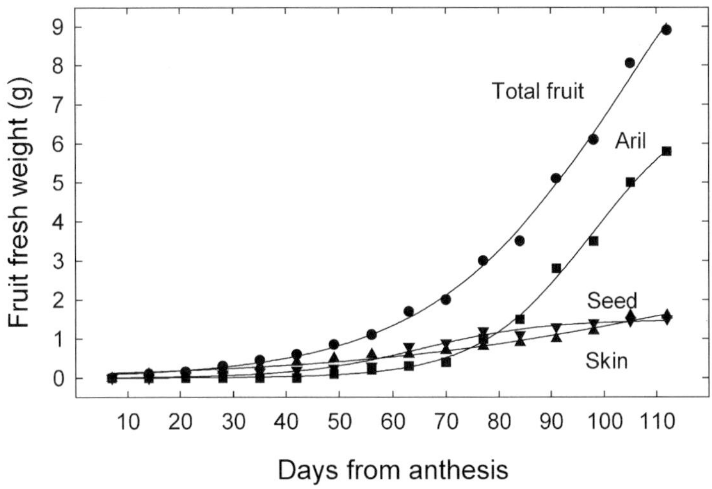

Fig. 7.18. Growth of 'Dongbi' longans in southern China (Ke *et al.*, 1992).

average fruit weight in Florida (Campbell and Malo, 1981). In China, about 40% of the flowers from an inflorescence are removed in April in an 'on-year', when the flower panicles are 12–15 cm long (Liu and Ma, 2001).

Conclusions

Longan and litchi have a similar pattern of fruit development, with growth soon after fruit set dominated by the differentiation of the seed and skin, and the aril developing later. The longan inflorescence is much larger than that of litchi, with a heavier set and a lower rate of fruit abscission. In many cultivars, the heavy productivity is at the expense of average fruit weight. Longan also has fewer cultivars with shrivelled, shrunken or aborted seeds.

References

Campbell, C.W. and Malo, S.E. (1981) Evaluation of the longan as a potential crop for Florida. *Proceedings of the Florida State Horticultural Society* 94, 307–309.

Chen, Q.X., Liao, J.S. and Hu, Y.L. (1995) The growth curve of longan fruit and the correlative analysis of its tissues. *Journal of Fujian Agricultural University* 24, 19–22.

Ke, G.W., Wang, C.C. and Huang, J.H. (1992) The aril initiation and ontogenesis of longan fruit. *Journal of Fujian Academy of Agricultural Sciences* 7, 22–26.

Knight, R.J., Jr, Manis, W.E., Kosel, G.W. and White, C.A. (1968) Evaluation of longan and lychee introductions. *Proceedings of the Florida State Horticultural Society* 81, 314–318.

Li, J.Z. (1984) Studies on fertilization of *Dimocarpus longan* Lour. *Acta Horticulturae Sinica* 11, 243–247.

Liu, X.H. and Ma, C.L. (2001) Production and research of longan in China. *Acta Horticulturae* 558, 73–82.

Morton, J. (1987) *Fruits of Warm Climates.* Creative Resources Systems, Winterville, North Carolina, 505 pp.

Subhadrabandhu, S. (1990) *Lychee and Longan Cultivation in Thailand.* Rumthai Publishers, Bangkok, Thailand, 40 pp.

Tindall, H.D. (1994) Sapindaceous fruits: botany and horticulture. *Horticultural Reviews* 16, 143–196.

Wangsin, N. and Pankasemsuk, T. (2003) Effect of potassium chlorate on flowering, total nitrogen, total nonstructural carbohydrate, C/N ratio, cytokinin-like and gibberellin-like substances in stem apex of longan cv. Daw. In: *Proceedings of the Second International Symposium on Lychee, Longan, Rambutan and Other Sapindaceae Plants.* Chiang Mai, Thailand (Abstract), p. 33.

Watson, B.J. (1984) Longan. In: Page, P. (ed.) *Tropical Fruits for Australia.* Queensland Government Printer, Brisbane, Australia, pp. 192–197.

Xu, X.D., Zheng, S.Q., Huang, J.S., Xu, J.H., Lin, Y.Q. and Liu, H.Y. (1997) Studies on fruit development of extremely late maturing species of longan. II. Dynamic changes of fresh and dry weight, the requirement and distribution of moisture and solute content in fruits. *Journal of Fujian Academy of Agricultural Sciences* 12, 19–23.

Yaacob, O. and Subhadrabandhu, S. (1995) *The Production of Economic Fruits in South-East Asia.* Oxford University Press, Oxford, UK, 419 pp.

Zee, F.T.P., Chan, H.T., Jr and Yen, C.R. (1998) Lychee, longan, rambutan and pulasan. In: Shaw, P.E., Chan, H.T., Jr and Nagy, S. (eds) *Tropical and Subtropical Fruits.* Agscience, Auburndale, Florida, pp. 290–335.

Zheng, S.Q., Huang, J.S. and Xu, X.D. (1994) Studies on fruit development of aborted-seeded longan: correlative analysis on fruit growing type and its characters. *Journal of Fujian Academy of Agricultural Sciences* 9, 22–25.

Zheng, S.Q. Huang, J.S., Xu, X.D., Huang, J.H. and Zheng, S.K. (1996) Observation on the development characteristics of aborted longan seed. *Acta Horticulturae Sinica* 23, 83–85.

Zhou, B.Y., Li, J.G., Huang, X.M. and Zhou, X.J. (1999) Changes of ZRs, GA and IAA content during fruit development of longan. *Journal of South China Agricultural University* 20, 50–53.

8 Fruit Disorders

Xuming Huang
College of Horticulture, South China Agricultural University, Guangzhou 510642, China

Introduction

Litchi and longan suffer from a range of disorders that reduce returns to commercial growers. The disorders, which may be manifest from fruit set to fruit maturity, include fruit cracking or splitting, sunburn and stunted fruit development. They are more common in certain cultivars and growing areas than others, while their incidence and severity also vary from season to season. The physiological mechanisms behind these disorders are not completely understood, although they often appear to be related to the weather or tree agronomy.

Fruit cracking

Fruit cracking is a major issue in litchi, but is less important in longan. Cracking in litchi is most common in China (Chen and Huang, 2001) and India (Mitra and Ghosh, 1991), the two largest producers in the world. The disorder also occurs in Thailand (Sethpakdee, 2002), Nepal (Budathoki, 2002) and South Africa (Joubert, 1970, 1986). In China, the annual loss caused by fruit cracking in 'No Mai Chee' ('Nuomici') (Fig. 8.1) alone is US$2.5 million (Li *et al.*, 1992). In India, one-third of the crop may be lost in susceptible cultivars such as 'Muzaffarpur' and 'Dehra Dun' (Kanwar *et al.*, 1972b).

Fig. 8.1. Fruit cracking in 'No Mai Chee' ('Nuomici') litchi in China.

Sharma and Ray (1987) and Li *et al.* (2001) have reviewed the problem in litchi. Studies have examined the susceptibility of different cultivars to cracking and the roles of temperature, water, plant growth regulators and nutrients. There have also been studies on the morphology and mechanical properties of the cell walls in the skin or pericarp. These studies have yielded valuable information about the underlying causes of cracking and how to reduce the problem.

Occurrence of the disorder

Fruit growth in litchi and longan consists of two distinct phases: growth of the pericarp, or skin, and the seed, followed by growth of the aril, or flesh (Kanwar *et al.*, 1972a; Huang *et al.*, 1983; Huang, 2001). Field observations showed that, in litchi, cracking occurs mainly after the fruit begin to colour, coinciding with the start of rapid aril growth (Kanwar *et al.*, 1972a; Kanwar and Nijjar, 1976; Li *et al.*, 1992; Huang *et al.*, 1999; Wang *et al.*, 2000) (Fig. 8.2). Huang and Xu (1983) proposed a 'ball skin versus bladder effect' theory to describe the relationship between a pre-grown skin and a growing aril. This theory suggested that cracking results from a lack of coordination between the 'ball skin' and the 'bladder' when stress from the 'inflating bladder' exceeds the tensile strength of the skin (Kanwar *et al.*, 1972a; Li and Huang, 1995). Abnormal development of the skin during early fruit growth is the basis of the disorder (Joubert, 1970); hence, several studies have examined the development and mechanical properties of the skin.

Methodology

Critical cracking turgor (CCT) is the maximum turgor that the skin can tolerate before it cracks, and can be determined in a pressure chamber (Ou, 1988; Huang *et al.*, 1999) as shown in Fig. 8.3. An injection needle is fixed on a rubber pad located at the centre of the chamber lid. The needle, pointing upwards, is forced through the fruit pedicel into the aril. Increasing pressure from nitrogen gas in the chamber (0.02 MPa/s) simulates the rising turgor of the expanding aril. The pressure required to split the fruit is recorded as the critical cracking turgor. However, this value does not include the existing fruit turgor that the skin is bearing before the external turgor is imposed, although it reflects how much extra turgor will cause the skin to split. Results of tests conducted in a laboratory correlated well with the incidence of cracking in orchard trees (Table 8.1).

Skin tensile strength and extensibility indicate the mechanical properties of the pericarp associated with cracking resistance, with an instron-like device used to measure these parameters in litchi (Huang and Huang, 1998). Strain–stress curves are developed as samples of fixed length and width are stretched until they break (Fig. 8.4). The maximum elongation (strain) and the maximum stress carried by the skin indicate extensibility and tensile strength. Results of these measurements indicate that resistant 'Wai Chee' ('Huaizhi') has higher tensile strength and extensibility than susceptible 'No Mai Chee' ('Nuomici') (Fig. 8.4).

Susceptibility of different cultivars

The incidence of cracking varies significantly amongst cultivars, suggesting a strong genetic link. Sanyal *et al.* (1990) reported that in India cracking was greatest in 'Deshi' (12%), followed by 'Early Large Red' (10%), 'Muzaffarpur' (7%), 'Nafarpal' (3%), 'Elachi' (2%), 'Bombai' (1%), 'McLean' (1%) and 'Bedana' (0.7%). Their report also suggested that early-ripening cultivars were more susceptible than later ones. Ou (1988) found that the relative order of susceptibility in China was 'No Mai Chee' > 'Kwai May' ('Guiwei') > 'Wai Chee' in both laboratory and field experiments. 'No Mai Chee' and 'Wai Chee' are the major late cultivars in China. Up to 80% of 'No Mai Chee' fruit may crack (Fig. 8.1), whereas the later 'Wai Chee' rarely cracks.

Skin morphology

Skin morphology varies amongst different cultivars. Kanwar *et al.* (1972b) suggested that

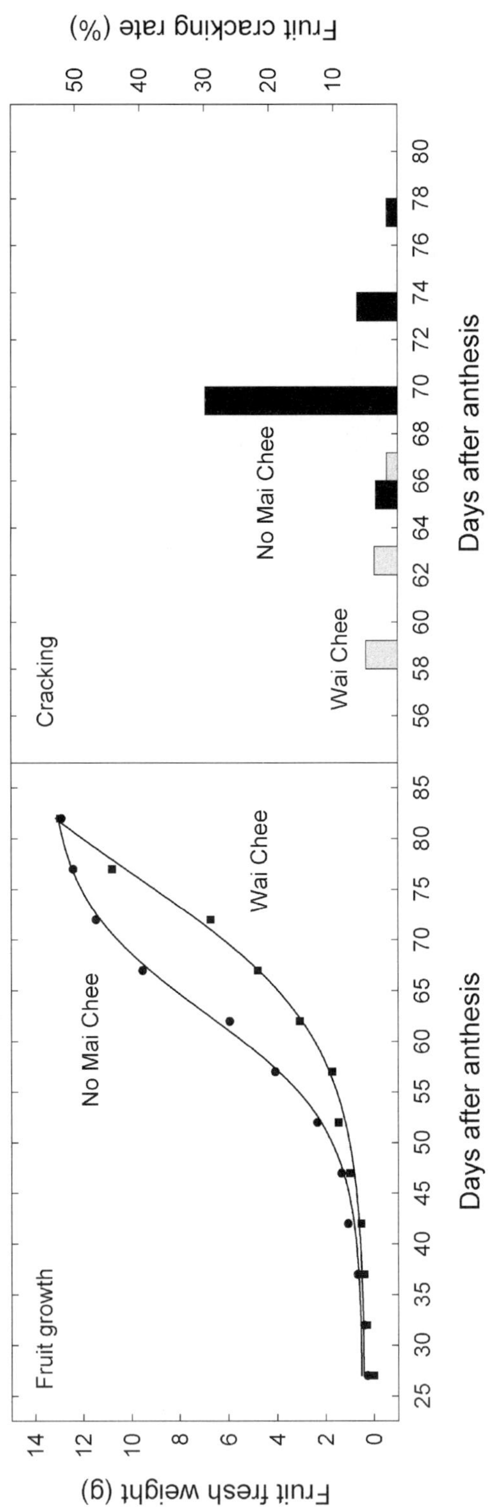

Fig. 8.2. Fruit growth and fruit cracking rate in 'No Mai Chee' ('Nuomici') and 'Wai Chee' ('Huaizhi') litchi in Guangdong, southern China (Huang *et al.*, 1999).

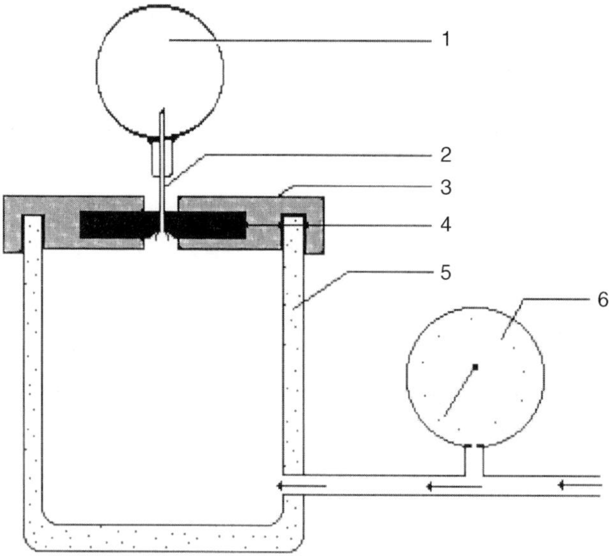

Fig. 8.3. Pressure chamber adapted to determine the critical cracking turgor. 1 = fruit sample; 2 = injection needle; 3 = chamber lid; 4 = rubber pad; 5 = chamber; and 6 = pressure gauge. Arrow shows flow of high-pressure nitrogen gas (Huang *et al.*, 1999).

Table 8.1. Critical cracking turgor (CCT) and cracking rate in different litchi cultivars in China. Fruit samples were taken during rapid aril growth.

	Ou (1988)		Huang *et al.* (1999)	
Cultivar	CCT (MPa)	Cracking rate (%)	CCT (MPa)	Cracking rate (%)
Wai Chee	0.39	14.8	0.44	6.5
Kwai May	0.28	43.3	–	–
No Mai Chee	0.26	50.8	0.31	34.0

cultivars with fewer tubercles per unit area of skin were less susceptible to cracking. However, Li and Huang (1995) drew a different conclusion based on scanning electron microscopy (SEM), with 'No Mai Chee' having broader and shallower tubercles than 'Wai Chee'. Their data agree with the results obtained by Joubert (1986). The relationship between properties of the skin and the size and density of tubercles needs further investigation. Kanwar *et al.* (1972b) also suggested that cultivars with thick

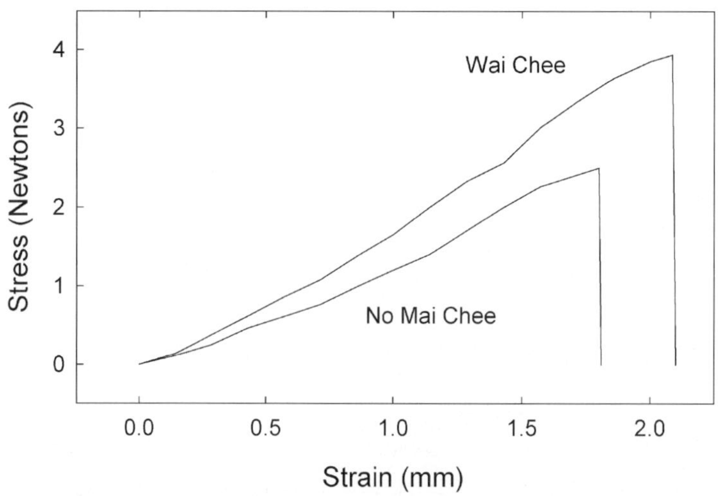

Fig. 8.4. Strain–stress curves of 'No Mai Chee' ('Nuomici') and 'Wai Chee' ('Huaizhi') litchi skin strips (10 mm long).

skins were less susceptible to cracking than those with thin skins. Wang (1998) and Huang *et al.* (1999) showed that cracking coincided with thinning of the skin during aril expansion; however, the peak of cracking did not occur at harvest when the skin was thinnest. Resistant 'Wai Chee' had an even thinner pericarp than 'No Mai Chee' at the critical period when the fruit started to turn red (Wang, 1998). Therefore, differences between cultivars could not simply be attributed to the thickness of the skin. Sanyal *et al.* (1990) examined aril-to-skin ratio and found it was lower in cracked fruit than in the normal fruit (3.4 vs. 4.9 in 'Deshi'), whereas Li and Huang (2001) showed the opposite trend.

The detailed structure of the skin has been studied in relation to cracking. Joubert (1986) suggested that the ratio of tangential to radial length of the cells at the base of the epicarp was higher in susceptible 'Mauritius' than in resistant 'Brewster' ('Chenzi') and 'Haak Yip' ('Heiye'). The skin of 'Wai Chee' has more regular mesocarp cells and better developed spongy tissues in the inner mesocarp than 'No Mai Chee' (Wang *et al.*, 2000; Fig. 8.5). The spongy tissue possibly buffers against the expanding aril and assists in extension of the skin. The larger spongy tissue of 'Wai Chee' may contribute to its high pericarp extensibility. Higher tensile strength may be related to a more elastic endocarp (Ou, 1988); however, recent observations did not

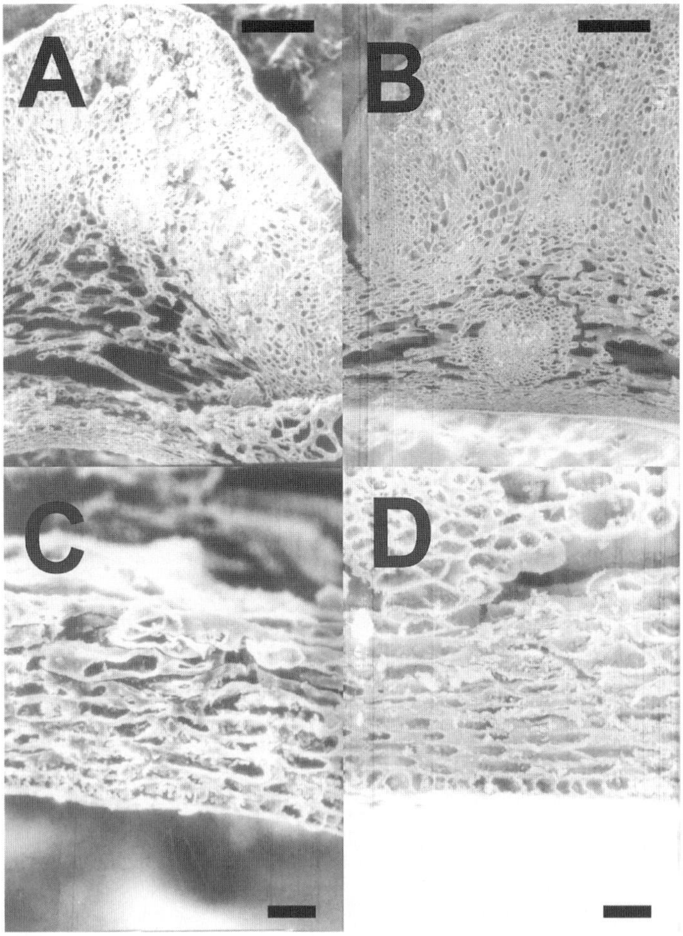

Fig. 8.5. Structure of the skin in 'Wai Chee' ('Huaizhi') (A and C) and 'No Mai Chee' ('Nuomici') (B and D) litchi prior to rapid aril expansion in Guangdong, southern China (56 days after anthesis). Exocarp (A and B, bar = 100 μm) and endocarp (C and D, bar = 10 μm).

reveal differences in the endocarp of the two cultivars (Fig. 8.5). The structure of the skin only partially explains differences in its mechanical properties and cracking in litchi cultivars. Other mechanisms need to be explored.

Cell walls

The mechanical strength of the skin is largely due to the cell walls. Load-carrying materials in the walls include cellulose integrated into microfibril networks and hemicellulose and pectin within the wall matrix (Brett and Waldron, 1990). Huang *et al.* (1999, 2001) found that higher concentrations of cellulose, hemicellulose and water-insoluble pectin in 'Wai Chee' than in 'No Mai Chee' contributed to cracking resistance. Corresponding assays carried out by Li (2001) showed that wall-degrading enzymes such as cellulase and pectinase were more active in the skin of 'No Mai Chee' than in 'Wai Chee'. Applications of calcium and gibberellin, which reduce the activity of cellulase, reduced cracking (Peng *et al.*, 2001).

Plant growth substances

Various authors have studied the relationship between cracking and endogenous growth substances. Sharma and Dhillon (1986, 1988) found that the concentrations of abscisic acid (ABA) were higher in the pericarp, seed and aril of cracked fruit than in normal fruit, while the concentrations of gibberellin (GA_3) were higher in the seeds at the critical period of cracking. A number of studies (Suryanarayana and Das, 1971; Kanwar and Nijjar, 1976; Sharma and Dhillon, 1987; Chandel and Kumar, 1995; Bhat *et al.*, 1997) suggested that auxins (2,4-dichlorophenoxyacetic acid and naphthyl acetic acid) at concentrations lower than 40 mg/l reduced cracking. Gibberellin (GA_3) at 10–50 mg/l was partially effective (Suryanarayana and Das, 1971; Sharma and Dhillon, 1987; Sinha *et al.*, 1999; Peng *et al.*, 2001). Ou (1988) suggested that GA increased the critical cracking turgor and pericarp tensile strength in 'Nuomici'. The results of Peng *et al.* (2001) indicated that GA reduced cracking by lowering the activity of cellulase. Shrestha (1981) reported that ethephon sprays at 10 mg/l reduced cracking from 12% (control) to 6% in 'Early Large Red'.

Weather

High temperatures or drought during fruit growth were associated with abnormal skin development and increased cracking in India (Kanwar *et al.*, 1972a,b; Kanwar and Nijjar, 1984a,b; Sharma and Ray, 1987; Chandel and Sharma, 1992) and South Africa (Joubert, 1970). The lower incidence of cracking in late cultivars, such as 'Hong Kong' in India, was attributed to aril growth occurring after the main period of high temperatures (Kanwar and Nijjar, 1984b). Menzel *et al.* (1995) reported that drought reduced fruit set and increased fruit cracking by 30–40% in South Africa. It has also been noted that the skin cracked after continuous rain or heavy irrigation, when fruit had absorbed a large volume of water (Joubert, 1970). In China, litchis develop during the rainy season, and fruit crack during periods of high humidity and heavy rain when the aril is expanding (Fig. 8.6). It was suggested that higher water availability led to excessive water uptake by the aril, which increased fruit turgor and thus cracking (Li *et al.*, 1992, 2001; Li, 1999).

Cracking is common when droughted trees are re-watered by irrigation or rain (Joubert, 1970; Huang, 1989; Huang *et al.*, 1994; Li and Huang, 1994; Menzel *et al.*, 1995). Joubert (1970) suggested that drought during the early stages of fruit development led to abnormal skin development, resulting in a pericarp that cannot stretch to accommodate the expanding aril. Drought also hardened and reduced the elasticity of the skin (Kanwar *et al.*, 1972b; Li and Huang, 1994). These changes were related to the loss of calcium from the pericarp and a direct inhibition of pericarp development (Li and Huang, 1994).

The author's recent study showed that the skin of 'No Mai Chee' had a higher activity of wall-bound peroxidase than 'Wai Chee'. This enzyme irreversibly catalyses the formation of phenolic cross-links between wall polymers, and leads to the loss of cell wall extensibility (Brett

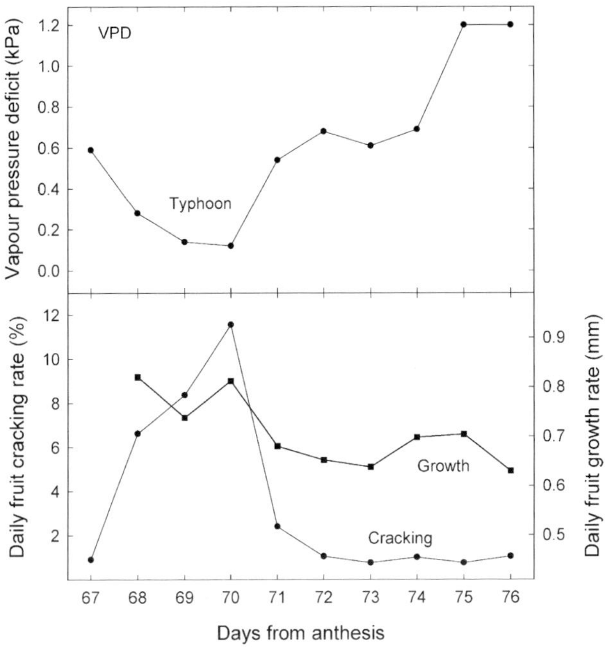

Fig. 8.6. Changes in daily cracking rate and fruit growth rate in 'No Mai Chee' ('Nuomici') litchi in Guangdong, southern China. Vapour pressure deficit (VPD) is also shown. A typhoon occurred between days 67 and 71. Daily cracking rate is the number of cracked fruit on day_n divided by the number of fruit on day_{n-1} (Li *et al.*, 1992).

and Waldron, 1990). Drought also increased the concentration of soluble solutes in the aril, which increased the growth potential of the aril through lower osmotic potential (Li and Huang, 1994). Heavy rain after a drought caused rapid growth of the aril, exerting a higher turgor against the less extensible skin (Li and Huang, 1994).

Nutrients

Deficiencies in nutrients can affect fruit development in litchi (Joubert, 1970). Much of the information about nutritional aspects of cracking has come from comparative analyses between cracked and normal fruit and from responses to nutrient applications.

Calcium plays an important role in cell wall structure and provides ionic bonds between pectins, which contribute to the cohesiveness between cells and the strength of cell walls (Brett and Waldron, 1990). Several studies have shown that the skin of cracked fruit had lower concentrations of calcium than those of normal fruit in the same cultivar (Sanyal *et al.*, 1990; Li and Huang, 1995; Li *et al.*, 1999). Orchards with a high incidence of cracking had lower concentrations of exchangeable calcium in the soil. In contrast, cracking in 'No Mai Chee' was inversely related to leaf calcium (Li *et al.*, 1999). In the laboratory, calcium hydroxide at 4 g/l increased critical cracking turgor (0.25 MPa vs. 0.17 MPa) and pericarp tensile strength (111 g/mm^2 vs. 78 g/mm^2) in 'No Mai Chee' (Ou, 1988). Other sources of calcium also reduced cracking in orchard trees (Li and Huang, 1995; Li *et al.*, 1999; Peng *et al.*, 2001). However, Sanyal *et al.* (1990) found no relationship between cracking and the concentration of calcium in the skin of different cultivars.

In situ X-ray microanalysis of cell walls revealed that the concentrations of calcium in the walls of the endocarp, mesocarp and epicarp of 'Wai Chee' were higher than in 'No Mai Chee' (Huang *et al.*, 1999). However, the skin of 'No Mai Chee' had a large number of cells in the epidermis and a high concentration of calcium in the cytoplasm (Huang *et al.*, 2001). This suggested that the lower concentration of cell wall calcium in 'No Mai Chee' was not due to a shortage of calcium, but to poor remobilization.

The concentration of boron was also lower in the pericarp of cracked fruit than in normal fruit, whereas there was no relationship with leaf boron (Li and Huang, 1995). Responses to boron applications have been variable (Misra and Khan, 1981; Sinha *et al.*, 1999). Applications of zinc and copper reduced cracking in India (Misra and Khan, 1981; Sharma and Dhillon, 1987; Sinha *et al.*, 1999). Cracking was negatively correlated with the concentration of nitrogen in the leaves, whereas the opposite was true for nitrogen in the soil. Phosphorus and magnesium were also found to be lower in orchards with severe cracking problems (Li and Huang, 1995). The roles of these nutrients in fruit cracking require further research.

Pests and diseases

Injury to the skin after early cell division will induce cracking during aril expansion (Joubert, 1970). Damage by pests such as fruit fly, *Ceratitis capitata* and *C. rosa*, and litchi moth, *Cryptophlebia peltastica*, induced cracking in South Africa (Villiers and de Villiers, 1990; Grove, 1999). Fungal diseases such as downy mildew (Zhang *et al.*, 1997) and litchi pepper spot, *Colletotrichum gloeosporioides*, also increase cracking.

Orchard management

The physiological causes of cracking are yet to be fully explained. Research has shown a strong relationship between cracking and the onset of drought after fruit set. Low concentrations of calcium in the skin are also associated with cracking. There are variations between different cultivars, areas, seasons and weather conditions. It is recommended that trees should not be subjected to shortages of water and nutrients after flowering. High plant calcium concentrations should also be maintained, excessive irrigation during aril growth should be avoided, and fruit pests and diseases should be controlled. In areas where the incidence of cracking is severe, only cracking-resistant cultivars should be grown.

Sunburn

Sunburn (lesion browning, pericarp necrosis) occurs in litchis grown in India, South Africa, Australia and Thailand (Kanwar *et al.*, 1972b; Joubert, 1986; Sanyal *et al.*, 1990; Menzel *et al.*, 2002; S. Rasananda, 2002, Thailand, personal communication) but is rare in China. It is more common in off-season than in traditional longan crops in Thailand (S. Rasananda and S. Kooaiyakul, 2002, Thailand, personal communication) and China. Joubert (1986) found that sunburn was initiated in the mesocarp and then spread to the epicarp and endocarp. The pericarp of affected fruit had higher activities of polyphenol oxidase, peroxidase and ascorbic acid oxidase than normal fruit, suggesting a possible physiological link.

Varieties differ markedly in susceptibility to sunburn (Chadha and Rajpoot, 1969; Kanwar *et al.*, 1972a; Sharma and Roy, 1987; Sanyal *et al.*, 1990). In India, early cultivars such as 'Deshi', 'Early Large Red' and 'Muzaffarpur' are more affected than late cultivars such as 'Bedana' and 'Nafarpal' (Sanyal *et al.*, 1990). Fruit on shaded branches also suffers less damage than those in the sun (Kanwar *et al.*, 1972b).

A higher incidence of sunburn was associated with temperatures above 38°C and relative humidities below 60% in India (Kanwar *et al.*, 1972b; Joubert, 1986). In China, litchi fruit development coincides with the wet season and sunburn is rare. The traditional crop of longan in Thailand grows under similar conditions and experiences less sunburn than off-season fruit (S. Kooaiyakul, 2002, Thailand, personal communication). Affected fruit had lower concentrations of calcium in the pericarp (22 vs. 29 mg/g dry weight in 'Deshi') (Sanyal *et al.*, 1990), although the benefits from calcium sprays are yet to be determined. Irrigation may keep the tree and fruit cool and might be beneficial under some circumstances (Menzel *et al.*, 2002). Paper bags covering the fruit clusters increased humidity, cooled the fruit and decreased the incidence of sunburn in South Africa (Oosthuizen, 1989).

Stunted fruit disorder

This disorder has not yet been fully documented, but is found in many orchards in South-east Asia. It is especially severe, affecting more than 50% of the off-season longan crop in China, when fruit develop during the hot, dry season.

Stunted fruit are significantly smaller than normal ones. In litchi, normal fruit turn green soon after the flowers are shed, whereas affected fruit remain green-yellow. In longan, the stunted fruit become hardened, with a rigid skin. Affected fruit in both crops often crack during aril growth. The cause of stunted fruit is not known.

Conclusions

The incidence of fruit disorders in litchi and longan varies with cultivar, season and agronomy. A full understanding of the physiology of the problems is yet to be developed. Fruit cracking occurs in litchi when the pressure of the expanding aril exceeds the tensile strength and the extensibility of the pericarp. The latter is related to the structure of the pericarp and cell-wall components, which differ between cultivars. Shortage of nutrients, especially of calcium, and drought and high temperatures, which retard the growth of the skin, increase cracking. High humidity and heavy rain during aril growth lead to excessive absorption of water by the aril and may aggravate cracking. Both species can be affected by sunburn and stunted fruit disorder, which can be significant problems in some orchards in South-east Asia. Little information is available regarding the underlying causes of these problems or methods that might overcome them.

Acknowledgements

I thank Huibai Huang, Raffi Stern, Shmuel Gazit, Prasert Anupunt, Sainarong Rasananda and Surachart Kooaiyakul for information on fruit disorders. The author's studies on litchi cracking were supported by the National Natural Science Foundation of China (No. 39770529 and 30000104) and the Guangdong Natural Science Foundation (No. 980156 and 990687).

References

Bhat, S.K., Raina, B.L., Chogtu, S.K. and Muthoo, A.K. (1997) Effect of exogenous auxin application on fruit drop and cracking in litchi (*Litchi chinensis* Sonn.) cv. Dehradun. *Advances in Plant Sciences* 10, 83–86.

Brett, C. and Waldron, K. (1990) *Physiology and Biochemistry of Plant Cell Walls.* Unwin Hyman, London, 194 pp.

Budathoki, K. (2002) Lychee production in Nepal. In: Papademetriou, M.K. and Dent, F.J. (eds) *Lychee Production in the Asia-Pacific Region.* Food and Agricultural Organization of the United Nations, Bangkok, Thailand, pp. 81–93.

Chadha, K.L. and Rajpoot, M.S. (1969) Studies on floral biology, fruit set and its retention and quality of some litchi varieties. *Indian Journal of Horticulture* 16, 124–129.

Chandel, S.K. and Kumar, G. (1995) Effect of irrigation frequencies and foliar spray of NAA and micronutrient solutions on yield and quality of litchi (*Litchi chinensis* Sonn.) cv. Rose Scented. *Advances in Plant Sciences* 8, 284–288.

Chandel, J.S. and Sharma, N.K. (1992) Extent of fruit cracking in litchi and its control measures in Kangra Valley of Himachal Pradesh. *South Indian Horticulture* 40, 74–78.

Chen, H.B. and Huang, H.B. (2001) China litchi industry: development, achievement and problems. *Acta Horticulturae* 558, 31–39.

Grove, T. (1999) The litchi moth: a literature review. *Cryptophlebia peltastica* (Meyrick) (Lepidoptera: Tortricidae). *Neltropika Bulletin* 305, 30–31.

Huang, H.B. (1989) The water relations problem of fruit trees in Guangdong. *Journal of South China Agricultural University* 10, 33–39.

Huang, H.B. (2001) Towards a better insight into the development of the arillate fruit of litchi and longan. *Acta Horticulturae* 558, 185–192.

Huang, H.B. and Xu, J.K. (1983) The developmental patterns of fruit tissues and their correlative relationships in *Litchi chinensis* Sonn. *Scientia Horticulturae* 19, 335–342.

Huang, H.B., Jiang, S.Y. and Xie, C. (1983) The initiation of aril and ontogeny of fruit in *Litchi chinensis* Sonn. *Journal of South China Agricultural College* 4, 78–83.

Huang, H.B., Gao, F.F. and Li, J.G. (1994) Fruit expansion growth in relation to water uptake under climatic changes. *Acta Horticulturae* 21, 124–128.

Huang, X.M. and Huang, H.B. (1998) Using wet strength tester to measure the mechanical strength of plant cell walls. *Plant Physiology Communications* 34, 207–209.

Huang, X.M., Wang, H.C., Gao, F.F. and Huang, H.B. (1999) A comparative study of the pericarp of litchi cultivars susceptible and resistant to fruit-cracking. *Journal of Horticultural Science and Biotechnology* 74, 351–354.

Huang, X.M., Li, J.G., Wang, H.C., Huang, H.B. and Gao, F.F. (2001) The relationship between fruit cracking and calcium in litchi pericarp. *Acta Horticulturae* 558, 209–216.

Joubert, A.J. (1970) *The Litchi.* Government Printer, Pretoria, South Africa, 21 pp.

Joubert, A.J. (1986) Litchi. In: Monselise S.P. (ed.) *Handbook of Fruit Set and Development.* CRC Press, Boca Raton, Florida, pp. 233–246.

Kanwar, J.S. and Nijjar, G.S. (1976) Fruit development in relation to cracking in litchi (*Litchi chinensis* Sonn.) at Gurdaspur (Punjab). *Haryana Journal of Horticultural Sciences* 5, 164–168.

Kanwar, J.S. and Nijjar, G.S. (1984a) Comparative evaluation of fruit growth in relation to cracking of fruits in some litchi cultivars. *Punjab Horticultural Journal* 24, 79–82.

Kanwar, J.S. and Nijjar, G.S. (1984b) A note on fruit growth in relation to cracking in litchi (*Litchi chinensis* Sonn.) cv. 'Hong Kong'. *Haryana Journal of Horticultural Sciences* 11, 72–73.

Kanwar, J.S., Nijjar, G.S. and Rajpoot, M.S. (1972a) Fruit growth studies in litchi (*Litchi chinensis* Sonn.) at Gurdaspur (Punjab). *Punjab Horticultural Journal* 12, 146–151.

Kanwar, J.S., Rajpoot, M.S. and Bajwa, M.S. (1972b) Sun-burning and skin-cracking in some varieties of litchi (*Litchi chinensis* Sonn.) and the factors affecting them. *Indian Journal of Agricultural Sciences* 42, 772–775.

Li, J.G. (1999) *Litchi Cultivation for High Yield, High Quality and High Profit.* China Agricultural University Press, Beijing, China, 186 pp.

Li, J.G. (2001) Ontogenetic studies of fruit development and minimizing cracking in *Litchi chinensis* Sonn. PhD Thesis, South China Agricultural University, Guangzhou, China.

Li, J.G. and Huang, H.B. (1994) Analysing the cause of litchi fruit cracking increased by rain after a long period of drought. *Advances in Horticultural Science* 1, 361–364.

Li, J.G. and Huang, H.B. (1995) Physico-chemical properties and peel morphology in relation to fruit-cracking susceptibility in litchi fruit. *Journal of South China Agricultural University* 16, 84–89.

Li, J.G. and Huang, X.M. (2001) Effects of anthesis time on fruit development, drop and cracking. *Guangdong Horticulture* 2, 13–15.

Li, J.G., Huang, H.B., Yuan, R.C. and Gao, F.F. (1992) Litchi fruit cracking in relation to fruit growth and water-uptake kinetics. *Journal of South China Agricultural University* 13, 129–135.

Li, J.G., Gao, F.F., Huang, H.B., Tan, Y.W. and Luo, J.T. (1999) Preliminary studies on the relationship between calcium and fruit cracking in litchi fruit. *Journal of South China Agricultural University* 20, 45–49.

Li, J.G., Huang, H.B., Gao, F.F., Huang, X.M. and Wang, H.C. (2001) An overview of litchi fruit cracking. *Acta Horticulturae* 558, 205–208.

Menzel, C.M., Oosthuizen, J.H., Roe, D.J. and Doogan, V.J. (1995) Water deficits at anthesis reduce CO_2 assimilation and yield of lychee (*Litchi chinensis* Sonn.) trees. *Tree Physiology* 15, 611–617.

Menzel, C.M., Bagshaw, J., Campbell, T., Greer, N., Noller, J., Olesen T., Waite, G.K., Kernot, I., Chapman, L. and Ridgen, P. (2002) *Lychee Information Kit.* Queensland Department of Primary Industries, Nambour, Australia.

Misra, R.S. and Khan, I. (1981) Effect of 2,4,5-trichlorophenoxyacetic acid and micronutrients on fruit size, cracking, maturity and quality of litchi cv. Rose Scented. *Progressive Horticulture* 13, 87–90.

Mitra, S. and Ghosh, B. (1991) Description and performance of some litchi cultivars in West Bengal, India. *Yearbook of the Australia Lychee Growers' Association* 1, 64–79.

Oosthuizen, J.H. (1989) Is covering of litchi clusters worthwhile? *Information Bulletin of the Citrus and Subtropical Fruit Research Institute* 199, 10–12.

Ou, L.X. (1988) Studies on fruit cracking in *Litchi chinensis* Sonn. MS thesis, South China Agricultural University, Guangzhou, China.

Peng, J., Xi, J.B., Tang, X.D., Wang, Y.G., Si, X.M. and Chen, J.S. (2001) Effect of Ca $(NO_3)_2$ and GA spray on leaves on the fruit cracking of 'Nuomici' litchi. *Acta Horticulturae Sinica* 28, 348–350.

Sanyal, S.J.R., Hasan, A., Ghosh, B. and Mitra, S.K. (1990) Studies on sun-burning and skin cracking in some varieties of litchi. *Indian Agriculturist* 34, 19–23.

Sethpakdee, R. (2002) Lychee production in Thailand. In: Papademetriou, M.K. and Dent, F.J. (eds) *Lychee Production in the Asia-Pacific Region.* Food and Agricultural Organization of the United Nations, Bangkok, Thailand, pp. 106–113.

Sharma, S.B. and Dhillon, B.S. (1986) Endogenous level of gibberellins in relation to fruit cracking in litchi (*Litchi chinensis* Sonn.). *Journal of Research Punjab Agricultural University* 23, 432–434.

Sharma, S.B and Dhillon, B.S. (1987) Effect of zinc sulphate and growth regulators on fruit cracking in litchi. *Indian Journal of Horticulture* 44, 57–59.

Sharma, S.B. and Dhillon, B.S. (1988) Endogenous levels of abscisic acid in relation to fruit cracking in litchi (*Litchi chinensis* Sonn.). *Agricultural Science Digest* 8, 55–58.

Sharma, S.B. and Ray, P.K. (1987) Fruit cracking in lychee: a review. *Haryana Journal of Horticultural Sciences* 16, 11–15.

Sharma, S.B. and Roy, P.K. (1987) Flowering and fruiting behaviour of some litchi cultivars. *Haryana Journal of Horticultural Sciences* 16, 168–174.

Shrestha, G.K. (1981) Effects of ethephon on fruit cracking of lychee (*Litchi chinensis* Sonn.). *HortScience* 16, 498.

Sinha, A.K., Singh, C. and Jain, B.P. (1999) Effect of plant growth substances and micronutrients on fruit set, fruit drop, fruit retention and cracking of litchi cv. Purbi. *Indian Journal of Horticulture* 56, 309–311.

Suryanarayana, V. and Das, R.C. (1971) Effect of 2,4-D, NAA, GA and 2,4,5-T on fruit weight and fruit-cracking in litchi (*Litchi chinensis* Sonn.) var. Muzaffarpur. *Plant Science* 3, 126–128.

Villiers, E.A. and de Villiers, E.A. (1990) The control of fruit flies on litchi. *Yearbook of the South African Litchi Growers' Association* 3, 29–30.

Wang, H.C. (1998) Studies on the cracking susceptibility of litchi (*Litchi chinensis* Sonn.) pericarp. MS thesis, South China Agricultural University, Guangzhou, China.

Wang, H.C., Wei, B.W., Gao, F.F. and Huang, H.B. (2000) Studies on the relation among fruit skin structure, cell division and fruit cracking in litchi (*Litchi chinensis* Sonn.). *Journal of South China Agricultural University* 21, 10–13.

Zhang, Z.W., Yuan, P.Y., Wang, B.Q. and Qui, Y.P. (1997) *Litchi Pictorial Narration of Cultivation.* Pomology Research Institute, Guangdong Academy of Agricultural Sciences, Guangzhou, China, 189 pp.

9 Photosynthesis and Productivity

Christopher M. Menzel
Department of Primary Industries and Fisheries, PO Box 5083, SCMC, Nambour, Queensland 4560, Australia

Introduction

The leaves of litchi and longan capture light and use it in the process of photosynthesis to produce the carbohydrates needed for growth. Other green tissues such as the young fruit can also fix CO_2, although not at the rate achieved in the leaves. Productivity is driven by the amount and distribution of light and nitrogen in the tree, along with water supply and temperature. A proportion of the carbohydrates produced in photosynthesis are used immediately in respiration and in the growth of new tissues, and the rest is stored in the branches to be used later. Litchi and longan obtain most of the carbohydrates needed for fruit growth from photosynthesis that takes place in the leaves situated just below the inflorescences, with the balance mainly coming from assimilates stored in the branches. These assimilates can be sourced from some distance in the tree if there are local shortages of CO_2 fixation.

Prolonged periods of overcast weather, extremes of temperature, and droughts can reduce photosynthesis and productivity in some growing areas. Until recently, most orchards were planted at low density and the trees were left to develop into large specimens. However, most new orchards are planted at high densities and are pruned regularly. These technologies have been widely adopted even though there is still little understanding of tree physiology. Experience with temperate fruit trees suggests that studies on the relationship between yield and the interception of light by the canopy should lead to the development of new management techniques and increased productivity. Higher leaf nitrogen concentrations, which increase CO_2 fixation, may also increase yields, along with the application of certain growth regulators and trunk girdling.

Experimental approaches

Changes in CO_2 fixation during leaf expansion have been tracked to determine when the young leaves begin to contribute assimilates to the tree. There has also been some effort to determine the relative contribution of photosynthesis that takes place in the fruit. Experiments have examined the effects of light and temperature on CO_2 fixation in single leaves and whole trees. These factors are the main environmental variables driving canopy photosynthesis. Girdling, defoliation and fruit thinning have been used to study the relationship between yield, photosynthesis and stored reserves. There have been efforts to develop pruning strategies for high-density orchards that have the potential to maximize light interception. More information is available for litchi than for longan, although it is not known whether the two crops have similar potential productivity. Limited information suggests that well-grown litchi orchards can be just as productive as other tropical fruit enterprises.

Productivity

Detailed information on productivity is not readily available for either crop. Yields can be presented on a tree basis or, preferably, on an area basis. Some statistics are available from China, Viet Nam and Thailand, showing the area under cultivation and total production, with yields per hectare calculated. There are problems with this approach when the data include orchards that have not come into full bearing, although a few industries list bearing and non-bearing trees separately.

Average yields for litchi in China in 1998 in the three main growing areas ranged from 1.3 to 3.0 t/ha, with Fujian being more productive than Guangdong or Guangxi (Table 9.1). Young non-bearing orchards were excluded from this analysis. Average productivity of longan was about half that of litchi, although the data included non-bearing orchards. A separate analysis by Liu and Ma (2001) showed that average yields of longan from 1987 to 1995 in Guangdong, Guangxi and Fujian ranged from 2.2 to 11.9 t/ha, with Fujian being slightly more productive. Yields increased over time, reflecting larger trees or possibly better agronomy. These data did not indicate biennial bearing across the whole industry.

Average yields for the two crops in Thailand in 1993 ranged from 3.2 to 6.8 t/ha, with longan more productive than litchi (Table 9.1). Chiang Mai produced the highest litchi yields, while Nan was the most productive longan-growing district. Subhadrabandhu and Yapwattanaphun (2001) showed that average yields for litchi for 1991–1995 ranged from 2.8 to 3.9 t/ha, with a mean of 3.4 t/ha. A variation of 20% around the mean indicated biennial bearing. Longan was slightly more productive, with yields for 1989–1998 ranging from 2.5 to 7.3 t/ha. There was a 50% fluctuation around the general mean of 5.1 t/ha, indicating substantial biennial bearing. Variations in whole-industry yields are likely to reflect seasonal weather such as frost or warm temperatures at flower initiation, whereas differences within an orchard probably reflect changes in tree agronomy, especially the incidence of major pests.

Choo (2000) presented data on the productivity of longan in Taiwan, with 53,385 t produced from 11,808 ha in 1998, indicating an average yield of 4.5 t/ha. There was strong biennial bearing, with yields of up to three times higher than this value in some years.

In Viet Nam, average yields of litchi in 1999 were 1.4 t/ha from 35,000 ha (Papademetriou and Dent, 2002). Average yields for longan were 9.3 t/ha in the Mekong Delta, 10.0 t/ha around Hanoi and 5.2 t/ha in the South-East Region (Choo, 2000). Total production was 265,000 t from 41,600 ha, equivalent to a mean yield of 6.5 t/ha.

Productivity can also be estimated from published research (Table 9.2). For 10-year-old litchis, average yields ranged from 32 to

Table 9.1. Productivity of litchi and longan in China and Thailand. Data for litchi in China from 1998 include bearing area only (Papademetriou and Dent, 2002). Data for longan in China from 1997 include non-bearing and bearing orchards (Liu and Ma, 2001). Data for litchi and longan in Thailand from 1993 include non-bearing and bearing orchards (Subhadrabandhu and Yapwattanaphun, 2001).

	Litchi			Longan		
	Area (ha)	Production (t)	Yield (t/ha)	Area (ha)	Production (t)	Yield (t/ha)
China						
Guangdong	110,700	327,000	3.0	119,600	124,700	1.0
Guangxi	80,260	106,700	1.3	210,100	133,300	0.6
Fujian	17,210	91,900	5.3	90,300	104,800	1.2
Thailand						
Lamphun				11,008	44,031	3.7
Chiang Mai	4,873	28,873	5.9	6,436	40,932	6.8
Chiang Rai	2,116	8,554	4.0	2,252	11,700	5.2
Phayao	488	2,735	5.6	922	4,161	4.5
Nan	157	502	3.2	1,129	9,486	8.4

Table 9.2. Productivity of litchi in India, Israel, South Africa and Australia.

Experiment	Age of trees (years)	No. trees/ha	Yield (kg/tree)	Yield (t/ha)
Nutrition expt in India (Lal *et al.*, 1999)	14	100	71	7.1
Growth regulator expt in Israel (Stern *et al.*, 1995)	11	278	70	19.5
Growth regulator expt in Israel (Stern *et al.*, 2000)	8	500	32	16.0
Nutrition expt in South Africa (Koen *et al.*, 1981a)	12	66	55	3.6
Nutrition expt in South Africa (Koen *et al.*, 1981b)	12	66	67	4.4
Irrigation expt in South Africa (Menzel *et al.*, 1995a)	10	100	51	5.1
Nutrition expt in Australia (Menzel *et al.*, 1994)	10	125	56	7.0
Pruning expt in Australia (Hieke *et al.*, 2002a)	10	300	52	15.6
Defoliation expt in Australia (Hieke *et al.*, 2002b)	8	300	36	10.8

71 kg/tree or 3.6–19.5 t/ha. High yields in Israel are related to intensive tree management, including strategic droughting, pruning and growth regulators. These yields are higher than the averages estimated from total production and the total area under cultivation, even taking into account non-bearing orchards (Table 9.1). Research is required to raise productivity in many countries in order to bridge the large gap between actual and potential yields.

Photosynthesis and plant development

Higher plants capture light through photosynthesis in green tissues containing chlorophyll, and use it to convert CO_2 and water into sugars and other compounds required for growth and development. Leaves are the main source of photosynthesis, but it also occurs in green stems, flowers and fruit. Respiration usually exceeds gross photosynthesis in young leaves, so these tissues must obtain their carbohydrates from the rest of the plant. The rate of photosynthesis increases as the leaves expand, with the mature leaves eventually contributing to the carbon economy of the plant. In temperate fruit trees, the maximum rate of CO_2 assimilation occurs when the leaves are fully expanded, whereas in tropical fruit species it occurs much later (see Kennedy and Johnson, 1981, in apple *Malus domestica*; Schaffer *et al.*, 1991, in avocado *Persea americana*). Photosynthesis is then maintained at this maximum rate before it declines when the leaves start to senesce. This period lasts for several weeks in temperate fruit and for several years in some tropical species.

The relationship between gas exchange and shoot development in litchi was investigated by studying the growth of 'Wai Chee' seedlings in a glasshouse (Hieke *et al.*, 2002c). Growth in the seedlings was confined to a single axis with leaves, stem and roots, and without flowers and fruit, which would affect the relationship between photosynthesis and leaf expansion in orchard trees. Shoots and leaves expanded in a sigmoid pattern to 80 mm and 500 cm^2 over 20 days in the glasshouse. Net CO_2 assimilation was 0.3 $\mu mol/m^2/s$ in soft red leaves, 1.6 $\mu mol/m^2/s$ in red-green leaves, 2.7 $\mu mol/m^2/s$ in light-green leaves and 5.8 $\mu mol/m^2/s$ in dark-green leaves. The light-green leaves were 75% expanded but had only half the photosynthesis of dark-green leaves. These differences in CO_2 fixation were associated with higher concentrations of chlorophyll in the older leaves than in the younger leaves.

Hieke *et al.* (2002d) also investigated the changes in photosynthesis during leaf expansion in orchard trees. In the first experiment, changes in leaf area and gas exchange of 'Bengal' were recorded as the leaves expanded over 50 days in spring, while in the second experiment, CO_2 fixation was studied on different leaves on terminal branches over a few days. Leaves expanded in a sigmoid pattern over 50 days and were half-expanded after 30 days. Net CO_2 assimilation, *A*, increased from –4.1 to 8.3 μmol/m²/s as the leaves changed from soft and red, to soft and light green, and then to hard and dark green. At the same time, dark respiration, R_d, decreased from 5.0 to 2.0 μmol $CO_2/m^2/s$. Net CO_2 assimilation was above zero when the leaves were half fully expanded. Chlorophyll concentrations increased from 0.7 mg/g fresh weight in young red leaves to 10.3 mg/g fresh weight in dark-green leaves. Stomatal conductance, g_s, changed similarly from 0.09 to 0.47 mol $H_2O/m^2/s$. Results from the first experiment are shown in Fig. 9.1.

There was a strong correlation between *A* and chlorophyll (r^2 = 99%) and g_s (r^2 = 81%), indicating that gas exchange followed the course of chlorophyll synthesis and stomatal opening, both of which lagged behind leaf expansion. Gross photosynthesis, P_g, calculated by summing *A* and R_d, was 2 μmol $CO_2/m^2/s$ in young leaves and 11 μmol $CO_2/m^2/s$ in old leaves. The results were similar when leaves were sampled over the season or for just a few days. It was concluded that litchi leaves do not contribute energy to the tree until they are about half fully expanded, and do not reach maximum rates of photosynthesis until after they are fully expanded and dark green.

In other experiments, Hieke (2000) investigated whether the position of the leaves affected the relationship between photosynthesis and leaf expansion in 'Bengal' trees. Measurements were made on three different shoots: leaves red to light green (Stage 1); leaves light green-brown to light green (Stage 2); and leaves dark green and mature (Stage 3). Leaves were sampled from the bottom (position 1) to the top (position 6) of each shoot. The effects of leaf position on gas exchange were smaller than the effects of shoot development (Table 9.3). Pooled across shoot development, *A*, g_s and chlorophyll were greater in leaves in positions one and two (lower, older leaves) than those in positions five and six (upper, younger leaves). Values were also higher in the shoots with dark-green leaves than those with red or light-green leaves. The shoots did not contribute to the tree until the leaves were light green and 50% of their final size.

Studies on longan in China showed that young leaves began to export photoassimilates when they were 7–10 days old, although maximum rates of CO_2 fixation did not occur for another 20 days (Liao *et al.*, 1996). Photosynthesis remained at close to the maximum rate (6–8 μmol $CO_2/m^2/s$) for at least another 60 days.

The presence or absence of fruit can influence the rate of leaf photosynthesis (Roe *et al.*, 1997). These authors investigated the effects of different numbers of leaves and fruit on gas exchange in girdled 'Tai So' branches in South Africa. Each branch had 0, 5, 10, 20 or 30 leaves and 0, 5, 10, 20 or 30 fruit. Photosynthesis peaked half way through fruit development and was higher when there were few leaves supporting the fruit, indicating a strong demand for carbohydrates by the crop. Net CO_2 assimilation was 18 μmol/m²/s in branches with 0.2 leaves per fruit and 5 μmol/m²/s in branches with no fruit. Averaged across the season, gas exchange varied by a factor of four across the range of treatments. Similar results were shown for avocado in Florida, where CO_2 fixation was halved when fruit were removed from the branches (Schaffer *et al.*, 1987).

Stomata are present in the outer layers of many fruit, but usually with only 1–10% of the frequency of that in leaves (Blanke and Lenz, 1989). The number of stomata is set at anthesis and remains constant, so stomatal density decreases as the fruit expand. Many species also contain chlorophyll in the outer layers of the fruit skin, which permits photosynthesis when they are exposed to sunlight. The rate of CO_2 assimilation is usually greatest at fruit set and declines as the fruit mature, with fruit contributing between 5% and 15% of their own carbon requirements (Whiley *et al.*, 1992).

Litchi fruit contribute only a small proportion of the carbon needed for their growth (Hieke *et al.*, 2002d). Fruit growth in 'Kwai May Pink' and 'Wai Chee' in subtropical Australia was

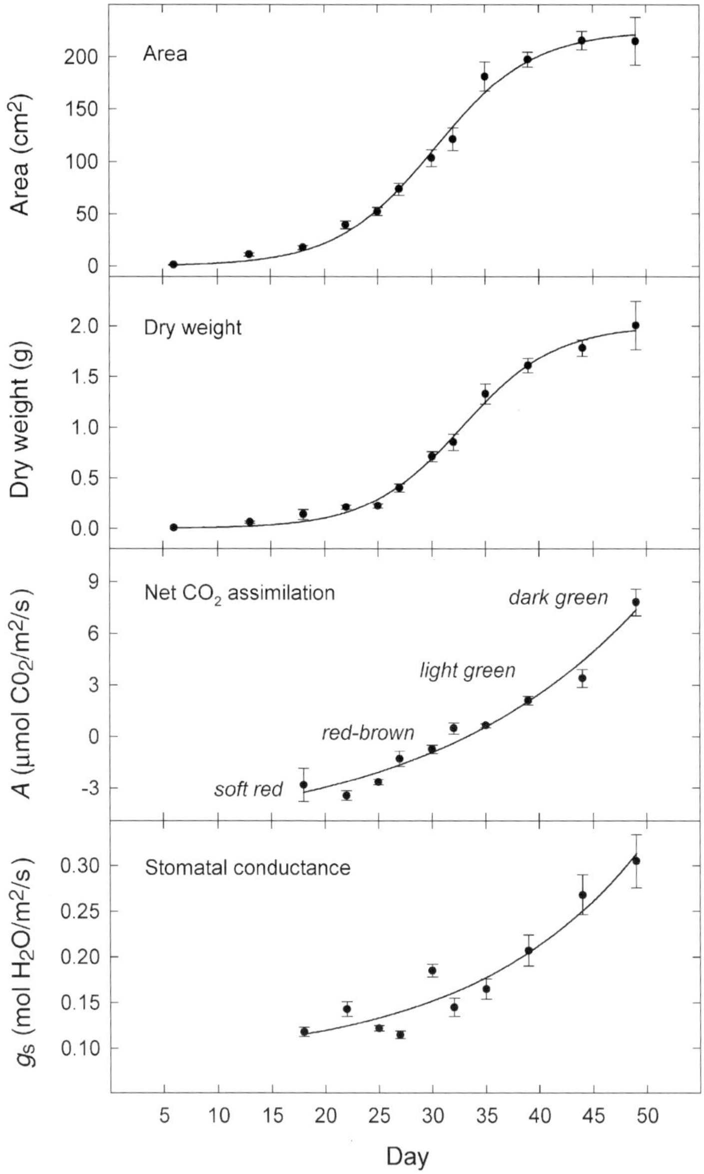

Fig. 9.1. Gas exchange during leaf expansion of 'Bengal' litchi in Australia. General appearance of the leaves indicated (from Hieke *et al.*, 2002d, with permission).

sigmoidal, with maximum fresh weight (29 g), dry weight (6 g) and fruit surface area (39 cm^2) attained after 97–115 days (Fig. 9.2). Fruit CO_2 output in the light, R_l, decreased from fruit set to fruit maturation (10 to 3 μmol/m^2/s), along with CO_2 output in the dark, R_d, (20 to 3 μmol/m^2/s). Photosynthesis never exceeded respiration; however, gross photosynthesis, P_g, the difference between R_l and R_d (4–8 μmol CO_2/m^2/s), was greatest in young green fruit. About 90% of the carbon required for fruit growth was accounted for in the dry matter of the fruit, with the remainder used in respiration. A carbon balance constructed by calculating total respiration and gross photosynthesis indicated that fruit photosynthesis contributed 3% of the

assessed in terms of the changes in dark respiration (R_d), light compensation point (LCP), apparent quantum yield (Φ), maximum rate of CO_2 assimilation (A_{max}) and light saturation point ($PPFD_{sat}$). Changes in g_s and C_i were measured to determine whether the lower photosynthesis under shade was due to closure of the stomata or an effect on the chemistry of CO_2 fixation. Light compensation point was the value of PPFD where R_d equalled A, whereas apparent quantum yield, a measure of photosynthesis efficiency, was determined from the slope of A vs. PPFD from 0 to 100 μmol quanta/m²/s. Light saturation point was calculated as the PPFD associated with 90% of A_{max}. The PPFD associated with a 50% reduction in A_{max} ($PPFD_{50\%}$) was also calculated. Changes in CO_2 assimilation with light were modelled as:

$$A = A_{max}\,[1 - \exp(-b * PPFD)] \qquad (3)$$

where A is net CO_2 assimilation, A_{max} is maximum rate of net CO_2 assimilation, PPFD is irradiance or photosynthetic photon flux density, and b is a constant.

There were three main components of the light-response curve: a linear part at low light where photosynthesis was proportional to irradiance, a non-linear part above the initial slope but below light saturation, and the light-saturated region (Fig. 9.3). Maximum CO_2 assimilation was similar in the two cultivars, whereas it was 50% lower in the potted plants. The irradiance required to saturate photosynthesis in potted plants was only 83% of that required for the orchard trees. The PPFD associated with a 50% reduction in maximum photosynthesis, along with R_d and Φ, was slightly higher in orchard trees than in the potted plants. The average LCP was 23 μmol quanta/m²/s. Whiley *et al.* (1999) quoted values of LCP of 30 μmol quanta/m²/s for field-grown avocado and mango (*Mangifera indica*), although these calculations did not take into account respiration in the rest of the tree. Values of Φ for litchi were similar to those recorded for avocado and mango. Whiley *et al.* (1999) reported lower CO_2 assimilation in potted plants than in orchard trees and suggested that root restriction in the plants led to the accumulation of starch in the leaves and end-product inhibition of photosynthesis. This mechanism probably operates in litchi.

Light can influence photosynthesis by affecting the entry of CO_2 into the leaf or the actual fixation of CO_2 within the cells. Changes in C_i during shading indicate which process is more sensitive. In the experiments of Hieke *et al.* (2002e), C_i increased as PPFD decreased, whereas there was no clear response in g_s. Thus, even under shade there was unrestricted diffusion of CO_2 through the stomata. The effects of shade on photosynthesis were mainly due to changes in leaf biochemistry.

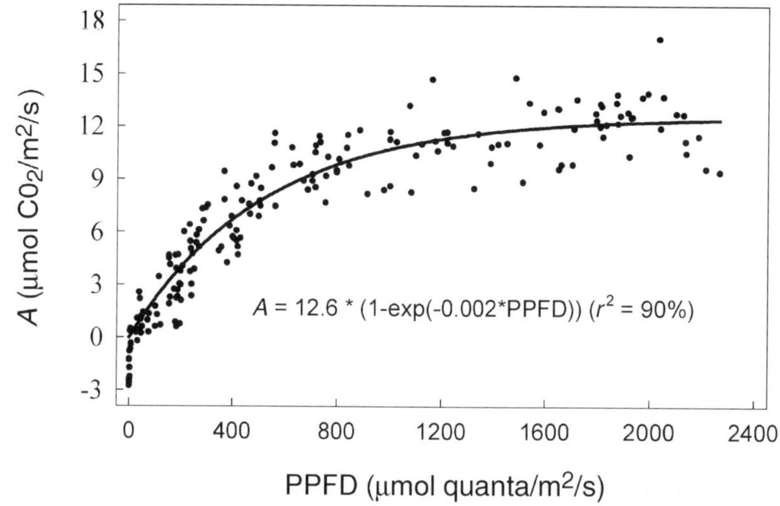

Fig. 9.3. Light response curve for net CO_2 assimilation (A) in 'Kwai May Pink' trees in Australia. Trees were shaded for 1 h before gas exchange was determined (from Hieke *et al.*, 2002e, with permission).

Riddoch *et al.* (1991) indicated that photosynthesis in *Nauclea diderrichi*, a pioneer tree from West Africa, was saturated at a PPFD of 1000 μmol quanta/m^2/s, whereas *Entandrophragma angolense*, a climax species, was saturated at 300 μmol quanta/m^2/s. Photosynthesis in mangosteen, *Garcinia mangostana*, from the shaded understorey forests of Malaysia, grown in 20% full sun, was saturated at 555 μmol quanta/m^2/s (Wiebel *et al.*, 1993). Litchi and *Nauclea diderrichi* exist on the forest floor as shaded seedlings, but require higher light levels for significant CO_2 fixation.

Only limited information is available on photosynthesis in longan. Laboratory experiments in China showed that LCP was 20–30 μmol quanta/m^2/s, while $PPFD_{sat}$ was 600–800 μmol quanta/m^2/s in 20- and 60-day-old leaves (Liao *et al.*, 1996). Photosynthesis in the field peaked at 1000 h, with a midday trough under clear skies, and a second peak at 1600 h, but with no midday depression during cloudy weather. There were seasonal changes, with peak photosynthesis between April and August associated with higher PPFDs, temperatures and soil water contents than at other times of the year.

Assimilate production and distribution

Carbohydrates are important in the growth of woody plants and account for over 65% of the dry matter in tree crops (Kozlowski and Keller, 1966). A proportion of the carbohydrates produced from photosynthesis is used immediately in respiration and growth, while the remainder accumulates as reserves during periods of excess production. The concentrations of carbohydrates vary within the tree and are often higher in the stem and roots than in the leaves. There are also strong seasonal influences in many species, reflecting leaf expansion, flowering and fruit development.

Litchi and longan do not produce new shoots continuously, even in a favourable environment. There is generally a rapid period of shoot elongation and leaf expansion followed by a period of leaf maturation before the next period of shoot growth. In litchi, the duration and interval of growth are strongly related to temperature and irradiation (Batten and Lahav, 1994; Olesen *et al.*, 2002). Under optimum conditions, trees may flush every 6 weeks. Orchard trees usually produce six to eight leaves on each flush, whereas seedlings produce four or five leaves. The last leaf is usually smaller than the others, suggesting competition within the new shoot. There are different optima for leaf initiation and expansion, with more leaves initiated at 25°C than at 30°C (Menzel and Simpson, 1988).

Hieke *et al.* (2002c) studied shoot development in litchi seedlings in a heated glasshouse. 'Wai Chee' seedlings were used rather than orchard trees because growth was limited to roots, stems or leaves without the complication of flowers and fruit. Shoot growth was rhythmic in the glasshouse, with 20 days of flushing and an interval of 10 days over three cycles. The growth of the shoots (S_{length}) and leaves (L_{area}) was described by a logistic model (Turner *et al.*, 1996):

$$S_{length} = S_{max}/\{1 + \exp[-k(t - m)]\} \quad (4)$$

$$L_{area} = L_{max}/\{1 + \exp[-k(t - m)]\} \quad (5)$$

where S_{max} is maximum shoot length (mm), L_{max} is maximum area of the leaves (cm^2), k is a rate constant (per day), m describes the time to reach maximum growth rate, and t is time in days. A high value of the coefficient k indicates that the shoot or leaf reached S_{max} or L_{max} earlier, while a small value of the coefficient m indicates earlier completion of growth. These models were used to assess the impact of treatments applied to alter the production and demand for assimilates in the seedlings (see below).

The concentrations of starch in the lower stem and roots of the seedlings decreased as the young red leaves expanded, and increased as the fully expanded leaves turned dark green. In contrast, the concentrations of starch in the other plant parts did not change with shoot development. If growth exceeds assimilation, starch reserves should decline, whereas if growth is less than assimilation, starch concentrations should increase. Thus, the demand for assimilates appeared to exceed supply when the young leaves were expanding.

Hieke *et al.* (2002c) removed leaves and roots from the seedlings and studied the impact on shoot development. The removal of 50% of the area of each fully expanded leaf had little effect on the expansion of younger leaves, although total plant dry weight was only 60% of that of unpruned seedlings (Table 9.4). Growth of the leaves was at the expense of the stems and roots in the defoliated plants. In contrast, removing half the roots just before bud swelling reduced leaf area by 80%. The young litchi shoots had low rates of CO_2 assimilation until the leaves were fully expanded, and were dependent on photosynthates from the rest of the plant. During leaf expansion, the translocation of carbohydrates to the shoot was at the expense of the roots.

Menzel *et al.* (1995b) studied the concentration of starch in orchard trees in Australia. The concentrations of starch in small and medium branches from 1 to 5 cm in diameter declined during leaf expansion from February to June, reached a maximum before anthesis in September, and then declined during fruit development from October to December. Changes in the leaves, twigs, trunk and roots were much smaller. It was concluded that starch reserves in the small and medium branches were used for leaf and fruit growth. Hieke *et al.* (2002c) also showed that the small and medium branches had the highest concentrations of starch. Starch was higher in branches with fully expanded dark-green leaves than those with young red leaves or fruit.

Small changes in the concentration of starch are likely to be important if they contribute to large changes in absolute weights of starch. It is therefore important to measure the dry weight of a tissue, along with the concentration of starch (Menzel *et al.*, 1995b). About half the starch reserves in a litchi tree were found in the small and medium branches (Table 9.5). This was due to their high concentration of starch, since they accounted for 35% of the tree's dry matter. In contrast, leaves accounted for about 25% of the tree's biomass, but contributed only 8% of the starch. A different pattern has been found for some other fruit trees, with about 40% of the carbohydrate reserves in mandarin, *Citrus reticulata*, in the roots (Goldschmidt and Golomb, 1982).

The relationship between fruit growth and leaf expansion

Leaves and fruit compete for resources in temperate trees, but whether this occurs in tropical species is less clear. Quinlan and Preston (1971) showed that fruit set was increased in apple, *Malus domestica*, when the new shoots were removed 5 days after full bloom, compared with unpruned trees. However, final yields were only 35% of the controls. In contrast, pinching the shoots, by removing the stem above the youngest unrolled terminal leaf, 15 days after full bloom increased yield by 60% compared with controls. This effect was reversed if the shoots were defoliated at the same time. Competition during flowering and early fruit growth limited

Table 9.4. Effects of defoliation on the growth of 'Wai Chee' seedling litchis in a glasshouse. Fifty per cent of the area of each fully expanded leaf was removed in the defoliation treatment. From Hieke *et al.* (2002c).

	Control	Defoliated
Plant dry weight (g)		
Leaves (previous flush)	8.6	4.3
Stem (previous flush)	1.8	1.6
Leaves (older flushes)	27.0	14.0
Stem (older flushes)	27.5	20.3
Root	20.4	11.4
Total plant	85.3	51.6
Max. shoot length in third flush (mm)	167	141
Max. leaf area in third flush (cm^2)	1305	1549

Table 9.5. Changes in plant dry weight, concentration of starch and weight of starch in litchi in Australia. Data are the means of eight cultivars. From Menzel *et al.* (1995b).

	Dry weight (kg/tree)	Concentration of starch (% dry weight)	Weight of starch (g/tree)
Leaves	8.4	2.2	205
Twigs	4.4	7.1	316
Small branches	9.7	9.6	930
Medium branches	4.7	8.1	405
Large branches	3.8	7.8	306
Trunk	2.2	4.1	101
Large roots	3.3	2.8	103

initial set, but the leaves were required for later fruit development.

Regulated deficit irrigation during Stage II of fruit growth reduced shoot extension and increased yield in deciduous fruit crops such as pear, *Pyrus communis*, in southern Australia (Mitchell *et al.*, 1986). Extension growth was reduced by up to 70% compared with well-watered trees, while yields were increased by up to 30%, suggesting competition between the fruit and young leaves. Batten *et al.* (1994) proposed that drought had a similar effect in litchi, redirecting resources from the leaves to the fruit, since trees that were droughted from flowering had twice as many fruit as trees from well-watered plots.

Hieke *et al.* (2002a) tested the ideas of Batten *et al.* by pruning inflorescences to encourage new leaves on half the branches of each tree during fruit development. Some of the trees were pruned every week to prevent all vegetative growth, in order to test the effect of flower removal alone without subsequent leaf growth. In other experiments, the effect of pruning half the branches was studied, but the pruning was scattered across the tree rather than on one side. A range of cultivars was evaluated, including 'Tai So' ('Dazao'), 'Bengal', 'Kwai May Pink', 'Salathiel' and 'Wai Chee' ('Huaizhi'). It was proposed that new leaves would induce fruit drop and that the yields of the pruned trees would be less than 50% of the yields of unpruned trees.

It was found that the yields of the pruned trees with (pruned once) or without new leaves (pruned weekly) were more than 50% of the yields of the controls in 10 out of 11 orchards, indicating that new leaves on one branch did not compete with fruit on adjacent branches (Table 9.6). The expanding leaves used carbohydrates produced or stored in the older leaves and branches, but they eventually contributed to the tree. The competition reported by Batten *et al.* (1994) is yet to be explained, although it is possible that drought had a direct effect on the physiology of the fruit.

Table 9.6. Effects of pruning on the yield of litchi (kg/tree) in Australia. In year 1, panicles were removed on one side of each tree (pruned trees). In year 2, trees were pruned weekly to remove all new growth or once to remove panicles on one side of each tree. In year 3, 50% of panicles scattered around the tree were removed. Pruning once at flowering encouraged new leaves to be produced. From Hieke *et al.* (2002a).

Experiment	Control	Pruned once	Pruned weekly
Year 1			
Tai So	54.8	22.1	
Bengal	82.3	56.6	
Kwai May Pink	23.2	12.6	
Salathiel	17.1	11.1	
Year 2			
Tai So	48.9	40.9	36.8
Bengal	72.9	48.9	50.4
Wai Chee	23.3	24.2	18.4
Year 3			
Tai So	55.9	44.1	
Bengal	29.1	43.8	
Kwai May Pink	51.6	43.0	
Wai Chee	85.8	60.8	

The relationship between yield and assimilate supply

Litchi and longan fruit can obtain carbohydrates from CO_2 assimilation in the fruit itself, from reserves in the branches, or from photosynthesis in the leaves. Assimilates in the leaves and branches can come from the shoots next to the inflorescences or from shoots some distance away.

McConchie and Batten (1991) studied fruit set in 'Bengal' litchi in northern New South Wales and found that 9% of female flowers that were hand pollinated produced fruit. Hand pollination was employed to reduce the impact of pollen production and transfer on fruit retention. In other experiments in Australia without hand pollination, up to 8% of female flowers carried a fruit to harvest (Menzel and Simpson, 1992). Yields in these two studies were much higher than those reported in India, where only 0.1–0.2% of female flowers carried fruit to harvest (Chadha and Rajpoot, 1969). Litchi trees typically produce many more flowers than they can carry fruit to maturation.

The productivity of temperate trees has been well researched, but there is less information for tropical species. In pear and related crops, the young fruit are dependent on stored reserves, whereas later growth is a function of CO_2 assimilation in the shoots (Teng *et al.*,

1998). Studies on macadamia (*Macadamia integrifolia* × *M. tetraphylla*) demonstrated a clear relationship between nut retention and leaf area in girdled branches, whereas nuts on defoliated, ungirdled branches were able to draw on resources from the rest of the tree (Trueman and Turnbull, 1994). No nuts were retained after 8 weeks on defoliated girdled branches.

Yuan and Huang (1988) studied the effects of defoliation on yield of 'H-1224' litchi in Guangdong. Branches about 1.5 cm in diameter were girdled at full bloom and 0, 2, 50 or 100 leaves were retained on the girdled branches. After 63 days, no fruit were retained on branches with 0 or 2 leaves, whereas branches with 50 or 100 leaves had 2.4 and 4.9 fruit. These experiments demonstrate the importance of the leaves for fruit production in girdled branches, although the number of fruit retained at harvest even in the branches with full leaf cover was lower than that recorded in South Africa and Australia (see below).

Roe *et al.* (1997) investigated the roles of stored carbohydrates and current CO_2 assimilation in 'Tai So' litchi in South Africa. Branches were girdled, with different proportions of leaves and fruit retained. Girdling isolated the fruit from the rest of the tree, with the crop being dependent on assimilates produced in the branch. Small and large branches were used, bearing the same number or a different number of leaves, to determine whether the amount of stored reserves affected fruit set. About 5% of fruit were retained in branches without leaves, and 60% of fruit in branches with three or six leaves per fruit. Large branches were no more productive than small branches if they had the same number of leaves, suggesting that a larger pool of reserves conferred no benefit in the absence of a larger leaf area. These results showed that the fruit were mainly dependent on current CO_2 fixation.

Hieke *et al.* (2002b) studied the relationship between yield and carbohydrates in litchi in a series of experiments in subtropical Australia. The trials were different from earlier research, with the treatments applied to whole trees. In the first experiment, all the leaves of the last flush or all the leaves of the previous flush (about eight leaves per terminal shoot) or all the old leaves were removed from 8-year-old 'Kwai May Pink' trees. Undefoliated trees acted as controls. About 4–5 kg of leaves were stripped from the trees in the first two defoliation treatments, compared with 8 kg for the third treatment. Fruit production was about 40% lower than controls after defoliation, with similar yields when the younger or older leaves were removed. The trees with the younger leaves removed had leaf area indices of 2.3 or 2.8, respectively, compared with 1.7 when all the older leaves were removed, indicating that the younger leaves were relatively more important for fruit production. These leaves also had higher rates of CO_2 fixation than the older leaves.

In other experiments, medium or large branches of 'Bengal' trees were girdled and defoliated after fruit set, and fruit retention was compared with ungirdled and undefoliated branches (Hieke *et al.*, 2002b). Defoliation had no effect on fruit production in non-girdled branches, whereas defoliated, girdled branches produced 30% (medium branches) or 56% (large branches) of the yields of non-defoliated, non-girdled branches. The results for the large branches are summarized in Table 9.7. The medium branches accounted for one-third of the carbohydrates required for fruit growth, and the large branches just over half. Girdling alone had no effect on the medium branches, but increased the yield of large branches compared with the controls. Fruit on non-girdled branches were apparently able to draw on resources from the rest of the tree, similar to the response in macadamia (Trueman and Turnbull, 1994).

Other studies by Hieke *et al.* (2002b) showed that the fruit were especially sensitive to assimilate supply in the first 20 days after anthesis. Yield was reduced by three-quarters in 'Bengal' trees when branches were girdled at flowering and defoliated 1 or 20 days after flowering, compared with fully leaved, girdled

Table 9.7. The effects of defoliation and girdling on fruit production in 'Bengal' litchi in Australia. Branches 5–10 cm in diameter were girdled and defoliated after fruit set. From Hieke *et al.* (2002b).

Treatment	Yield (g/panicle)	Av. fruit wt (g)
Control	922	18.5
Defoliated	1087	17.2
Girdled	1130	18.3
Defoliated + girdled	410	16.7

branches, and was reduced by half when they were defoliated on days 40 or 61. In contrast, defoliation 14 days before harvest on day 83 had no effect on fruit yield.

Fruit thinning experiments were conducted on 10-year-old 'Kwai May Pink' trees with 0, 20, 50 or 80% of the inflorescences stripped by hand just after fruit set. Thinned trees had similar yields to unthinned trees (65–82 kg), with thinning increasing fruit retention in the remaining clusters under a higher leaf-to-fruit ratio. Thinning had no effect on average fruit weight, with a general mean of 23 g amongst the four treatments (Hieke *et al.*, 2002b). The responses from whole trees were similar to those obtained with girdled branches (Roe *et al.*, 1997).

In the Australian experiments, the concentrations of starch were higher in the branches and lower in the leaves and roots (Hieke *et al.*, 2002b). There were differences during the season, with higher values before flowering and lower values after flowering. Defoliation decreased the concentrations of starch, while girdling and fruit thinning increased them, although the differences were generally small. These results suggest that litchi fruit depend more on current CO_2 assimilation than on reserves. Most of the photoassimilates contributed to the fruit come from the leaves just behind the inflorescences. The branches generally act independently, but can supply assimilates to fruit some distance away if there are severe shortages of fixed CO_2.

Girdling and growth regulators have been used to increase litchi production in Israel, South Africa, Australia and China (Stern *et al.*, 1995, 1997, 2000, 2001; Stern and Gazit, 1997; Roe *et al.*, 1997; Hieke, 2000). Girdling resulted in the accumulation of carbohydrates above the cut by redirecting them towards the fruit (Fig. 9.4). The effects of auxin sprays are not fully understood but they probably increase the strength of the developing fruit as sinks at a critical period. These aspects are discussed in further detail in Chapter 7, this volume.

Li *et al.* (2001) studied the relationship between yield and shoot production in longan in China. Yield and quality in trees with many summer shoots were superior to those in trees with few summer shoots. A study conducted with 20-year-old trees of 'Wulongling' in Fujian indicated that shoots that flushed in summer and autumn produced higher yields than shoots that

Fig. 9.4. Girdling of the lower trunk in litchi (from Menzel *et al.*, 2002, with permission).

flushed only in autumn (Lin, 1997). Presumably the summer shoots provided the photoassimilates required for fruit development. It was also possible that the trees with summer flushes were healthier.

Relationship between fruit size and yield

The relationship between yield and crop load in temperate species has been established by several studies (see Forshey and Elfving, 1977, for apple). Generally, fruit thinning increased average fruit size, although this was at the expense of yield. Fewer studies have been conducted in tropical trees. Galliani *et al.* (cited in Goldschmidt and Koch, 1996) showed that average fruit dry weight in mandarin, *Citrus reticulata*, was 10, 12 and 20 g, and yields were 25, 15 and 4 kg/tree, under crop loads of 3000, 1500 and 500 fruit/tree, respectively.

Hieke *et al.* (2002a) showed that the removal of 50% of the potential crop at flowering had no effect on average fruit weight in several litchi cultivars in subtropical Queensland. Increasing the leaf-to-fruit ratio increased relative fruit retention, but did not lead to larger fruit. Roe *et al.* (1997) obtained similar results in girdling and defoliation experiments in South Africa. Ray and Sharma (1988) reported a weak correlation between average fruit weight and yield per branch (r^2 = 36%) or yield per unit of trunk diameter (r^2 = 50%) in 'Shahi' litchi in India. The relationship between yield and number of fruit on each tree was as follows:

Yield (kg/tree) = 27.54 + 0.017 * number of fruit/tree (r^2 = 81%) (6)

Yields ranged from 60 to 145 kg/tree, with about 2000–7000 fruit/tree. These data suggest an average fruit weight of about 17 g across different productivities, although the regression line would be expected to pass through the origin (zero fruit, zero yield).

Relationship between yield and tree size

Little information is available for either crop on the relationship between yield, leaf area and tree size. Researchers in Australia investigated the relative productivity of ten small 'Kwai May Pink' litchi trees in Bundaberg (Menzel *et al.*, 2000). The trees ranged from 6 to 9 years of age and were grown at 0.5 m spacings within rows. The total number of leaves on each tree was recorded in July, while the leaf area of each tree was calculated from the relationship between leaf area and leaf weight determined at harvest in January. At the same time, the heights and widths of the canopies were used to calculate canopy surface area. Yield per tree was determined at harvest.

The total number of leaves on each tree varied from 2700 to 9200, while canopy surface area ranged from 24 to 54 m^2/tree and total leaf area from 11 to 41 m^2/tree. There was a linear relationship between total leaf area and canopy surface area passing through the origin:

Total leaf area (m^2/tree) = 0.67 * canopy surface area (m^2/tree) (7)

Relative leaf area index, the ratio of leaf area to canopy surface area, was essentially independent of tree size, and ranged from 0.48 to 0.83. Yield was related to canopy surface area and total leaf area:

Number of fruit/tree = 19.2 (± 1.6) * canopy surface area (m^2/tree) (8)

Number of fruit/tree = 30.7 (± 2.6) * total leaf area (m^2/tree) (9)

Yield was related to the leaf area supporting the crop, consistent with the trees being small and widely spaced, with little shading and similar partitioning of dry matter. It was concluded that once trees begin to bear, fruit production is simply a function of the effective canopy surface area. About seven leaves were required to fill each 'Kwai May Pink' fruit (area of an individual leaf was 45.7 ± 1.8 cm^2) compared with two to three leaves for 'Tai So' ('Dazao') and one to two leaves for 'Souey Tung' ('Shuidong') (Roe *et al.*, 1997). However, these two experiments are not directly comparable. Menzel *et al.* (2000) used whole trees, whereas the estimates in South Africa were based on girdled branches, where assimilates are stored and do not contribute to the rest of the tree. The leaves of 'Kwai May Pink' are also smaller than those of 'Tai So'.

Few authors have investigated the relationship between yield and leaf area in longan. Huang *et al.* (1992) studied the relative productivity of old orchards in Fujian with different-sized canopies and showed that yield was strongly correlated with the diameter of the tree crowns. Trees with canopies 3.3, 4.0 and 4.4 m in diameter had relative yields of 3.2, 4.5 and 6.6 kg/m^2 leaf area. It was concluded that the best yields (1400 kg/tree) were associated with trees that had two to four main branches and a trunk girth of 1.0 m. It was suggested that 30 leaves were required on each terminal fruiting branch to produce high yields.

Effects of light on growth and productivity

Light has strong effects on CO_2 fixation in leaves, dry matter production and plant development. The effects on yield depend on the actual light regime during different stages of plant development, and the relative sensitivity of the species. The actual light environment during growth can also determine the response of plants. Some species adapt to low light by diverting more dry matter into leaf expansion or produce higher concentrations of chlorophyll. Species are broadly classified into sun and shade plants. Sun plants typically have smaller thicker leaves, higher rates of water and CO_2 exchange, lower leaf area per unit of plant dry matter (lower leaf area ratios), higher root-to-shoot ratios, and higher LCPs than shade plants (Atwell *et al.*, 1999).

In their native environment, litchi and longan are dominant species in the upper canopies of the rainforests, but spend some time as seedlings until a gap in the canopy allows them to exploit the improved light environment on the forest floor. Research has shown that photosynthesis of litchi is saturated at about half full sun, whereas maximum leaf expansion occurs at higher light levels. Low light levels also reduce fruit growth.

The effects of light on leaf expansion and growth are usually described in terms of photosynthetic photon flux density, PPFD, or total solar (short-wave) radiation or irradiance. Total radiation is the energy of the radiation from 300 to 3000 nm (PPFD is from 300 to 700 nm), usually expressed as $MJ/m^2/day$. About half of the short-wave radiation received at the Earth's surface is photosynthetically active light. In subtropical Australia, solar radiation on clear days ranges from 16 MJ/m^2 in winter to 32 MJ/m^2 in summer, and is about 80% lower during overcast weather (Fig. 9.5). Average values range from 12 to 24 $MJ/m^2/day$. In the tropics there is less seasonal variation, with values dependent on cloud cover. Continuous overcast weather reduces incoming (extraterrestrial) solar radiation by half. Measurements of short-wave radiation are not widely recorded in litchi- and longan-growing areas, as the necessary instrumentation is complex and expensive. Models relating solar radiation and hours of bright sunshine at a given site are available, but they tend to be unreliable under intermittent cloud cover. The relationship for different sites in South Africa (D. Roe, Nelspruit, 1993, personal communication) was:

$$I_o = I_e\,[0.25 + 0.50\,(n/N)] \qquad (10)$$

where I_o is irradiance at the Earth's surface ($MJ/m^2/day$), I_e is irradiance outside the Earth's atmosphere ($MJ/m^2/day$), n is the hours of bright sunshine each day, and N is the potential hours of bright sunshine each day or daylength.

Hieke *et al.* (2002e) investigated the effects of radiation on shoot extension and leaf expansion of 'Kwai May Pink' litchi in subtropical Queensland. In the first experiment, seedlings were grown in a heated greenhouse in 20% full sun, equivalent to maximum noon PPFD of 200 μmol quanta/m^2/s. Their growth was compared with that of seedlings growing in full sun with a maximum PPFD of 1080 μmol quanta/m^2/s. Shoot and leaf growth were recorded also on plants grown outdoors in artificial shade providing 20, 40, 70 or 100% of full sun, equivalent to maximum PPFDs of 500, 900, 1400 or 2000 μmol quanta/m^2/s. In these experiments, shoots and leaves expanded in a sigmoidal pattern to 69 mm and 497 cm^2 for each flush in full sun compared with 27 mm and 189 cm^2 in shaded seedlings. Seedlings were 1.7 times heavier in full sun compared with those under heavy shade. In the shade, leaf growth continued at the expense of the roots, so shaded seedlings had higher shoot-to-root ratios than

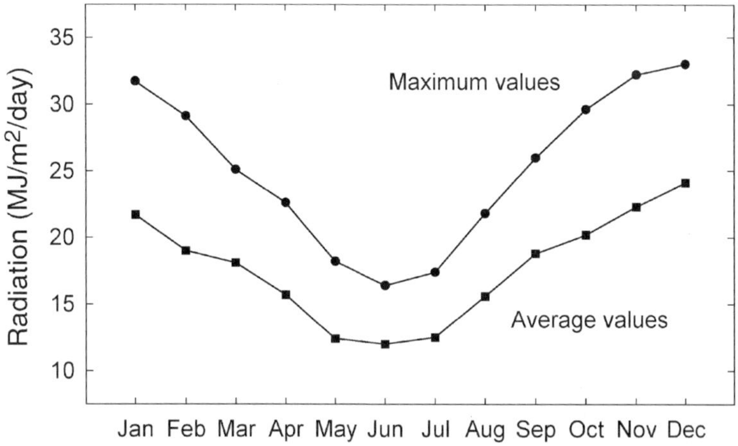

Fig. 9.5. Seasonal changes in solar radiation (I_o) at Nambour in southern Queensland, Australia (latitude 27°S).

plants grown in full sun (3.7 vs. 3.1). In the potted plants grown outdoors in 20, 40, 70 or 100% of full sun, final leaf area per shoot was 44, 143, 251 and 362 cm^2, respectively. Plants grown in shade also had shorter shoots (66 mm) than plants grown in full sun (101 mm).

Rainforest trees vary greatly in their adaptation to light. Some species are shade-sensitive and only establish when there are large gaps in the canopy, whereas others are tolerant or adaptable, persist as shade-tolerant seedlings, and exploit small gaps in the canopy. A 6-week shade treatment (PPFD of 725 μmol quanta/m^2/s) had no effect on shoot growth in avocado, where full sun at noon was equivalent to a PPFD of 1350 μmol quanta/m^2/s (Chirachint and Turner, 1988). In contrast, mangosteen seedlings grown in 50% or 80% full sun were larger than those grown in 20% or 100% full sun (Wiebel *et al.*, 1994). The substorey avocado would be expected to be more responsive to light than the shade-adapted understorey mangosteen. The data collected by Hieke *et al.* (2002e) suggest that higher light levels are required for growth and photosynthesis in litchi than for survival of seedlings on the forest floor.

Fruit development in litchi is also sensitive to light conditions (Yuan and Huang, 1988). These authors girdled branches (1.5 cm in diameter) of 'H-1224' trees in Guangdong and shaded all the leaves to reduce light interception by 90%. Control branches were girdled but left in the full sun. The shade was applied for 7 days at full bloom, or for 7 days 3 weeks after full bloom, with the pattern of fruit drop followed for another 3 weeks. In the first experiment with shade at full bloom, shaded panicles bore an average of 0.2 fruit compared with 8.5 fruit for the control panicles. In the second experiment with shade during fruit development, there was an average of 0.8 fruit on each shaded panicle compared with 2.2 fruit on each control panicle. These data suggest that a week of overcast weather just after fruit set could increase fruit drop in litchi.

Effects of temperature on growth and productivity

Temperature has a major effect on plants. Subtropical species usually start growing at 10°C, with maximum growth at about 30°C. New leaves and flowers are killed at temperatures just below freezing, while still lower temperatures generally kill whole trees. Temperatures from 10°C to 30°C affect the rate of CO_2 fixation in the leaves and various aspects of plant development.

Temperatures vary with latitude, elevation and local geography. Average temperatures fall by about 0.5°C for each degree increase in latitude, and about 0.6°C for each 100 m rise in elevation above 500 m (Figs 9.6 and 9.7).

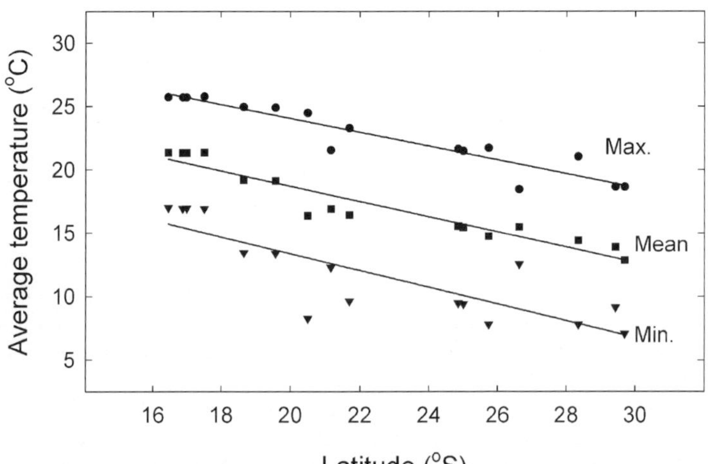

Fig. 9.6. The relationship between average temperatures in July, and latitude along the east coast of northern Australia (from Menzel *et al.*, 2000, with permission).

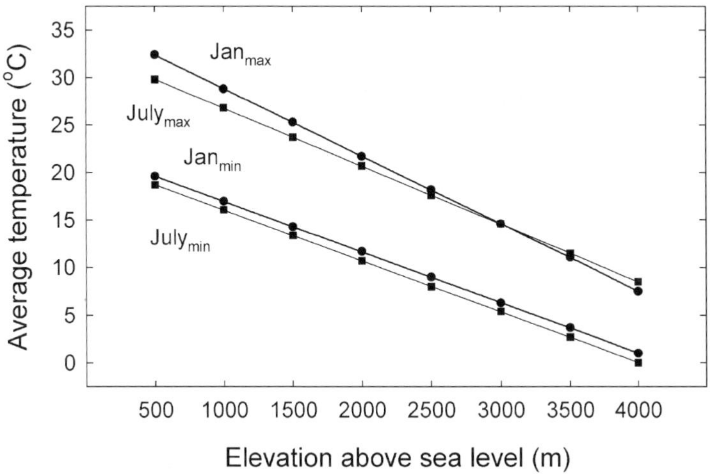

Fig. 9.7. The relationship between average maximum and minimum temperatures in January and July, and elevation in Papua New Guinea (from McAlpine *et al.*, 1983).

Average monthly temperatures are more relevant for predicting plant growth than mean annual values. Temperature affects leaf expansion, flower initiation, flower development, pollination and fruit development in litchi and longan; however, only aspects relating to tree growth are discussed here. The effects on flower and fruit development are reviewed elsewhere in this volume.

Menzel and Paxton (1985) investigated the effects of temperature on the growth of six seedling litchi cultivars ('Bengal', 'Haak Yip', 'Kwai May', 'Gee Kee', 'Tai So' and 'Wai Chee') in a glasshouse in subtropical Australia. Stem extension, node production and leaf production increased with increasing day/night regimes from 15/10°C to 20/15°, 25/20° and 30/25°C. There was a significant correlation between node production and mean daily temperature (day temperature plus night temperature divided by two):

$$\text{No. nodes} = -8.05 + 0.73 * \text{average temperature } (r^2 = 96\%) \quad (11)$$

The base temperature for shoot growth was 11°C, although no upper threshold was determined. Plant dry weight increased with increasing temperatures, with a 4.8-fold increase in leaf dry weight, a 4.5-fold increase in stem dry weight, and a 2.5-fold increase in root dry weight over the range tested. High temperatures increased the proportion of leaf in the seedlings (58% vs. 48%) and decreased the proportion of roots (25% vs. 38%). The colour of the new leaves changed from red to pink to green with increasing temperature. These results indicated that the best vegetative growth in litchi occurs above 25°C.

Batten and Lahav (1994) examined the effects of constant 10, 15, 20, 30 or 35°C on shoot growth of 'Haak Yip', 'Tai So' and 'Bengal' litchi in controlled-environment glasshouses. A modified Richards' function described leaflet elongation over time:

$$L_{length} = L_{max}/\{1 + a * \exp[-b\,(t - m)]\}^{1/a} \quad (12)$$

where L_{length} is leaflet length at a given time, L_{max} is maximum leaflet length, m is a time offset parameter, a defines the asymmetry of the curve ($a = 1$ gives the symmetrical logistic), b defines the rate of rise of the curve, and t is time in days.

The minimum temperature at which leaflet elongation ceased varied from 8 to 13°C in the three cultivars, whereas the optimum for leaf elongation was 35°C. 'Tai So' took 16 days to attain full leaf elongation at 25°C and 23 days at 20°C; however, the time between flushes almost doubled from 32 to 58 days. A flush took 150 days to develop at 15°C. A model was developed to predict the rate of flush development per day for 'Tai So':

$$P = [(T - 12.9)^{0.9577}]/[365 + 21.28 (T - 12.9)^{0.9577}] \quad (13)$$

where P is the rate of flush development per day and T is average daily temperature. The value 12.9 is the base temperature for leaflet elongation in 'Tai So'. The model was used to predict the rate of shoot development, with growth complete when the sum from the start of a flush equalled one.

Batten and Lahav (1994) also looked at the impacts of temperature on dry matter production. The base temperature for shoot dry matter production was between 10 and 14°C, while the optimum temperature was between 28 and 34°C. The optimum for leaf elongation was greater than the optimum for shoot production. Other data collected by Batten *et al.* (1992) indicated that the optimum for stomatal opening was about 37°C. In contrast, flowering generally only occurs below 20°C (Menzel and Simpson, 1995). The analysis of Batten and Lahav (1994) suggests that it may not be possible for trees to complete a vegetative flush between harvest and the expected time of floral initiation in cool subtropical locations, or at elevation in the tropics (see Fig. 9.6).

Seasonal changes in maximum and minimum temperatures for various litchi- and longan-growing areas are shown in Fig. 9.8. Average maxima/minima in Guangzhou in China range from 19.1/8.7°C in January to 32.8/25.2°C in August. It is much warmer in Chiang Mai in Thailand than in Guangzhou in winter, with regimes in January of 29.0/13.0°C. Temperatures during summer are similar (31.0/23.0°C in Chiang Mai). Galilee in Israel has a typical Mediterranean climate with cool, wet winters and hot, dry summers. Temperatures in January (17.9/9.2°C) are similar to those in Guangzhou; however, summer minima are about 4°C cooler (32.5/20.9°C).

Cairns in northern Queensland has similar temperatures to Guangzhou during summer (31.4/23.6°C in January), but winters are much warmer (25.6/17.0°C in July). Nambour in southern Queensland and Nelspruit in South Africa have cool winters and warm summers. Maxima in January are 1–2°C cooler than in Cairns, and maxima in July, 2–4°C cooler. Minima are 4°C cooler in summer in Nambour and Nelspruit than in Cairns, and 10°C cooler in winter. Temperatures are ideal for flower initiation in Guangzhou, Galilee, Nelspruit and Nambour, intermediate in Chiang Mai, and too warm in Cairns. Temperatures during summer are ideal for maximum photosynthesis and plant growth (days of 30°C or higher) for most of the year in Chiang Mai, and for about 5 out of 12 months in Guangzhou, Galilee and Cairns. Temperatures at Nelspruit and Nambour are below the optimum for maximum CO_2 fixation and leaf expansion in all months.

Planting density and productivity

Monteith (1977) and others demonstrated a fundamental relationship between the production of dry matter in crops and seasonal light interception in the UK. These responses have been well documented in temperate fruit such as apple, but have received little attention in tropical species. In apple, maximum yields of 10 t/ha (dry matter) were achieved with a total intercepted PPFD of 1000 MJ/m^2 (Lakso, 1994). In other experiments, maximum yields of 90 t/ha (fresh weight) were associated with interception of 80% of the incident light received for the season.

Tree shape and planting density influence light interception and its distribution through the canopy. Various models of canopy photosynthesis and dry matter production have been developed in apple, and optimum tree shapes and planting densities have been determined for different environments. In recent years, the tropical fruit industries, including litchi and longan, have moved to closer plantings, although these developments have occurred in the absence of a sound knowledge of tree physiology. Data collected in Australia suggest that small litchi trees are as productive as large trees when yields are expressed on a leaf-area basis (Menzel *et al.*, 2000). Further experiments are required to define the optimum tree shape and size for these crops. Data collected by Hieke *et al.* (2002b) indicated a strong relationship between yield and the amount of CO_2 fixed by the canopy, which would be dependent on tree shape, spacing and light interception.

In China there is no standard layout for litchi orchards, although most farmers prefer

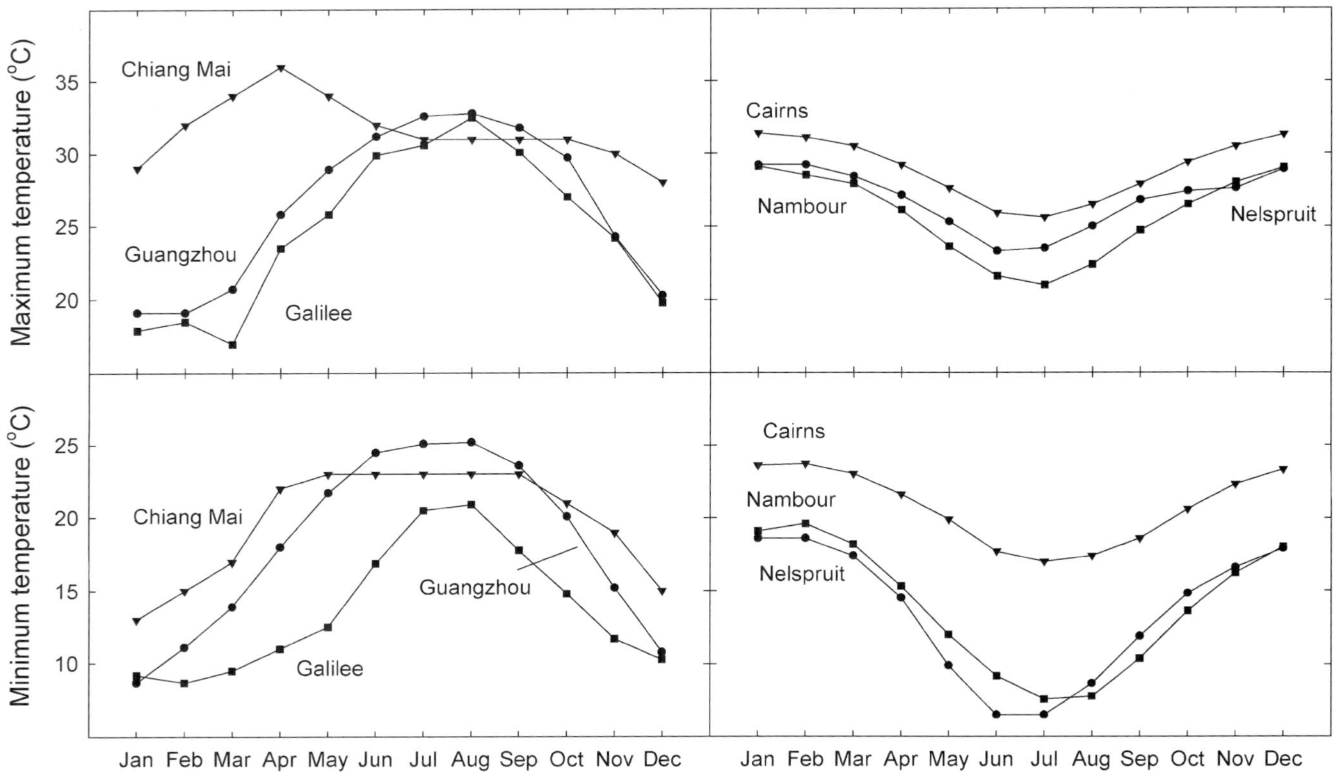

Fig. 9.8. Seasonal changes in temperature for various litchi- and longan-growing areas. Data from Chiang Mai, Thailand (lat. 19°N, elevation 300 m), Guangzhou, China (lat. 23°N, elevation 18 m), Galilee, Israel (lat. 32°N, elevation 200 m below sea level), Nelspruit, South Africa (lat. 25°S, elevation 660 m), and Cairns (lat. 17°S, elevation 3 m) and Nambour (lat. 27°S, elevation 29 m), Australia.

plantings of 2.5–3.0 × 3.5–4 m, equivalent to 825–1100 trees/ha (Menzel, 2002). Other crops, such as bean, groundnut, sweet potato, vegetables, pineapple and papaya, are planted in the interrows and the orchard thinned to 300 trees/ha after a few years. Some sections of the industry have adopted high-density plantings of up to 1500 trees/ha. These are often based on the popular early cultivar, 'Fay Zee Siu', and are dependent on regular pruning, girdling, watering and fertilizing.

In Viet Nam, the normal tree spacing is 7 or 8 m, depending on the fertility of the soil and orchard topography (Fig. 9.9). There are few high-density orchards. Tree spacing in Thailand ranges from 3 to 8 m, with the closer spacings requiring a higher level of orchard management than in traditional plantings.

The traditional litchi growers in India use a spacing of 9–10 m, equivalent to 100 trees/ha planted in a square system (Menzel *et al.*, 2002). Old trees in these orchards may be 10 or 12 m high. There are experimental plantings at 4.5 × 4.5 × 9 m (329 trees/ha), in double hedgerows. The closer plantings provide greater fruit production per hectare and fruit quality equal to that of traditional plantings. A light pruning is recommended after harvest. In Bangladesh, old orchards are planted at 7–12 m and new plantings at 4 m. Traditional spacings of 10–12 m are still used in Nepal, with the interrows planted to vegetables or other crops. These are removed after about 8 years. Planting distances in the Philippines are 7 or 8 m, equivalent to 150–200 trees/ha.

In South Africa, litchi orchards older than 10 years of age are planted mostly at 10–12 m, equivalent to 70–100 trees/ha (Milne, 1999). Orchards younger than this are planted at 6 or 7 m, equivalent to 200–280 trees/ha. The development of pruning technologies in Israel has fostered the development of high-density orchards at 3 × 5 m or 4 × 5 m, equivalent to 500–667 trees/ha (Goren and Gazit, 1996). Older orchards contain only 278 trees/ha.

Suggested row and tree spacings for litchi trees in Australia range from 6 × 4 m to 10 × 4 m for trees grown as hedges (Menzel *et al.*, 2002). The correct spacing depends on the planned height of the trees, the cultivar and the growing region. Vigorous, spreading cultivars grow more vigorously in tropical Cairns compared with subtropical Nambour. Trees are maintained at a maximum height of 5 m after pruning. These spacings and tree sizes make good use of the land and provide maximum yield and cash flow early in the life of the orchard, while facilitating harvest from relatively low trees (Fig. 9.10).

Recommended spacings for longan in China are 4.5–5.0 m between trees and 5.0–6.0 m between rows, equivalent to 330–450

Fig. 9.9. A litchi orchard in Viet Nam, interplanted with pineapples (photograph courtesy of Christopher Menzel).

trees/ha (Li *et al.*, 1998; Huang, 1999). For high-density planting, growers use 660–900 trees/ha (Fig. 9.11). There are experimental plantings of dwarf, heavily pruned 'Shixia' longan in Guangdong and elsewhere at 1500–1800 trees/ha (Fig. 9.12). In Viet Nam, the normal spacing adopted is 7 or 8 m, with very few high-density orchards. Tree spacings in Thailand range from 6 to 12 m. Many of the older orchards have only 50 trees/ha.

Effects of pruning on productivity

Litchi and longan produce flowers and fruit on terminal shoots. Research has shown that strategic pruning can be used to alter the timing of the harvest as well as yield, and to maintain tree height and production in high-density plantings.

Young (1977) was the first to document the response of litchi to pruning. Branches were removed from the sides and tops of large

Fig. 9.10. A high-density litchi orchard in southern Queensland, Australia (photograph courtesy of Christopher Menzel).

Fig. 9.11. A high-density longan orchard in Guangxi, southern China (photograph courtesy of Christopher Menzel).

Fig. 9.12. A pruned longan orchard in Guangxi, southern China (photograph courtesy of Christopher Menzel).

'Brewster' ('Chenzi') trees in Florida during winter. By the third year after pruning, the heights and widths of the pruned trees were only about half of those of the unpruned trees. There was no flowering or crop the year after pruning, while production in year 2 was about 30% of unpruned trees, and about 80% in year 3. Results with less severe and more strategic pruning in Taiwan were more encouraging, with trees lightly pruned after harvest in a single year (Yen and Tien, 1985). Pruning had no effect on the proportion of branches flowering in 'Yook Ho Pow' (87% vs. 90%) and 'Kang Wei' (63% vs. 63%), but reduced flowering in 'Sah Keng' (26% vs. 61%).

In Israel, 'Tai So', 'Bengal' and 'Floridian' trees were tip-pruned between September and February (Goren, 1990). Flowering was generally best with pruning from September to November, with the trees initiating inflorescences in December and January. Pruning and droughting before flowering form the basis of regular production in Israel, with yields of 10 t/ha achieved using these technologies (Goren and Gazit, 1992). Menzel *et al.* (1996) investigated the effects of tip-pruning on the productivity of 'Bengal', 'Kwai May Pink' and 'Wai Chee' litchi over 3 years in subtropical Queensland and showed that flowering and yield were as good after winter pruning as with summer pruning, although both were less than in unpruned trees.

Campbell (1994) studied the relationship between flowering and shoot development in 5-year-old 'Haak Yip' trees in Florida. Trees were pruned to remove all leaf flushes emerging after 31 August, 30 September or 30 November. When a leaf flush emerged after this time, the branch was pruned back to the last mature leaf of the previous flush. Control trees were left unpruned. Pruning in August, September or November produced an average of 3.1, 3.5, and 1.2 inflorescences per shoot, whereas control trees had 1.2 inflorescences per shoot. Fruit set was negligible in all the treatments. These results suggest that trees with leaves emerging after November are unlikely to flower in Florida; however, they do not indicate the optimum time for pruning.

Olesen *et al.* (2002) developed a model to predict the optimum time for pruning litchis along the east coast of Australia. Trees of 'Kwai May Pink' were pruned over spring and summer at Alstonville (lat. 29°S), and subsequent shoot elongation over one, two or three cycles measured. Leaves were produced when the mean temperature during early flush development was above 17–19°C, and flowers were produced when temperatures were below this range. Trees with successive flushes in February (early autumn) and June (early winter) flowered more than trees with successive flushes in April and August. The rate of flush development (per day) was related to the average product of mean daily temperature and total daily irradiation (400–1100 nm) (x, °C.MJ/m^2/day) during flush development:

$$\text{Flushing rate} = 0.017/\{[1 + e^{0.055\,(492 - x)}]\ 0.055\} \qquad (14)$$

This regression was used with long-term weather records to estimate the date a flush needed to commence for one or two flushes to be completed by the winter solstice at different latitudes (Fig. 9.13). The earliest date for the completion of one flush ranged from 16 February in northern New South Wales (lat. 30°S) to 13 March in northern Queensland (lat. 17°S). In a separate trial at Mareeba (lat. 17°S), trees pruned on 10 February, for which it was estimated that they would produce about 1.5 flushes prior to winter (i.e. flushes in late autumn and early spring, but not in winter), flowered and yielded poorly. In contrast, trees pruned on 11 March, with an estimated single flush up to winter, flowered well and produced good yields. Non-pruned trees had a similar pattern of flushing as the trees pruned on 11 March and were just as productive. These data suggest that strategic pruning can be used to manipulate flushing times and cropping (Fig. 9.14).

Long-term weather records were also used to estimate the number of days per year with mean daily temperatures below 20°C, as a guide to potential flowering (Menzel and Simpson, 1995). There was a strong latitudinal gradient, with more than 100 inductive days but a median of 17 days in northern New South Wales, and a minimum of 0 days for tropical Cairns (lat. 17°S).

Menzel *et al.* (2000) conducted pruning experiments on longan in Australia. On seven occasions from November to January, trees of 'Kohala' and 'Chompoo' were pruned at Nambour (lat. 27°S), and flushing and weather were recorded as reported by Olesen *et al.* (2002). Linear regressions were fitted to the relationship between flush interval and flushing rate against the product of irradiation and temperature (x, °C.MJ/m^2/day), with the two cultivars giving similar responses. The relationship between flushing rate (per day) and weather in 'Chompoo' was:

$$(\text{Flushing rate})^{1.5} = [(0.000059{*}x) - 0.015]^{1.5} \qquad (15)$$

The results for longan were broadly consistent with those for litchi (Olesen *et al.*, 2002). The trees had a minimum cycle of 7 weeks compared with 8–9 weeks for litchi, but this was slower under cool, cloudy weather. Trees pruned at different times had different flowering, suggesting a similar physiology to litchi.

Small pruning experiments were carried out on 10-year-old 'Kohala' trees on the Atherton

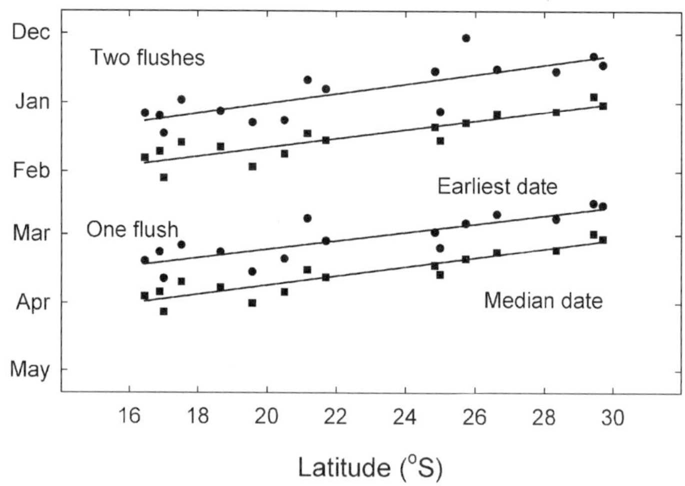

Fig. 9.13. The relationship between the date on which a flush must emerge for one or two flushes to be completed before the winter solstice in litchi and latitude along the east coast of northern Australia (from Olesen *et al.*, 2002, with permission).

Fig. 9.14. Heavy yield of litchis in northern Queensland, Australia, after pruning in January (photograph courtesy of Christopher Menzel).

Tableland and the impacts on flowering were determined (Menzel *et al.*, 2000). In the first experiment, trees were pruned in late March and their performance was compared with unpruned trees. About 0.5 m was removed from each of the terminal branches. Average yields at harvest were 83 kg/tree for the controls and 40 kg/tree for the pruned trees. In the second experiment, selected cropping branches were pruned at harvest in mid-March or in early April, and compared with non-cropping branches. The non-cropping branches produced leaves in April and flowered in June, while those that were pruned at harvest produced leaves in March and leaves and flowers in July. The branches that were pruned in April produced leaves in April and a mixture of leaves and flowers in August. The proportion of branches flowering in the different treatments in October was 76, 38 and 29%, respectively. It is apparent that pruning of longan in northern Queensland should be carried out as soon as possible after harvest.

In some seasons, litchis may flush during late autumn. These flushes are unlikely to produce flowers. Research in Australia has shown that the young flushes can be removed with a hedge trimmer or burnt off with an application of ethephon (Menzel *et al.*, 2000). The trees produce new shoots, and flower if the weather is cool. A similar strategy would probably be effective in longan.

Practical aspects of pruning in litchi

Menzel *et al.* (2002) described two major types of pruning for litchi in Australia. Thinning in winter involves selective limb removal, allowing light to penetrate to the lower branches to maintain productivity in the lower canopy (Fig. 9.15). Pruning also allows the wind to move unimpeded through the tree, reducing the twisting effect from strong winds, and allows better penetration of sprays. Shaping and strategic pruning involves cutting back the outside of the canopy to a hedgerow or individual tree, after harvest (Fig. 9.16). The pruned trees are smaller and easier to spray, harvest and net for the exclusion of pests (Fig. 9.17).

With selective limb removal, about 10–20% of the branches are removed from inside the canopy. After pruning, some light can be seen through the canopy. Long branches are removed first, as this helps to control the height of the tree. These limbs should be removed at a point close to the main trunk. In vigorous cultivars, long limbs that fail to branch are cut back to produce a more compact tree and to increase the number of potential fruiting sites.

Strategic pruning helps to control tree size but it must be carried out at the right time to ensure flowering and cropping (Olesen *et al.*, 2002). Most growers in Australia prefer hedgerows to individual trees. For hedgerows, a

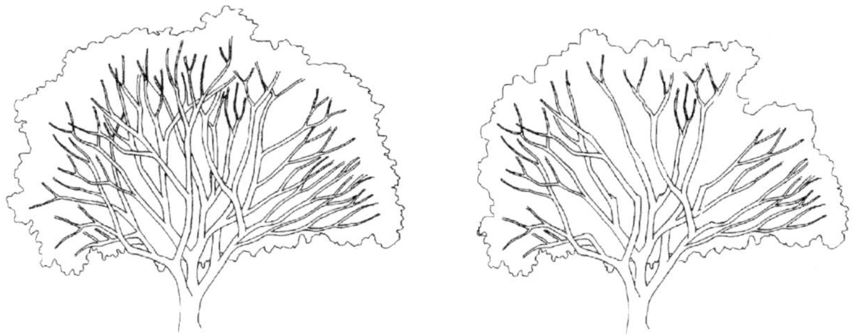

Fig. 9.15. Branch removal in litchis in Australia, with an unpruned tree on the left, and a thinned tree on the right (from Menzel *et al.*, 2002, with permission).

Fig. 9.16. Shaping of litchis in Australia, with flat tops on the left, and angled tops on the right (from Menzel *et al.*, 2002, with permission).

Fig. 9.17. Litchi trees in southern Queensland, Australia with individual tree netting (photograph courtesy of Neil Greer).

regular programme of trimming the tops and sides of the trees to maintain tree size from year 5 is recommended (Fig. 9.18). The trees should be pruned into a cone or rounded cone. Trees are shaped when the gap across the interrow is less than 2 m. The foliage is cut back to produce a laneway space of about 4 m. Hedging after harvest allows the tree to flush once or twice before winter, and then to flower. Selective limb removal is sometimes carried out at the same time. Young trees are typically skirted to keep the lower branches off the ground (Fig. 9.19).

Large trees can be reduced to a manageable size with judicious pruning over 2–3 years. The recommended approach is to remove whole branches rather than to prune just the top of the tree, with no more than 20% of the branches being removed. Any regrowth is removed before flowering. Although regular pruning is preferred, tree removal is sometimes the only option available in old orchards. Light penetration is improved in overgrown orchards by removing every second tree within the rows. Exposed limbs should be treated with a water-based paint to prevent sunburn.

Practical aspects of pruning in longan

A brief description of strategies used in Thailand and China is provided, with the emphasis on improving tree structure, fruit size and the regularity of bearing.

Grafted trees can be pruned to a workable height of 6–8 m. In Thailand, water shoots and small branches from the interior of the tree are removed, with the trees skirted to 1–2 m. In China, one strong branch is retained after every growth flush to form a natural round-shaped crown with six to ten main branches. Trees in Thailand are cut to a 1.2 m high trunk and three to four vigorous shoots selected to form the main frame of the tree. These laterals are forced into wide angles from the trunk with the aid of sticks. Two branches are left on each lateral shoot. Similarly, two sub-branches are left on each branch

Fig. 9.18. Pruning of litchi at Bundaberg in southern Queensland, Australia.

Fig. 9.19. Skirting of young litchis in Queensland, Australia (from Menzel *et al.*, 2002, with permission).

and so on. Finally, after 3–4 years, the canopy consists of 24–32 terminal branches. The tree is quite short, with good light penetration through the canopy.

Chinese growers often remove 40% of the flower spikes and 30% of the young fruit in heavily cropping trees, leaving more than 500 g fruit on each terminal branch at harvest. The flowering panicles are pruned when they are about 10–12 cm long, and 'weak' and 'strong' branches can be easily distinguished. Thinning increases fruit size, with fruit greater than 2.5 cm in diameter (18 g) attracting a premium price. Fruit are usually thinned 4–6 weeks after fruit set when they are the size of a pea. Fruit thinning is essential after flower thinning because of the high rate of fruit set and greater competition amongst developing fruit. Thai growers remove half the flowers on each flower spike before fruit set, and 10% of the fruit after fruit set in an 'on' year. Small spikes may carry up to 30 fruit or 600 g at harvest and large spikes up to 180 fruit or 3600 g. The fruit should be at least 1 cm apart on the panicle.

Conclusions

The productivity of litchi and longan orchards varies dramatically between years and growing areas. Average yields are 2–5 t/ha, although yields up to four times this have been achieved in research plots. Litchi, and especially longan, are also subject to biennial bearing, with the yields in 'off' years being only 50–80% of the yield in 'on' years. This suggests that there is a large gap between potential and actual productivity.

Average yields are low in some orchards because of heavy shading, drought or poor pest control. Environmental conditions during flower initiation and fruit set can also reduce production. Opportunities exist for increasing productivity by increasing the interception and distribution of light by the leaves through the development of more efficient tree shapes and sizes. The optimum time to prune the trees after harvest also needs to be established for different growing areas. Experiments have shown a strong relationship between yield and photosynthesis in the leaves behind the fruit clusters, suggesting that high nitrogen applications, which increase canopy CO_2 fixation, will probably increase production.

Acknowledgements

Comments on the manuscript by Trevor Olesen from NSW Department of Primary Industries in Alstonville are appreciated. Many thanks to colleagues who contributed to the litchi research programme at Maroochy, especially Don Simpson and Susanna Hieke.

References

Atwell, B.J., Kriedemann, P.E. and Turnbull, C.A. (1999) *Plants in Action*. Macmillan Education Australia, South Yarra, Victoria, Australia, 664 pp.

Batten, D.J. and Lahav, E. (1994) Base temperature for growth processes of lychee, a recurrently flushing tree, are similar but optima differ. *Australian Journal of Plant Physiology* 21, 589–602.

Batten, D.J., Lloyd, J. and McConchie, C.A. (1992) Seasonal variation in stomatal response of two cultivars of lychee (*Litchi chinensis* Sonn.). *Australian Journal of Plant Physiology* 21, 589–602.

Batten, D.J., McConchie, C.A. and Lloyd, J. (1994) Effects of soil water deficit on gas exchange characteristics and water relations of orchard lychee (*Litchi chinensis* Sonn.) trees. *Tree Physiology* 14, 1177–1189.

Blanke, M.M. and Lenz, F. (1989) Fruit photosynthesis. *Plant Cell and Environment* 12, 31–46.

Campbell, R.J. (1994) Fall pruning induces blooming in young lychee trees. *Proceedings of the Florida State Horticultural Society* 107, 348–350.

Chadha, K.L. and Rajpoot, M.S. (1969) Studies on floral biology, fruit set, and its retention and quality of some litchi varieties. *Indian Journal of Horticulture* 26, 124–129.

Chirachint, W. and Turner, D.W. (1988) Shade reduces the foliar symptoms of 'Fuerte' avocado affected by salt, without significantly changing the concentrations of Na, K or Cl in the leaves. *Scientia Horticulturae* 36, 1–15.

Choo, W.K. (2000) *Longan Production in Asia*. Food and Agricultural Organization of the United Nations, Bangkok, Thailand, 44 pp.

Forshey, C.G. and Elfving, D.C. (1977) Fruit numbers, fruit size, and yield relationships in 'McIntosh' apples. *Journal of the American Society for Horticultural Science* 102, 399–402.

Goldschmidt, E.E. and Golomb, A. (1982) The carbohydrate balance of alternate-bearing citrus trees and the significance of reserves for flowering and fruiting. *Journal of the American Society for Horticultural Science* 107, 206–208.

Goldschmidt, E.E. and Koch, K.E. (1996) Citrus. In: Zamski, E. and Schaffer, A.A. (eds) *Photo-assimilate Distribution in Plants and Crops: Source–Sink Relationships*. Marcel Dekker, New York, pp. 797–823.

Goren, M. (1990) High density litchi orchards by reducing tree height. *Alon Hanotea* 44, 699–704.

Goren, M. and Gazit, S. (1992) Small stature litchi orchards: a new approach to the growing of litchi. *Acta Horticulturae* 349, 69–72.

Goren, M. and Gazit, S. (1996) Management of lychee orchards in Israel. In: Welch, T. and Ferguson, J. (eds) *Proceedings of the Fourth National Lychee Seminar.* Australian Lychee Growers' Association, Yeppoon, Queensland, pp. 74–77.

Hieke, S. (2000) Physiology of leaf and fruit growth in lychee (*Litchi chinensis* Sonn.). Doctoral thesis, Humboldt University, Berlin, Germany.

Hieke, S., Menzel, C.M., Doogan, V.J. and Lüdders, P. (2002a) The relationship between fruit and leaf growth in lychee (*Litchi chinensis* Sonn.). *Journal of Horticultural Science & Biotechnology* 77, 322–325.

Hieke, S., Menzel, C.M., Doogan, V.J. and Lüdders, P. (2002b) The relationship between yield and assimilate supply in lychee (*Litchi chinensis* Sonn.). *Journal of Horticultural Science & Biotechnology* 77, 326–332.

Hieke, S., Menzel, C.M. and Lüdders, P. (2002c) Shoot development, chlorophyll, gas exchange and carbohydrates in lychee seedlings (*Litchi chinensis*). *Tree Physiology* 22, 947–953.

Hieke, S., Menzel, C.M. and Lüdders, P. (2002d) Effects of leaf, shoot and fruit development on photosynthesis of lychee trees (*Litchi chinensis*). *Tree Physiology* 22, 955–961.

Hieke, S., Menzel, C.M. and Lüdders, P. (2002e) Effects of light availability on leaf gas exchange and expansion in lychee (*Litchi chinensis*). *Tree Physiology* 22, 1249–1256.

Huang, H.S. (1999) Techniques for early bearing and high production of young longan trees. *South China Fruits* 28 (4), 29–30.

Huang, Y.H., Su, J.F. and Wang, Z.X. (1992) Discussion on the relationship between longan tree structure and yield. *Fujian Agricultural Science and Technology* 1, 17–18.

Kennedy, R.A. and Johnson, R.S. (1981) Changes in photosynthetic characteristics during leaf development in apple. *Photosynthesis Research* 2, 213–223.

Koen, T.J., Langenegger, W. and Smart, G. (1981a) Determination of the fertiliser requirements of litchi trees. *Information Bulletin of the Citrus and Sub-Tropical Fruit Research Institute* 103, 9–12.

Koen, T.J., Langenegger, W. and Smart, G. (1981b) Nitrogen fertilization of litchi trees. *Information Bulletin of the Citrus and Sub-Tropical Fruit Research Institute* 107, 9–11.

Kozlowski, T.T. and Keller, T. (1966) Food relations of woody plants. *Botanical Review* 32, 293–382.

Lakso, A.N. (1994) Apple. In: Schaffer, B. and Andersen, P.C. (eds) *The Handbook of Environmental Physiology of Fruit Crops*, Vol. I: *Temperate Crops.* CRC Press, Boca Raton, Florida, pp. 3–42.

Lal, R.L., Tiwari, J.P. and Misra, K.K. (1999) Response of litchi (*Litchi chinensis* Sonn.) cv. Rose Scented to nitrogen and potassium fertilization. *Progressive Horticulture* 31, 79–83.

Li, J.G., Pan, X.W., Li, R. and Dai, H.F. (2001) Study on measures for raising longan fruit quality. *South China Fruits* 30 (4), 29.

Li, R., Pan, X.W. and Li, J.G. (1998) Some important problems for growing longan fruit. *South China Fruits* 27 (4), 28.

Liao, J.S., Liu, S., Chen, Q.X. and Zheng, G.H. (1996) A study on photosynthesis of longan (*Dimocarpus longan* Lour.) and its influential factors. *Acta Horticulturae Sinica* 23, 1–7.

Lin, R.S. (1997) The relationship between fruiting branch and flower cluster quality. *South China Fruits* 26 (4), 29.

Liu, X.H. and Ma, C.L. (2001) Production and research of longan in China. *Acta Horticulturae* 558, 73–92.

Lloyd, J., Wong, S.C., Styles, J.M., Batten, D., Priddle, R., Turnbull, C. and McConchie, C.A. (1995) Measuring and modeling whole-tree gas exchange. *Australian Journal of Plant Physiology* 22, 987–1000.

McAlpine, J.R., Keig, G. and Falls, R. (1983) *Climate of Papua New Guinea.* Commonwealth Scientific and Industrial Research Organization, Canberra, Australia, p. 95.

McConchie, C.A. and Batten, D.J. (1991) Fruit set in lychee (*Litchi chinensis* Sonn.):

variation between flowers, panicles and trees. *Australian Journal of Agricultural Research* 42, 1163–1172.

Menzel, C.M. (2002) *The Lychee Crop in Asia and the Pacific.* Food and Agricultural Organization of the United Nations, Bangkok, Thailand, 108 pp.

Menzel, C.M. and Paxton, B.F. (1985) The effect of temperature on growth and dry matter production of lychee seedlings. *Scientia Horticulturae* 26, 17–23.

Menzel, C.M. and Simpson, D.R. (1988) Effect of temperature on growth and flowering of litchi (*Litchi chinensis* Sonn.). *Journal of Horticultural Science* 66, 335–344.

Menzel, C.M. and Simpson, D.R. (1992) Flowering and fruit set in lychee (*Litchi chinensis* Sonn.) in subtropical Queensland. *Australian Journal of Experimental Agriculture* 32, 105–111.

Menzel, C.M. and Simpson, D.R. (1995) Temperatures above 20°C reduce flowering in lychee (*Litchi chinensis* Sonn.). *Journal of Horticultural Science* 70, 981–987.

Menzel, C.M., Haydon, G.F., Doogan, V.J. and Simpson, D.R. (1994) Time of nitrogen application and yield of Bengal lychee on a sandy loam soil in subtropical Queensland. *Australian Journal of Experimental Agriculture* 34, 803–811.

Menzel, C.M., Oosthuizen, J.H., Roe, D.J. and Doogan, V.J. (1995a) Water deficits at anthesis reduce CO_2 assimilation and yield of lychee (*Litchi chinensis* Sonn.) trees. *Tree Physiology* 15, 611–617.

Menzel, C.M., Rasmussen, T.R. and Simpson, D.R. (1995b) Carbohydrate reserves in lychee trees (*Litchi chinensis* Sonn.). *Journal of Horticultural Science* 70, 245–255.

Menzel, C.M., Simpson, D.R. and Doogan, V.J. (1996) Preliminary observations on growth, flowering and yield of pruned lychee trees. *Journal of the Southern African Society for Horticultural Sciences* 6, 16–19.

Menzel, C.M., Olesen, T., McConchie, C., Wiltshire, N., Diczbalis, Y. and Wicks, C. (2000) *Lychee, Longan and Rambutan: Optimum Canopy Management.* Rural Industries Research and Development Corporation, Canberra, Australia, 92 pp.

Menzel, C.M., Bagshaw, J., Campbell, T., Greer, N., Noller, J., Olesen, T. and Waite, G.K. (2002) *Lychee Information Kit.* Queensland Department of Primary Industries, Nambour, Australia.

Milne, D.L. (1999) Lychee production and research in South Africa. In: Menzel, C.M., Waite, G.K. and Bronson, R. (eds) *Proceedings of the Fifth National Lychee Conference.* Australian Lychee Growers' Association, Twin Waters, Queensland, pp. 25–31.

Mitchell, P.D., Chalmers, D.J., Jerie, P.H. and Burge, G. (1986) The use of initial withholding of irrigation and tree spacing to enhance the effect of regulated deficit irrigation on pear trees. *Journal of the American Society for Horticultural Science* 111, 858–861.

Monteith, J.L. (1977) Climate and the efficiency of crop production in Britain. *Philosophical Transactions of the Royal Society of London Series B* 281, 277–294.

Olesen, T., Menzel, C.M., Wiltshire, N. and McConchie, C.A. (2002) Flowering and shoot elongation of lychee in eastern Australia. *Australian Journal of Agricultural Research* 53, 977–983.

Papademetriou, M.K. and Dent, F.J. (2002) *Lychee Production in the Asia-Pacific Region.* Food and Agricultural Organization of the United Nations, Bangkok, Thailand, 128 pp.

Quinlan, J.D. and Preston, A.P. (1971) The influence of shoot competition on fruit retention and cropping of apple. *Journal of Horticultural Science* 46, 525–534.

Ray, P.K. and Sharma, S.B. (1988) Fruit growth dynamics and yield relationships in 'Shahi' litchi. *Indian Journal of Horticulture* 45, 171–179.

Riddoch, I., Lehto, T. and Grace, J. (1991) Photosynthesis of tropical tree seedlings in relation to light and nutrient supply. *New Phytologist* 119, 137–147.

Roe, D.J., Menzel, C.M., Oosthuizen, J.H. and Doogan, V.J. (1997) Effects of current CO_2 assimilation and stored reserves on lychee fruit growth. *Journal of Horticultural Science* 72, 397–405.

Schaffer, B., Ramos, L. and Lara, S.P. (1987) Effect of fruit removal on net gas exchange of avocado leaves. *HortScience* 22, 925–927.

Schaffer, B., Whiley, A.W. and Kholi, R.R. (1991) Effects of leaf age on gas exchange characteristics of avocado (*Persea americana* Mill.). *Scientia Horticulturae* 48, 21–28.

Stern, R.A. and Gazit, S. (1997) Effect of 3,5,6-trichloro-2-pyridyl-oxyacetic acid on fruitlet abscission and yield of 'Mauritius' litchi (*Litchi chinensis* Sonn.). *Journal of Horticultural Science* 72, 659–663.

Stern, R.A., Kigel, J., Tomer, E. and Gazit, S. (1995) 'Mauritius' lychee fruit development and reduced abscission after treatment with the auxin 2,4,5-TP. *Journal of the American Society for Horticultural Science* 120, 65–70.

Stern, R.A., Nadler, M. and Gazit, S. (1997) 'Floridian' litchi yields increased by 2,4,5-TP spray. *Journal of Horticultural Science* 72, 609–615.

Stern, R.A., Stern, D., Harpaz, M. and Gazit, S. (2000) Applications of 2,4,5-TP, 3,5,6-TPA, and combinations thereof increase lychee fruit size and yield. *HortScience* 35, 661–664.

Stern, R.A., Stern, D., Miller, H., Xu, H.F. and Gazit, S. (2001) The effects of the synthetic auxins 2,4,5-TP and 3,5,6-TPA on yield and fruit size of young 'Fei zi xiao' and 'Hei ye' litchi trees in Guangxi Province, China. *Acta Horticulturae* 558, 285–288.

Subhadrabandhu, S. and Yapwattanaphun, C. (2001) Lychee and longan production in Thailand. *Acta Horticulturae* 558, 49–57.

Syvertsen, J.P. and Lloyd, J. (1994) Citrus. In: Schaffer, B. and Andersen, P.C. (eds) *The Handbook of Environmental Physiology of Fruit Crops*, Vol. II: *Sub-tropical and Tropical Crops*. CRC Press, Boca Raton, Florida, pp. 65–99.

Teng, Y., Tanabe, K., Tamura, F. and Itai, A. (1998) Effects of spur defoliation and girdling on fruit growth, fruit quality, and leaf and shoot carbohydrate levels on the spurs of different ages in 'Nijisseiki' pear. *Journal of the Japanese Society for Horticultural Science* 67, 643–650.

Trueman, S.J. and Turnbull, C.G.N. (1994) Fruit set, abscission and dry matter accumulation on girdled branches of macadamia. *Annals of Botany* 74, 667–674.

Turner, D.W. (1994) Bananas and plantains. In: Schaffer, B. and Andersen, P.C. (eds) *The Handbook of Environmental Physiology of Fruit Crops*, Vol. II: *Sub-tropical and Tropical Crops*. CRC Press, Boca Raton, Florida, pp. 37–64.

Turner, D.W., Menzel, C.M. and Simpson, D.R. (1996) Short-term drying of half the root system reduces growth but not water status or photosynthesis in leaves of passionfruit (*Passiflora* sp.). *Scientia Horticulturae* 65, 25–36.

Whiley, A.W., Schaffer, B. and Lara, S.P. (1992) Carbon dioxide exchange of developing avocado (*Persea americana* Mill.) fruit. *Tree Physiology* 11, 85–94.

Whiley, W.A., Searle, C., Schaffer, B. and Wolstenholme, B.N. (1999) Cool orchard temperatures or growing trees in containers can inhibit leaf gas exchange of avocado and mango. *Journal of the American Society for Horticultural Science* 124, 46–51.

Wiebel, J., Eamus, D., Chacko, E.K., Downton, W.J.S. and Lüdders, P. (1993) Gas exchange characteristics of mangosteen (*Garcinia mangostana* L.) leaves. *Tree Physiology* 13, 55–69.

Wiebel, J., Chacko, E.K., Downton, W.J.S. and Lüdders, P. (1994) Influence of irradiance on photosynthesis, morphology and growth of mangosteen (*Garcinia mangostana* L.) seedlings. *Tree Physiology* 14, 263–274.

Yen, C.R. and Tien, Y.J. (1985) Effect of pruning on the growth and yield of litchi. 1. Effect of harvest pruning on the flushing and flowering in varieties of different vigor. *Journal of the Chinese Society for Horticultural Science* 31, 40–43.

Young, T.W. (1977) Effect of branch girdling on yield of severely pruned 'Brewster' lychee trees. *Proceedings of the Florida State Horticultural Society* 90, 251–253.

Yuan, R.C. and Huang, H.B. (1988) Litchi fruit abscission: its patterns, effect of shading and relation to endogenous abscisic acid. *Scientia Horticulturae* 36, 281–292.

10 Plant Water Relations and Irrigation

Christopher M. Menzel
Department of Primary Industries and Fisheries, PO Box 5083, SCMC, Nambour, Queensland 4560, Australia

Introduction

Litchis and longans are native to the tropical rainforests of South-east Asia, where the trees are subject to seasonal droughts, and where most of the rain falls between April and September. However, both species have a deep root system and can survive long dry periods (Menzel, 1991; Menzel and Simpson, 1994). Cloudy weather and high humidity, even during the dry season, also tend to reduce the impact of drought on the trees. Irrigation of these crops in Asia is rare because of its cost or lack of infrastructure, whereas supplementary watering is required to produce commercial crops in South Africa, Israel and Australia.

Water deficits affect many aspects of plant physiology. In most species, droughts close the stomata and reduce CO_2 assimilation, stem extension, leaf expansion and fruit growth. Some plant functions or tissues are more sensitive than others to drought. Differences in the sensitivity of the shoots and fruit to drought have been used to increase yields and returns in pear, *Pyrus communis*, and similar crops (Mitchell and Chalmers, 1982). However, these ideas have been little explored in tropical species, and the use of regulated deficit irrigation to increase productivity is rare.

Both crops are particularly sensitive to water supply during the period from flowering to early fruit development. Yield and quality in areas with access to supplementary watering are generally superior to those in orchards dependent on rainfall. For litchi, where irrigation is available, a watering of 60–80 mm every second or third week is usually sufficient for satisfactory production. Experiments have shown that soil water deficits can be used to manipulate flushing patterns and increase flowering and possibly fruit production, under some circumstances. Less information is available for longan.

Experimental approaches

Most of the information on the water relations of litchi has been collected in South Africa, Israel and Australia, where the orchards are typically irrigated. Fewer data sets are available from South-east Asia, where most of the trees depend on rain. Only limited research has been conducted on longan, and it is not known whether the two species have similar water requirements. Researchers have been interested in determining the impacts of drought on plant water relations and gas exchange under laboratory, glasshouse and field conditions through attempts to model the opening and closing of the stomata. Various authors have studied the effects of drought on shoot growth, flowering, fruit production and quality. The relationship between plant water use and weather has been studied and irrigation recommendations have been developed for commercial orchards. The responses of trees growing in the field have generally confirmed models developed from data obtained from potted plants.

Soil and plant water relationships

Soil water is often described in terms of water potential, Ψ, a measure of the free energy of water, with pure water without solutes having a Ψ value of zero. Soil water potential is negative. Water typically moves along a gradient from higher (wetter, less negative) to lower potential (drier, more negative) from the soil to root, to the leaf and then to the air. Soils are porous and hold water in the pores by capillary forces (Atwell *et al.*, 1999), and as plants withdraw water for use in transpiration, the soil dries and the larger pores drain. Water in the remaining pores is held even more tightly and is under suction or negative hydrostatic pressure. This suction, equivalent to soil water potential or Ψ_{soil} in most situations, typically ranges from 0.01 MPa at field capacity to more than 1.5 MPa in very dry soils.

The volume of water available for plant growth varies with soil type and rooting depth. Most authors present data in terms of the depth of water available between field capacity and wilting point, equivalent to Ψ_{soil} of –0.01 and –1.5 MPa. About 70 mm of water is available in a 1 m profile of sand at field capacity, 140 mm of water in a sandy loam, 220 mm of water in a loam, 140 mm of water in a clay loam, and 200 mm of water in a well-structured clay (Atwell *et al.*, 1999). Many authors present data as average soil water content, θ (in %), for a particular depth of soil. A sandy soil would have a θ of about 17% at field capacity and about 10% at wilting point, equivalent to 70 mm of water available in 1 m.

The water relations of plant tissues are usually described in terms of water potential or relative water content (Whiley and Schaffer, 1997). As in soil, Ψ in plants is the measure of the free energy of water, with pure water without solutes having a Ψ of zero. As solutes are added, Ψ declines and becomes more negative. Hydrostatic pressure (tension) also exerts a significant effect on the energy of water. Thus Ψ has two main components in plants, the osmotic or solute potential, Ψ_π, and hydrostatic or pressure (turgor) potential, Ψ_p:

$$\Psi = \Psi_\pi + \Psi_p \qquad (1)$$

There is also a third component, matric potential, Ψ_m, related to the adhesive forces between water molecules and the cell walls, although these are generally negligible in plants. In cells, Ψ_p is generally positive or equal to zero, while in xylem tissues of transpiring plants, Ψ_p is negative (under tension).

As with most plants, leaf water potential in litchi has typically been determined using a pressure chamber (Ritchie and Hinckley, 1975; Atwell *et al.*, 1999). When pressure is applied to an excised leaf sealed into the chamber, xylem sap is forced back to the cut surface. This point or the balancing pressure (BP) is taken to be numerically equivalent, although opposite in sign, to the original xylem tension, Ψ_{xylem}. At equilibrium, the following relationship applies:

$$-BP = \Psi_{xylem} = \Psi_{leaf} \qquad (2)$$

Well-watered trees typically have a Ψ_{leaf} of about –1.0 MPa in the early morning. Values of Ψ_{leaf} of –2.0 MPa indicate moderate water deficits, while values below –3.0 MPa indicate severe drought. Leaves often wilt when Ψ_{leaf} falls to –4.0 MPa, and they die at lower values.

Some authors use the relative water content (RWC) of the leaves as an index of plant water status. This is derived by taking fresh leaf samples, soaking them in water under low light for 24 h until they are fully turgid, and drying them (Turner, 1981). Relative water content can then be calculated from the original fresh weight (Wt_{fresh}), turgid weight (Wt_{turgid}) and dry weight (Wt_{dry}):

$$\text{RWC (\%)} = [(Wt_{fresh} - Wt_{dry})/(Wt_{turgid} - Wt_{dry})] \times 100 \qquad (3)$$

Net CO_2 assimilation and growth typically fall as RWC declines. Fully hydrated plants have a RWC of above 90%, while values between 70% and 80% represent a moderate drought. Values below 70% are associated with severe drought.

Irrigation requirements

The difference between evaporation and rainfall in a given month, along with the depth of water stored in the soil, can be used to indicate the likely need for supplementary watering. The actual evaporation rate provides an estimate of the potential water application at that time, in the absence of rainfall. Some examples are provided for sites in China, Israel, South Africa and Australia (Fig. 10.1a–d).

The different sites would be expected to experience drought at different times during the phenological cycle. It is dry for most of the year in Nelspruit, South Africa, and Bundaberg, Australia. In Guangzhou, China, the period from flower initiation to anthesis experiences a short drought, while Alstonville, Australia, has a longer drought from anthesis to early fruit growth. This extends from before flower initiation to harvest at Cairns and Mareeba, Australia. The dry period is prolonged in Galilee, Israel, and covers the period from anthesis until just before flower initiation.

Guangzhou (latitude 23°N; elevation 18 m), the centre of litchi and longan production in China, has a moist subtropical climate. Annual rainfall is 1694 mm, with a relatively dry period from October to March during flower initiation and development (Fig. 10.1a). Annual evaporation is similar to rainfall of 1716 mm, at about 150 mm/month from May to November. Evaporation exceeds rainfall for 6 months of the year, with about 100 mm of irrigation (evaporation – rain) required each month from October to December, leading up to flower initiation. Fruit are normally harvested during May and June in most districts, when the rainy season has started, so trees without supplementary watering experience water deficits from flower initiation to anthesis.

Litchi-growing regions in Israel have a characteristic Mediterranean climate with hot, dry summers and cool, wet winters. In Galilee in the north-east (lat. 32°N; 200 m below sea level), most of the rain falls between December and March, covering the period from floral initiation to fruit set (Fig. 10.1b). Annual rainfall is only 613 mm and annual evaporation is double that at 1877 mm. Evaporation exceeds rainfall for 9 out of 12 months, with 200 mm of irrigation (evaporation – rain) required each month from May to September. These seasonal droughts are used to manipulate flushing, with only irrigated trees producing new shoots after harvest. Fruit are harvested in early July in this environment.

Nelspruit in South Africa (lat. 25°S; elevation 660 m) has a similar annual rainfall (770 mm) to Galilee, but has a subtropical climate (Fig. 10.1b). Most of the rain falls between October and March (summer dominance), and is especially heavy during the latter part of fruit development. Annual evaporation is similar to Galilee at 1742 mm, with highest rates (160 mm/month) from September to March and evaporation exceeding rainfall in all months. There is a total irrigation requirement (evaporation – rain) of 972 mm compared with 1264 mm in Israel.

Alstonville in northern New South Wales (lat. 29°S) is at the southern range of litchi and longan production areas in Australia and has a wet subtropical climate. Annual rainfall is 1861 mm, with more rain in summer than in winter (Fig. 10.1c). Evaporation ranges from 75 mm in June to 186 mm in December. Rainfall exceeds evaporation for 7 months of the year, with a peak irrigation demand (evaporation – rainfall) of 86 mm in September.

Bundaberg in southern Queensland (lat. 25°S, 14 m elevation) is at the centre of commercial production in Australia, and has a dry subtropical climate (Fig. 10.1c). Annual rainfall is only 60% of that at Alstonville, with winter being much drier than summer. Evaporation ranges from 81 to 220 mm/month, 20% higher than at Alstonville. Evaporation exceeds rainfall in all months, with 85–113 mm (evaporation – rain) required each month between September and December.

Cairns and Mareeba in Queensland (lat. 17°S) are located at the northern end of commercial production in Australia. Cairns is situated on the coast (elevation of 3 m) and has a wet tropical climate, whereas Mareeba is further inland at an elevation of 404 m and is much drier (Fig. 10.1d). Annual rainfall is 2029 mm at Cairns and 925 mm at Mareeba. Both sites

Fig. 10.1a–d. (*see next 2 pages*) Seasonal changes in rainfall, evaporation and potential irrigation requirement (evaporation – rainfall) for various litchi- and longan-growing areas: (a) Guangzhou, China (lat. 23°N, elevation 18 m); (b) Nelspruit, South Africa (lat. 25°S, elevation 660 m); Galilee, Israel (lat. 32°N, elevation 200 m below sea level); (c) Alstonville, Australia (lat. 29°S, elevation 140 m); Bundaberg, Australia (lat. 25°S, elevation 14 m); (d) Cairns, Australia (lat. 17°S, elevation 3 m); Mareeba, Australia (lat. 17°S, elevation 404 m).

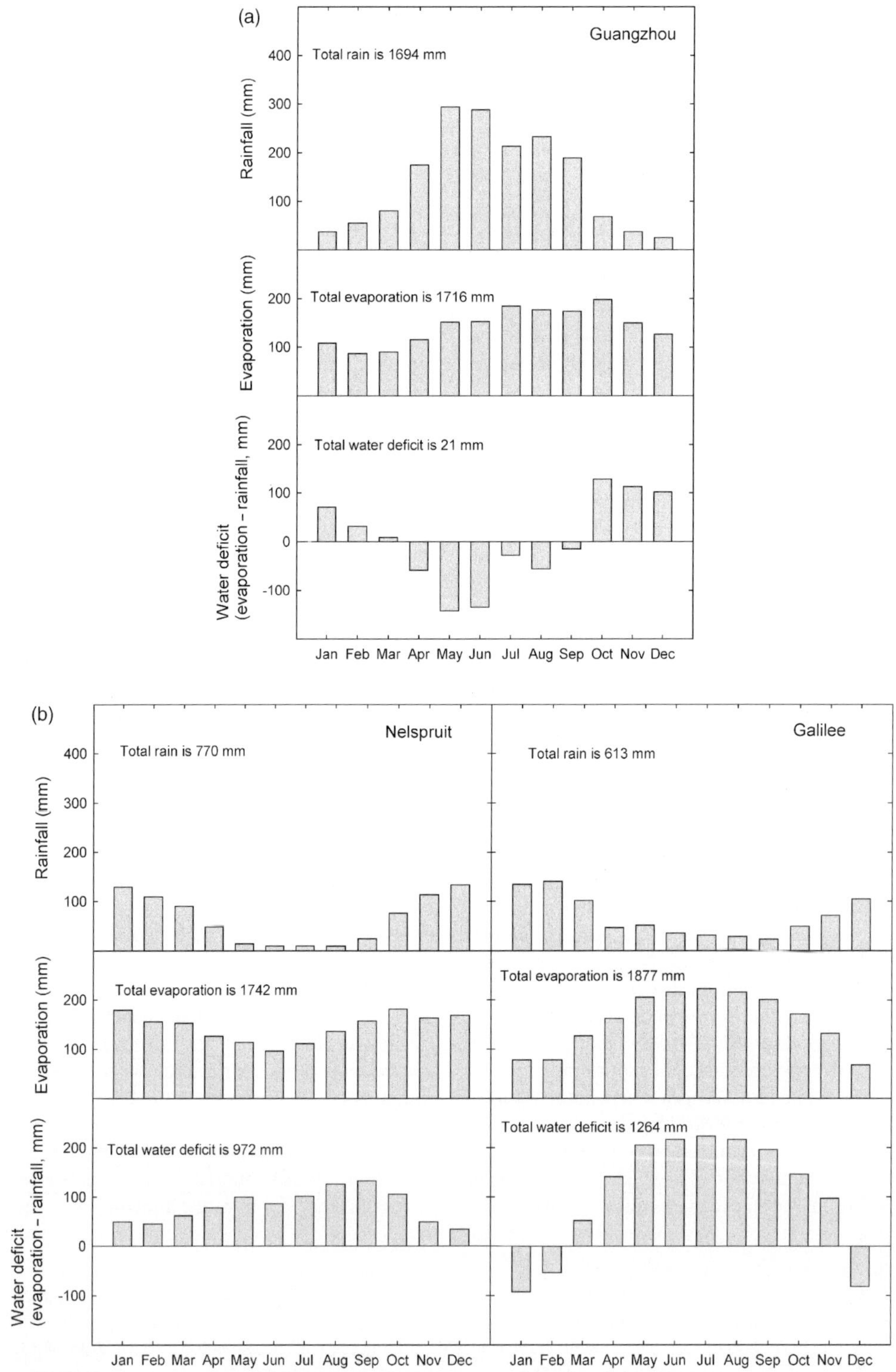
(a)
Guangzhou
Total rain is 1694 mm
Rainfall (mm)
Total evaporation is 1716 mm
Evaporation (mm)
Total water deficit is 21 mm
Water deficit (evaporation - rainfall, mm)
Jan Feb Mar Apr May Jun Jul Aug Sep Oct Nov Dec
(b)
Nelspruit
Total rain is 770 mm
Total evaporation is 1742 mm
Total water deficit is 972 mm
Galilee
Total rain is 613 mm
Total evaporation is 1877 mm
Total water deficit is 1264 mm
Rainfall (mm)
Evaporation (mm)
Water deficit (evaporation - rainfall, mm)
Jan Feb Mar Apr May Jun Jul Aug Sep Oct Nov Dec

Fig. 10.1a, b.

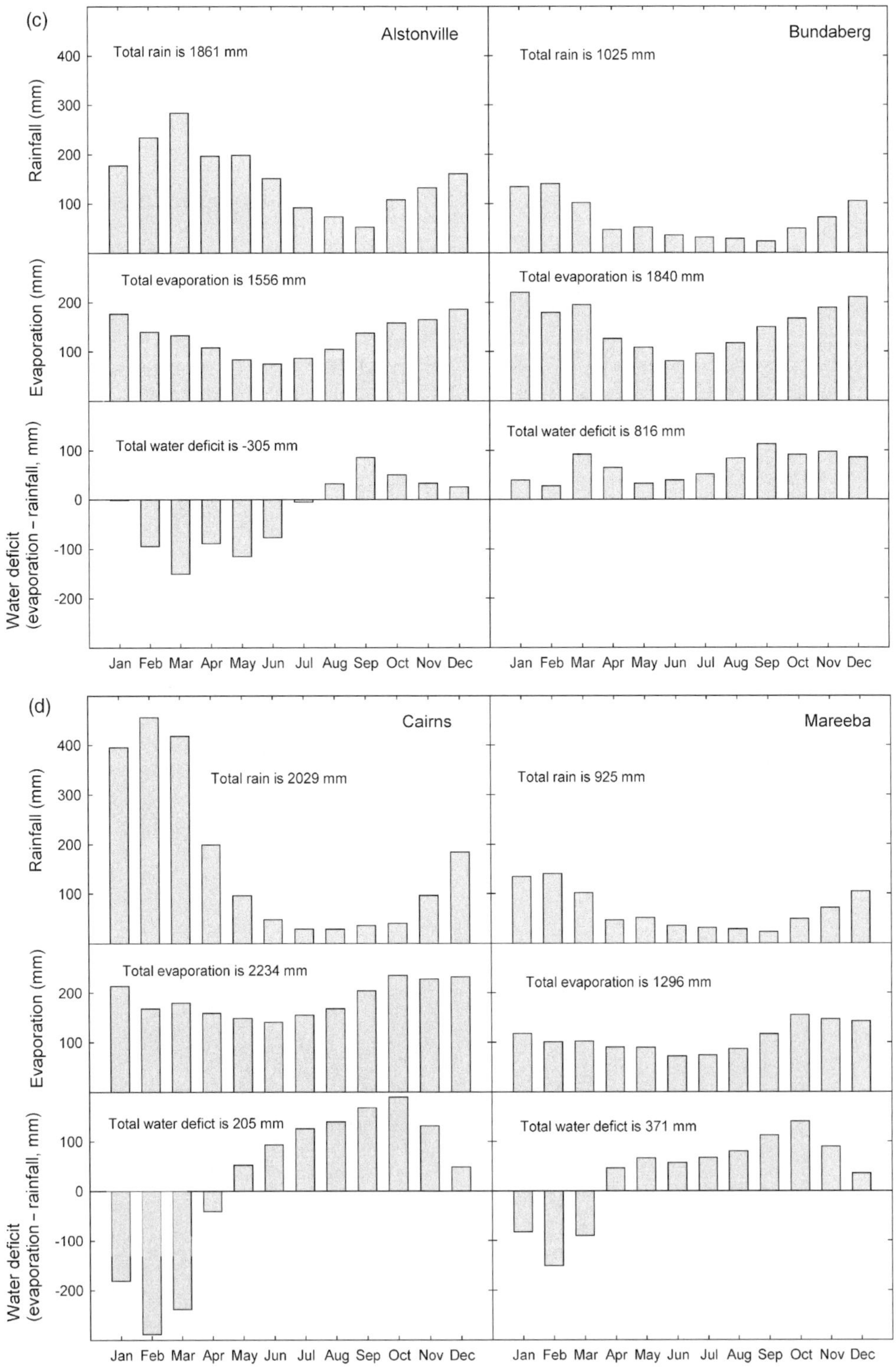
(c)
Alstonville
Bundaberg
Total rain is 1861 mm
Total rain is 1025 mm
Total evaporation is 1556 mm
Total evaporation is 1840 mm
Total water deficit is -305 mm
Total water deficit is 816 mm
Rainfall (mm)
Evaporation (mm)
Water deficit (evaporation – rainfall, mm)
(d)
Cairns
Mareeba
Total rain is 2029 mm
Total rain is 925 mm
Total evaporation is 2234 mm
Total evaporation is 1296 mm
Total water deficit is 205 mm
Total water deficit is 371 mm
Jan Feb Mar Apr May Jun Jul Aug Sep Oct Nov Dec

Fig. 10.1c, d.

receive most of the rain between December and March, with the rest of the year being relatively dry. Evaporation varies from 141 to 236 mm/month at Cairns and from 72 to 155 mm/month at Mareeba. The higher values for Cairns are related to strong coastal breezes. Evaporation exceeds rainfall for 8 or 9 months of the year, with peak irrigation demand in September and October.

The accumulated water deficit (evaporation – rainfall) during the year is greatest at Bundaberg (816 mm), followed by Mareeba (371 mm) and Cairns (205 mm). Rainfall exceeds evaporation at Alstonville by 305 mm. This analysis suggests that irrigation is essential for commercial production in Bundaberg, Mareeba and Cairns, but is less important at Alstonville.

Bundaberg has a similar latitude to Guangzhou (23–24°), and similar annual evaporation (about 1800 mm). However, Bundaberg's rainfall is only 60% of that at Guangzhou (1025 mm vs. 1694 mm). The rainfall is also less uniformly distributed during the year. It can be concluded that unirrigated trees face greater water deficits in Bundaberg than in Guangzhou. The weather at Guangzhou matches that at Alstonville more closely, even though the latter has a higher latitude (29°). Alstonville is slightly wetter, with more consistent rainfall during the year. These data suggest that irrigation is required for commercial production in Israel, South Africa and China, but is more critical at Galilee, since it is very dry leading up to harvest. Nelspruit is similar to Bundaberg in Australia, but is drier than Alstonville, Cairns and Mareeba.

Irrigation in different countries

Most of the litchi and longan orchards in China are not irrigated (Fig. 10.2), although some trees planted along rivers and streams have access to water at depth in the soil (Menzel, 2002). Only a few of the new orchards planted away from the rivers at elevation are irrigated, since water resources are generally reserved for rice. It is usually dry from October to March, during flowering and fruit set in Guangzhou, Guangxi and Fujian. Huang (1989) suggested that drought reduced fruit production in Guangdong in unirrigated orchards, with severe dry weather resulting in small and split fruit and excessive leaf fall. Drought is also reported to reduce production of these crops in Taiwan (Liao, 1997).

Litchi orchards in Viet Nam are reliant on rainfall, since irrigation is not readily available in the elevated production areas (Menzel, 2002). Most of the litchi orchards in Thailand are located in the northern hills on steep slopes and are not easily irrigated (Subhadrabandhu, 1990; Menzel, 2002). In some years, severe droughts reduce production. Mini-sprinklers

Fig. 10.2. A longan orchard in Guangxi, southern China. Most litchi and longan plantings in South-east Asia are not irrigated.

are sometimes used in the larger commercial plantings in the lower areas. Longans in Thailand also experience drought where supplementary watering is not available.

Experiments in India showed that regular irrigation was required for commercial litchi production (Menzel, 2002). Supplementary watering reduced the incidence of fruit-splitting, which can be quite severe in some districts. Most orchards are watered by hand or are flood-irrigated, even though drippers are more efficient. Trees in Nepal, Bangladesh and the Philippines also suffer water deficits during fruit development. Irrigation is generally recommended for commercial production in South Africa, Israel and Australia (Anonymous, 1992; Goren and Gazit, 1996; Menzel and McConchie, 1998).

Tree water use

Water use in mature trees has been studied in relation to the weather through simple estimates of crop evapotranspiration, ET_{crop}. This relies on direct measurements of evaporation from a Class A pan evaporimeter, E_{pan}, and the application of an appropriate crop factor, k_{crop}, where:

$$ET_{crop} = E_{pan} \times k_{crop} \quad (4)$$

Crop factors vary with the species and growing environment and range from 0.2 to 1.3, with values declining as the soil dries (Atwell *et al.*, 1999). Evapotranspiration includes transpiration from the crop as well as evaporation from the soil. There are no data separating evaporation from the soil and transpiration from the tree in these crops, although in a mature litchi or longan orchard with thick leaf mulch and a long irrigation cycle, evaporation from the soil would be expected to account for only a small proportion of ET_{crop}.

An alternative approach to using an evaporimeter is to rely on estimates of potential evapotranspiration under non-limiting conditions, using local values of radiation, temperature, wind and vapour pressure, which govern water loss from a canopy. To estimate ET_{crop}, the daily reference crop evapotranspiration value, E_{ref}, is modified by an empirical crop coefficient, k_{crop}, where:

$$ET_{crop} = E_{ref} \times k_{crop} \quad (5)$$

There are several different models used to calculate $E_{ref,}$ including the Penman–Monteith equations that use a reference crop, usually lucerne or a grass (Atwell *et al.*, 1999). Different crop coefficients are required if lucerne or grass is used as the reference crop. Values normally range from 0.2 in bare soil to > 1 in dense canopies. Many earlier reports refer to potential evapotranspiration rather than to reference evapotranspiration. The original term was used to describe transpiration from a plant surface without any limitation imposed by water availability or transpiration. These various relationships have been investigated in litchi.

In Israel, litchi orchards are irrigated to replace 50–60% of E_{pan} (Stern *et al.*, 1993). The dry season lasts 6 months, with an annual rainfall of only 600 mm, making irrigation particularly important during the summer. South African researchers suggested an allowable depletion of 50% of available soil water from August to March before irrigation applications are necessary, and 90% depletion from April to July, with an effective rooting depth of 0.4 m (Anonymous, 1992). This is a difficult strategy to implement unless actual soil water levels are determined. The suggested k_{crop} was 0.85, with water use calculated to be 3.3 mm/day in July and 5.6 mm/day in March. These were theoretical calculations only and were not verified by actual tree water use. In drought experiments in South Africa (Fig. 10.3), average ET_{crop} from panicle emergence in July to harvest in January was 3.7 mm/day, with E_{pan} ranging from 3 to 10 mm/day (Menzel *et al.*, 1995). This resulted in k_{crop} ranging from 0.4 to 1.2, with no seasonal trend, but decreasing with E_{pan} (Fig. 10.4).

In India, ET_{crop} was related to ET_{ref} (Hasan and Chattopadhyay, 1992). Values of k_{crop} ranged from 0.9 to 1.5, with an ET_{crop} of 2 mm/day in winter and 5 mm/day in summer. The results of the experiments in South Africa and India show that tree water use does not increase linearly with increasing E_{pan} or $ET_{ref,}$ presumably because of the partial closure of the stomata during dry weather. Estimates of tree water use calculated from k_{crop} appear unreliable for scheduling irrigation in litchi, and probably also in longan.

Fig. 10.3. Stand of 10-year-old 'Tai So' ('Mauritius') litchis used to study the water relations of orchards in South Africa (photograph courtesy of Johan Oosthuizen).

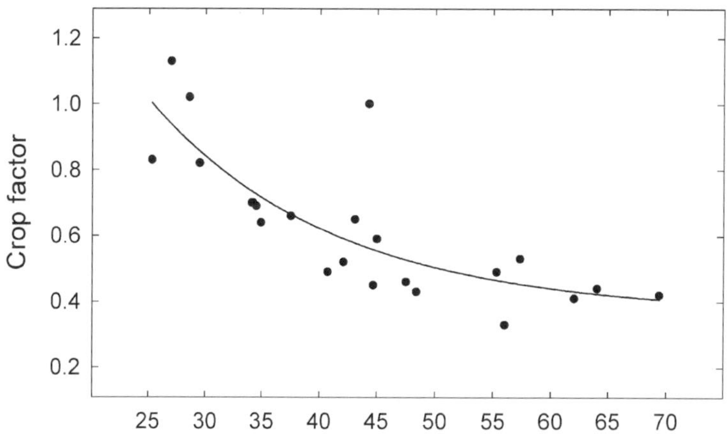

Fig. 10.4. The relationship between the crop factor, k_{crop}, and evaporation from a Class A pan evaporimeter (E_{pan}) in a 'Tai So' litchi orchard in Nelspruit, South Africa. Trees irrigated weekly (from Menzel *et al.*, 1995, with permission).

Extraction of soil water

Soil water extraction patterns have been described for litchi in a few locations, and they probably also apply to longan. In Australia, the roots of mature trees were found to a depth of 1 m in many soils, although most roots were located in the 0.0–0.6 m horizon (Menzel *et al.*, 1990). Other research indicated that the depth of rooting might not necessarily indicate water extraction over that depth. Eight-year-old 'Bengal' trees growing on a red clay loam in northern New South Wales extracted soil water from below 1 m (Batten *et al.*, 1994). Calculations of ET_{crop} during a drought indicated a total water deficit of 420 mm for the non-irrigated plants, with 180 mm available to a soil water potential, ψ_{soil}, of −1.0 MPa in the top 1 m, and a total of 330 mm available in the top 2 m. That is, the trees extracted about 240 mm below 1 m, and about 90 mm below 2 m.

In South Africa, mature 'Tai So' trees extracted water from below 1.5 m in a sandy loam (Menzel *et al.*, 1995). The average soil water content, θ, in well-watered plots was 14.5% during the experiment (Fig. 10.5). In droughted plots, θ declined from 14.2% at week 1, to 8.9% at week 10, and varied between 7.6% and 9.5% from week 11 to week 26 at the end of the drought. There was little water use from 0 to 1.5 m after week 16, indicating that the trees were extracting soil water below this depth. Net depletion in the top 1.5 m from weeks 1 to 16 was 126 mm. Soil water was used from all depths, with 28% of measured water use from 0 to 1.5 m extracted from 0–0.3 m, 23% from 0.3–0.6 m, 17% from 0.6–0.9 m, 19% from 0.9–1.2 m, and 13% from 1.2–1.5 m. It can be concluded that mature trees are able to extract soil water from considerable depths, and that a short-term drought may not necessarily reduce photosynthesis and growth.

Relationship between gas exchange and tree water status

Plants take up water from the soil and transpire it through the leaves, with a small fraction used in metabolism. Leaf water potential declines (becomes more negative) during the day as the rate of water loss by the leaves exceeds the rate of water uptake by the roots. Leaf water potential also declines during periods of drought. Initially, the plants recover during the night, but eventually Ψ_{leaf} reaches critical values that reduce metabolism and growth. Stem extension and leaf expansion typically cease before net CO_2 assimilation (*A*) becomes zero.

Diurnal changes in leaf gas exchange in well-watered and droughted 'Tai So' trees were studied in South Africa (C.M. Menzel, J.H. Oosthuizen and D. Roe, unpublished data, Nelspruit, 1993, cited in Menzel and Simpson, 1994). Leaf water potential declined from pre-dawn to early in the afternoon, with a near full recovery in the late afternoon in well-watered controls, but not in droughted plants (Fig. 10.6). Stomatal conductance, g_s, and *A* reached maximum values at 0700–0800 h, and were lower in droughted trees than in the controls for most of the day.

Liao *et al.* (1996) studied the gas exchange of longan in Fujian, China. There was a broad optimum between April and August, reflecting higher light, temperature and soil water contents. These authors also studied diurnal variations in gas exchange during summer. On fine sunny days, *A* reached maximum values at 1000 h, declined during the middle of the day, and then recovered late in the afternoon. In contrast, there was no midday depression during cloudy weather. Maximum CO_2 assimilation was more

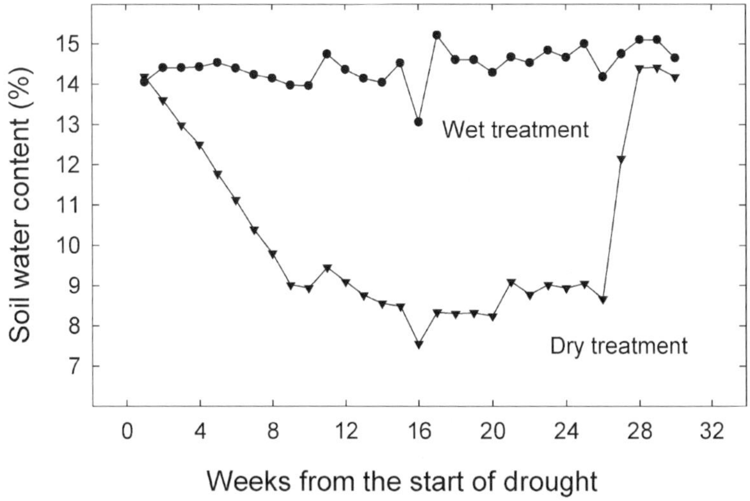

Fig. 10.5. Effects of drought on changes in volumetric soil water content, θ (0–1.5 m), in a 'Tai So' litchi orchard in Nelspruit, South Africa. Irrigation withdrawn from half the plots from panicle emergence on 28 July (week 1). Trees re-watered after harvest in January (week 26). Trees in the wet treatment irrigated weekly (from Menzel *et al.*, 1995, with permission).

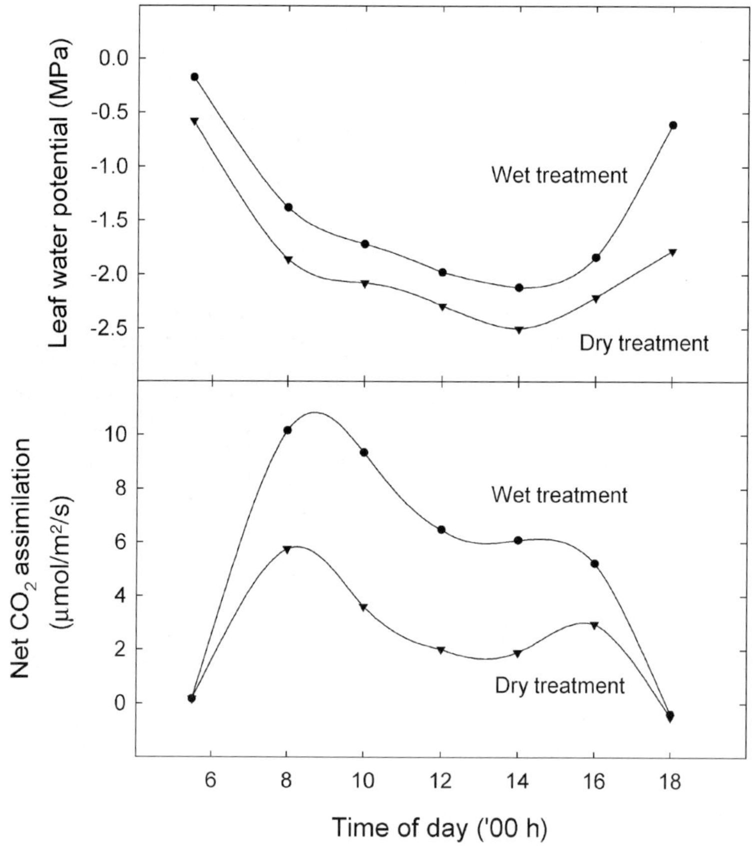

Fig. 10.6. Diurnal changes in leaf water potential, Ψ_{leaf}, and net CO_2 assimilation, *A*, in October in a 'Tai So' litchi orchard in Nelspruit, South Africa. Irrigation withdrawn from half the plots from panicle emergence in July. Trees re-watered after harvest in January (week 26). Trees in the wet treatment irrigated weekly. Volumetric soil water content, θ (0–1.5 m), in the wet treatment was 14.4% and 9.1% in the dry treatment (data from Christopher Menzel and Johan Oosthuizen, South Africa, 1992).

than double under clear skies (9 µmol/m²/s) than under cloudy weather (4 µmol/m²/s).

Unirrigated 8–10-year-old 'Bengal' trees growing in a deep, well-drained clay that was dried out from flowering until harvest were compared with trees irrigated weekly to replace 85% of potential evapotranspiration in drought experiments at Alstonville in northern New South Wales (Batten *et al.*, 1994). Potential evaporation (same term as reference evaporation or ET_{ref} mentioned above) is the water use of a well-watered grass sward. For a Class A pan evaporimeter with a wire bird cover and surrounded by grass, potential evapotranspiration of the grass is about 85% of E_{pan}. Consequently, the irrigated trees were watered to replace 72% of the pan evaporation (pan factor of 0.85 and a k_{crop} of 0.85).

Pre-dawn and noon Ψ_{leaf} declined to –0.9 and –2.4 MPa in unirrigated trees, while corresponding minimum values in the controls were –0.4 and –2.0 MPa. There were no differences in tree water status between the two groups of plants for the first 6 weeks of the experiment, indicating that the unirrigated plants were able to extract sufficient soil water to replace transpiration. Calculated whole-tree conductance did not change with decreasing plant water status, suggesting that the drought was not severe enough to cause xylem cavitations and embolisms. In contrast, daytime g_s declined as tree water status fell (r^2 = 64%), indicating an effect on gas exchange. These experiments suggest a long irrigation cycle is not likely to reduce dry matter production in moist subtropical Australia. The impacts of drought on flowering and fruit growth recorded in these experiments are discussed in a later section.

The relative drought tolerance of litchi was also demonstrated in experiments in South Africa (see Figs 10.7 and 10.8) (Menzel *et al.*, 1995). In well-watered 'Tai So' trees, Ψ_{leaf} at 0900 h ranged from –0.6 to –1.5 MPa. There was more variation in the early afternoon, with Ψ_{leaf} ranging from –0.6 to –2.2 MPa. Leaf water potential in droughted trees ranged from –0.8 to –2.6 MPa in the morning, and from –0.9 to –2.8 MPa in the afternoon. The difference in tree water status between the two groups of plants increased from week 7 to week 20, when Ψ_{leaf} was about 1.0 MPa lower in the droughted plots. There was a slow decline in tree water status as average θ fell from 14.5% to 9.8%, and then a more rapid decline below 9.8% when the trees had used two-thirds of the water available from 0–1.5 m (Fig. 10.9). This response is similar to that in India, where yields did not decline until the trees had used half of the soil water available in the major root-zone (Hasan and Chattopadhyay, 1991; Chandel, 1995).

Because the trees dried out slowly in South Africa (Menzel *et al.*, 1995), *A* did not respond

Fig. 10.8. Measuring leaf photosynthesis in litchi with a leaf chamber in drought experiments in South Africa (photograph courtesy of Johan Oosthuizen).

Fig. 10.7. Measuring leaf water potential in litchi with a pressure chamber in drought experiments in South Africa (photograph courtesy of Johan Oosthuizen).

immediately water was withdrawn, but fell to about 25% of maximum values after 10 weeks, when Ψ_{leaf} fell to –2.5 MPa in the morning (Fig. 10.10). Average values across the season were only half of those in well-watered plants. Tree water status recovered within a week of re-watering, whereas gas exchange remained lower after 5 weeks. It was concluded that drought had a long-term effect on the chemistry of the leaves.

Researchers in Israel considered that mid-day stem water potential, Ψ_{stem}, was a better indicator of tree water status and gas exchange than Ψ_{leaf} in a range of fruit crops including litchi (Stern *et al.*, 1998; Naor, 2000). They found that the correlation between g_s and Ψ_{stem} was higher than that with Ψ_{leaf}. Stem water potential was measured by enclosing a leaf still attached to the tree in a plastic bag covered with foil, and waiting 1.5 h before taking the measurement in the

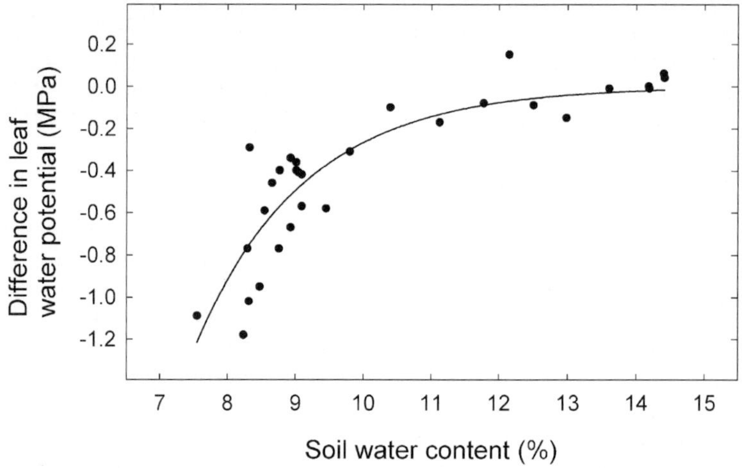

Fig. 10.9. The relationship between the difference in leaf water potential, Ψ_{leaf}, between wet and dry treatments ($\Psi_{leaf\ wet} - \Psi_{leaf\ dry}$), and volumetric soil water content, θ (0–1.5 m), in a 'Tai So' litchi orchard in Nelspruit, South Africa (from Menzel *et al.*, 1995, with permission).

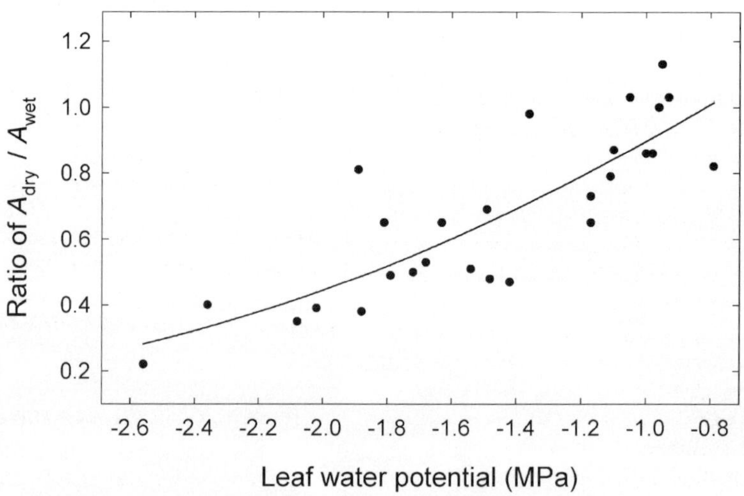

Fig. 10.10. The relationship between relative net CO_2 assimilation, *A* (A_{dry}/A_{wet}), and leaf water potential, Ψ_{leaf}, in a 'Tai So' litchi orchard in Nelspruit, South Africa. Irrigation withdrawn from half the plots from panicle emergence in July. Data collected at 0900 h. Trees re-watered after harvest in January (week 26). Trees in the wet treatment irrigated weekly (from Menzel *et al.*, 1995, with permission).

pressure chamber. Naor (2000) found that there was a higher correlation between yield and Ψ_{stem} than with Ψ_{leaf} in apple, *Malus domestica*, although these ideas were not evaluated in litchi. It was concluded that Ψ_{stem} could be used to schedule irrigation, but the effects of vapour pressure deficit (VPD) on the values needed to be taken into account. These issues are less important when Ψ_{leaf} is measured early in the morning under low VPDs.

Williams and Araujo (2002) studied the changes in Ψ_{leaf} and Ψ_{stem} in two cultivars of grapevine, *Vitis vinifera*, during a drought in the Napa Valley in California. Leaf water potential was measured before dawn (0330 h) and at midday (1230–1330 h) and stem water potential was measured at midday. All three measurements were strongly related to each other, and equally related to soil water content and gas exchange. Leaf water potential at noon was about 0.7 MPa lower than at dawn, while noon leaf water potential was 0.4 MPa lower than noon stem water potential. Their data suggest that all three measurements were valid estimates of vine water status.

CO_2 assimilation is often independent of plant water status until a threshold is reached, and approaches zero when leaves wilt. This critical value can be used to estimate the required range in plant water status for the production of maximum yields. Studies in South Africa showed that the relationship between A and Ψ_{leaf} in 'Tai So' litchi shifted with the rate of soil drying and the number of drought cycles (Roe *et al.*, 1995). Potted plants were grown in sand (quick drying) or clay (slow drying), and with one or two cycles of drought. Asymptotic relationships were shown between relative A (A_{dry}/A_{wet}) and Ψ_{leaf}. The values of Ψ_{leaf} for a relative A of 0.5 and zero were –1.5 and –2.5 MPa for a short drying cycle in sand, shifting to –2.0 and –3.0 MPa in a subsequent drying cycle, and to –2.4 and –3.5 MPa for a long drying cycle in a clay. It was suggested that results collected from potted plants may not necessarily predict the response of field-grown trees, with slow drying and intermittent drought.

Roe *et al.* (1995) also determined whether the effects of drought on photosynthesis were due to the stomata closing or to changes in leaf biochemistry. Both g_s and A declined as Ψ_{leaf} in the morning fell from –1.4 to –3.2 MPa (Table 10.1). Net CO_2 assimilation decreased to 18% of values in well-watered plants when the leaves wilted. The concentration of CO_2 in the leaves (C_i) increased in the droughted plants, suggesting that there was little effect on the entry of CO_2 into the stomata. Plant water status recovered within a day of the trees being re-watered, whereas the resumption of photosynthesis took a few days longer. These data suggest that drought affected the biochemistry of CO_2 fixation. Similar experiments have also been conducted in Australia (Fig. 10.11).

Table 10.1. Effects of drought on leaf water potential (Ψ_{leaf}), stomatal conductance (g_s), net CO_2 assimilation (A) and internal CO_2 concentration (C_i) in potted 'Tai So' litchi in South Africa. From Roe *et al.* (1995), reproduced with permission from the *Journal of Horticultural Science & Biotechnology*.

Drying cycle (days)	Ψ_{leaf} (MPa)	g_s (mmol $H_2O/m^2/s$)	A (μmol $CO_2/m^2/s$)	C_i (ppm)
0	–1.4	170	6.8	286
5	–1.6	125	5.2	282
15	–2.2	100	3.8	294
19	–3.0	80	2.1	323
22	–3.2	70	1.2	334

Little information is available on the effects of drought on gas exchange in longan, although Liao *et al.* (1996) studied the response of potted trees under controlled conditions in Fujian. Net CO_2 assimilation declined when relative soil water content declined below 50% and RWC declined below 90%. Gas exchange was zero when relative soil water content declined to below 30% and RWC declined to below 50%. A relative soil water content of 30% was close to the limit of water extraction by the plants. These results suggest that drought is not likely to affect photosynthesis in longan until the plants have used about 70% of the water available in the soil.

Models of leaf gas exchange

The structure of the leaf includes a barrier or cuticle to restrict water loss so that the bulk of transpiration occurs through the stomatal pores. There is a strong relationship between potential yield and water use in plants, with between

Fig. 10.11. Measuring leaf photosynthesis in drought experiments with potted litchis in southern Queensland, Australia (photograph courtesy of Don Simpson).

100 and 1000 g of water used per gram of dry matter produced (Atwell *et al.*, 1999). Plant physiologists have thus been interested in studying gas exchange in leaves growing in different environments. The optimum conditions for photosynthesis may not necessarily reflect the environment where a plant species evolved, since survival may be more related to optimizing leaf area or reproductive development rather than the rate of photosynthesis per unit leaf area.

Studies on potted 'Bengal' and 'Kwai May Pink' plants in a laboratory indicated that litchi stomata are sensitive to changes in light, temperature and vapour pressure (Batten *et al.*, 1992). In these experiments, environmental parameters were changed independently, whereas this control is not readily available in the field. When photosynthetic photon flux density, PPFD, was 1200 μmol quanta/m^2/s and leaf temperature was 35°C, g_s declined from 120 to 80 mmol H_2O/m^2/s as partial pressure of water vapour, Δe, increased from 20 to 37 mmol/mol. At 25°C, g_s declined from 140 to 90 mmol H_2O/m^2/s as Δe increased from 5 to 20 mmol/mol. Reductions in A with changes in Δe were less pronounced and usually were less than 20%. High vapour pressure deficits could therefore account for reductions in gas exchange during the middle of the day.

In the experiments of Batten *et al.* (1992), maximum values of g_s were obtained with a PPFD of 1200 μmol quanta/m^2/s for 'Kwai May Pink' at all leaf temperatures, and at 13 and 29°C for 'Bengal'. Stomatal conductance was still increasing at the higher irradiance in 'Bengal' at 35°C. In 'Bengal', g_s continued to increase with increasing leaf temperature to 35°C, indicating that the optimum temperature was greater than this value. At a PPFD of 1200 μmol quanta/m^2/s, g_s was 100 mmol H_2O/m^2/s at 23°C and 170 mmol H_2O/m^2/s at 35°C.

Lloyd *et al.* (1995) investigated the diurnal changes in canopy photosynthesis in northern New South Wales by enclosing litchis in whole-tree gas exchange chambers. The responses to light, temperature and VPD were similar to those recorded for single leaves in the laboratory (Batten *et al.*, 1992). However, above a PPFD of 500 μmol quanta/m^2/s, gas exchange was higher under overcast conditions than under clear sky conditions, probably due to lower leaf temperatures and VPDs under the cloud cover. There was also more diffuse light during cloudy weather, with the lower leaves receiving more radiation than when it was sunny. The results of these two experiments suggest that maximum gas exchange in litchi occurs under warm, sunny, humid conditions.

Changes in soil temperature also influence water use in litchi. Average daytime g_s in 'Bengal' doubled from August to December in northern New South Wales as soil temperatures increased from 13 to 26°C (Batten *et al.*, 1992).

The increase in g_s was correlated with soil temperature (r^2 = 63%), but less so with air temperature (r^2 = 41%). Higher average g_s combined with higher E_{pan} during warm weather lead to lower Ψ_{leaf} in summer than in winter. As might be expected, tree water use was greater in summer than in winter.

Liao *et al.* (1996) investigated the changes in gas exchange of longan under laboratory and field conditions in Fujian. Net CO_2 assimilation was saturated with a PPFD of 600–800 μmol quanta/m^2/s in 20- and 60-day-old leaves. The optimum leaf temperature for *A* was 25°C for leaves grown at 22°C, and 28°C for leaves grown at 27°C. Photosynthesis was also higher for leaves grown at the higher temperature. These data suggest that the optimum temperature and light conditions for photosynthesis are lower in longan than in litchi, although the maximum values of *A* seem similar for the two species. Other experiments in China showed that the optimum leaf temperature for gas exchange ranged from 22 to 28°C during the experiment, and was related to the temperature that the leaves had been exposed to in the previous 5–10 days (Liu *et al.*, 1997). The optimum temperature for gas exchange was higher than the prevailing growth temperatures in spring and autumn, and lower in summer. The optimum leaf temperature increased as PPFD increased from 200 to 800 μmol quanta/m^2/s.

Researchers in Japan examined the effects of high temperatures on gas exchange and chlorophyll fluorescence of longan seedlings in the laboratory (Yamada *et al.*, 1996a). The leaves were exposed to a PPFD of 2000 μmol quanta/m^2/s at 30, 33 or 36°C, and a relative humidity of 70%. The leaves were very sensitive to temperature under high light, with *A* at 36°C being only 19% of the value at 30°C. Stomatal conductance also decreased from 204 to 127 mmol H_2O/m^2/s with increasing temperature. Transpiration was not substantially affected, whereas C_i was higher at 36°C than at 30°C. Results from other experiments showed that chlorophyll fluorescence, F_v/F_m, was lower at 40°C (0.808) than at 25°C (0.827). These data suggest that lower photosynthesis was related to higher respiration and photo-inhibition rather than to closure of the stomata. In the same experiments, 'Irwin' mango (*Mangifera indica*) was more tolerant than longan to high temperatures. Yamada *et al.* (1996b) investigated the heat tolerance of several tropical species using the same experimental approach (fluorescence at 25 and 45°C) and found that longan was more sensitive to high temperatures than were the other species.

Relationship between growth and weather

Various authors have attempted to assess the relationship between growth and weather; however, the results of many of the experiments are difficult to interpret because of the correlation between radiation, temperature and rainfall (see Menzel, 1983, 1984; Menzel and Simpson, 1994). Research has indicated that a period of drought can assist flowering, at least in litchi, but is not essential. There is some evidence to suggest that prolonged periods of overcast or rainy weather during anthesis reduce fruit set.

Research over 4 years in Hawaii suggested a link between flowering and water supply in two 'Brewster' orchards (Nakata, 1955). In one experiment, 53% of terminal branches flowered when trees were irrigated and fertilized in August compared with 7% flowering in trees irrigated and fertilized in August and September. It is difficult to separate the effects of water and nutrition in these experiments, although the effects of fertilizers in other experiments were generally small. Nakata's data show a weak inverse relationship between flowering in the orchards and rain during October and November (r^2 = 55%). A closer relationship might have been expected between shoot growth and ψ_{leaf}. In other studies in Florida with the same cultivar, there was little correlation between flowering and soil water status (Young and Harkness, 1961; Young, 1970).

Observations in Kaohsiung, Taiwan (lat. 23°N) suggest that low humidity reduced litchi fruit set (Batten, 1986). There was a strong correlation (r^2 = 99%) between yield and mean relative humidity at 0500 and 2100 h in March, during flowering. Yield ranged from 10 to 40 kg/tree as mean RH increased over the small range from 75% to 85%. In contrast, at Alstonville in Australia, there was no relationship between yield and humidity during female flowering for 'Tai So' over 7 years (Batten, 1986). Mean yield was 36 kg/tree and mean RH at 0900 h was 73%. No information was provided on the range of these parameters. Rain extending over several days during anthesis could cause crop failure because of damage to the flowers. Flowers can be ruined by water, but wet flowers caused by showers that last for less than a day or wetting with insecticides do not generally reduce fruit set (Batten and McConchie, 1992).

Overcast conditions during flowering are common in southern China and other parts of South-east Asia, but are rare in Israel, South Africa and Australia. There was no correlation between the proportion of female flowers setting fruit (19–26%) and daily maxima from 25 to 35°C, or maximum VPDs from 1.5 to 3.5 kPa in northern New South Wales (McConchie and Batten, 1991). Further research is required to define the optimum conditions for pollination and fruit set. Small decreases in fruit set are not likely to reduce yields because most trees set far more fruit than they can carry to harvest (Hieke *et al.*, 2002b). The results presented for these types of experiments highlight the difficulty of using correlations to analyse field data.

Effects of drought on the growth of young trees

Little information is available on the irrigation requirements of young orchards, although Singh and Pathak (1983) studied the effects of drought on 7-year-old 'Calcuttia' trees over 5 years in Uttar Pradesh. The trees were irrigated weekly or fortnightly between March and June (summer), and fortnightly or monthly between October and February (winter). Average rainfall during the two periods was 49 and 162 mm, respectively. The best tree heights and trunk girths were obtained with weekly irrigation in summer and monthly irrigation in winter. Growth was only slightly less with 25 mm applications at each irrigation compared with 75 mm. Tree water use at the lower application rates would be equivalent to 4 and 1 mm/day during summer and winter, respectively.

Effects of drought on flowering in potted plants

Litchi shoot growth is very sensitive to changes in tree water status. Menzel *et al.* (1989) examined the response of 'Kwai May Pink' in a glasshouse at 30°/25°C and found that leaf production, stem extension and shoot dry weight decreased as the level and duration of drought increased. Shoot growth ceased when ψ_{leaf} fell to –1.8 MPa in the morning (Table 10.2). The trees produced new leaves when they were re-watered and ψ_{leaf} rose above –0.7 MPa, but they flowered after being transferred to a glasshouse at 15°/10°C. Chaikiattiyos *et al.* (1994) recorded similar data for other cultivars. Drought does not appear to have a direct effect on litchi flowering, with temperature being more important (Menzel and Simpson, 1988, 1995). A drought may assist flowering by altering flushing patterns in the terminal branches. If

Table 10.2. Effects of drought on shoot growth of 'Kwai May Pink' litchis in a glasshouse. Shoot growth measured after 12 weeks. Data from Menzel *et al.* (1989), reproduced with permission from the *Journal of Horticultural Science & Biotechnology.*

Ψ_{leaf}	Irrigation cycle	Stem extension (cm)	No. of leaves	Leaf area (cm^2/plant)	Leaf wt (g/plant)	Stem wt (g/plant)
Above –0.6 MPa (control)	Every day	30.6	13.3	617	6.5	1.6
From –0.6 to –1.0 MPa (moderate cyclical)	Every day after threshold reached	25.1	8.4	527	4.5	0.8
From –0.9 to –2.1 MPa (severe cyclical)	Every 2–4 days	15.0	6.0	371	2.9	0.4
From –0.9 to –1.1 MPa (moderate constant)	Every day after threshold reached	7.8	3.6	191	1.6	0.2
From –1.8 to –2.0 MPa (severe cyclical)	Every 4–6 days	0	0	0	0	0

drought delays flushing until a period of cool weather, the trees will flower. Strategic pruning and girdling probably operate in a similar manner (Olesen *et al.*, 2002). This is of practical significance in areas, such as Israel, that experience dry weather in winter.

Once inflorescences have initiated, the best fruit set is achieved when the plants are well-watered. A cyclic drought (pre-dawn ψ_{leaf} of –2.0 MPa), achieved by watering potted plants every 4–7 days, reduced panicle growth and the number of flowers per shoot compared with plants watered daily with a ψ_{leaf} above –0.7 MPa (Menzel and Simpson, 1991). Only a few flowers (mostly male) were retained on the droughted plants. Regular irrigation after flower initiation is recommended in order to optimize potential production.

Effects of drought on flowering in orchard trees

Several experiments have demonstrated that drought can increase flowering in litchi. However, the desired response is only achieved if the water deficit is followed by cool weather. Drought cannot be used to induce flowering in tropical environments. Other research results suggest that drought during actual flower development or anthesis reduces yields.

Nakata and Suehisa (1969), using 8-year-old 'Tai So' trees in Hawaii, were the first to study the effects of irrigation on growth and flowering of litchi. Soil water potential in the 'wet' treatment was above –0.03 MPa from June to February, with panicles emerging in December. In the 'dry' treatment, ψ_{soil} was –0.5 MPa from June to August and –1.5 MPa from September to December. Heavy rain occurred in December and ψ_{soil} rose to about –0.03 MPa. Soil water potential in the 'covered' treatment declined to –0.9 MPa in January, and then increased to –0.03 MPa in March after irrigation. The 'dry' treatment had a long, severe drought and the 'covered' treatment a short, mild drought, which included the time after floral initiation. The 'wet', 'covered' and 'dry' treatments produced 1.5, 1.2 and 0.7 leaf flushes from June to December. Only 50% of branches flowered in the 'wet' plots compared with more than 80% in the other treatments.

Israeli workers showed that water deficits increased flowering and production under a Mediterranean climate with winter rainfall. Stern *et al.* (1993) conducted a series of drought experiments over 4 years using the two main commercial varieties, 'Tai So' and 'Floridian'. The trials were conducted at two sites in eastern Galilee on a heavy clay, and at one coastal site on a sandy loam. In the first experiment, droughted trees received only 50% of their normal irrigation (25–30% of E_{pan}) for 7 days from 1 October. Since the trees did not show any symptoms of droughting, irrigation was withdrawn for another 3 weeks until some of the leaves started to turn brown (1 November). The trees were then given limited irrigation of 1 mm/day until it started to rain on 20 November. Control trees were given 50–60% of E_{pan} during the experiment. Drought increased flowering and yield by two- to threefold compared with the values for the well-watered trees.

In other experiments in Israel (Stern *et al.*, 1993), trees were droughted for 35 days from 14 September (early drought) or 29 September (late drought), or were fully irrigated. Droughted trees received no irrigation for the first 35 days and then 1 mm/day until the beginning of the rainy season in mid-November. The control trees received 50–60% of E_{pan}. Drought increased flowering and yield compared with well-watered trees (Table 10.3). In the third experiment, irrigation was cut off on 20 September (Kfar Hitim) or 1 October (Palmahim) for 35 days. The trees were then given 1 mm/day until it rained in mid-December. Once again, the droughted trees had greater flowering and yield than the well-watered trees. For instance, at Kfar Hitim, the 'Tai So' trees had a flowering score of 2.1 in the controls and 3.0 in the droughted plots (scored from 0 to 3), which were reflected in yields of 23.7 and 60.7 kg/tree, respectively. In all four experiments, the trees produced one or two leaf flushes after harvest, before irrigation was withdrawn. The control trees produced another flush (leaves) in November, whereas the droughted trees did not flush until it started to rain in November or December. This flush was invariably an inflorescence.

Further drought experiments in Israel examined what degree of water deficit was

Table 10.3. Effects of drought on flowering and yield of 11-year-old litchis in Israel. Trees droughted from 14 September (early drought) or 29 September (late drought), with the rainy season commencing in mid-November. Flowering scored on a scale of 0 (no flowering) to 5 (heavy flowering). Data from Stern *et al.* (1993).

	'Tai So'		'Floridian'	
Treatment	Flowering (0–5)	Yield (kg/tree)	Flowering (0–5)	Yield (kg/tree)
Control	2.5	22.2	1.1	6.1
Early drought	4.7	40.4	4.3	19.8
Late drought	4.5	39.7	4.0	19.7

required to achieve the best flowering and production (Stern *et al.*, 1998). Trees were irrigated over 2 years and received 0, 25 or 50% of E_{pan} from early October until rain fell in November or December. Control trees were given 100% of E_{pan}. The three drought treatments effectively prevented shoot extension, suggesting that a water application to replace 50% of E_{pan} was sufficient to prevent vegetative growth. Flowering was greater in the droughted trees than in the controls, with no differences amongst the three drought treatments. Yields were also greater after droughting; however, there were differences between the different treatments in the second year. The zero treatment produced lower yields than the 25% and 50% treatments, reflecting a smaller tree canopy. It was concluded that an irrigation cycle to replace 50% of E_{pan} was sufficient to control shoot growth and induce flowering in Israel.

Researchers in West Bengal showed that drought during actual flowering reduced fruit production (Hasan and Chattopadhyay, 1991). These workers investigated the response of 10-year-old 'Bombai' trees to four irrigation regimes, with trees irrigated when they had used 30, 45 or 60% of the water available between field capacity and wilting point, or not irrigated (control plots). The experiment was conducted from November to May, covering the period from panicle initiation to fruit harvest. Trees irrigated after they had used 30% of the available soil water produced more flowers and flowering branches than the other treatments. However, increased flower production did not always translate into higher yields, with the 30% and 45% treatments having similar numbers of fruit per panicle. These results suggest that an irrigation cycle allowing for a 45% depletion of available soil water was adequate during flower development in this environment. This research confirms the effect of drought on flower development in potted plants recorded by Menzel and Simpson (1991).

Little information is available on the effects of drought on flower initiation and development in longan. Analysis of weather data in Fujian revealed an inverse correlation between yield and temperatures during winter (Ke *et al.*, 1998). In other studies in Fujian, poor flower initiation (fewer than 20% of terminal branches flowering) was associated with temperatures above 15°C in January (Zhuang, 1999). In some years, leafy inflorescences are produced, with long, thin flower clusters (Li and Li, 1999). Inflorescences with small leaves did not set many fruit (Pan *et al.*, 2000). Many of these authors suggested that the trees should be droughted in the period leading up to flower initiation, and then irrigated during flower development. However, no data were presented showing the relationship between plant development and water deficits.

Effects of drought on fruit growth and quality

Experiments in India, South Africa and Australia indicated that drought may have different effects on yield and quality depending on the level and timing of the water deficit. Mature trees can tolerate quite long periods between irrigations before production is affected.

Experiments in West Bengal indicated that regular irrigation is required for commercial litchi production (Hasan and Chattopadhyay, 1990). 'Bombai' trees were irrigated when the volume of soil water in the main root-zone available to the plants had declined by 30, 45 or 60%. Control trees received no irrigation. Increases in tree height, stem diameter and canopy spread, yield and average fruit weight were greater with the two higher watering regimes compared with the other treatments. It was suggested that commercial orchards should be irrigated from November to May when available soil water had declined by 45%, equivalent to an irrigation every 35 days. Other data collected by Hasan and Chattopadhyay (1992) indicated similar yields after a 30% or 45% depletion of available soil water.

Chandel (1995) found that 'Dehra Dun' litchi trees irrigated in Himachal Pradesh (430 m elevation) when available soil water had declined by 20% or 40% had higher fruit retention, fruit weight and yield than trees that were irrigated after a 60% depletion or were not irrigated. Yields were 52, 50, 46 and 40 kg/tree in the different treatments, respectively. Chandel and Kumar (1995) also studied the response in 'Rose Scented' in Uttar Pradesh, although the experiments are not directly comparable. Surprisingly, trees irrigated every 15 days had slightly higher yields (78 kg) than those irrigated every 7 days (67 kg) and both yielded higher than non-irrigated trees (56 kg). Trees growing in this area appear to require more frequent irrigation than those in West Bengal (Hasan and Chattopadhyay, 1990).

Australian research has demonstrated that litchis are relatively drought hardy when grown on deep clay soils (Batten *et al.*, 1994). In a simple experiment, a set of non-irrigated 'Bengal' trees was compared with a set of trees that were irrigated weekly from flowering to replace 85% of potential evapotranspiration at Alstonville (see earlier section). Fruit were about 10% smaller on the non-irrigated trees than on the irrigated trees, but fruit production was doubled in the dry treatment (Table 10.4). Greater fruit set was attributed to less competition between new leaves and fruit, although no shoot growth data were presented. Other studies in Australia indicated that young leaves developing on terminal shoots did not reduce fruit set in adjacent branches (Hieke *et al.*, 2002a).

Litchi orchards are irrigated in South Africa. In Nelspruit (lat. 25°S), Menzel *et al.* (1995) compared a group of 'Tai So' trees irrigated weekly to replace evapotranspiration, and another set that

Table 10.4. Effects of drought on fruit growth (fresh weight) in 'Bengal' litchi in northern New South Wales. Trees droughted from flowering. From Batten *et al.* (1994).

	Irrigated trees	Droughted trees
Fruit wt (g)	23.3	19.6
Skin wt (g)	4.4	4.2
Seed wt (g)	2.8	3.1
Aril wt (g)	16.0	12.3
No. fruit per panicle	12.2	23.2
Yield per panicle (g)	284	461

was allowed to dry out gradually over 6 months from panicle emergence. The experiments were similar to those conducted by Batten *et al.* (1994) in Australia, except that the trees were droughted from panicle emergence. The orchard was growing on a sandy loam with about 126 mm of soil water available in the 1.5 m profile. In contrast, the trees in Australia were growing on a clay soil, with 180 mm available to 1 m, and 330 mm available to 2 m. In South Africa, drought reduced initial fruit set by 30% and final set by 70% compared with well-watered trees. Water deficits did not alter the pattern of fruit growth, but reduced yield from 51 kg/tree in well-watered plots to 7 kg/tree in droughted plots (Fig. 10.12). About 15% of terminal branches initiated new leaves after fruit set in the well-watered plants, whereas leaf growth ceased in the droughted trees. These results suggest that leaf production is more sensitive to drought than is fruit production, although the difference in the results compared with those recorded by Batten *et al.* (1994) need to be resolved. The trees in South Africa were subject to an earlier, longer and more severe drought than those in Australia. It is possible that there was a direct effect of the drought on the growth of fruit in the work by Batten *et al.*

Not all parts of the fruit develop at the same time (Menzel, 2002). During the first 7–8 weeks after fertilization, the fruit skin or pericarp, the embryo and the seed skin are formed. At the end of this stage, the aril or flesh forms only a negligible portion of the fruit. During the next 2–3 weeks, the cotyledons (or seed leaves), which comprise most of the seed, are formed, and the development of the aril begins. At the end of this

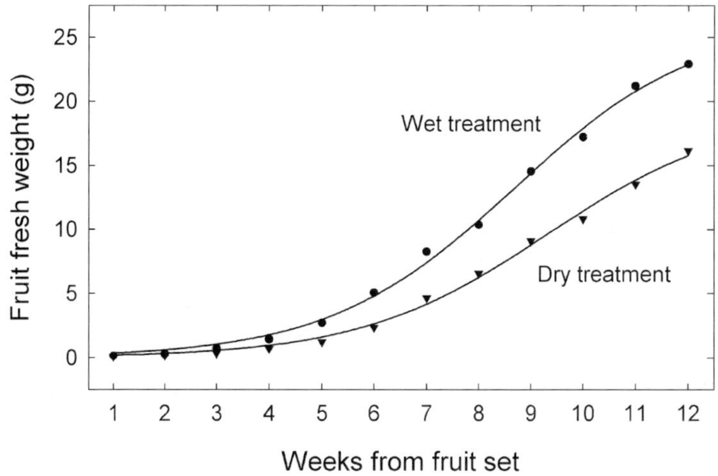

Fig. 10.12. Effects of drought on fruit growth in a 'Tai So' litchi orchard in Nelspruit, South Africa. Irrigation withdrawn from half the plots from panicle emergence on July 28. Fruit set was 11 weeks later. Trees re-watered after harvest in January (week 26). Trees in the wet treatment irrigated weekly (from Menzel *et al.*, 1995, with permission).

stage, the aril comprises about one-third of fruit fresh weight. The final period of fruit growth is dominated by expansion of the aril, which accounts for 65–75% of final fruit fresh weight. Drought affects the various components of the fruit differently, with quality suffering if the growth of the skin or aril is affected.

In Australia, water deficits decreased total fruit fresh weight in 'Bengal' by 16% compared with well-watered trees (Batten *et al.*, 1994). This was due to a 23% reduction in the weight of the aril, while seed weight increased (Table 10.4). There was no effect on the weight of the pericarp. In South Africa, where trees were exposed to an earlier and more severe water deficit, fruit weight in 'Tai So' was reduced by 60% compared with control plots (Menzel *et al.*, 1995). As in Australia, this was mainly due to the production of smaller arils in the droughted trees. Drought also increased the amount of split fruit, which at harvest accounted for 41% of the crop on the droughted trees and only 10% in the well-watered plots. The droughted trees were not re-watered until after harvest; however, it is possible that fruit water status in those trees improved under cloudy, humid weather. These results suggest that a period of drought soon after fruit set can reduce quality and increase cracking. These aspects are discussed in further detail by Xuming Huang (see Chapter 8, this volume).

Irrigation management

In the absence of irrigation, an annual rainfall requirement of 1500 mm is suggested for both crops. Prolonged dry periods during fruit development are likely to reduce commercial production.

Studies in India, Australia and South Africa (Hasan and Chattopadhyay, 1990; Batten *et al.*, 1994; Menzel *et al.*, 1995) showed that the trees can extract water at considerable depths in most soils and produce acceptable yields under fairly long irrigation cycles. A watering every 2 weeks is sufficient in a sandy loam, and every 3–4 weeks in clays, which have greater water-holding capacity. Irrigation in a sandy loam before 50% of the available soil water is used would maintain tree water status in the acceptable range if the soil profile is brought back to field capacity with each irrigation. This strategy is dependent on the trees being well-grown with a deep root system, and the soil having a good structure. Orchards growing on shallow soils with limited roots at depth will need more frequent watering. Very sandy soils can also dry out quickly during hot weather.

Maximum water requirements for trees of different ages can be calculated from seasonal values of evaporation, E_{pan}, and tree canopy cover:

Water application (l/tree/day) = E_{pan} (mm/day) × tree canopy cover (m^2) (6)

This analysis assumes a k_{crop} of 1.0. Suggested water applications for trees in southern Queensland, a humid subtropical location, are shown in Table 10.5. It would be much drier in Israel, South Africa and India. Since k_{crop} varies throughout the year, the only effective way to irrigate is to monitor actual changes in soil water under the trees. This approach takes into account run-off and free drainage after heavy rain or irrigation, whereas these factors are not taken into account in methods based on E_{pan} or E_{ref}. Crop specialists can provide advice on irrigation systems and application rates for different orchards.

Irrigation systems

The irrigation system should have sufficient capacity to deliver between 2 and 8 Ml/ha of good-quality water each year (Menzel *et al.*, 2002). If the water contains algae or iron bacteria, special filtering systems are required. Salinity, as measured by electrical conductivity, EC_w, should not exceed 0.5 dS/m or 320 mg/l of soluble salts. The system should be capable of delivering water to the major root-zone, which could be 1.5 m deep and extend from the trunk to just past the dripline of the canopy. Consultants can prepare an irrigation plan for each soil type and block.

Under-tree mini-sprinklers are common in South Africa and Australia (Fig. 10.13), whereas trickle tape is popular in Israel. Mini-sprinklers with an output of 80–250 l/h are recommended, with the micro-spray feature used for the first 2 years to limit water throw. One row of trickle tape is usually sufficient for young trees, with a second line being installed after 3 or 4 years. Trickle systems require sand filters to prevent large particles blocking the small holes in the tape. Trickle irrigation uses less water and provides more efficient wetting of the root-zone than sprinklers, but is not suitable for sands because the water travels straight down the profile with little lateral movement, and trees can dry out in hot weather.

The irrigation system should deliver water uniformly to each tree in the orchard. Sprinklers operating at low pressure dump water at the end of the sprinklers' throw, or produce a fine mist if the pressure is too high. Some trees may also receive more water than others and low branches may impede its uniform distribution. Distribution uniformity (DU) can be determined by positioning a grid of at least 35 collecting cans under the sprinkler area and measuring the volume collected in each can. Distribution uniformity is:

DU (%) = (Av. of the lowest 25 percentile of readings/Av. of all readings) × 100 (7)

A DU of 75% or higher is very good, whereas values less than 67% are unacceptable. Sprinklers should also be set at the height recommended by the manufacturer and replaced when worn. They should also be checked regularly for blockages caused by ants or particles, or for damage caused by birds, rats and other animals.

Table 10.5. Suggested weekly irrigation rates (l/tree) for orchards of different ages in southern Queensland. Water application (l/week) = E_{pan} (mm/day) × tree canopy cover (m^2) × 7. The analysis assumes a k_{crop} of 1.0. From Menzel *et al.* (2002), with permission.

			Irrigation (l/tree)		
			Year 5	Year 10	Year 15
		Mean E_{pan}	Canopy area (m^2/tree)		
Time of year	Stage of growth	(mm/day)	2	16	32
May–June	Pre-flowering	2.5	35	280	560
July–September	Flowering	4.0	56	448	896
October–February	Fruit growth	4.5	63	504	1008
March–April	Leaf growth	3.5	49	392	784

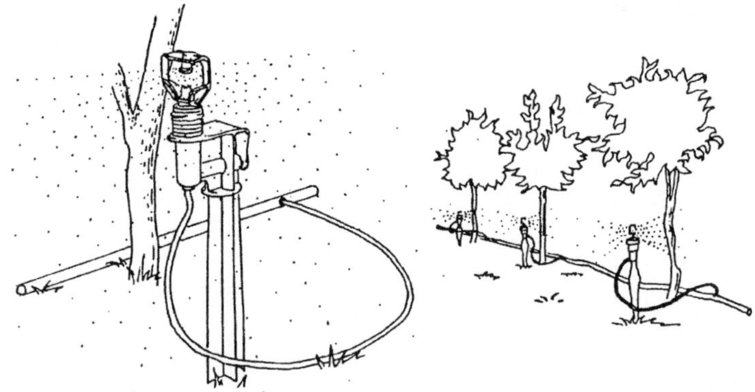

Fig. 10.13. Under-tree mini-sprinklers used in litchi and longan orchards in Australia (from Menzel *et al.*, 2002, with permission).

Monitoring tree water use

Several techniques are available for monitoring tree water use; however, soil-based systems, such as non-portable capacitance probes, are preferred (Menzel *et al.*, 2002). These devices measure the volumetric water content of the soil by determining the dielectric constant next to an access tube. For most soils, the dielectric constant is proportional to its water content. The device consists of multiple sensors mounted on probes with slots every 0.1 m to accommodate snap-in sensors. The probes are placed within vertical PVC access tubes installed throughout the orchard (Fig. 10.14). These are left in place for the season and can be moved to other sites as required.

Measurements from the sensors, generally positioned at depths of 0.2, 0.4, 0.6 and 1.0 m, are relayed at regular intervals, via a cable, to a datalogger and are downloaded to a computer to map daily tree water use. One probe is needed for each block and this should be located in a position that reflects the water use of all of the trees (this needs to be determined initially for each site). The current cost of a starter kit consisting of logger, solar panel, two 1 m probes and software is US$8000. The equipment can also be hired. Consultants can be used to set up the system and manage it, or at least to provide some initial advice.

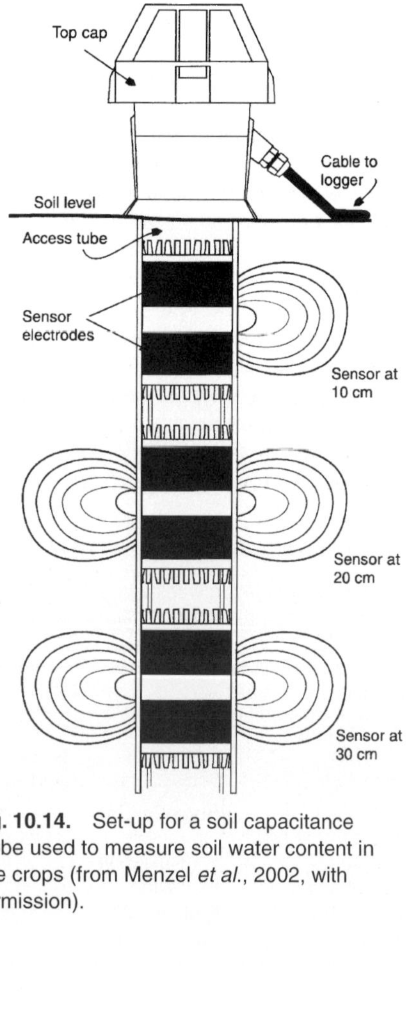

Fig. 10.14. Set-up for a soil capacitance probe used to measure soil water content in tree crops (from Menzel *et al.*, 2002, with permission).

Conclusion

Most litchi and longan orchards in South-east Asia depend on regular rainfall, with irrigation being either too expensive or not available. Research has shown that drought affects litchi production in South Africa, Israel, India, Australia and Hawaii, but its importance in China, Thailand and Viet Nam has not been quantified. The role of drought in longan production is also not well understood. In South-east Asia, it is usually dry during litchi and longan flowering and early fruit development. Mulching and cover crops can assist water conservation; however, it is recommended that, in addition to these measures, new orchards should be irrigated if possible. In the absence of irrigation, an annual rainfall of 1200–1500 mm is required for regular production.

Studies have been undertaken to help in understanding the effects of drought on the physiology of litchi and how these impact on commercial production, whereas only limited information is available for longan. Mature orchards extract water from below 1.5 m, with tree water status not affected until 40% or 50% of the available soil water has been used.

Gas exchange and growth continue at normal rates until a threshold soil or plant water deficit is reached. Net CO_2 assimilation starts to decline when leaf water potential, Ψ_{leaf}, in the morning falls below –2.0 MPa and approaches zero when the leaves wilt and Ψ_{leaf} falls below –3.5 MPa. Tree water status recovers within a few days of re-watering, whereas gas exchange is affected for much longer. Shoot growth is more sensitive to drought than photosynthesis or fruit growth. Leaves cease to expand once Ψ_{leaf} falls below –2.5 MPa, whereas dry matter accumulation in the fruit continues. The sensitivity of shoot growth to water deficits is used to increase flowering in Israel, where it is cool and dry before flower initiation. Whether this approach works consistently in South-east Asia is not known. Tropical lowland environments are too warm for consistent successful flowering. Mild water deficits after flowering shift resources in favour of the fruit, whereas severe droughts reduce yield and fruit quality.

Tree water use is best determined by measuring changes in soil water with non-portable capacitance probes or similar devices, whereas estimates of crop water use from weather data or reference crops are unreliable. An irrigation applied every 2–4 weeks is generally adequate for well-grown orchards.

Acknowledgements

Comments on the manuscript by David Turner from the University of Western Australia are appreciated. Many thanks to colleagues who contributed to the litchi research programme at Maroochy.

References

Anonymous (1992) *The Cultivation of Litchi.* Agricultural Research Council, Pretoria, South Africa, 61 pp.

Atwell, B.J., Kriedemann, P.E. and Turnbull, C.A. (1999) *Plants in Action.* Macmillan Education Australia, South Yarra, Victoria, Australia, 664 pp.

Batten, D.J. (1986) Towards an understanding of reproductive failure in lychee (*Litchi chinensis* Sonn.). *Acta Horticulturae* 175, 79–83.

Batten, D.J. and McConchie, C.A. (1992) Pollination in lychee. In: Fullelove, G.D. (ed.) *Proceedings of the Third National Lychee Seminar.* Australian Lychee Growers' Association, Bundaberg, pp. 23–28.

Batten, D.J., Lloyd, J. and McConchie, C. (1992) Seasonal variation in stomatal responses of two cultivars of lychee (*Litchi chinensis* Sonn.). *Australian Journal of Plant Physiology* 19, 317–329.

Batten, D.J., McConchie, C.A. and Lloyd, J. (1994) Effects of soil water deficit on gas exchange characteristics and water relations of orchard lychee (*Litchi chinensis* Sonn.) trees. *Tree Physiology* 14, 1177–1189.

Chaikiattiyos, S., Menzel, C.M. and Rasmussen, T.S. (1994) Floral induction in tropical fruit trees: effects of temperature and water supply. *Journal of Horticultural Science* 69, 397–415.

Chandel, J.S. (1995) Influence of different irrigation regimes alone and in combination with sprays of growth regulator and nutrient on fruiting and quality of litchi cv 'Dehra Dun'. *Advances in Horticulture and Forestry* 4, 23–27.

Chandel, J.S. and Kumar, G. (1995) Effect of irrigation frequencies and foliar spray of NAA and micronutrient solutions on yield and quality of litchi (*Litchi chinensis* Sonn.) cv. Rose Scented. *Advances in Plant Sciences* 8, 284–288.

Goren, M. and Gazit, S. (1996) Management of lychee orchards in Israel. In: Welch, T. and Ferguson, J. (eds) *Proceedings of the Fourth National Lychee Seminar.* Australian Lychee Growers' Association, Yeppoon, Queensland, pp. 74–77.

Hasan, M.A. and Chattopadhyay, P.K. (1990) Effect of different soil moisture regimes on growth and yield of litchi cv. Bombai. *Indian Journal of Horticulture* 47, 397–400.

Hasan, M.A. and Chattopadhyay, P.K. (1991) Flowering behaviour of litchi as affected by different soil moisture regimes. *South Indian Horticulture* 39, 67–70.

Hasan, M.A. and Chattopadhyay, P.K. (1992) Note on effect of different soil moisture regimes on water requirement, crop coefficient and water-use efficiency of litchi. *Indian Journal of Horticulture* 49, 155–157.

Hieke, S., Menzel, C.M., Doogan, V.J. and Lüdders, P. (2002a) The relationship between fruit and leaf growth in lychee (*Litchi chinensis* Sonn.). *Journal of Horticultural Science & Biotechnology* 77, 322–325.

Hieke, S., Menzel, C.M., Doogan, V.J. and Lüdders, P. (2002b) The relationship between yield and assimilate supply in lychee (*Litchi chinensis* Sonn.). *Journal of Horticultural Science & Biotechnology* 77, 326–332.

Huang, H.B. (1989) The water relations problem of fruit trees in Guangdong. *Journal of South China Agricultural University* 10, 33–39.

Ke, G.W., Tang, J.F. and Liu, R.F. (1998) Study on the reasons causing low yield of longan fruit trees and possible improvements. *South China Fruits* 27 (1), 25–26.

Li, H.Y. and Li, C.F. (1999) The early high quality and high production techniques for longan trees. *South China Fruits* 28 (6), 30–31.

Liao, J.S., Liu, S., Chen, Q.X. and Zheng, G.H. (1996) A study on photosynthesis of longan (*Dimocarpus longan* Lour.) and its influential factors. *Acta Horticulturae Sinica* 23, 1–7.

Liao, W.C. (1997) *A Manual of the Prevention of Damage to Fruit Trees by Extreme Environmental Conditions.* Special Publication of the Taichung District Agricultural Improvement Station, Taiwan, Number 39, 162 pp.

Liu, S., Liao, J.S., Chen, Q.X. and Zheng, G.H. (1997) Photosynthetic response to surrounding temperature in longan leaves. *Journal of Fujian Agricultural University* 26, 407–410.

Lloyd, J., Wong, S.C., Styles, J.M., Batten, D.J., Priddle, R., Turnbull, C. and McConchie, C.A. (1995) Measuring and modelling whole-tree gas exchange. *Australian Journal of Plant Physiology* 22, 987–1000.

McConchie, C.A. and Batten, D.J. (1991) Fruit set in lychee (*Litchi chinensis* Sonn.): variation between flowers, panicles and trees. *Australian Journal of Agricultural Research* 42, 1163–1172.

Menzel, C.M. (1983) The control of floral initiation in lychee: a review. *Scientia Horticulturae* 21, 201–215.

Menzel, C.M. (1984) The pattern and control of reproductive development in lychee: a review. *Scientia Horticulturae* 22, 333–345.

Menzel, C.M. (1991) *Litchi chinensis* Sonn. In: Verheij, E.W.M. and Coronel, R.E. (eds) *Plant Resources of South-East Asia No. 2: Edible Fruit and Nuts.* Pudoc, Wageningen, The Netherlands, pp. 191–195.

Menzel, C.M. (2002) *The Lychee Crop in Asia and the Pacific.* Food and Agricultural Organization of the United Nations, Bangkok, Thailand, 108 pp.

Menzel, C.M. and McConchie, C.A. (1998) Lychee and longan. In: Hyde, K.W. (ed.) *The New Rural Industries.* Rural Industries Research and Development Corporation, Canberra, Australia, pp. 288–295.

Menzel, C.M. and Simpson, D.R. (1988) Effect of temperature on growth and flowering of litchi (*Litchi chinensis* Sonn.) cultivars. *Journal of Horticultural Science* 63, 347–358.

Menzel, C.M. and Simpson, D.R. (1991) Effects of temperature and leaf water stress on panicle and flower development of litchi (*Litchi chinensis* Sonn.). *Journal of Horticultural Science* 66, 335–344.

Menzel, C.M. and Simpson, D.R. (1994) Lychee. In: Schaffer, B. and Andersen, P.C. (eds) *Handbook of Environmental Physiology of Fruit Crops* Vol. II: *Sub-tropical and Tropical Crops.* CRC Press, Boca Raton, Florida, pp. 123–145.

Menzel, C.M. and Simpson, D.R. (1995) Temperatures above 20°C reduce flowering in lychee (*Litchi chinensis* Sonn.). *Journal of Horticultural Science* 70, 981–987.

Menzel, C.M., Rasmussen, T.S. and Simpson, D.R. (1989) Effects of temperature and leaf water stress on growth and flowering of litchi (*Litchi chinensis* Sonn.). *Journal of Horticultural Science* 64, 739–752.

Menzel, C.M., Aitken, R.L., Dowling, A.W. and Simpson, D.R. (1990) Root distribution of lychee trees growing in acid soils of subtropical Queensland. *Australian Journal of Experimental Agriculture* 30, 699–705.

Menzel, C.M., Oosthuizen, J.H., Roe, D.J. and Doogan, V.J. (1995) Water deficits at anthesis reduce CO_2 assimilation and yield of lychee (*Litchi chinensis* Sonn.) trees. *Tree Physiology* 15, 611–617.

Menzel, C.M., Bagshaw, J., Campbell, T., Greer, N., Noller, J., Olesen, T. and Waite, G.K. (2002) *Lychee Information Kit.* Queensland Department of Primary Industries, Nambour, Australia.

Mitchell, P.D. and Chalmers, D.J. (1982) The effect of reduced water supply on peach tree growth and yields. *Journal of the American Society for Horticultural Science* 107, 853–856.

Nakata, S. (1955) Floral initiation and fruit set in lychee, with special reference to the effects of sodium naphthalene acetate. *Botanical Gazette* 117, 126–134.

Nakata, S. and Suehisa, R. (1969) Growth and development of *Litchi chinensis* as affected by soil moisture stress. *American Journal of Botany* 56, 1121–1126.

Naor, A. (2000) Midday stem water potential as a plant water stress indicator for irrigation scheduling in fruit trees. *Acta Horticulturae* 537, 447–454.

Olesen, T., Menzel, C.M., Wiltshire, N. and McConchie, C.A. (2002) Flowering and shoot elongation of lychee in eastern Australia. *Australian Journal of Agricultural Research* 53, 977–983.

Pan, X.W., Li, J.G., Li, R., Gu, H.X. and Dai, H.F. (2000) Study on the techniques for promoting the formation of flower clusters on longan trees. *South China Fruits* 29 (1), 26–27.

Ritchie, G.A. and Hinckley, T.M. (1975) The pressure chamber as an instrument for ecological research. *Advances in Ecological Research* 9, 165–254.

Roe, D.J., Oosthuizen, J.H. and Menzel, C.M. (1995) Rate of soil drying and previous water deficits influence the relationship between CO_2 assimilation and tree water status in potted lychee (*Litchi chinensis* Sonn.). *Journal of Horticultural Science* 70, 15–24.

Singh, S.N. and Pathak, R.A. (1983) Effect of irrigation intensity, its frequency and manures on the growth of litchi, cv. Calcuttia. *Punjab Horticultural Journal* 23, 197–202.

Stern, R.A., Adato, I., Goren, M., Eisenstein, D. and Gazit, S. (1993) Effects of autumnal water stress on litchi flowering and yield in Israel. *Scientia Horticulturae* 54, 295–302.

Stern, R.A., Meron, M., Naor, A., Wallach, R., Bravdo, B. and Gazit, S. (1998) Effect of fall irrigation level in 'Mauritius' and 'Floridian' lychee on soil and plant water status, flowering intensity, and yield. *Journal of the American Society for Horticultural Science* 123, 150–155.

Subhadrabandhu, S. (1990) *Lychee and Longan Cultivation in Thailand.* Rumthai Publishers, Bangkok, Thailand, 40 pp.

Turner, N.C. (1981) Techniques and experimental approaches for the measurement of plant water status. *Plant and Soil* 58, 339–366.

Williams, L.E. and Araujo, F.J. (2002) Correlations among predawn leaf, midday leaf, and midday stem water potential and their correlations with other measures of soil and plant water status in *Vitis vinifera. Journal of the American Society for Horticultural Science* 127, 448–454.

Whiley, A.W. and Schaffer, B. (1997) Stress physiology. In: Litz, R.E. (ed.) *The Mango: Botany, Production and Uses.* CAB International, Wallingford, UK, pp. 147–173.

Yamada, M., Fukumachi, H. and Hidaka, T. (1996a) Photosynthesis in longan and mango as influenced by high temperatures under high irradiance. *Journal of the Japanese Society for Horticultural Science* 64, 749–756.

Yamada, M., Hidaka, T. and Fukumachi, H. (1996b) Heat tolerance in leaves of tropical fruit crops as measured by chlorophyll fluorescence. *Scientia Horticulturae* 67, 39–48.

Young, T.W. (1970) Some climatic effects on flowering and fruiting of 'Brewster' lychees in Florida. *Proceedings of the Florida State Horticultural Society* 83, 362–367.

Young, T.W. and Harkness, R.W. (1961) Flowering and fruiting behaviour of Brewster lychees in Florida. *Proceedings of the Florida State Horticultural Society* 74, 358–363.

Zhuang, W.B. (1999) Study on the reason for the low flowering rate of longan and its control. *South China Fruits* 28 (5), 30.

11 Plant Nutrition and Fertilizing

Christopher M. Menzel
Department of Primary Industries and Fisheries, PO Box 5083, SCMC, Nambour, Queensland 4560, Australia

Introduction

Elemental nutrients represent only a small fraction of total plant biomass, but are essential for the normal functioning of all cells. More than 90 elements can be found in different species, although only about 14 of these appear to be crucial for plant growth and development (Atwell *et al.*, 1999). In litchi and longan, nitrogen (N) is the most important nutrient, followed by phosphorus (P), potassium (K), calcium (Ca) and magnesium (Mg). The micronutrients iron (Fe), boron (B), copper (Cu), zinc (Zn) and manganese (Mn), as their name implies, are required only in very small amounts. These nutrients are involved in cell metabolism or are incorporated into the structure of the cells themselves. The nutrients must be obtained from the soil via a relatively small volume of plant roots compared with the volume of soil that they exploit. To capture nutrients in the soil, the plant must continually produce new roots that have an active nutrient-absorbing system. The energy for this growth and metabolism comes from photosynthesis in the leaves. Nutrient uptake by plants thus has a substantial biological cost.

Litchi and longan are common species in the rainforests of southern China and other parts of South-east Asia. The trees are long-lived and often dominate the upper canopies of the forests. In these environments, there is considerable recycling of plant nutrients from fallen leaves, twigs, flowers and fruit, with the tree growing and taking up nutrients throughout the year. The proportion of dry matter and nutrients diverted to the reproductive structures each year is relatively small compared with the total pool of nutrients available in the tree, leaf litter and surrounding soil.

Yields begin to decline when the concentration of one or more of these nutrients reaches abnormally low or high levels, with characteristic symptoms appearing in the leaves, stems, flowers or fruit. Mature trees may take a long time to exhibit these responses, although once nutrient concentrations fall below critical values, it may take several years for the trees to recover after a deficiency has been corrected. In most well-managed orchards, the application of fertilizers has little impact on commercial production, because weather and pests are more important.

General principles

Plants grow and absorb nutrients from the soil continuously and best growth is usually associated with a particular internal nutrient concentration. The task of plant physiologists is to define the range of nutrient concentrations associated with maximum growth (Atwell *et al.*, 1999). The internal nutrient requirement is the minimum concentration of a nutrient that is required within plant cells to sustain growth and metabolism. The external nutrient requirement is the amount of a nutrient, either naturally occurring or applied, that the plant needs to find

 Litchi and Longan: Botany, Production and Uses (eds C.M. Menzel and G.K. Waite)

in the soil for it to meet its internal requirement. Estimates of internal nutrient requirements have been made for many plants and are usually based on the concentration of the nutrients present in recently matured leaves expressed on a dry weight basis. However, the actual concentration of individual nutrients required for optimum growth probably varies between roots, leaves, stems, flowers and fruit. Internal nutrient requirements are generally similar for different plants, although there are some differences between grasses and dicots, and amongst species and cultivars within these groups. The ability of some species to fix nitrogen can also affect their internal nutrient requirements. External nutrient requirements vary with soil type, drought, pH, cation exchange capacity of the soil, and other factors. There are also interactions between some nutrients, such as potassium, calcium, magnesium and sodium.

Research approaches

A vast array of research has been conducted on litchi nutrition, although the results of much of it are of limited use to commercial producers. There has been less work on longan. Longan is possibly more productive than litchi (see Menzel, Chapter 9, this volume) and thus has a greater nutrient requirement. The principles developed for one crop probably apply to the other.

Leaf nutrient standards have been developed for litchi in South Africa and Australia and for litchi and longan in China, but they are based on different times of sampling. The suggested standards are usually similar when these differences in phenology at sampling are taken into account, and are similar to those developed for other tree crops. These standards form the basis of nutrition management in the absence of calibrations between yield and leaf nutrient concentrations. The philosophy behind this approach is that acceptable yields can be obtained over a fairly wide range of in-leaf concentrations for most nutrients.

Soil nutrient concentrations associated with commercial production have been determined in various orchards, but the recommendations derived generally only apply to trees grown in similar soils. Differences in analytical methods between laboratories may also affect nutrient diagnoses and the corrective recommendations made. Regular soil analyses can help monitor soil pH, salinity and other problems that are not directly related to leaf nutrient composition. Surveys indicate the relative importance of nutrients affecting productivity or quality in different growing areas, but do not indicate the optimum nutrient applications required for different orchard sites.

Symptoms of nutrient deficiencies have been determined by growing plants in sand culture in the absence of one or more of the studied nutrients. It is much easier to induce these symptoms under glasshouse conditions than in the field, as visible symptoms are relatively rare in orchard trees. However, the reliance on nutrient deficiency symptoms appearing in a tree is not a sound basis for fertilizer management, since growth, yield and quality are often reduced before obvious symptoms become visible.

Calibration curves between growth/yield and leaf nutrient concentrations have been derived from various pot and field experiments. Normally, the relationship between growth and leaf nutrient concentration developed at one site applies to other sites, provided that the appropriate tissue is sampled for nutrient analysis. In contrast, the relationship between plant performance and soil nutrient concentrations is site-specific.

Significant responses to nitrogen have been reported, with smaller increases in yield resulting from the application of phosphorus, potassium, calcium, magnesium and some of the micronutrients. Much of this research is of limited practical value because of the confounding of treatments (e.g. a single N + P + K treatment versus a nil fertilizer control treatment) or no data on leaf or soil nutrient concentrations. Many researchers have also mixed young and old leaves, which have a different nutrient status, in the same sample. In addition, some of these experiments have been conducted for only one year or with a single cultivar or soil type.

Little attention has been paid to nutrient balances to ensure that fertilizer applications reflect the ratio of plant uptake. There is little evidence to suggest that nitrogen can be used to manipulate flowering in tropical locations or that the time of nutrient application affects productivity.

Soil type

Litchis and longans are grown on a range of soils including alluvial sands, rocky soils, loams, heavy clays, and soils with a high content of organic matter or lime (Menzel, 1987, 1991, 2002). The major soil types utilized are sands and alluvial soils, medium to heavy clays, and calcareous soils. Trees perform best on well-drained clay loams of medium to high fertility that have a minimum of 1 m of well-drained topsoil. Slopes greater than 15% are best avoided, as they do not allow the safe use of machinery, and may erode.

Many old orchards in China were established on alluvial sands close to rivers, where they had good drainage and access to the water table. The majority of the longans in Viet Nam are grown in the southern areas in the fertile alluvial soils of the Mekong Delta. In India, well-drained alluvial soils that provide access to the water table are also considered ideal for litchi cultivation. In South Africa, litchis grow well on sandy soils in the cooler subtropical areas, with the trees developing roots to 1 m depth (Anonymous, 1992). Sandy soils are also utilized in Israel and Australia (Menzel *et al.*, 1990; Goren and Gazit, 1996; Menzel and McConchie, 1998). Many of these soils have low concentrations of phosphorus, potassium, magnesium and boron, and regular applications of fertilizer are required in order to maintain growth and yield. The major problem with very sandy soils is that they may dry out and subject the trees to drought during hot weather if irrigation is not available or is inadequate.

Medium to heavy clay soils are common in China (Figs 11.1 and 11.2) and Viet Nam, and are also used in India, Thailand, Israel, South Africa and Australia. These soils typically contain 10–40% clay, have an acid or alkaline pH, and possess a range of clay chemistry, fertility and colour, which ranges through red, brown, yellow, grey and black. Soil pH and fertility can usually be adjusted relatively easily. Some clay soils in South-east Asia have low concentrations of organic matter, phosphorus and potassium, but they are generally more fertile than sands and are less likely to dry out. Root growth can be a problem in heavily compacted sites that become waterlogged during the wet season, which reduces production and may kill trees. Drains or mounds may be installed before planting and organic matter applied to improve soil structure.

Calcareous soils dominate plantings of these crops in Florida, Israel and India, with some sites having up to 30% free lime and a pH above 8.5. These soils can cause problems with potential deficiencies of iron, zinc and manganese. Orchards growing on sites with pH above 7.0 need to be carefully managed. Some of the orchards in Hawaii are grown in volcanic rock, which requires special care to manage watering and fertilizing (Fig. 11.3).

Fig. 11.1. Planting site for a litchi orchard in Guangdong, southern China, showing a heavy clay soil.

Fig. 11.2. A young longan orchard in Guangxi, southern China, showing a heavy clay soil (photograph courtesy of Christopher Menzel).

Soil pH

Little critical information exists regarding the optimum pH favoured by either crop. Commercial orchards can be found on acid soils with a pH of 5.0–6.0 in China and Australia, and on calcareous soils with a pH of 7.0–8.5 in India and Florida (Menzel, 1987; Menzel *et al.*, 1990). At pHs below 4 and above 8, the availability of some micronutrients to the trees may be restricted (Fig. 11.4; Menzel *et al.*, 2002). For litchi and longan, a pH of 5.5–6.0 (1:5 water test) is suggested for most situations. Various fertilizers and amendments can be applied for overcoming the problems associated with a low or a high pH. The amendments that can be used depend on the pH of the soil and whether there are any associated shortages of calcium or magnesium. If pH and calcium are low, lime or calcium carbonate should be used. If pH and magnesium are low, dolomite or calcium and magnesium carbonate is recommended. If pH is above 6 but calcium is low, gypsum or calcium sulphate should be used, since it does not raise the pH of the soil. If pH is above 6 but magnesium is low, magnesium sulphate or the fairly insoluble magnesium oxide is recommended.

Since the lime requirement of a soil is also influenced by its buffering capacity or cation

Fig. 11.3. A young litchi tree growing in a volcanic rock soil in Hawaii (photograph courtesy of Christopher Menzel).

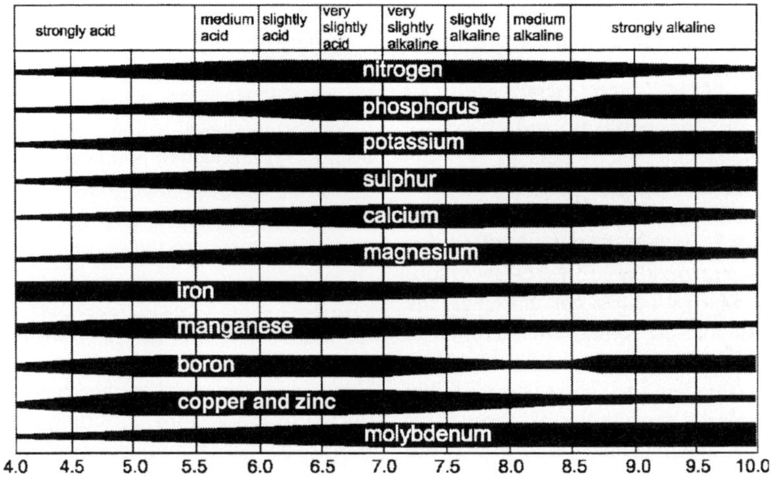

Fig. 11.4. The availability of plant nutrients changes with soil pH (from Menzel *et al.*, 2002, with permission).

exchange capacity (CEC) dependent on the amount and type of clay and organic matter, many scientists prefer to base calculations of liming on the buffer pH determined by extraction in calcium chloride (Table 11.1). Both the buffer and water pH are used to determine lime requirements.

Table 11.1. Lime requirement (t/ha at 0.1 m depth) for soils with different pH. Only apply lime when the soil pH (water) is lower than the target pH. Data from Phil Moody and Bob Aitken, Queensland Department of Natural Resources and Mines.

Mehlich soil buffer pH	Lime required to bring soil pH (water) to 5.5	Lime required to bring soil pH (water) to 6.5
4.5	8.6	18.0
4.7	7.1	15.0
4.9	5.7	12.0
5.1	4.4	10.0
5.3	3.2	7.5
5.5	2.2	5.7
5.7	1.3	4.1
5.9	0.7	2.8

Salinity

Responses to excess salts have not been well defined for either crop. Litchi appears to be less sensitive to salt than is avocado, *Persea americana*, or macadamia, *Macadamia integrifolia* × *M. tetraphylla*, but is still in the low-tolerance class of plants (Whiley and Saranah, 1980). The use of irrigation water with an electrical conductivity, EC_w, of more than 0.6–1.0 dS/m or total dissolved salts, TDS, of more than 500 mg/l may reduce productivity in commercial orchards.

Mycorrhizas

Mycorrhizal associations occur in about 90% of higher plants (Atwell *et al.*, 1999). The plants supply carbon to the symbiotic fungus, which assists the plants with nutrient uptake from the soil. Arbuscular mycorrhizas are the most common fungi associated with plants and they colonize about 80% of all species. The two other mycorrhizal types, ectomycorrhizas and ericoid mycorrhizas, are found only in a few families, including some trees.

Coville (cited in Groff, 1921) was the first to study the relationship between litchi and mycorrhizas. Fungi were isolated from root tubercles of seedlings grown in peat and sand, whereas they were absent on plants grown in the standard mix of loam, sand and manure. Seedlings bearing tubercles were larger and had more roots than plants lacking the fungi. Experiments in India showed that seedlings grown in peat plus mycorrhizal soil were larger than those grown in peat plus regular soil (Kadman and Slor, 1974).

Pandey and Misra (1971, 1975) described the taxonomy, morphology and habit of *Rhizophagus litchi*, which belongs to the vesicular-arbuscular group of mycorrhizae (VAM). The endophytes survived on living roots but could not be cultured on artificial media. Mycorrhizas were found only on short-lived sub-lateral roots, which the fungi penetrated via the epidermal cells into the cortex, whereas the root hairs, endodermis and vascular tissue were free of infection. Pandey and Misra showed that infection increased the uptake of phosphorus and potassium, leading to increased growth and yield compared with non-inoculated trees.

In many of the above experiments, it was not always determined that the control plants were free of the mycorrhizas, which could only be achieved if the soil was sterilized or if soil-free media were used. It is possible that the greater growth recorded by Coville and others was due to improved fertility or better pH of the mycorrhizal medium. Some of these concerns have been addressed by research conducted in Florida. Janos *et al.* (2001) grew 'Tai So' ('Dazao') plants for 469 days in 95-litre containers of soil-free medium with and without inoculation, and with low and high phosphorus applications. The mycorrhizas, mainly *Glomus coremiodes*, were obtained from the roots of bahiagrass, *Paspalum notatum*. High concentrations of phosphorus in half the containers were achieved by applying 1.3 g of triple superphosphate (19% P).

The application of phosphorus had no effect on the mycorrhiza, net CO_2 assimilation rate or plant growth, whereas the presence of the fungi increased leaf growth as early as 120 days after inoculation, compared with control plants. The mycorrhizas colonized only 7% of the total root surface of the plants, but at harvest, shoot dry weight was 39% higher than in the controls. The inoculated plants had a greater proportion of fine roots and higher concentrations of phosphorus, potassium, copper and zinc. It was concluded that *Glomus* enhanced plant growth even when soil phosphorus concentrations were high.

Limited information from China suggests that mycorrhizas are also important in longan (Liang, 1992, 1995). Seedlings inoculated with *Glomus caledonium*, *G. citricola* and *G. mosseae* were taller and heavier than non-inoculated, sterilized plants. The presence of fungi also increased leaf chlorophyll concentration, CO_2 assimilation and phosphorus uptake. Further research is required to determine the role of these fungi in commercial production.

Factors affecting leaf composition

Various physiological studies have been conducted in these crops, including investigations of the changes in nutrient concentrations with season and leaf age, as well as the effects of fruiting and the position of leaves within the tree. The variation in nutrient concentrations in different cultivars has also been studied. This research has been used to develop appropriate sampling techniques to estimate tree nutrient status.

The physiological age of a leaf has a strong effect on its nutrient composition. Studies in Australia on 'Bengal' litchi showed that the concentrations of phosphorus (Fig. 11.5), potassium and zinc decreased over 52 weeks, whereas those of calcium, magnesium, iron, manganese and boron increased (Menzel *et al.*, 1987). The responses of nitrogen and copper were variable and depended on the level of supply. It was suggested that sampling and analysis of the mature leaves of a current flush, or the top one or two leaves of the previous flush, would provide a reliable estimate of tree nutrient status. Studies in India provided similar data for the local cultivar 'China' (Kotur and Singh, 1993). Typically, nitrogen (from 1.9% to 1.6%), phosphorus (from 0.12% to 0.08%) and potassium (from 0.81% to 0.56%) decreased over 9 months, while calcium (from 2.1% to 3.2%), magnesium (from 0.26% to 0.60%), iron (from 138 to 252 μg/g) and manganese (from 76 to 192 μg/g) increased.

Other studies with litchi in Australia showed that the presence or absence of fruit influenced nutrient concentrations in the leaves (Menzel *et al.*, 1988a). Leaves were sampled from fruiting and non-fruiting branches from flowering through to fruit harvest in each of three 'Tai So' and 'Bengal' orchards. The major effect of fruiting was to reduce leaf potassium in two out of the three orchards. Nitrogen, phosphorus and zinc concentrations were lower in fruiting branches in a third of the orchards, while calcium, magnesium, manganese and boron were higher. There were also strong seasonal effects, reflecting the relative physiological ages of the leaves. Leaves from fruiting branches were a better indicator of the nutrient reserves available for fruit production than leaves on non-fruiting branches.

Changes in leaf composition with the season can be large enough to affect the diagnosis of nutrient deficiencies. Menzel *et al.* (1992b) sampled recently matured leaves from vegetative flushes or behind fruiting clusters from seven 'Tai So', 'Haak Yip' ('Heiye') and 'Wai Chee' ('Huaizhi') orchards over 3 years in subtropical Queensland. The concentrations of the mobile nutrients such as nitrogen (Fig. 11.6), phosphorus and potassium reached a maximum at flowering, and then declined during fruit

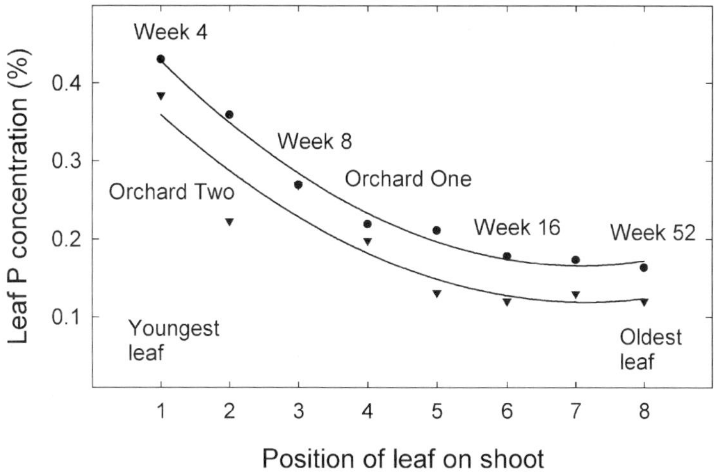

Fig. 11.5. Relationship between phosphorus concentration and leaves in different positions on terminal branches of 'Bengal' litchi in Australia. Relative ages of selected leaves indicated. Redrawn from Menzel *et al.* (1987), with permission from the *Journal of Horticultural Science & Biotechnology*. The optimum leaf phosphorus after panicle emergence is 0.14–0.22%.

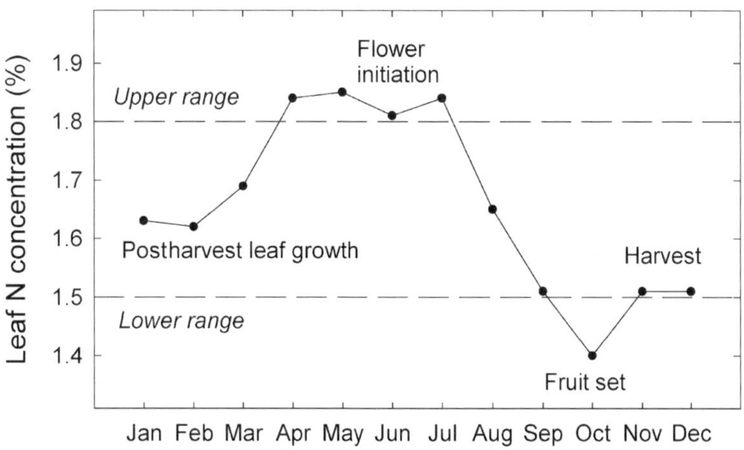

Fig. 11.6. Seasonal changes in leaf nitrogen in 'Tai So' litchi in Australia. Redrawn from Menzel *et al.* (1992b), with permission from the *Journal of Horticultural Science & Biotechnology*. The optimum concentration after panicle emergence is 1.50–1.80% N.

development. In contrast, leaf calcium, magnesium, iron, copper, boron and manganese were at a minimum at flowering, and then increased. It was concluded that the choice of sampling time would influence the diagnosis for nitrogen, potassium, calcium, iron and boron in these orchards. Overall, the concentrations of most nutrients were stable at panicle emergence, the suggested time for taking leaf tests. In contrast, the standards developed in South Africa (Koen and du Plessis, 1993) did not take these seasonal influences into account.

Leaf nutrient concentrations required for normal growth have been determined for the major litchi cultivars in Australia (Menzel *et al.*, 1992b,c). Leaf nitrogen, phosphorus and potassium ranged from 1.44 to 1.58%, 0.18 to 0.24% and 0.50 to 0.90% of leaf dry matter, respectively. The importance of these differences is not known, although the range in nutrient concentration was not large enough to influence agronomic diagnosis. Suggested nutrient concentrations for litchi in Australia were 1.50–1.80% for N, 0.14–0.22% for P and 0.70–1.10% for K (Menzel *et al.*, 1992b).

The nutrient composition of longan also changes with leaf and fruit development. The concentrations of nitrogen and potassium decreased as the leaves aged, whereas those of magnesium and calcium increased (Zhuang *et al.*, 1984). There were smaller changes in leaf phosphorus. The concentrations of all nutrients, except calcium, were lower in an 'on' year than in an 'off' year. Other studies in China reported lower concentrations of nitrogen (1.3–1.6% vs. 1.4–2.0%), phosphorus (0.12–0.19% vs. 0.14–0.28%) and potassium (0.6–0.9% vs. 0.7–1.3%) in an 'on' year compared with values in an 'off' year (Zhao, 1991).

Liu *et al.* (1986) studied changes in nutrient concentrations in longan cultivars in Fujian and showed that they were affected by the season and the type of tissue sampled. The concentration of nitrogen was higher in vegetative (1.9%) than in fruiting shoots (1.3%), and was higher in summer (1.9%) than in autumn (1.6%). In contrast, the first, second and third leaves on a terminal branch had similar nitrogen concentrations (1.8%). Leaf nitrogen concentrations decreased during flowering and fruit development, and increased after harvest. They suggested that leaves should be sampled in January, before flowering, when there was a correlation between yield and leaf nitrogen (r^2 = 60–80%).

The effects of environment on the uptake and translocation of nutrients has received little attention, apart from some work in litchi. In Hawaii, drought reduced phosphorus concentrations in shoots compared with those in well-watered trees ('Tai So'), reflecting the poor root growth of the plants in dry soil (Nakata and Suehisa, 1969). Calcium and magnesium uptake was also affected, whereas there was no effect on nitrogen and potassium. These authors did not indicate whether these differences were large enough to influence the diagnosis of deficiencies in droughted orchards. Studies on young trees grown in glasshouses in Australia found that temperature generally had no effect on leaf nutrient composition (Menzel and Simpson, 1988). However, shoot potassium was lower and shoot calcium higher above 15°/10°C. Shoot manganese was also higher at 25°/20°C and 30°/25°C than at 15°/10°C and 20°/15°C. None of the nutrients were deficient.

Experiments have shown that leaf age, fruiting and season affect nutrient concentrations. These effects are large enough to influence nutrient diagnosis in many orchards, suggesting that the protocol for sampling leaf nutrient concentrations needs to be standardized across different growing areas.

Leaf and soil analyses

Research has resulted in the development of leaf nutrient standards for litchi in South Africa, Australia and China, while tentative standards have been suggested for longan in China.

South African workers have determined optimum nitrogen, phosphorus and potassium concentrations for litchi from fertilizer experiments, with the samples collected 6–8 weeks after fruit set in October or November (Koen and du Plessis, 1993). The two central leaflets from the second compound leaf under the fruiting cluster were selected for the sample (Fig. 11.7). The range suggested for nitrogen (see Table 11.2) was somewhat lower than the concentrations associated with maximum yields in other experiments conducted at Nelspruit (Koen *et al.*, 1981a,b).

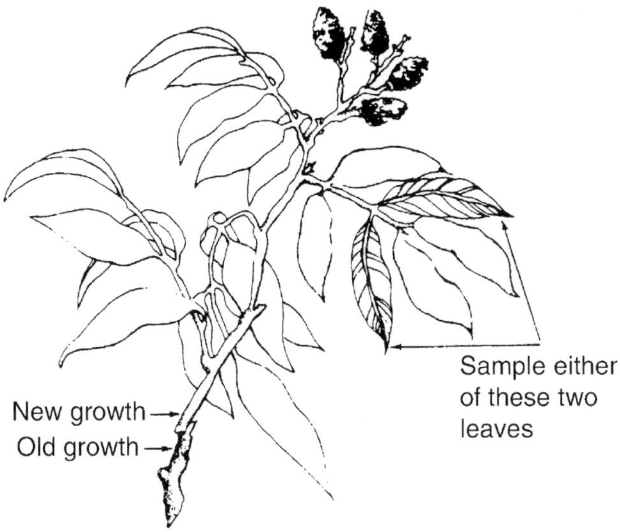

Fig. 11.7. Leaves for tissue testing of litchi in South Africa are sampled after fruit set (redrawn from Anonymous, 1992).

Table 11.2. Suggested optimum leaf nutrient concentrations for litchi and longan, on a dry weight basis.

Nutrient	Litchi Australia (Menzel *et al.*, 1992b)	Litchi South Africa (Koen and du Plessis, 1993)	Litchi Taiwan (Huang *et al.*, 1998)	Litchi Guangdong (Chen *et al.*, 1998)	Longan Fujian (Liu *et al.*, 1986)	Longan Fujian (Wang *et al.*, 1994)	Longan Fujian (Zhuang *et al.*, 1995)	Longan Taiwan (Chen, 1997)
N (%)	1.50–1.80	1.26–1.46	1.60–1.90	1.60–1.75	> 1.70	1.21–1.73	1.40–1.90	1.47–1.79
P (%)	0.14–0.22	0.15–0.20	0.12–0.27	0.11–0.14	0.12–0.20	0.17–0.25	0.10–0.18	0.11–0.19
K (%)	0.70–1.10	0.90–1.06	–	0.54–0.79	0.60–0.80	0.52–1.02	0.50–0.90	0.89–1.77
Ca (%)	0.60–1.00	–	0.60–1.00	–	1.52–2.50	0.59–1.33	0.90–2.00	0.76–1.12
Mg (%)	0.30–0.50	–	0.30–0.50	–	0.20–0.30	0.09–0.23	0.13–0.30	0.24–0.47
Fe (μg/g)	50–100	–	50–100	–	–	–	30–100	100–120
Mn (μg/g)	100–250	–	100–200	–	–	–	40–200	200–300
Zn (μg/g)	15–30	–	15–30	–	–	–	10–40	20–28
Cu (μg/g)	10–25	–	10–25	–	–	–	4–10	15–25
B (μg/g)	25–60	–	25–60	–	–	–	15–40	40–60
Na (μg/g)	< 500	–	–	–	–	–	–	–
Cl (%)	< 0.25	–	–	–	–	–	–	–
Time of sampling	After panicle emergence	After fruit set	Anthesis	Before flower initiation	After panicle emergence	Before flower initiation	Before flower initiation	Before flowering

Standards developed for litchi in Australia were based on surveys of high-yielding orchards in southern Queensland, with samples collected 1–2 weeks after panicle emergence between May and August (Menzel *et al.*, 1992b). The equivalent period in South-east Asia is from October to December. The first leaf under the panicle was sampled from eight branches uniformly distributed around each tree (Fig. 11.8). The leaf sample was accompanied by a soil sample taken from 0 to 0.15 m depth each year. One such sample normally represented a planting of no more than 3 ha, with separate samples recommended for each soil type, block or cultivar. Approximately 20 uniform trees that were well separated throughout the orchard were selected. The recommended nutrient concentrations (Table 11.2) were similar to the

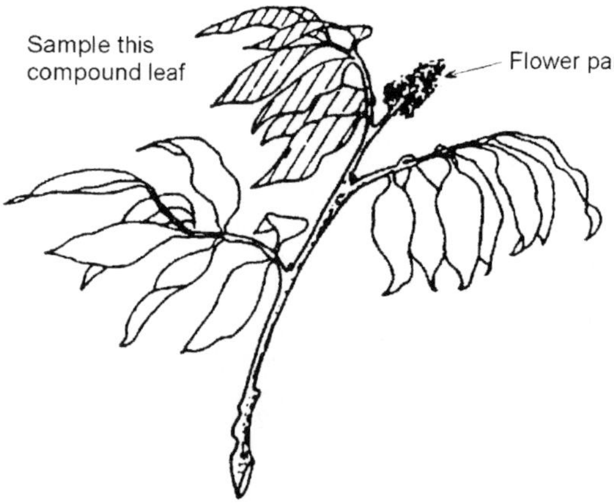

Fig. 11.8. Leaves for tissue testing of litchi in Australia are sampled after panicle emergence (redrawn from Menzel *et al.*, 2002, with permission).

values suggested in South Africa, even though the leaves were sampled earlier.

Changes in soil fertility, leaf nutrient composition and yield were evaluated in 30 'Haak Yip' orchards in central Taiwan (Huang *et al.*, 1998). Most of the soils had a pH of < 5 and low concentrations of calcium and magnesium. The suggested optima for these nutrients at anthesis are similar to those developed in Australia (Table 11.2). The leaves were sampled later than in Australia and would be expected to have lower concentrations of nitrogen (0.2%), phosphorus (0.01%) and potassium (0.1%), and higher concentrations of calcium (0.1%) and magnesium (0.05%). Chen *et al.* (1998) set optimum concentrations for nitrogen, phosphorus and potassium from fertilizer experiments conducted in Guangdong, with the leaves collected in autumn before flower initiation (Table 11.2). Nitrogen and phosphorus concentrations were similar to values found in Australia, whereas potassium concentrations were 0.3–0.4% lower.

Liu *et al.* (1986) investigated the nutritional status of leaves of 'Wulongling' and 'Fuyan' longan in Fujian over 3 years, and determined the concentrations associated with higher yields (Table 11.2). Leaf samples were collected in January. Wang *et al.* (1994) carried out studies on 30- to 50-year-old trees of 'Shuizhang' in Fujian, applying fairly high rates of nitrogen (440 kg/ha), phosphorus (146 kg/ha) and potassium (306 kg/ha). Other studies in Fujian showed that nutrient status varied with year and locality (Zhuang *et al.*, 1995). It was suggested that samples for analysis be collected in autumn before flower initiation. Chen (1997) studied seasonal changes in nutrient concentrations in seven orchards in Taiwan and concluded that leaves just behind the flower clusters constituted the most appropriate sample. Suggested optima from these studies in China are indicated in Table 11.2.

Soil analysis can be used to ensure that a particular site meets the fertility status considered to be adequate for a specific crop and soil. Regular tests can help diagnose potential problems relating to soil pH, salinity and certain nutrient interactions that are not directly related to plant composition.

Tentative standards were developed for litchi in Australia from surveys of high-yielding orchards (Menzel *et al.*, 1992a). Soil samples were taken at the same time as were leaves for tissue analysis, just after panicle emergence and before the application of fertilizer. Collection of samples just after fertilizer application was avoided to preclude sampling errors that might be associated with the uneven distribution of the fertilizers. Research showed that the feeder roots and nutrients under the trees were concentrated in the topsoil (Menzel *et al.*, 1990). Consequently, sampling of the 0–0.15 m layer provided the most reliable estimate of nutrient reserves in the soil. Leaf mulch was removed before each soil sample was taken and this was accompanied by a leaf sample collected from the associated trees, with the sample taken from halfway between the trunk and the dripline or edge of the canopy (Fig. 11.9). Separate samples were taken for each block, soil and cultivar.

Soil optima recommended by Menzel *et al.* (1992a) for maximum yields in Australia were as

Fig. 11.9. The procedure for collecting soil samples for chemical analyses (redrawn from Menzel *et al.*, 2002, with permission).

follows: pH, 5.5–6.0; organic carbon, 1.0–3.0%; electrical conductivity, < 0.20 dS/m; Cl, < 250 mg/kg; Na, < 1.0 cmol (+)/kg; NO_3-N, 10 mg/kg; P, 100–300 mg/kg; K, 0.5–1.0 cmol (+)/kg; Ca, 3.0–5.0 cmol (+)/kg; Mg, 2.0–4.0 cmol (+)/kg; Cu, 1.0–3.0 mg/kg; Zn, 2–15 mg/kg; Mn, 10–50 mg/kg; and B, 1.0–2.0 mg/kg. It is not known whether these optima apply to soils in South-east Asia and other locations.

Orchard surveys

Information has been collected on leaf and soil concentrations in commercial orchards, sometimes along with records of tree performance. Data sets based on many different soil types and cultivars are more valuable than those based on only a few orchards. If properly conducted, these surveys can highlight potential nutrient problems.

The relative importance of different nutrients affecting litchi in Australia was determined at 13 sites in coastal subtropical Queensland (Menzel *et al.*, 1992c). Cultivars included in the study were 'Tai So' (n = 11), 'Bengal' (n = 18), 'Kwai May Pink' (n = 6), 'Kwai May Red' ('Guiwei', n = 3), 'Salathiel' (n = 3) and 'Wai Chee' (n = 5). Leaf nitrogen, potassium, calcium, magnesium and iron concentrations were lower, and leaf phosphorus, manganese, zinc and copper higher than the Australian standards in some orchards. It was suggested that nitrogen, potassium, calcium, magnesium and iron should be applied to some orchards, while it was desirable to reduce phosphorus applications in other instances. High leaf manganese, zinc and copper were related to contamination by sprays or dust.

Jiang *et al.* (1999) investigated the nutrient status of various fruit orchards including litchi, in 12 villages in Fujian. The soils were quite acidic, with the pH ranging from 4.1 to 5.0, and low in organic matter (1.5%), total nitrogen (0.05%), phosphorus, potassium, magnesium and boron.

In a survey of litchi in India, leaf samples were collected from 25 'Rose Scented' orchards in Uttar Pradesh (Kunwar and Singh, 1993). The trees were growing on sandy and clay loams that ranged in pH from 7.0 to 8.2, typical of the area. Leaf nitrogen ranged from 0.9% to 2.1%, and was below 1.8% in 80% of the orchards. Leaf phosphorus was low in more than 80% of the orchards (0.03–0.22%), whereas potassium (0.55–1.30%), calcium (0.40–0.95%) and magnesium (0.24–0.60%) were adequate. The application of additional nitrogen and phosphorus was indicated for commercial production.

A similar survey of 15 orchards in north-eastern Haryana showed that half the plantings had low soil carbon, three-quarters had low soil nitrogen, and one in ten had low phosphorus (Joon *et al.*, 1997). Soil pH, electrical conductivity and potassium were adequate. Leaf nitrogen ranged from 1.4% to 2.4%, phosphorus from 0.18% to 0.31%, potassium from 0.60% to 0.97%, calcium from 1.20% to 2.85% and magnesium from 0.07% to 0.40%. The relative productivity of these orchards was not indicated, although some nitrogen and magnesium concentrations were low (Menzel *et al.*, 1992b). An analysis of 45 orchards growing on sands and loams in the Punjab (mean pH of 7.5) found that many plantings had low concentrations of nitrogen, potassium, magnesium, zinc and manganese (Hundal and Arora, 1993).

Zhang *et al.* (1999) examined variations in soil fertility in 14 longan orchards in the Pearl River Delta in Guangdong. All the orchards had low soil nitrogen concentrations, while 36% were deficient in phosphorus and 93% deficient in potassium, calcium and magnesium. Applications of these nutrients were recommended by the authors.

The many surveys that have been conducted in litchi and longan can help to identify potential nutrient problems, but they are not able to indicate the amounts of the different nutrients required to restore or maintain productivity under various management systems and for different soils.

Nutrient uptake

Destructive harvest of trees in Australia showed that the greatest reserves of nutrients in 'Bengal' litchi occurred in the leaves, twigs and small branches. These structures accounted for 75% of the total reserves of the tree, whereas the other plant parts usually contained less than 5% (Menzel *et al.*, 1992d). The high reserves present in the leaves, twigs and small branches were accounted for in their large contribution to the plant's weight, although the concentrations of nutrients were also higher in these tissues than in the other plant parts (Fig. 11.10). The concentrations of nutrients in the leaves reflected the reserves in the rest of the plant, indicating that they were a reliable index of the tree's nutrient status.

In Australia, average nutrient concentrations on a dry matter basis in 'Bengal' fruit were as follows: N, 0.85%; P, 0.19%; K, 1.04%; Ca, 0.10%; Mg, 0.18%; Mn, 29 μg/g; Zn, 34 μg/g; Cu, 36 μg/g; and B, 15 μg/g (Menzel *et al.*, 1992d). A 50 kg crop removed the following nutrients (g/tree) in the fruit: N, 98; P, 22; K, 120; Ca, 12; Mg, 21; Mn, 0.3; Zn, 0.4; Cu, 0.4; and B, 0.2. Thus the fruit used more potassium than nitrogen. The amounts of nutrients needed for new leaves, stems, roots and flowers were not included in these calculations. Some of the nutrients would be recycled from leaf litter and fallen twigs, flowers and fruit. If rates of fertilizer application are several times higher than those removed by the crop, then an orchard is being over-fertilized.

Nitrogen

The literature concerning nitrogen nutrition in litchi is fairly extensive. Symptoms of deficiency have been described and calibrations developed, between growth and leaf concentrations in potted plants, and between yield and leaf concentrations in orchard trees. There has been some attempt to determine the impact of strategic applications of nitrogen during the year, and the responses to different types of fertilizer. Less information is available for longan.

Nitrogen is the most important nutrient affecting growth and productivity, with deficiency symptoms appearing when leaf concentrations fall below 1.3% in dry matter. Because nitrogen moves from old to young leaves when its concentration in plants is low, the older leaves are the first to turn yellow (Goldweber, 1959; Mallik and Singh, 1965; Thomas *et al.*, 1995; Lei *et al.*, 1998). When the deficiency is severe, leaf expansion, flowering and fruit set are severely affected (Plate 30). Fruit are small with a low flesh recovery and poor eating quality (Goldweber, 1959; Mallik and Singh, 1965; Thomas *et al.*, 1995).

The effects of nitrogen on shoot and root growth of litchi were investigated in seedlings in sand culture (Menzel *et al.*, 1995a), because symptoms of deficiency usually occur

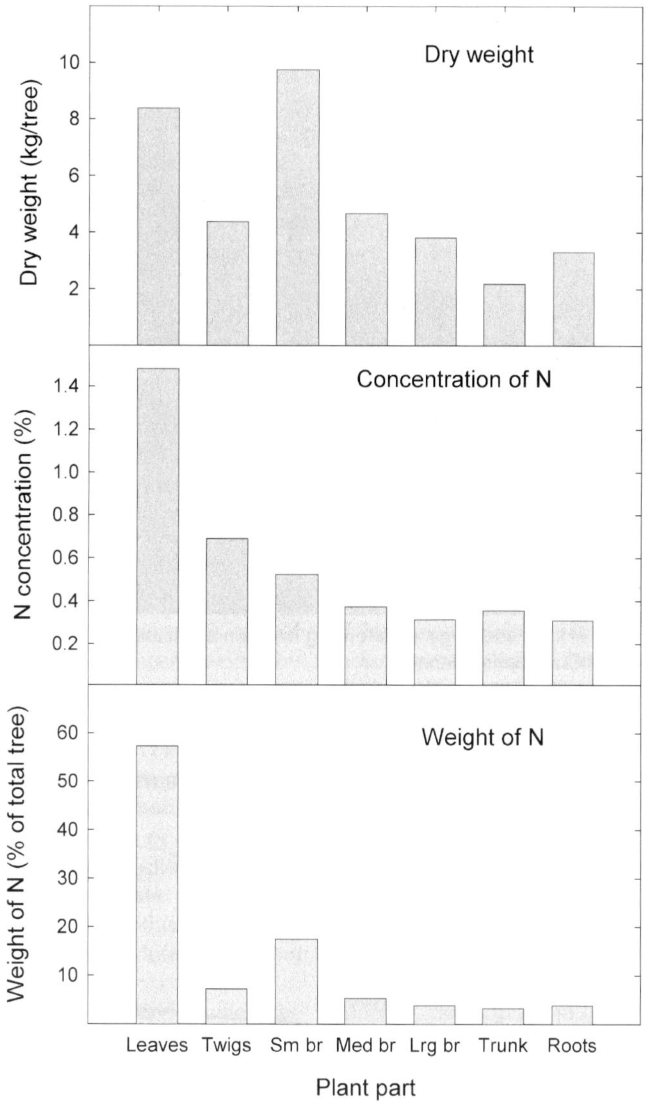

Fig. 11.10. Distribution of dry matter and nitrogen, and concentration of nitrogen in 'Bengal' litchi in Australia. Twigs are less than 1 cm in diameter, small branches (Sm br) are 1–3 cm in diameter, medium branches (Med br) are 3–5 cm in diameter, large branches (Lrg br) are 5–10 cm in diameter, and trunks are greater than 10 cm in diameter. Redrawn from Menzel *et al.* (1992d), with permission from the *Journal of Horticultural Science & Biotechnology*.

much earlier under these conditions than when plants are grown in the field (Fig. 11.11). High rates of nitrogen application increased the concentration of the element in the plant (Fig. 11.12). Shoot growth increased with nitrogen, up to the highest leaf nitrogen obtained (about 1.4%), indicating that the leaf nitrogen required for maximum shoot growth was greater than this value (Fig. 11.13). In contrast, 95% of maximum root growth occurred with a leaf nitrogen concentration of 1.0–1.2%. Differences in shoot growth were recorded above and below 1.2% nitrogen in the leaves, but before large differences in leaf colour were observed. These results suggested that productivity would be affected long before the leaves turn yellow.

Li *et al.* (2001c) revisited some of these ideas using 6-year-old trees of 'Tai So' and 'Brewster', which are the dominant cultivars in Florida. Treatments were 181 g N/tree applied at: panicle differentiation (January); panicle differentiation and flowering (March); panicle differentiation, flowering and after harvest (June); or at the previous stages and in November. The concentrations of nitrogen in the leaves ranged from 1.1% to 2.7% in 'Tai So' and from 1.1% to 2.4% in 'Brewster', with the lower values being deficient (Menzel *et al.*, 1995a). Trees receiving the highest rate of fertilizer did not always have higher leaf nitrogen concentrations than trees receiving lower amounts. Flowering was greater in the trees receiving 181 and 362 g N (82% and 94% for 'Tai So') compared with trees receiving 543 and 724 g N (40% and 20% for 'Tai So'), but yields were lower than 15 kg/tree in all treatments.

Koen *et al.* (1981a) investigated the effects of fertilizers on the yield of 'Tai So' over 8 years in South Africa. The highest nitrogen application (140 g increasing to 1000 g/tree) gave the highest yield (mean of three seasons). The concentration of nitrogen in the leaves sampled after fruit set increased from 1.35% to 1.52% with increasing nitrogen applications. In other experiments, maximum yields were obtained with a leaf nitrogen concentration of 1.47%, with lower yields at 1.42% or 1.52% (Koen *et al.*, 1981b). At the end of the experiment, the 12-year-old trees with the best yield had received 840 g N/year. Nitrogen also increased fruit weight and sugar-to-acid ratio and reduced the proportion of malformed fruit. The leaf concentrations associated with maximum yields were similar to those recorded in Australia (Menzel *et al.*, 1992b, 1994), even though the samples were collected after fruit set, compared with before flowering in Queensland.

Ghosh *et al.* (cited in Mitra, 1988) found that the highest yields of 'Bombai' achieved over 3 years in India were obtained with the highest application of nitrogen used (600 g N applied to 6-year-old trees). The optimum leaf nitrogen concentration was 1.33% before flowering, 1.47% at flowering, and 1.15% after harvest. Lal *et al.* (1999) studied the effects of fertilizer (0–1.8 kg N/tree) on growth, yield and leaf composition in 14-year-old 'Rose Scented' trees over 2 years in Uttar Pradesh. Leaf nitrogen concentrations increased from 0.91% to 1.42% with increasing fertilizer application rate. The best growth and yield were obtained with 1.2 kg N/tree, equivalent to a leaf nitrogen concentration of 1.36% before flowering.

Chen *et al.* (1998) studied the response of 21-year-old trees growing on an acid clay loam in Guangdong, with a pH of 4.4, organic matter of 1.4–1.8%, and total nitrogen of 0.07%. Application of 0.8 kg N/tree increased yield compared with unfertilized controls, the higher yields being associated with a leaf nitrogen concentration in autumn of 1.60–1.75%. These values are similar to those reported for maximum yields in Australia (Menzel *et al.*, 1994). It can be concluded from these studies that low nitrogen concentrations can limit litchi production.

The time of fertilizer application did not affect fruit production in 6-year-old 'Bengal' trees growing in a sandy loam containing 2.8 mg NO_3-N/kg in southern Queensland (Menzel *et al.*, 1994). Applications equivalent to 750 kg N/ha were made in year 4 after panicle emergence in July, after harvest in January, or split between the two periods. Control trees received no nitrogen. The time of application had little impact on leaf nitrogen values, which varied by only 0.1% in the different treatments.

It took unfertilized trees 4 years to show significant reductions in leaf nitrogen concentrations (Fig. 11.14) and yield compared with fertilized trees in experiments in Australia. Yield increased as leaf nitrogen in August in year 4 increased from 0.95% to 1.56%, with lower yields in control trees associated with poor leaf growth in the previous 2 years, and lower CO_2 assimilation compared with trees receiving nitrogen. The 12-year-old trees of Koen *et al.* (1981a,b) achieved maximum yields with an application of 0.84 kg N/tree or about 300 kg N/ha broadcast, whereas the trees in Australia received two or three times this amount.

Figure 11.15 shows the relationship between yield and leaf nitrogen in litchi measured in various experiments. Maximum yields were associated with a leaf nitrogen concentration of 1.4% in India, and about 1.5% in South Africa and Australia. There was some evidence that yield continued to increase with further nitrogen applications in the latter studies, suggesting that the optimum leaf nitrogen is higher than 1.5%. The optimum value in South Africa would be expected to be about 0.3 units lower

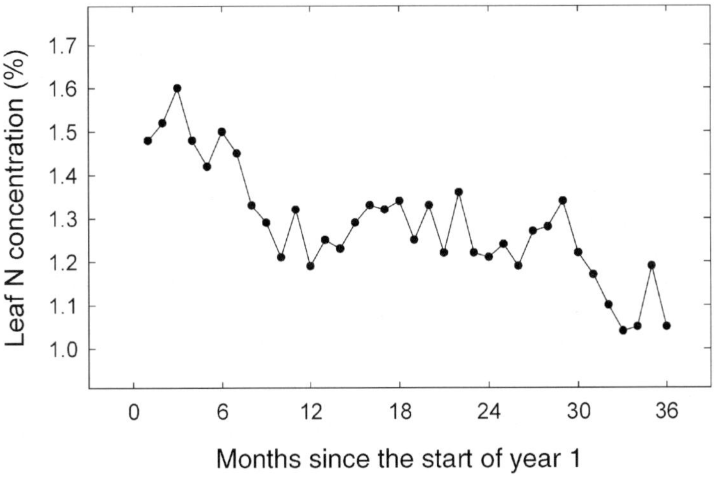

Fig. 11.14. Changes in leaf nitrogen concentration in 'Bengal' litchi over 3 years in Australia. The trees received no nitrogen fertilizer from the October before year 1 (redrawn from Menzel *et al.*, 1994, with permission). The optimum is 1.50–1.80% N.

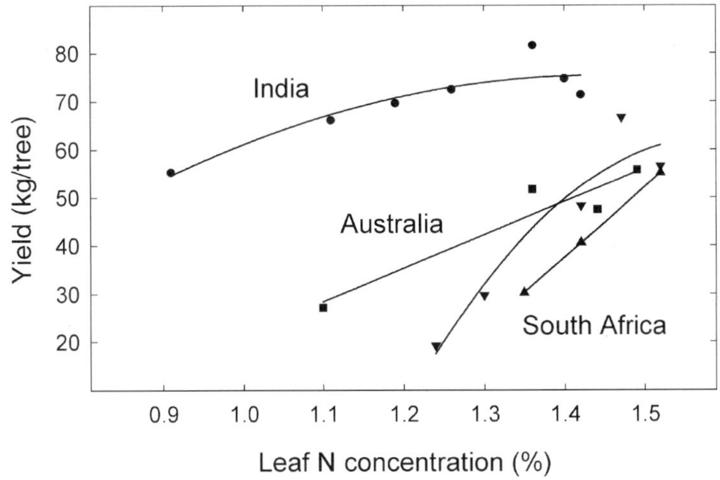

Fig. 11.15. Relationship between yield and leaf nitrogen concentration in litchi. Samples in India, Australia and South Africa were collected before flower initiation, after panicle emergence and after fruit set, respectively (redrawn from Lal *et al.*, 1999; Menzel *et al.*, 1994; and Koen *et al.*, 1981a,b).

(% dry matter) than in Australia because the samples were taken later in the phenological cycle (Fig. 11.6).

Nitrogen is generally applied as urea (46% N), ammonium nitrate (34% N) or sulphate of ammonia (21% N), with potassium nitrate (13% N and 38% K), calcium nitrate (15% N and 18% Ca) and calcium ammonium nitrate (27 N% and 8% Ca) being less important. Continuous applications of ammonium and some other nitrogen fertilizers can acidify the soil, and regular applications of lime or dolomite may be required to correct this.

Nitrogen is probably the main nutrient affecting production in orchard trees and the literature in litchi is extensive. Research has characterized the symptoms of deficiency and tree response in terms of growth and yield. Maximum yields are associated with leaf nitrogen of about 1.5%, although there is some evidence that higher concentrations might provide higher yields. The manipulation of trees to attain low nitrogen concentrations prior to flowering does not increase fruit production.

Phosphorus

Symptoms of phosphorus deficiency have been induced in potted plants but they are rare in the field. Only limited responses to phosphorus applications have been recorded, since most orchards have high concentrations of this nutrient.

Various symptoms of deficiency have been described for plants growing in sand culture (Goldweber, 1959; Mallik and Singh, 1965; Lei *et al.*, 1998). Dead patches developed on the tips and margins of mature leaves and progressed towards the midrib. The necrosis was typically copper-brown. Eventually, the leaves curled, dried out and fell. Symptoms of deficiency may occur in trees growing on acid or alkaline soils that are well supplied with the nutrient because of its reduced availability at extreme pH values, although some sandy and organic sites are naturally low in the nutrient.

'Brewster' and 'Purbi' litchis grown in pots under low or nil phosphorus produced no obvious symptoms and resembled control plants (Goldweber, 1959; Mallik and Singh, 1965). Shoot and root growth were normal, except for an increase in trunk girth and premature leaf drop. Goldweber also noted that plants grown in the absence of phosphorus fruited, although the fruit were smaller than in the controls. There was no relationship between leaf phosphorus and nutrient supply in the studies by Mallik and Singh (1965). Phosphorus concentrations at the end of the experiment when deficiency symptoms were visible, were 0.31% for the 1/8 P treatment, 0.11% for the control and 0.51% for the 2P treatment. This suggests that leaves other than the recently matured leaf were used for analysis. Leaf concentrations above 0.08% are considered adequate in Australia (Menzel *et al.*, 1992b).

Only small responses to phosphorus have been recorded for litchi in South Africa. Koen *et al.* (1981a) studied the effect of fertilizers (0, 300 and 600 g P/tree increasing to 0, 750 and 1500 g P/tree over 8 years) in 9–12-year-old 'Tai So' orchards. Average yields from three seasons (1976, 1978 and 1979) were 37.8, 42.3 and 45.8 kg/tree, respectively. No explanation was given for the absence of fruiting during the other seasons. Fruit quality was not affected. Soil phosphorus determined by a resin extract, which is different from the method used in Australia, ranged from 2.7 to 8.9 mg/kg, while leaf phosphorus ranged from 0.12% to 0.18%. A reaction was not expected unless soil concentrations were below 6 mg/kg. The leaf phosphorus concentrations were similar to those suggested in Australia (Menzel *et al.*, 1992b).

Menzel *et al.* (1995b) studied the effect of phosphorus on litchi production in subtropical Queensland. These authors were interested to see whether deficiencies would appear in trees not fertilized for 3 years, and if excessive rates of application had any detrimental effects on production. The 'Tai So' and 'Kwai May Pink' trees were growing in a sandy loam and a red clay loam, respectively, which differed in their ability to buffer against sudden changes in external nutrient supply. The sites were selected on the basis that they had fertility levels common to many orchards in Australia. Fertilizer applications were equivalent to 0–2.4 t P/ha. In the first 2 years, there was no effect on leaf phosphorus in the trees receiving fertilizer, while in year 3, phosphorus accumulated in the leaves of trees given high rates of fertilizer (up to 0.26% and 0.44% P). In contrast, leaf phosphorus in the controls fell from 0.33% to 0.28% over 3 years at one site and was similar over that time at the other site (0.18%).

In the experiments of Menzel *et al.* (1994), fruit production was similar over the range in leaf phosphorus of 0.18–0.44%, compared with the Australian standard of 0.14–0.22% (Menzel *et al.*, 1992b). The buffering capacities of the soil and tree were thus indicated. These results suggest that annual applications may not be required, indicating savings for growers equivalent to US$70/ha for 15-year-old trees. The leaf standard for phosphorus may also need to be reviewed. The 12-year-old trees of Koen *et al.* (1981a) would have had a canopy diameter of about 6 m and a canopy cover of about 28 m^2. Maximum yields were achieved with 0.1 t P/ha, whereas in Australia there was no increase in yield with 0.2 t P/ha.

Where mixed fertilizers are applied regularly, low phosphorus concentrations are unlikely to occur because the nutrient is not readily leached from most soils. On the other hand, regular applications of phosphorus may lead to deficiencies of other nutrients such as iron and zinc. Phosphorus is available as superphosphate (9% P, 11% S and 20% Ca), triple superphosphate (19% P, 2% S and 18.5% Ca), diammonium phosphate (18% N, 20% P and 2% S) and monoammonium phosphate (12% N, 22% P and 3% S). When required, a single application of any of these applied to the soil is usually sufficient for two or three seasons.

Symptoms of phosphorus deficiency have been described for potted plants, but are rarely observed in orchard trees. Responses to phosphorus applications are usually smaller than those recorded for nitrogen. Most orchards have adequate concentrations of the nutrient.

Potassium

Symptoms of potassium deficiency have been noted in plants growing under glasshouse conditions. Limited calibration experiments have related growth and yield to leaf concentrations. Responses to potassium are relatively rare in orchard trees.

Many litchi orchards in South-east Asia and Australia have leaf potassium concentrations below 0.80% (Menzel, 2002). This can occur late in the season when potassium is translocated to developing fruit, after heavy nitrogen applications (Mallik and Singh, 1965), or after heavy rain. This is more likely to occur on sands than on clay loams or heavy clays.

The first symptom of potassium deficiency is a general yellowing of the leaves, followed by the death of the leaf tips and later the bases of the leaves (Goldweber, 1959; Mallik and Singh, 1965; Thomas *et al.*, 1995; Lei *et al.*, 1998). The old leaves are eventually shed so that the canopy consists of small terminal clusters of new leaves. Severe deficiency in sand culture restricted shoot and root development, flowering and fruit set, and killed trees. Deficiency in the experiments of Mallik and Singh was not associated with reduced leaf concentrations (1.15–1.26% K), possibly because physiologically young leaves were used for the analyses. In the experiments of Thomas *et al.*, deficient plants had leaf potassium of 0.38% compared with 1.28% in the controls. The value for the controls was similar to that obtained from trees in South Africa and Australia (see below).

Joiner and Dickey (1961) irrigated 'Brewster' plants in sand culture with a nutrient solution containing 8, 32 or 180 mg K/l. The fertilizer increased leaf potassium from 1.11% to 2.24%, but reduced shoot extension and trunk growth, especially under high nitrogen applications. These responses were associated with a lower concentration of magnesium, which fell from 0.57% to 0.45%. The concentration of magnesium in the roots was only 0.01% under high potassium and nitrogen, whereas plants with low rates of fertilizers had 0.13% Mg.

It has been reported that potassium affects litchi photosynthesis (Deng *et al.*, 1994). Net CO_2 assimilation was higher under high irradiance and high ambient CO_2 concentrations with higher applications of potassium (K_0, K_1 and K_2 treatments). For instance, at saturating photosynthetic photon flux densities, or PPFDs, of 900 μmol quanta/m^2/s, net CO_2 assimilation was 2.0, 3.5 and 3.0 μmol quanta/m^2/s, for the respective potassium treatments. These results suggest that potassium can increase production by increasing photosynthesis in the leaves.

Lal *et al.* (1999) investigated the influence of fertilizers on the production of litchi in Uttar Pradesh. There were seven rates of potassium up to 900 g/tree, applied over 2 years to 14-year-old trees of 'Rose Scented'. Potassium concentrations increased from 0.96% to 1.39% in leaves collected before flowering in October, but there was no effect on yield (52–61 kg/tree). These responses were associated with higher vegetative growth at the intermediate rates of application. No soil concentrations were presented, although it appears that these orchards had sufficient potassium to support satisfactory production for at least 2 years. The authors reported a significant response to nitrogen in the same experiment.

There was no response to potassium in fertilizer experiments in South Africa (Koen *et al.*, 1981a). Rates up to 500 g K/tree were applied to 9–12-year-old 'Tai So' orchards, resulting in yields from 41.0 to 43.5 kg/tree. Soil concentrations ranged from 50 to 61 mg K/kg and leaf concentrations from 0.91% to 1.06% K in the different treatments, with a response expected only when initial soil concentrations were below 56 mg K/kg. A different analysis technique was used compared with that employed in Australia, but the optimum leaf concentrations were similar to values recorded in Australia (Menzel *et al.*, 1992b).

Similar experiments on 'Tai So' and 'Kwai May Pink' trees conducted over 3 years in Queensland produced no symptoms of deficiency or excess, with applications of 0–3.2 t K/ha, which is about ten times industry recommendations (Menzel *et al.*, 1995b). Leaf potassium fell only slightly over time in trees given no fertilizer. Fruit production was similar with leaf potassium of 0.75–1.37% compared with the industry standard of 0.70–1.10% (Menzel *et al.*, 1992b). These results suggest that annual applications may not be required in orchards that already have large amounts of soil-extractable potassium.

Potassium is applied to orchards as potassium sulphate (41% K and 16.5% S), potassium chloride (50% K and 50% Cl) or potassium nitrate (38% K and 13% N). The fertilizers can be applied in conjunction with nitrogen in two or three applications at panicle emergence, fruit set and harvest.

Symptoms of potassium deficiency are well documented in potted plants and field-grown trees; however, only small responses to potassium have been recorded. Field calibration experiments have provided tentative optimum leaf nutrient concentrations.

Calcium

Limited information is available on calcium nutrition. Symptoms of deficiency have been achieved by growing plants in sand culture, but are rare in the field. Calcium is reported to affect fruit quality, at least in litchi. Calcium is applied to both crops when soil or leaf calcium concentrations or soil pH decline, even though no optimum leaf and soil concentrations have been determined.

Litchis and longans grown in sand without calcium produced smaller leaves than those fertilized with calcium (Goldweber, 1959; Lei *et al.*, 1998). Necrotic areas also developed along the margins of the leaves, which were then shed. Eventually the leaves, stems and roots ceased to grow. Plants flowered, but did not set fruit.

The role of calcium in litchi fruit cracking (fruit splitting) has been studied extensively in China. There was a strong correlation between cracking and total calcium concentration in different tissues, with less calcium present in the skin of cracked fruit (Li and Huang, 1995). Further studies by Huang *et al.* (2001) showed that the concentration of the nutrient in the cell wall as pectin-bound calcium was higher in the skin of the resistant cultivar 'Wai Chee' ('Huaizhi') than in the skin of the susceptible cultivar 'No Mai Chee' ('Nuomici'). X-ray microscopic analysis revealed a large number of calcium-rich cells in the epidermis of 'No Mai Chee', whereas there were relatively few of these cells in 'Wai Chee'. These results suggested that the lower cell-wall calcium concentrations in 'No Mai Chee' skin were related to lower calcium mobilization rather than to a shortage of the nutrient. Further research is required to resolve this issue. In the interim, foliar sprays of calcium are recommended to reduce cracking (Saha *et al.*, 1993; Li *et al.*, 2001a).

Calcium toxicity is rare, although high soil concentrations may reduce the uptake of manganese, zinc, boron, copper and phosphorus. Many of the soils in India and Florida have high concentrations of free lime, and thus must be managed carefully to avoid micronutrient deficiencies.

The nutrient is found in calcium sulphate (18–20% Ca, and 14–18% S), calcium nitrate (18–19% Ca and 15% N), calcium carbonate or lime (35–40% Ca), calcium and magnesium carbonates (12–15% Ca and 8–13% Mg), and calcium ammonium nitrate (8% Ca and 27% N). Calcium is also available in superphosphate (20% Ca) and triple superphosphate (18.5% Ca).

Because of its influence on soil acidity, management of calcium involves analysis for calcium, magnesium and pH. Lime, dolomite and gypsum are used to improve the calcium status of soils; however, care must be taken not to overuse these materials, as they may alter soil chemistry. The choice of product depends on the effect required. Lime is used when pH and calcium are low, while dolomite is used when pH, calcium and magnesium are low. Gypsum at the rate of 500 g/m^2 tree canopy is normally used when the pH is within the desired range, but calcium is low. The amendments can be applied in early autumn or spring.

Research suggests a strong relationship between fruit cracking and low calcium in litchi; however, no relationships have been established between growth/yield and leaf calcium concentrations. The role of this nutrient in commercial production has yet to be determined.

Magnesium

Research on magnesium nutrition has been limited to potted plants, from which leaf concentrations associated with deficiency have been determined. No field experiments have been conducted.

The concentration of magnesium in orchards is often below 0.30%, especially when trees are grown on sands that are readily leached. It is not known whether these low values reduce productivity or fruit quality. Magnesium is not readily transported from the old leaves in litchi and longan; hence symptoms occur first in the young leaves. Plants grown in sand in the absence of the nutrient had small leaves (Goldweber, 1959; Lei *et al.*, 1998). The tips and centres of the leaves died and they were eventually shed (Thomas *et al.*, 1995). Flowering was suppressed when leaf concentrations fell below 0.25% (Joiner and Dickey, 1961). Deficiencies can be induced by heavy applications of nitrogen and potassium. Tomer and Kadman (1985), for instance, noted that growth of 'Tai So', 'Bengal' and 'Floridian' in Israel was reduced as nitrogen in the irrigation water increased from 25 to 100 mg/l. It is possible that the high nitrogen may have interfered with magnesium uptake, although no leaf samples were analysed.

In sand culture experiments in Florida (Thomas *et al.*, 1995), symptoms of deficiency were not associated with lower leaf concentrations compared with well-fertilized 'Tai So' and 'Brewster' plants. Average leaf concentrations were 0.41% Mg, possibly because old and young leaves were mixed in the samples.

Li *et al.* (2001b) studied the effect of magnesium on the growth of longan seedlings under hydroponics. Magnesium deficiency resulted in decreased growth of both roots and shoots, with the roots being particularly sensitive. Seedlings with low magnesium concentrations had a higher shoot-to-leaf ratio compared with seedlings with sufficient nutrient. Concentrations of leaf magnesium below 0.25% were considered to be deficient.

The nutrient is available as magnesium sulphate (10% Mg), calcium and magnesium carbonates and magnesium oxide (54% Mg). Dolomite applied to correct pH may not supply enough magnesium if soil concentrations are very low. Annual applications of magnesium sulphate after panicle emergence at the rate of 40 g/m^2 tree canopy are suggested.

Zinc

Symptoms of zinc deficiency have been described for plants grown in pots. The response of trees in the field to foliar sprays applied at flowering has also been studied.

Deficiency occurred on acid, leached soils where native zinc was low, or on alkaline soils where zinc was not readily available to plants, especially after heavy nitrogen applications. Symptoms include general bronzing or mottling of the leaves (Plate 31) and the production of small shoots and fruit. Litchi branches die when zinc concentrations fall below 10 μg Zn/g dry weight in the leaves. Plants grown in sand without the nutrient in Florida had lower leaf concentrations (19 μg/g) than those fertilized with zinc (54 μg/g), but did not show symptoms of deficiency (Thomas *et al.*, 1995). These leaf concentrations are much higher than those reported in Australia (Menzel *et al.*, 1992b). The absence of deficiency symptoms and high leaf concentrations suggest that the plants had access to other sources of zinc.

Dutt (1962) sprayed litchi trees of unspecified age and cultivar in India, with 3.1 g Cu/l, 2.7 g Zn/l or 0.2 g B/l before flowering. The number of fruit per tree at harvest was 40, 200, 100 and 70 for the control, copper, zinc and boron treatments, respectively. All fruit produced by the controls were immature. Those produced after boron were partially immature but all the fruit produced after zinc or copper were large and sweet. No soil or leaf nutrient concentrations were determined. Many other studies have demonstrated only a slight effect of zinc sprays on productivity in India (Menzel, 1987).

Various zinc fertilizers are available, including zinc sulphate heptahydrate (23% Zn), zinc sulphate monohydrate (36% Zn), and the less soluble zinc oxide (80% Zn). It is necessary to monitor both leaf and soil concentrations to determine orchard zinc requirements. Soil applications of zinc sulphate heptahydrate at 5 g/m^2 tree canopy are preferred. If, after such applications there is little or no response and leaf concentrations are still below optimum, two applications of zinc sulphate heptahydrate at 1 g/l can be made to the expanding leaves after harvest.

Iron

Research on the role of iron in nutrition is limited in both crops. Symptoms of deficiency have been described along with suggested remedial actions; however, the relationship between plant growth and leaf concentrations is not known.

Orchards established on sands often have low iron concentrations, and iron deficiencies may be further exacerbated by excessive phosphorus, which interferes with iron uptake by the roots. Problems may also arise in alkaline soils with pHs above 7.0 or after excessive lime applications. Plants showing deficiency have a general yellowing of the leaves, which spreads from the young to the older growth (Thomas *et al.*, 1995). When the deficiency is severe, the leaves turn almost white and the branches die (Plate 33). Concentrations below 40 μg Fe/g cause problems in Australia, although both fertilized (36 μg Fe/g) and deficient plants (22 μg Fe/g) had concentrations below this value in Florida (Thomas *et al.*, 1995).

Malcolm (1953) corrected iron deficiency in young 'Brewster' trees growing in a calcareous soil in Florida by applying Fe-EDTA to the leaves at the rate of 4 g Fe/l. Leaf colour returned within a few days in trees that were slightly affected, while new shoots appeared after 8 weeks in 60% of trees that were severely affected.

Iron is available as iron sulphate (23% Fe) for use in all situations, and iron chelate or iron EDDHA (Sequestrene, 5–15% Fe) where the pH is greater than 7. Concentrations of iron in the leaves are a good indication of the success of any programme, with values below 40 μg/g considered deficient. Deficiencies can be corrected by applying iron sulphate to the soil at 10 g/m^2 tree canopy, although foliar sprays at 5 g/l are required for orchards growing on soils with 30% free lime or phosphorus concentrations above 500 mg/kg.

Copper

Little research has been conducted on copper nutrition in these species, although tentative leaf nutrient concentrations associated with deficiency have been proposed in Australia. Horticulturists in India have also examined the effects of copper amendments on litchi fruit set. Deficiencies are fairly rare.

Deficiency was most likely in sands with naturally low concentrations of copper, or in soils with high concentrations of organic matter or limestone. Relatively high concentrations of nitrogen, phosphorus and zinc exacerbated the problem. In Australia, the leaves curled along their margins and died when leaf concentrations fell below 6 μg Cu/g (Plate 32). Fruit were small with reduced pulp recovery (Dutt, 1962). In some soils in Australia, copper and zinc deficiencies occur together, with the trees producing long shoots with small leaves that abscise.

Copper is available as bluestone or copper sulphate pentahydrate (25% Cu). A single application applied to the soil at 4 g/m^2 tree canopy is adequate in most situations. Copper sulphate can also be applied to the foliage at 2 g/l. Fungicides such as copper oxychloride and copper hydroxide supply the nutrient, although these are not routinely used for disease control in litchi or longan.

Boron

Optimum concentrations of boron have been established for high-yielding trees in Australia. Researchers in South-east Asia have also examined the response to soil and foliar applications.

Symptoms of deficiency have not been well described for either species, although typical symptoms in tropical tree crops include retarded root and shoot growth, leading to tree decline. The leaves also turn yellow with a shot-hole effect, while fruit are misshapen. Dutt (1962) reported that litchi trees in India with low boron concentrations produced small fruit with a low pulp recovery. Many orchards had low soil or plant boron concentrations. Deficiency was more likely to be found on acid sands where native boron had been leached, and on alkaline soils with free lime, and on soils with high concentrations of organic matter.

High boron concentrations are generally required for satisfactory fruit set in plants. Deficiencies reduced flowering and pollination, whereas vegetative growth was often unaffected (Hanson and Breen, 1985a). Fruit set was related to boron concentrations in the flowers rather than to that in the leaves, with limited remobilization occurring when concentrations were low (Hanson and Breen, 1985b). Both soil and foliar applications increased litchi production in China (Xu *et al.*, 1984). Trees (15–20-year-old specimens with a ground cover of 15–20 m^2) sprayed with 0.45 g B/l or given 20–25 g B as a soil dressing produced three to four times the yield of control plots. No leaf or soil boron concentrations were presented, nor were the times of boron application recorded. The actual tree yields were not indicated.

Boron is available as borax (11% B), solubor (22% B) or boric acid (17% B). One soil application of solubor per year at 2 g/m^2 tree canopy at panicle emergence is usually sufficient, although a second application is required after harvest if leaf concentrations are very low (Menzel *et al.*, 1992b). Solubor can be dissolved in water and sprayed evenly under the trees. Care needs to be taken when applying the nutrient through the irrigation, as the rates of application need to be adjusted to take into account the area covered by sprinklers. Foliar applications of solubor at 2 g/l may be required in some clays that restrict boron uptake by plant roots. Boron must be applied with care, as the margin between deficiency and toxicity is small. Initial symptoms of toxicity include salt-burn along the margins of the leaves, and some leaf drop (Plate 34). These symptoms may affect the trees for several leaf flushes, or for several years in severe cases. In litchi, toxicity occurs with leaf concentrations of 100 μg B/g or higher.

Manganese

No symptoms of manganese deficiency are described for these crops. In other fruit crops, the symptoms of deficiency are similar to those of zinc and iron, with yellowing between the veins of the leaves. Thomas *et al.* (1995) followed the growth of young 'Tai So' and 'Brewster' trees in sand lacking manganese over 30 weeks. The plants grown without the nutrient did not develop any visual symptoms of deficiency, even though they had lower leaf concentrations (40 μg Mn/g) than the controls (400 μg Mn/g). Values of 100–250 μg Mn/g dry weight are considered acceptable for trees grown on moderately acid soils in Australia (Menzel *et al.*, 1992b).

Manganese deficiency may occur in soils with a pH greater than 7, or be induced by the indiscriminate use of lime. This is more likely to be an issue in India and Florida in orchards on calcareous soils. Toxicity may develop in soils with a pH below 5.5 and soil manganese concentrations above 50 mg/kg. The pH to a depth of 0.15 m should be raised to 6.5 under these circumstances. In most soils, deficiency can be corrected by one or two applications of manganese sulphate (33% Mn) at 5 g/m^2 tree canopy. Sands require annual applications, and timing is not critical. When the deficiency is chronic, foliar applications of manganese sulphate at 2 g/l are suggested (Menzel *et al.*, 2002).

Fertilizer programme for young orchards

A soil analysis conducted at least 6 months before a new orchard is planted in Australia is recommended (Fig. 11.16), as this allows plenty of time for the required fertilizers to be applied (Menzel *et al.*, 2002). Soil nutrient concentrations that are considered optimum vary depending on the laboratory conducting the analyses and their extraction procedures. The results of the analysis should be discussed with a horticulturist in order to determine what

Fig. 11.16. Collection of soil samples before the establishment of a litchi or longan orchard in southern Queensland, Australia. Redrawn from Menzel *et al.* (2002), with permission.

fertilizers are required. Insoluble fertilizers such as lime, dolomite, gypsum and superphosphate should be incorporated into the root-zone, whereas boron, potassium and nitrogen should be applied when the trees have established and start to grow. A broad fertilizer guide for litchi in Australia is given in a later section of this chapter.

Pre-plant applications normally correct any deficiencies in the soil and supply enough nutrients for initial tree growth. The programme for young trees consists of small, regular applications: in year 1, apply 30 g of general fertilizer (12 N:6 P:13 K:13 S) or similar to each tree every 3 months, increasing this to 40, 60 and 80 g in years 2, 3 and 4. Apply 8, 15 and 20 l/tree of organic matter in the spring of years 1, 2 and 3. The fertilizers should be distributed evenly under the canopy, and to 0.3 m past the dripline, keeping it at least 0.2 m away from the trunk to avoid injuring the tree.

Table 11.3. Annual fertilizer requirements to achieve acceptable yields of litchi in Australia, and longan in Taiwan (g/tree). From Menzel *et al.* (2002) and Zee *et al.* (1998).

Tree age (years)	Nitrogen	Phosphorus	Potassium
Litchi			
4–5	200	80	300
6–7	300	100	400
8–9	400	130	550
10–11	500	170	700
12–13	600	200	800
14–15	800	250	1200
15+	1000	300	1400
Longan			
3–4	100	90	50
5–6	150	150	100
7–8	200	210	150
9–10	300	300	300
11–12	500	450	625
13–14	700	750	875
15+	1000	1050	1250

Fertilizer programme for bearing orchards

In bearing orchards, the aim, through annual leaf and soil tests, is to maintain nutrient concentrations within a specified optimum range. Soil analysis is used primarily to monitor and adjust pH, calcium, magnesium, phosphorus and boron concentrations. Rates of the major nutrients required for trees in Australia and Taiwan are shown in Table 11.3. Urea and sulphate of potash, which supply a single major nutrient, are preferred to compound fertilizers, which supply too much phosphorus and insufficient magnesium, zinc and boron.

Trees carrying a heavy crop may require up to twice the amount of nitrogen and potassium indicated, so leaf and soil analyses should be used to adjust fertilizer application rates. For example in litchi, if 1.5 kg of potassium sulphate is applied and leaf potassium is 0.50% (the desired concentration is 0.70–1.10% in Australia), increase the rate in the coming year by 20% to 1.8 kg. If next year the leaf test indicates 0.70% K, apply an extra 10% (2.0 kg). If the leaf test exceeds 1.10% K, reduce the application rate by 5%. Nitrogen and potassium can be applied in two or three applications at panicle emergence (after leaf and soil testing), after fruit set and after harvest, to reduce losses from leaching. Phosphorus, if required, can be added to the first application.

If the range for micronutrients is within optimum values, use the recommended rates indicated earlier, but if the concentrations are below optimum, a second application is required. If the leaf test is above the standard value, apply nothing. Most micronutrients can be applied between spring and summer. Soil applications generally provide the best results, since uptake through the leaves via foliar sprays is very limited. However, foliar sprays of boron, copper and manganese can be applied to the mature summer and autumn leaves, and zinc applied to the expanding leaves if deficiencies are chronic.

Rates of fertilizer recommended for 10-year-old litchis carrying a 100 kg crop in Guangdong are as follows: 600 g N, 40 g P and 250 g K applied before flowering; 200 g N, 50 g P and 700 g K applied at full bloom; and 600 g N, 40 g P and 250 g K applied prior to harvest (Menzel, 2002). The fertilizer is normally applied in a trench dug around the tree, although foliar fertilizers can be used at flowering. The amount of phosphorus applied is much higher than that recommended in Australia.

In India, 12–15-year-old trees receive 600–800 g N, 150–200 g P and 300–500 g K in two or three applications, with an emphasis on organic fertilizers (Menzel, 2002). Applications of foliar zinc, copper, manganese and boron are also suggested. In South Africa, 10-year-old trees receive 700 g N, 70 g P and 250 g K (Anonymous, 1992). The nitrogen and potassium are applied at flowering and harvest, and the phosphorus at harvest. Since most of the soils contain little zinc or boron, four foliar applications of zinc per year are recommended, along with boron every 2 years. Trees in India and South Africa receive similar amounts of nitrogen and phosphorus to orchards in Australia, but less potassium.

Fertigation

The application of soluble fertilizer through the irrigation water has many advantages over broadcasting solid fertilizers (Menzel *et al.*, 2002). Annual rates of nitrogen and potassium can often be reduced by 30–50% compared with standard under-tree fertilizing.

The fertilizer is dissolved in water in a drum or tank and sucked or injected through the watering system via a Venturi injection pump or a pressure differential system. Fertilizers must be highly soluble to avoid damaging the pump and lines, and must also be compatible to avoid precipitates that block the sprinklers. Good filtration is essential, as are sprinklers that deliver uniform volumes of water to each tree. Fertigation is normally carried out every 2 weeks in sands and every 4 weeks in clays, between panicle emergence and harvest.

High concentrations of iron can promote the growth of certain bacteria. To address these problems, a full water test is advised at the start. Specialists can provide advice on irrigation design and operation of the system. The soil pH in the wetted and dry areas should be monitored, as most fertilizers acidify the soil. Nitrate-based fertilizers such as potassium nitrate and calcium nitrate lower the pH less than ammonium-based fertilizers. Micro-fine lime can be applied as a suspension to correct pH, provided there is sufficient agitation in the fertilizer reservoir.

Conclusions

Symptoms of nutrient deficiencies have been induced by growing plants in sand culture that lacks one or more nutrients. Deficiencies in orchard trees can occur sometimes on sands, certain clays, and soils that are high in organic matter, acidity or lime. Responses to nutrient applications in orchard trees have generally been recorded only for nitrogen. Optimum leaf nutrient concentrations have been established for the two crops, and calibrations between yield and leaf nitrogen, phosphorus and potassium calculated. Acceptable yields and fruit quality are generally achieved over a wide range of leaf nutrient concentrations. Responses to particular fertilizer regimes are normally site-specific. Most orchards require regular applications of nitrogen, potassium, magnesium, zinc and boron, with fertilizer applications based on the results of leaf and soil tests. The fertilizer can be applied at any time during the year, with soil applications generally being adequate in most situations.

Acknowledgements

Comments on the manuscript by David Turner from the University of Western Australia are appreciated. Many thanks to colleagues who contributed to the litchi research programme at Maroochy.

References

Anonymous (1992) *The Cultivation of Litchis*. Agricultural Research Council, Pretoria, South Africa, 61 pp.

Atwell, B.J., Kriedemann, P.E. and Turnbull, C.A. (1999) *Plants in Action*. Macmillan Education Australia, South Yarra, Victoria, Australia, 664 pp.

Chen, W.S. (1997) Changes of nutrient content in longan leaves from flowering to fruit development. *Special Publication of the Taichung District Agricultural Improvement Station* 38, 115–120.

Chen, X.D., Chen, J.S., Zhang, F.B. and Yang, S.H. (1998) Effectiveness of nitrogen and potash fertilizers on litchi. *Guangdong Agricultural Sciences* 2, 27–29.

Chen, Y.H. (1985) An approach to the on-year and off-year in longan fruiting. *Journal of Fujian Agricultural College* 14, 331–338.

Deng, Y.C., Ni, Y.Y. and Chen, N.R. (1994) Studies on the effects of potassium on photosynthesis and respiration of litchi. *Journal of South China Agricultural University* 15, 80–84.

Dutt, A.K. (1962) Response of litchi to minor elements in laterite soils. *Fertilizer News* 7, 24–26.

Goldweber, S. (1959) Observations on lychees grown in pot culture. *Proceedings of the Florida State Horticultural Society* 72, 353–356.

Goren, M. and Gazit, S. (1996) Management of lychee orchards in Israel. In: Welch, T. and Ferguson, J. (eds) *Proceedings of the Fourth National Lychee Seminar*. Australian Lychee Growers' Association, Yeppoon, Queensland, pp. 74–77.

Groff, G.W. (1921) *The Lychee and Lungan*. Orange Judd, New York, 180 pp.

Hanson, E.J. and Breen, P.J. (1985a) Effects of fall boron sprays and environmental factors on fruit set and boron accumulation in 'Italian' prune flowers. *Journal of the American Society for Horticultural Science* 110, 389–392.

Hanson, E.J. and Breen, P.J. (1985b) Xylem differentiation and boron accumulation in 'Italian' prune flower buds. *Journal of the American Society for Horticultural Science* 110, 566–570.

Huang, W.T., Huang, Y.M., Hsiang, W.M., Chang, M.H., Lin, M.L., Wang, C.H. and Wu, W.L. (1998) Nutrition studies on lychee (*Litchi chinensis* Sonn.) orchards in central Taiwan. *Journal of Agricultural Research of China* 47, 388–407.

Huang, X.M., Li, J.G., Wang, H.C., Huang, H.B. and Gao, F.F. (2001) The relationship between fruit cracking and calcium in litchi pericarp. *Acta Horticulturae* 558, 209–211.

Hundal, H.S. and Arora, C.L. (1993) Nutrient status of litchi (*Litchi chinensis* Sonn.) orchards in submontaneous area of Punjab. *Journal of the Indian Society of Soil Science* 41, 778–779.

Janos, D.P., Schroeder, M.S., Schaffer, B. and Crane, J.H. (2001) Inoculation with arbuscular mycorrhizal fungi enhances growth of *Litchi chinensis* Sonn. trees after propagation by air-layering. *Plant and Soil* 233, 85–94.

Jiang, R.H., Mao, Y.L., Chen, T.S., Yan, L., Xing, S.H. and Lin, S.Q. (1999) Soil fertility of orchards and nutrient situation of fruit trees in Changtai town. *Journal of Fujian Agricultural University* 28, 466–470.

Joiner, J.N. and Dickey, R.D. (1961) Effects of nitrogen, potassium and magnesium levels on growth and chemical composition of lychee (*Litchi chinensis* Sonn.) grown in sand culture. *Proceedings of the American Society for Horticultural Science* 77, 270–277.

Joon, M.S., Dahiya, S.S. and Singh, S. (1997) Nutritional survey of litchi orchards of the north-eastern Haryana, India. *International Journal of Tropical Agriculture* 15, 275–278.

Kadman, A. and Slor, E. (1974) Experiments with propagation of the litchi (*Litchi chinensis*) in Israel. *Indian Journal of Horticulture* 31, 28–33.

Koen, T.J. and du Plessis, S.F. (1993) Optimal leaf analysis norms for litchis. *Information Bulletin of the Institute for Tropical and Subtropical Crops* 248, 19–21.

Koen, T.J., Langenegger, W. and Smart, G. (1981a) Determination of the fertiliser requirements of litchi trees. *Information Bulletin of the Citrus and Subtropical Fruit Research Institute* 103, 9–12.

Koen, T.J., Langenegger, W. and Smart, G. (1981b) Nitrogen fertilization of litchi trees. *Information Bulletin of the Citrus and Subtropical Fruit Research Institute* 107, 9–11.

Kotur, S.C. and Singh, H.P. (1993) Leaf-sampling technique in litchi (*Litchi chinensis*). *Indian Journal of Agricultural Sciences* 63, 632–638.

Kunwar, R. and Singh, R. (1993) Note on the nutritional survey of litchi orchards in Doon Valley of

Garhwal Hills of Uttar Pradesh. *Progressive Horticulture* 25, 164–165.

Lal, R.L., Tiwari, J.P. and Misra, K.K. (1999) Response of litchi (*Litchi chinensis* Sonn.) cv. Rose Scented to nitrogen and potassium fertilization. *Progressive Horticulture* 31, 79–83.

Lei, D.C., Liang, C.P., Chen, Y.Z. and Huang, F.Y. (1998) Preliminary report on sand culture experiment for Dawuyuan longan variety for identifying the shortage of mineral element. *South China Fruits* 27 (3), 29–30.

Li, J.G. and Huang, H.B. (1995) Physio-chemical properties and peel morphology in relation to fruit cracking susceptibility in litchi. *Journal of South China Agricultural University* 16, 84–89.

Li, J.G., Huang, H.B., Gao, F.F., Huang, X.M. and Wang, H.C. (2001a) An overview of litchi fruit cracking. *Acta Horticulturae* 558, 205–207.

Li, Y., Liu, X.H. and Zhuang, W.M. (2001b) Effects of Mg on longan seedling growth and Mg nutrition diagnosis. *Journal of Fujian Agricultural University* 30, 394–397.

Li, Y.C., Davenport, T.L., Rao, R. and Zheng, Q. (2001c) Nitrogen, flowering and production of lychee in Florida. *Acta Horticulturae* 558, 221–224.

Liang, X.T. (1992) Effects of V-A mycorrhiza on growth of seedlings of mango, longan and lychee. *Journal of Guangxi Agricultural University* 11, 96–101.

Liang, X.T. (1995) Inoculation of forest and fruit trees with vesicular-arbuscular mycorrhizal fungi in Guangxi Province, China. In: Brundett, M., Dell, B., Malajczuk, N. and Gong, M.Q. (eds) *Proceedings of an International Symposium and Workshop on Mycorrhizas for Plantation Forestry in Asia.* Australian Centre for International Agricultural Research, Canberra, pp. 114–118.

Liu, X.H., Zheng, J.J., Pan, D.M. and Xie, H.C. (1986) An investigation on the leaf nutritional diagnosis criteria of longan (*Dimocarpus longan* Lour.). *Journal of Fujian Agricultural College* 15, 237–245.

Malcolm, J.L. (1953) Chelates for the correction of iron chlorosis in subtropical plants. *Proceedings of the Florida State Horticultural Society* 66, 179–184.

Mallik, P.C. and Singh, D.L. (1965) Hunger signs in litchi (*Litchi chinensis*). *Indian Agriculturalist* 9, 127–132.

Menzel, C.M. (1987) Lychee nutrition: a review. *Scientia Horticulturae* 31, 195–224.

Menzel, C.M. (1991) Litchi. In: Verheij, E.W.M and Coronel, R.E. (eds) *Plant Resources of South-East Asia* Vol. 2: *Edible Fruit and Nuts.* Pudoc, Wageningen, The Netherlands, pp. 191–195.

Menzel, C.M. (2002) *The Lychee Crop in Asia and the Pacific.* Food and Agricultural Organization of the United Nations, Bangkok, Thailand, 108 pp.

Menzel, C.M. and McConchie, C.A. (1998) Lychee and longan. In: Hyde, K.W. (ed.) *The New Rural Industries.* Rural Industries Research and Development Corporation, Canberra, Australia, pp. 288–295.

Menzel, C.M. and Simpson, D.R. (1988) Effect of temperature on growth and flowering of litchi (*Litchi chinensis* Sonn.) cultivars. *Journal of Horticultural Science* 63, 347–358.

Menzel, C.M., Carseldine, M.L. and Simpson, D.R. (1987) The effect of leaf age on nutrient composition of non-fruiting litchi (*Litchi chinensis* Sonn.). *Journal of Horticultural Science* 62, 273–279.

Menzel, C.M., Carseldine, M.L. and Simpson, D.R. (1988a) The effect of fruiting status on nutrient composition of litchi (*Litchi chinensis* Sonn.) during the flowering and fruiting season. *Journal of Horticultural Science* 63, 547–556.

Menzel, C.M., Carseldine, M.L. and Simpson, D.R. (1988b) Crop development and leaf nitrogen in lychee in subtropical Queensland. *Australian Journal of Experimental Agriculture* 28, 793–800.

Menzel, C.M., Aitken, R.L., Dowling, A.W. and Simpson, D.R. (1990) Root distribution of lychee trees growing in acid soils of subtropical Queensland. *Australian Journal of Experimental Agriculture* 30, 699–705.

Menzel, C.M., Barry, G.A. and Simpson, D.R. (1992a) Observations on the concentrations of soil nutrients in lychee orchards in subtropical Australia. *Journal of the Southern African Society for Horticultural Science* 2, 36–40.

Menzel, C.M., Carseldine, M.L., Haydon, G.F. and Simpson, D.R. (1992b) A review of existing and proposed new leaf nutrient standards for lychee. *Scientia Horticulturae* 49, 33–53.

Menzel, C.M., Haydon, G.F., Doogan, V.J. and Simpson, D.R. (1992c) Observations on the leaf nutrient status of lychee in Australia. *Journal of the Southern African Society for Horticultural Science* 2, 86–88.

Menzel, C.M., Haydon, G.F. and Simpson, D.R. (1992d) Mineral nutrient reserves in bearing litchi trees (*Litchi chinensis* Sonn.). *Journal of Horticultural Science* 67, 149–160.

Menzel, C.M., Haydon, G.F., Doogan, V.J. and Simpson, D.R. (1994) Time of nitrogen application and yield of Bengal lychee on a sandy loam soil in subtropical Queensland. *Australian Journal of Experimental Agriculture* 34, 803–811.

Menzel, C.M., Haydon, G.F. and Simpson, D.R. (1995a) Growth of lychee (*Litchi chinensis* Sonn.) in sand culture under variable nitrogen supply. *Journal of Horticultural Science* 70, 757–767.

Menzel, C.M., Simpson, D.R., Haydon, G.F. and Doogan, V.J. (1995b) Phosphorus and potassium fertilization of lychee. *Journal of the Southern African Society for Horticultural Science* 5, 97–99.

Menzel, C.M., Bagshaw, J., Campbell, T., Greer, N., Noller, J., Olesen, T. and Waite, G.K. (2002) *Lychee Information Kit.* Queensland Department of Primary Industries, Nambour, Australia.

Mitra, S.K. (1988) Litchi. In: Bose, T.K., Mitra, S.K. and Sadhu, M.K. (eds) *Mineral Nutrition of Fruit Crops.* Naya Prokask, Calcutta, India, pp. 366–371.

Nakata, S. (1955) Floral initiation and fruit set in lychee, with special reference to the effects of sodium naphthalenacetate. *Botanical Gazette* 117, 126–134.

Nakata, S. and Suehisa, R. (1969) Growth and development of *Litchi chinensis* as affected by soil moisture stress. *American Journal of Botany* 56, 1121–1126.

Pandey, S. and Misra, A.F. (1971) *Rhizophagus* in mycorrhizal association with *Litchi chinensis* Sonn. *Mycopathology and Mycological Applications* 45, 337–354.

Pandey, S. and Misra, A.F. (1975) Mycorrhiza in relation to growth and fruiting of *Litchi chinensis* Sonn. *Journal of the Indian Botanical Society* 54, 280–293.

Saha, D.P., Sharma, R.K., Kumar, R. and Thankur, S. (1993) Effect of preharvest application of calcium salts on fruit cracking and physio-chemical composition of fruit in litchi (*Litchi chinensis* Sonn.). *Orissa Journal of Horticulture* 21, 53–57.

Thomas, M.B., Ferguson, J. and Crane, J.H. (1995) Identification of N, K, Mg, Mn, Zn and Fe deficiency symptoms of carambola, lychee, and papaya grown in sand culture. *Proceedings of the Florida State Horticultural Society* 108, 370–373.

Tomer, E. and Kadman, A. (1985) The effect of various amounts of fertilizers on the growth and fruiting of litchi trees in the Negev region. *Alon Hanotea* 39, 447–450.

Wang, R.J., Zhuang, Y.M., Chen, I.X., Wu, W.B. and Huang, Z. (1994) Studies on nutrient status of a productive longan orchard. *China Fruits* 3, 13–16.

Whiley, A.W. and Saranah, J.B. (1980) Salinity tolerance of some avocado, litchi and macadamia nut trees. *Biennial Report of Maroochy Horticultural Research Station, Australia* 2, 16–17.

Xu, R.Y., Chen, R.M., Chen, J.L., Wang, K.G. and Kang, Z.L. (1984) A study of the effect of microelement boron on litchi production. *Journal of Fujian Agricultural College* 13, 305–311.

Young, T.W. and Harkness, R.W. (1961) Flowering and fruiting behaviour of Brewster lychees in Florida. *Proceedings of the Florida State Horticultural Society* 74, 358–363.

Young, T.W. and Koo, R.C.J. (1964) Influence of nitrogen source and rate of fertilization on performance of Brewster lychees. *Proceedings of the Florida State Horticultural Society* 77, 406–410.

Zee, F.T.P., Chan, H.T., Jr and Yen, C.R. (1998) Lychee, longan, rambutan and pulasan. In: Shaw, P.E., Chan, H.T., Jr and Nagy, S. (eds) *Tropical and Subtropical Fruits.* AgScience, Auburndale, Florida, pp. 290–335.

Zhang, F.B., Chen, J.S. and Chen, X.D. (1999) Diagnosis of soil nutrient limiting factors in longan orchards in Pearl River Delta. *Guangdong Agricultural Sciences* 3, 33–37.

Zhao, D.L. (1991) Studies on the mineral nutrient and requirement rule in 'on' or 'off' year of lonyan tree. *Journal of Fujian Academy of Agricultural Sciences* 6, 60–66.

Zhuang, Y.M., Li, L.R., Jiang, Y., Wang, R.J., Su, M.H. and Liu, H.W. (1984) Seasonal variation of the amounts of the major nutrient elements in the longan foliage and in the orchard soil. *Acta Horticulturae Sinica* 11, 165–170.

Zhuang, Y.M., Wang, R.J., Xie, Z.N. and Xu, W.B. (1995) Optimum range of mineral element contents in the leaves of Shuizhang longan. *Journal of Fujian Agricultural University* 24, 281–286.

12 Pests

Geoff K. Waite
Department of Primary Industries and Fisheries, PO Box 5083 SCMC, Nambour, Queensland 4560, Australia

Introduction

Insects and mites attack all parts of litchi and longan trees (Table 12.1). Some species cause severe damage by boring into and throughout the trunk and branches, while others devour or distort leaves and reduce photosynthetic capacity. Usually, the most economically important damage is that which is inflicted on the flowers and fruit. Even though other tree parts may be attacked and damaged by pests, it is the flowers and ultimately the fruit that provide an income for producers. Hence, pests of the fruit in particular, must be adequately managed in order to ensure a profitable harvest that is free from insects and insect damage that might contribute to postharvest disorders, or may represent a quarantine problem that causes rejection by importing countries.

Fruit borers

Caterpillars that attack the fruit cause direct crop loss. Those that tunnel through the flesh of the fruit are difficult to control because once inside the fruit they are safe from insecticides. The major pest of litchis and longans in China, Taiwan and Thailand is *Conopomorpha sinensis* Bradley, known as the litchi stem-end borer in China, and the litchi fruit borer in Thailand. This insect was at one time known as *Acrocercops cramerella* Snellen, until Bradley (1986) described three previously unrecognized species congeneric with the cocoa pod borer, *Conopomorpha cramerella* Snellen. Two of the new species described by Bradley, *C. sinensis* and *C. litchiella* Bradley, attack litchi and longan, while *C. cramerella* is restricted to rambutan and cocoa (Bradley, 1986). *C. sinensis* and *C. litchiella* are recorded from India and South-east Asia to China and Taiwan. *C. litchiella* was found recently in north Queensland.

C. sinensis lays small, yellow, scale-like eggs on the fruit of litchi and longan, any time after fruit set. Eggs may also be deposited on new leaves and shoots. On hatching, the larva bores into the fruit, leaf or shoot. One or more eggs may be laid on a fruit, but only one larva survives.

Mature larvae are 6–10 mm long, and are brown, or green if they have fed on leaves. They pupate on or under mature leaves in cream, oval cocoons. The moth is very small with long filiform antennae and narrow, fringed forewings that extend 8–11 mm when expanded. The pest may complete four or five generations during the fruiting season. In cooler regions, development continues over the winter period either in alternative hosts or in the terminals of the main hosts, litchi and longan. During the off-season, when fruit is not available to the pest, larvae can survive by feeding on young leaves and shoots, similar to *C. litchiella*. *C. sinensis* prefers litchis to longans.

In both crops, unsprayed orchards suffer extensive damage. Considerable damage can be

Table 12.1. Arthropod pests of litchi and longan.

Order	Family	Species	Geographical area	Plant part affected
Acari	Eriophyidae	*Aceria litchii* (Keiffer)	China, Taiwan, Australia, Hawaii, India, Pakistan	Litchi leaves, flowers, fruit
		Aceria dimocarpi (Kuang)	China	Longan leaves, twigs
		Aceria longana Boczek and Knihinicki	Thailand	Longan terminals, flowers
	Tetranychidae	*Oligonychus coffeae* (Nietner)	Australia	Litchi leaves
	Tarsonemidae	*Polyphagotarsonemus latus* (Banks)	Australia, China, Hawaii	Litchi and longan leaves, terminals
Coleoptera	Cerambycidae	*Aristobia testudo* (Voet)	China	Litchi, longan branches
		Anoplophora maculata (Thomson)	Taiwan	Litchi branches
		Uracanthus cryptophagus Olliff	Australia	Litchi branches
	Chrysomelidae	*Rhyparida discopunctulata* Blackburn	Australia	Litchi leaves
		Monolepta australis (Jacoby)	Australia	Litchi leaves
		Proctophana tomentosa Lacordaire	Brazil	Litchi leaves
	Curculionidae	*Cratopus angustatus* Boh.	Réunion	Litchi leaves, flowers, fruit, roots
		Cratopus humeralis Boh.	Réunion	Litchi leaves, flowers, fruit, roots
		Euthyrrhinus meditabundus Fab.	Australia	Litchi twigs
		Orthorrhinus klugii Boh.	Australia	Litchi twigs
	Scarabaeidae	*Xylotrupes gideon* (Linnaeus)	China, Australia	Litchi fruit
Diptera	Cecidomyiidae	*Litchiomyia chinensis* Yang and Luo	China	Litchi leaves
	Tephritidae	*Ceratitis capitata* (Weidemann)	South Africa, Réunion, Hawaii	Litchi fruit
		Ceratitis rosa Karsch	South Africa, Réunion	Litchi fruit
		Bactrocera dorsalis Hendel	Hawaii	Litchi fruit
		Bactrocera cucurbitae (Coquillett)	Hawaii	Litchi fruit
		Bactrocera tryoni (Froggatt)	Australia	Litchi fruit
Hemiptera	Tessaritomidae	*Tessaritoma papillosa* Drury	China, Viet Nam, Thailand, Myanmar, Philippines, India	Litchi and longan flowers, fruit
		Tessaritoma javanica Thunberg	India	Litchi flowers, fruit
		Tessaritoma quadrata Distant	India	Litchi flowers, fruit
		Lyramorpha rosea Westw.	Australia	Litchi fruit
	Coreidae	*Amblypelta nitida* Stål	Australia	Litchi and longan fruit
		Amblypelta lutescens lutescens (Distant)	Australia	Litchi and longan fruit
		Paradasynus longirostris Hsiao	China	Longan fruit
	Rhopalidae	*Leptocoris rufomarginata* (Fabricius)	Australia	Litchi fruit

Homoptera	Coccidae	*Pulvinaria* (*Chloropulvinaria*) *psidii* (Maskell)	China, Taiwan, Australia, Florida, India	Litchi and longan twigs, leaves, fruit
		Chloropulvinaria hesperidum Linnaeus	Australia	Litchi twigs
		Parasaissetia nigra (Nietner)	India	Litchi twigs
		Saissetia coffeae (Walker)	India	Litchi twigs
		Ceroplastes rubens Maskell	Australia	Longan leaves
		Ceroplastes ceriferus (Fabricius)	Taiwan	Litchi leaves
		Nipaecoccus vastator (Maskell)	Taiwan	Litchi leaves
		Icerya seychellarum West.	Réunion	Litchi twigs
		Kerria lacca Kerr	Taiwan	Litchi twigs
		Hemiberlesia lataniae (Signoret)	Australia	Litchi branches, twigs
		Fiorinia nephelii Maskell	India	Litchi twigs
		Parlatoria pseudopyri Kuwana	India	Litchi twigs
		Parlatoria cinerea Danne and Hadden	India	Litchi twigs
	Pseudococcidae	*Planococcus citri* (Risso)	Taiwan	Litchi fruit
Lepidoptera	Cossidae	*Zeuzera coffeae* Nietner	China, Taiwan	Litchi and longan branches
	Tortricidae	*Conopomorpha sinensis* Bradley	China, Taiwan, Thailand	Litchi and longan fruit, leaves, shoots
		Conopomorpha litchiella Bradley	China, Taiwan, Thailand	Litchi and longan fruit, leaves, shoots
		Cryptophlebia peltastica Meyr.	Mauritius, Seychelles, Réunion, South Africa, Madagascar	Litchi fruit
		Cryptophlebia leucotreta Meyr.	South Africa	Litchi fruit
		Cryptophlebia ombrodelta (Lower)	Thailand, China, Taiwan, Japan, Australia, Hawaii	Litchi and longan fruit
		Cryptophlebia illepida Butler	Hawaii	Litchi fruit
		Olethreutes praecedens Wals.	Réunion	Litchi leaves
		Olethreutes perdulata Meyr.	Australia	Litchi leaves
		Platypeplus aprobola (Meyrick)	China, India, Australia	Litchi leaves
		Epiphyas postvittana (Walker)	Australia, Hawaii	Litchi leaves
		Adoxophyes cyrtosema Meyr.	China	Litchi and longan leaves
		Homona coffearia Nietner	China, Thailand	Litchi and longan leaves
		Homona difficilis Meyrick	Thailand	Litchi and longan leaves
		Crocidosema litchivora sp. nov.	Florida	Litchi flowers

continued

Table 12.1. *Continued.*

Order	Family	Species	Geographical area	Plant part affected
	Lycaenidae	*Deudorix epijarbas* Moore	India	Litchi fruit
		Deudorix epijarbas amatius	China, Thailand	Litchi and longan fruit
		Deudorix diovis Hewitson	Australia	Litchi and longan fruit
	Noctuidae	*Eudocima fullonia* (Clerck)	Australia	Litchi and longan fruit
		Eudocima salaminia (Cramer)	Australia	Litchi and longan fruit
		Eudocima jordani (Holland)	Australia	Litchi and longan fruit
		Oxyodes tricolor Guen.	Australia	Litchi and longan leaves
		Oxyodes scrobiculata F.	Thailand	Litchi and longan leaves
		Achaea janata (L.)	Australia	Litchi and longan leaves
	Metarbelidae	*Salagena* sp.	South Africa	Litchi branches
		Indarbela quadrinotata Walker	India	Litchi branches
		Indarbela tetraonis Moore	India	Litchi branches
		Arbela dea Swinhoe	China	Litchi branches
	Yponomeutidae	*Comoritis albicapilla* Moriuti	China	Litchi bark
Thysanoptera	Phlaeothripidae	*Dolichothrips indicus* Hood	India	Litchi flowers
	Thripidae	*Megalurothrips distalis* Karny	India	Litchi leaves
		Scirtothrips dorsalis Hood	China	Litchi and longan shoots
		Thrips imaginis Bagnall	Australia	Litchi and longan flowers

caused even in sprayed orchards (Huang *et al.*, 1994). Most of the natural enemies of *C. sinensis* are tiny wasp parasitoids. *Phanerotoma* spp., *Colastes* spp. and *Pholestesor* spp. (Braconidae) and *Goryphus* spp. (Ichneumonidae) attack the larva, while *Goryphus* spp. may also attack the pupal stage (Dolsopon *et al.*, 1997a).

In Thailand, fruit are inspected weekly from immediately after fruit set, to detect eggs of *C. sinensis*. It is recommended that infested fruit be picked and destroyed while infestation levels remain at 1–2%. When the pest becomes more active, an insecticide such as permethrin is applied at weekly intervals, ceasing at least 2 weeks before harvest. Damaged fruit that has fallen from the tree should be inspected to assess the prevalence of natural enemies of the pest in the orchard. If parasitism is significant, allowance for this must be made in deciding when to spray. Moth populations have been assessed, based on the catch of males on sticky traps using pheromone baits. Fruit panicles may also be bagged to exclude the pest. This does not provide complete protection, although fruit colour and quality are improved in both crops. Litchis mature earlier than longans, and litchi trees are pruned after harvest to remove any remaining fruit that might harbour pupae of the pest and so present a threat to the developing longan crop.

All stages of the litchi leafminer, *C. litchiella*, are similar to those of the fruit borer. The female moth lays light-yellow eggs on new shoots. The creamy white larvae bore into the shoots, but they may also mine the leaf blade. Although mature larvae prefer to feed on the midrib and veins of young leaves, they may also bore into fruit. The larval stage occupies 10–14 days, after which the larva pupates on mature leaves. The pupal stage lasts 7–10 days and moths live for about a week.

The leafminer is attracted to leaf flushes, especially during the rainy season from June to October in Thailand. Litchi is more heavily infested than longan. Dolsopon *et al.* (1997b) found that the pest destroyed 75% of litchi shoots compared to 50% on longan. Sprays, if they are necessary, must be applied to the very young leaves of important flushes, especially the second flush. If 30–40% of larvae are parasitized, spraying is not recommended. The same species of parasitoids that attack the fruit borer effectively suppress *C. litchiella* in Thailand (Dolsopon *et al.*, 1997a).

Several species of the genus *Cryptophlebia* attack litchis and longans. In South Africa, and on the Indian Ocean islands of Madagascar, Mauritius, the Seychelles and Réunion, *Cryptophlebia peltastica* Meyr. is a major pest of litchis (Quilici *et al.*, 1988; de Villiers and Stander, 1989; Quilici, 1996). *Cryptophlebia leucotreta* Meyr. damages litchis in South Africa but the frequency of attack is insignificant compared with *C. peltastica* (Newton and Crause, 1990). *Cryptophlebia bactrachopa* Meyr. is a pest of the crop in Malawi (La Croix and Thindwa, 1986).

Cryptophlebia ombrodelta (Lower) occurs in Thailand, China, Japan, Taiwan, Hawaii and Australia. In addition to macadamia, it is the major pest of litchi and longan fruit in Australia and Hawaii. *Cryptophlebia illepida* Butler is also a significant pest of litchi and macadamia in Hawaii (Jones, 1994).

The larvae of all *Cryptophlebia* spp. feed internally in the fruit. In immature fruit a larva will bore directly into the seed, which is completely eaten (Plate 44). However, ovipositing moths are strongly attracted to more mature fruit that have larger seeds, especially when the fruit begin to colour (Rogers and Blair, 1981; Newton and Crause, 1990). The juice that oozes from wounds on mature fruit may stain neighbouring fruit on a panicle and cause further losses. This is particularly so in cultivars such as Bengal, which produces large panicles bearing many fruit in a tight cluster (Waite, 1992a).

Egg, larval and pupal parasitoids attack *Cryptophlebia* spp., but they are not always capable of preventing economic damage in orchards. In India, the egg parasitoid, *Trichogrammatoidea fulva* Nagaraja, was considered to be the most important of a number of natural enemies of *C. ombrodelta*, parasitizing up to 68% of eggs (Anonymous, 1978). *T. cryptophlebiae* Nagaraja is a particularly effective egg parasitoid in South Africa, where it has been used extensively for inundative release into citrus orchards for the control of *C. leucotreta* (Newton and Odendaal, 1990). It has parasitized 63% of the eggs of *C. peltastica* in litchis, comparable to that achieved against *C. leucotreta* in citrus under natural conditions (Newton and Crause, 1990). *T. cryptophlebiae* is now established in Australia, and although parasitism of *C. ombrodelta* eggs laid on litchis is low, 90% or more of eggs laid on the later-maturing longan crop may be parasitized.

To control *Cryptophlebia* spp. in South Africa, de Villiers (1992a) recommended the application of the insect growth regulator, triflumuron, as a single full-cover spray 40 days before harvest, or two sprays of teflubenzuron 2 weeks apart, commencing when the fruit are 10 mm in diameter. Alternatively, panicles may be covered with paper bags to exclude the pest. This procedure also improves fruit colour and quality (de Villiers, 1983b). In Australia, carbaryl and azinphos-methyl are used against *C. ombrodelta* with varying success. Synthetic pyrethrins are effective, but their frequent use may initiate outbreaks of scales and mites. Newer insecticides, including the insect growth regulator, tebufenozide (Mimic), provide better control, with less disruption of natural enemies. Follett and Sanxter (2001) concluded that a hot-water treatment of litchi and longan of 49°C for 20 min for disinfestation of fruit flies in Hawaii would also kill *Cryptophlebia* spp. Similarly, irradiation treatments that are effective against fruit flies will also kill *Cryptophlebia* spp. (Follett and Lower, 2000).

Species of *Deudorix* are minor pests of litchis and longans in India (*Deudorix epijarbas* Moore), China, Thailand (*D. epijarbas amatius*) and Australia (*D. diovis* Hewitson). The female butterfly lays single eggs on the fruit. On hatching, the larva bores inside, completely destroying the flesh and seed. Unlike *Cryptophlebia* spp., whose larvae usually feed on only one fruit, individual *Deudorix* larvae may damage up to four fruit during their development. In Queensland, longans suffer more damage than litchis and *Deudorix* is often more important than *Cryptophlebia*. Attendance of the larvae by several species of ants, including *Polyrhachis* spp., probably limits suppression of the pest by parasitoids.

Fruit-piercing moths attack a range of fruit including mango, carambola, citrus, mangosteen and stonefruit, as well as litchi and longan, throughout South-east Asia, the South Pacific and Australia (Banziger, 1982; Fay and Halfpapp, 1993). The species that cause most damage in Australia are *Eudocima* (*Othreis*) *fullonia* (Clerck), *E. salaminia* (Cramer) (Plate 45) and *E. jordani* (Holland) (Fay and Halfpapp, 1999), while in China, *Oraesia emarginata* Fabricius is one of a number of species that damage litchis.

Fruit-piercing moths are unusual in that the adult stage causes damage to the fruit. The larvae develop on *Erythrina* spp. in the Pacific Islands, and on vines belonging to the Family Menispermaceae in the forests of northern Queensland. The moth's proboscis can drill through the skin of fruit, allowing it to feed on the juice (Plate 46). Yeasts and bacteria that are introduced into the wound on the proboscis of the moth initiate fermentation of the fruit (Sands and Schotz, 1989). *Drosophila* spp. are attracted to the fermenting fruit and their activity hastens the process of deterioration. Within a few days a frothy exudate seeps from the hole. This may stain sound, undamaged fruit, causing further losses.

Growers in Australia and Thailand patrol orchards at night with spotlights and attempt to manually remove as many moths as possible. This may be achieved by catching the moths as they feed on the fruit, or by catching them in a net, or swatting them with a tennis racquet as they fly from tree to tree. However, when moths are very numerous, this becomes futile.

Moth traps can be made with shade cloth draped loosely over a frame of wire. These are baited with ripe fruit such as citrus and bananas. The moths are attracted to feed on the fruit and become entangled in the folds of cloth when they attempt to fly off. They are then killed each morning. Large numbers of traps are required to protect an entire orchard, and substantial damage may still result. In Thailand, ripe fruit of banana and pineapple are dipped in insecticide and hung in the trees to poison the feeding moths. In some countries, paper bags used to protect fruit from other pests are effective against these moths, and are specifically recommended for this purpose in Thailand. Recent research in Australia has resulted in the development of artificial baits that incorporate food lures and insecticides in an agar gel (Fay, 2002). These have reduced moth damage in citrus by 75%.

Parrots and fruit bats are a severe problem in litchi and longan in eastern Australia, and many growers have erected nets to protect their crops. Initially, these vertebrate pests were excluded by draping nets over individual trees.

Newer technologies have enabled the erection of tunnel nets that cover individual rows, or complete enclosures that cover the whole orchard. The high capital cost of the permanent nets is offset by their extended life and the ongoing protection afforded by them. The entire cost is often saved through the prevention of bird and bat damage in one season. These same nets also exclude fruit-piercing moths. A mesh size of 12 mm with double cross-hairs will also exclude smaller insect pests such as fruitspotting bugs and fruit borer moths.

Lepidopterous leaf feeders

The loopers, *Oxyodes scrobiculata* F. and *O. tricolor* Guen., feed on litchi and longan leaves in Thailand and Australia, respectively. In Australia, *O. tricolor* attacks litchis in south-east Queensland but is not a pest in northern Queensland (Plate 37). The castor oil looper, *Achaea janata* (L.), sporadically causes leaf damage in Queensland and is often present with *O. tricolor*. The main period of activity is during the postharvest flush, when young leaves are the preferred larval food.

In Thailand it is recommended that carbaryl be applied when infestations of *O. scrobiculata* reach two or three young larvae per leaflet. If 40% or more of larvae are parasitized, sprays should not be applied. In Queensland, *Bacillus thuringiensis* Berliner (Bt), endosulfan or carbaryl may be used to control loopers.

Several species of leafroller attack litchis. *Olethreutes praecedens* Wals. is a minor pest in Réunion (Vayssieres, 1997), while *Olethreutes perdulata* Meyr. is an occasional pest in Queensland (Waite, 1992a). *Platypeplus aprobola* (Meyrick) has been recorded on litchis in China (Anonymous, 1978) and India (Butani, 1977), and this species and *Epiphyas postvittana* (Walker) are recorded attacking the crop in Australia (Storey and Rogers, 1980) and Hawaii (Higgins, 1917). *Adoxophyes cyrtosema* Meyr. and *Homona coffearia* Nietner attack litchi and longan in Guangzhou and Fujian (Anonymous, 1978). *H. coffearea* and *H. difficilis* infest litchi, longan and rambutan in Thailand. The orange fruit borer, *Isotenes miserana* (Walker), is an omnivorous leafroller that also attacks flowers and fruit in Queensland. In Florida, *Crocidosema litchivora* sp. nov. is an apparently recent arrival that damages litchi inflorescences (Brown *et al.*, 2002).

In Australia, leafroller damage is generally tolerated. To minimize this in critical vegetative flushes, especially in young trees, sprays of carbaryl may be applied when 50% of the leaf flushes are infested (Waite and Elder, 1996). In India the manual removal of rolled leaves that contain larvae was recommended when infestations were light, with sprays of phosphamidon, fenitrothion or endosulfan applied to control severe infestations (Butani, 1977; Sahoo and Maiti, 1992).

Bugs

Several bugs belonging to the Family Tessaritomidae attack litchis and longans throughout China, South-east Asia and Australia. *Tessaritoma papillosa* Drury, the litchi stinkbug, occurs in southern China, Viet Nam, Thailand, Burma, the Philippines and India (Anonymous, 1978). Butani (1977) noted that *Tessaritoma javanica* Thunberg and *T. quadrata* Distant are found on litchi in India. In Australia the litchi stinkbug, *Lyramorpha rosea* Westw., damages fruit, but is rarely a problem in commercial orchards.

Apart from utilizing litchi and longan as hosts in China, *T. papillosa* attacks orange, pomelo, castor oil, pomegranate, eucalyptus, canna, loquat and rose. The annual life cycle of the bug is tied to seasonal conditions and tree phenology. Adults tend to aggregate and overwinter mostly on litchi and longan, but may also be found on other hosts, in areas out of the wind but with adequate sunshine. Bug activity commences in spring, coinciding with the production of litchi and longan inflorescences. The female bugs are especially attracted to trees with many flowers and new terminals, where they mate and lay up to 14 egg masses, each containing about 14 eggs. These are usually attached under leaves. Peak oviposition occurs in late March in Guangdong, but continues throughout summer until September (Anonymous, 1978). The eggs, which are round and creamy white, turn red just before hatching after about 13 days at 25°C (Unahawutti, 1990).

Newly hatched nymphs from each egg mass remain in a cluster for several hours after hatching, before they disperse (Plate 43). The nymphs develop through five instars, maturing to become light-brown adults after about 80 days. When disturbed, or during the heat of the day, first and second instar nymphs may drop to the ground, returning to the tree later when it has cooled. Nymphs can survive for periods of up to 12 days without feeding. Both adults and nymphs expel defensive odours when disturbed.

The first nymphs mature in June while there are still old adults in the trees. The new adults do not mate immediately, since their reproductive organs are immature. They overwinter in this state, and recommence the cycle in spring. In litchis and longans, adults and nymphs feed on terminals, which may be killed, and also on flowers and fruit, which fall. The physiology of stinkbug stings was studied in detail by Zhang (1997), who found that more immature fruit than mature fruit fall. According to Liu and Lai (1998), chemical control is only partially effective, and 20–30% of fruit is still damaged.

The main natural enemies of *T. papillosa* in Guangdong are the egg parasitoids, *Encyrtus* (*Ooencyrtus*) spp., *Anastatus* spp. and *Blastophaga* spp. These may parasitize 70–90% of eggs late in the season, with *Encyrtus* spp. being the most effective (Anonymous, 1978; Liu and Lai, 1998). Liu and Gu (1998) found that in an orchard under integrated pest management, combined parasitism rates by *Anastatus* spp. and *Ooencytus* spp. were 42–47% in June. In an orchard that relied on chemical control, parasitism was only 0–3%. During the 1970s, biological control of the litchi stinkbug in Guangdong was initiated using the egg parasitoid, *Anastatus japonicus* Ashmead, the flat venter wasp, after field trials in the late 1960s had demonstrated its value (Huang *et al.*, 1974) (see Fig. 12.1).

Mass releases of the egg parasitoid are effective in controlling litchi stinkbug and, at least during the time that Peoples' Communes were in place in China, were a cheaper form of control than chemicals (Huang *et al.*, 1974). After the break-up of the Communes, insecticides seemed to be the favoured method of control, although biological control using egg parasitoids is once again being considered in the context of integrated pest management (IPM). Timing of sprays is critical for successful control of the bug. It has been found to vary in its susceptibility to the commonly used chemical trichlorfon at different times of the year, depending on the bugs' body fat content and the nature of the fat.

Maximum resistance to trichlorfon occurred during the winter, when fat levels reached 20–25% and the proportion of unsaturated fat was high. Maximum susceptibility occurred from February until July, when fat levels were 4–12%, and the proportion of unsaturated fat was low. On this basis, one spray was applied in early March when resistance was low and before the first eggs were laid. In late April and early May, resistance in adults is still low and most of the

Fig. 12.1. Mass releases of the egg parasitoid *Anastatus japonicus* are effective in controlling litchi stinkbug, *Tessaritoma papillosa* (photograph courtesy of Geoff Waite).

new generation are in instars one and two, which are also susceptible. This is a good time to apply a second spray, if necessary, before the nymphs develop to the third instar, which is difficult to kill (Anonymous, 1978). Zeng *et al.* (2001) suggested a mixture of chlorpyrifos and cypermethrin as an alternative for better control, and to slow the rate of resistance development to trichlorfon. Good coverage, especially of old, very large trees, is essential to achieve a satisfactory kill.

In Guangxi, the main parasitoids of *T. papillosa* eggs are *Ooencyrtus* spp. and *Anastatus* spp., which together parasitize an average of 67% of eggs (Zhou and Xian, 1994). Mass releases of *Ooencyrtus* spp. into litchis and longans in Guangxi have given significant levels of control. Egg parasitoids of *T. papillosa* in Thailand are *Anastatus* sp. nr. *japonicus* and *Ooencyrtus phongi*. They are mass-reared in eggs of the wild silkworm, *Philosamia ricini* Hutt., and are released early in the season. Results are similar to those achieved in China, with the parasitoids accounting for 79% and 21%, respectively, of the bug eggs (Nanta, 1992).

The coreid bugs, *Amblypelta nitida* Stål, the fruitspotting bug (Plate 41), and *A. lutescens lutescens* (Distant), the banana-spotting bug, are major pests of tropical tree crops grown in Queensland (Waite, 1990). The adult bugs overwinter on citrus or non-crop hosts, which may be native plants or exotic ornamentals such as *Murraya paniculata*. In spring, when litchis and longans flower, the bugs migrate into the orchards. In litchis they prefer to feed on green fruit during the 6-week period following flowering (Plate 42), while longans remain attractive from fruit set until harvest. Orchards located adjacent to the natural rainforest breeding areas of the bugs are particularly susceptible to attack (Waite and Huwer, 1998). Female bugs lay individual eggs, mostly on leaves close to fruit. These take about 7 days to hatch, the nymphs developing through five instars before they become adults in about 42 days under average summer conditions.

In South Africa, the coconut bug, *Pseudotheraptus wayi* Brown, has become a problem in a variety of crops over the last 20 years (de Villiers, 1990b). It is not currently recognized as a pest of litchis, but its history of adapting to new hosts suggests that it could eventually become a problem. *Paradasynus longirostris* Hsiao has recently been described as an important pest of longans in Fujian, where the adults and nymphs attack the fruit (Huang *et al.*, 2000).

Coreid bugs feed on the developing seed within litchi and longan, and this initiates abscission a couple of days later. The feeding puncture is invisible on the surface of the fruit, and the only way to distinguish fruit damaged by bugs from naturally shed fruit is to dissect them to detect the typical lesion on the dark-brown seed testa. In south Queensland, *A. nitida* damaged 95% of fallen fruit on unsprayed litchis over three seasons, and in northern Queensland, *A. l. lutescens* damaged 98% of fruit (Waite, 1990). Comparable data are not available for longans, but a similar situation is likely. In litchis, adults may occasionally feed on mature fruit, but usually only nymphs, which are trapped on the trees because they are flightless, can be found feeding on such fruit.

Fruitspotting bugs have few natural predators apart from spiders, particularly those belonging to the family Thomisidae. However, egg parasitoids may have a significant effect late in the season. In northern Queensland, the egg parasitoids, *Ooencyrtus* spp., *Anastatus* spp. and *Gryon* spp., account for 30–60% of the eggs of the banana-spotting bug (Fay and Huwer, 1993). In southern Queensland, *Anastatus* spp. and *Gryon meridionis* parasitize eggs of *A. nitida* and *A. l. lutescens* to a similar extent (Waite and Petzl, 1997).

In most crops attacked by fruitspotting bugs, the insects are extremely difficult to detect, being well camouflaged in the canopy because of their colour. They also have good sensory perception, which enables them to escape predation. In litchis and longans, where the fruit are borne on the outside of the tree, the bugs are easily visible on the panicles. However, reliance on finding the bugs to determine when to spray is often misleading, particularly if monitoring is carried out irregularly. Dissection of fallen fruit is the best way to sample for the presence of bugs. Two sprays of endosulfan, applied 2 weeks apart during the first 6 weeks after fruit set, is usually sufficient to protect the crop. In longans, which are subject to attack over a longer period, further sprays may be necessary, but those applied for the control of *C. ombrodelta* also control the bugs.

Leptocoris rufomarginata (Fabricius) and *Leptocoris tagalica* Burm., which belong to the Family Rhopalidae, infest litchi trees in Queensland on a sporadic basis. They cause green fruit to fall, and may sometimes be as important as the fruitspotting bugs (Waite, 1992a). Damage symptoms are characteristic and easily distinguished from those of *Amblypelta* spp., with *Leptocoris* spp. leaving only a 'pinprick' on the seed compared with the extensive lesions produced by *Amblypelta*.

Soft scales

Pulvinaria (Chloropulvinaria) psidii (Maskell), the green shield scale infests litchis, and to a lesser extent longans, in China, Taiwan, Australia, Florida and India. In Queensland, crawlers are produced in early spring by adult scales that infest the leaves and twigs. Some of these crawlers move on to developing inflorescences, and later colonize the young fruit (Plate 40). Growers often mistake the female scales for mealybugs because the egg masses are enclosed by waxy filaments that often cover most of the scale.

P. psidii causes no damage through its feeding and is not considered to be important in India (Butani, 1977). However, when significant populations develop on the fruit as they often do in Florida (Butcher, 1954), China, Taiwan and Australia (Waite and Elder, 1996), the fruit is unmarketable because of the presence of the scales. The scales also produce honeydew, which supports the growth of sooty mould on infested fruit and on panicles situated below. This discoloration results in downgrading or rejection of the fruit in the market-place. Soft brown scale, *C. hesperidum* Linnaeus, causes similar problems (Plate 39).

The mealybug ladybird, *Cryptolaemus montrouzieri* Mulsant, and the green lacewing, *Mallada signata* (Shneider), are the most effective predators of *P. psidii* in Queensland. No significant parasitism has been noted in the scale on litchis in Queensland, but in China, *Anicetus ceroplastis* Ishii parasitizes immature and female scales. An unidentified coccinellid and several predatory mites also feed on the scales in China, and during the wet season an unspecified entomophagous fungus may cause significant mortality (Anonymous, 1978). In Queensland it is recommended that green shield scale populations be monitored just before and during the emergence of flower panicles to determine infestation levels on the twigs. If a significant proportion of twigs are infested, an oil spray should be applied to prevent infestation of the flowers and fruit (Waite and Elder, 1996).

C. hesperidum Linnaeus is an occasional pest of litchis in Queensland, where its parasitoids have been disrupted by chemical sprays, or it is protected from its natural enemies by ants (Waite, 1986). *Parasaissetia nigra* (Nietner) and *Saissetia coffeae* (Walker) infest litchis in India, but they are not important commercially (Butani, 1977).

Pink wax scale, *Ceroplastes rubens* Maskell, is present in most litchi- and longan-producing countries. It is a pest only on longans in Australia, infesting the leaves (Plate 38). The resulting heavy films of sooty mould probably reduce photosynthesis. On heavily infested trees, all of the fruit may be spoiled by mould. Corrective sprays of low viscosity oil or methidathion should be applied to limit contamination if scale populations are significant. The parasitoid *Anicetus beneficus* Ishii and Yasumatsu attacks pink wax scale in coastal Queensland, but does not always control it. *Ceroplastes ceriferus* (Fabricius), Indian white wax scale, and *Nipaecoccus vastator* (Maskell) are common on litchis in Taiwan, but are only minor pests. *N. vastator* is a minor pest of litchis in northern New South Wales, Australia. In Réunion, *Icerya seychellarum* West. and *Pulvinaria* spp. have been recorded on litchis, the former often attended by the ant, *Solenopsis* spp. (Vayssieres, 1997).

The lac insect, *Kerria lacca* Kerr, introduced into Taiwan from Thailand in 1940 for the production of shellac, is now a pest of many fruit and flower species, amongst them litchi and longan. Heavy infestations may cause twigs to wilt and die, reducing flowering and fruiting. The heavy production of honeydew encourages the growth of sooty mould (Hsieh and Hwang, 1981). Several parasitoids and predators have been recorded, but natural control is generally ineffective (Hwang and Hsieh, 1981). Sprays of dimethoate and fenthion, applied when crawlers are being produced, provide good control (Hsieh and Hwang, 1983).

Armoured scales

The diaspidid scales, *Hemiberlesia lataniae* (Signoret) and *Fiorinia* sp. nr. *nephelii* Maskell, occasionally infest litchis in Queeensland, but they are seldom a problem (Waite, 1992a). Similarly, *F. nephelii*, *Parlatoria pseudopyri* Kuwana, *P. cinerea* Danne and Hadden and *Aulacaspis* spp. are recorded from the crop in India, but are of no commercial significance (Butani, 1977). The hard scales generally infest twigs, and if allowed to multiply, the vitality of young trees may be reduced, and terminals killed.

Microhymenoptera undoubtedly suppress hard scale populations in the absence of disruptive chemical sprays, but the identity of natural enemies of the above species has not been determined, except for *H. lataniae* in Queensland. There, latania scale attacking avocados is controlled by a complex of natural enemies that includes *Aphytis* sp. *proclia* group and *Encarsia citrina* (Craw.) (Aphelinidae), as well as *Signiphora flavella* Girault and *S. perpauca* Girault (Signiphoridae). *Chrysopa oblatis* Banks (Chrysopidae) and *Rhizobius satellus* Blackburn also feed on the scales (Waite, 1988).

Branch borers

The litchi longicorn beetle, *Aristobia testudo* (Voet), is a serious pest of litchi and longan in Guangdong, China (Zhang *et al.*, 1997). The beetle has one generation per year. Adults emerge from June to August, and girdle branches by chewing off 10 mm strips of bark. The eggs are laid on the wound and are covered with an exudate. They hatch in late August, and the larvae live under the bark until January, when they bore into the xylem and create tunnels, up to 600 mm long, in the wood (Ho *et al.*, 1990). These tunnels have openings packed with frass, situated at regular intervals and opening to the exterior, for aeration. In June, the larva blocks the tunnel with wood fibre and frass just before it pupates. This tunnelling results in the death of the branch.

The white spotted longicorn beetle, *Anoplophora maculata* (Thomson), in Taiwan and the citrus branch borer, *Uracanthus cryptophagus* Olliff, in Australia, can kill branches up to 100 mm in diameter (Plate 36). In some orchards, damage from *U. cryptophagus* is increasing, especially under protective netting, and where insecticides are used less frequently to control other pests.

Recommendations for borer control in China include the manual removal of beetles by orchard workers during the period of adult activity. Eggs and young larvae can also be removed from accessible branches. Established larvae can be located through the presence of frass packed into the ends of tunnels, and 'fished out' with wire hooks or a knife. A skilled worker can kill 112 larvae in 2 hours (Ho *et al.*, 1990). Alternatively, dichlorvos may be injected into the tunnels, which are then sealed with clay (Zhang *et al.*, 1997). *A. testudo* has no known natural enemies, although experimental injections of the nematode *Steinernema carpocapsae* (Weiser) into larval tunnels has provided 73–100% control (Xu *et al.*, 1995).

The coffee leopard moth, *Zeuzera coffeae* Nietner, has two generations a year in mainland China and in Taiwan. Eggs are laid in groups of 20–30 in crevices in the bark of litchis and longans. On hatching, the larvae bore into the branches, which may be killed. They pupate just below the surface of the bark, the empty pupal case protruding from the exit hole when the moth emerges. The pest occurs mainly in abandoned orchards and is of minor importance.

Weevils

Cratopus angustatus Boh. and *Cratopus humeralis* Boh. attack litchis in Réunion. The weevils mate mostly at night, and the eggs are laid in irregular groups between leaves that are glued together with a mucilaginous secretion produced by the female. On hatching, the larvae migrate to the soil, where they feed on the roots of the tree. Sometimes the adults aggregate on certain trees, although the reason for this has not been ascertained. *Cratopus* spp.

adults feed on leaves, flower panicles and fruit less than 10 mm long, but the primary target is the fruit peduncle (Vayssieres, 1997). In Queensland, the weevils *Euthyrrhinus meditabundus* Fab. and *Orthorrhinus klugii* Boh. chew small patches of bark from twigs, in which they lay their eggs, but cause only minor damage. They also occasionally ring-bark and kill terminals, but the overall effect on trees is insignificant.

Chemicals can be applied when weevil infestations are noted on the trees. However, care needs to be taken since the pyrethroid cyhalothrin (Karate), applied to control *Cratopus* spp. in Réunion, eliminated the natural enemies of *Icerya* spp., resulting in an outbreak of that pest (Vayssieres, 1997).

Scarab beetles

The elephant beetle, *Xylotrupes gideon* (Linnaeus), damages litchi fruit in all production areas of Australia, and is also recorded from China. The larvae develop in the soil or in organic matter, where they feed on plant roots and humus material. They grow to a length of about 50–70 mm, pupate in the soil or compost, and the large sexually dimorphic adults emerge in spring.

The beetles are attracted to litchis by ripe fruit that have split or have been damaged by other pests, especially parrots or fruit bats. Once they have colonized a tree and start feeding on the damaged fruit, sound fruit may also be severely damaged (Plate 47). Regular inspection of orchards when the fruit are ripening is necessary to detect beetles. They are tolerant of many of the sprays applied for other pests, and high rates of carbaryl or chlorpyrifos are required to kill them. Alternatively they can be removed manually by knocking them into buckets with a stick, which is costly because of the labour involved.

Leaf-feeding beetles

Small black or brown leaf-eating beetles belonging to the genus *Rhyparida* often attack new flushes of litchis and longans in northern Queensland. The most damaging species is *Rhyparida discopunctulata* Blackburn, which emerges in swarms after spring rains. These swarms strip the leaves from any suitable host trees in their flight path. The red-shouldered leaf beetle, *Monolepta australis* (Jacoby), does the same when it swarms, especially in southern Queensland, although numerous non-swarming beetles may be attracted to split fruit without damage being inflicted. *Proctophana tomentosa* Lacordaire has recently been recorded damaging litchi leaves in Brazil (Sánchez Soto and Nakano, 2003).

Mites

The litchi erinose mite, *Aceria litchii* (Keiffer), also known in China as litchi hairy mite, hairy spider or dog ear mite, occurs throughout China and Taiwan (Huang *et al.*, 1990), India (Prasad and Singh, 1981), Pakistan (Alam and Wadud, 1963), Hawaii (Nishida and Holdaway, 1945) and Australia (Waite, 1986). It has not been recorded in South Africa, Israel, Canary Islands, Mauritius, Madagascar and Réunion.

Adult erinose mites are minute, 0.13 mm long, and pinkish white. Female mites lay eggs singly on the leaf surface, amongst the erineum or 'felt' induced by their feeding. The eggs are small (0.032 mm in diameter), spherical and translucent. A slow-moving protonymph hatches in 3–4 days. This stage is followed by two further nymphal stages or deutonymphs. The adult stage is reached about 13 days after oviposition (Alam and Wadud, 1963) with 13–15 overlapping generations produced each year in India and China (Prasad and Singh, 1981; Zhang *et al.*, 1997). Adult mites crawl from old infested leaves to the new growth flushes, where their feeding stimulates the production of the erineum in which they shelter and feed (Plate 35). Population abundance follows the flushing activity of the trees, with numbers booming on the summer flush, and moderating during winter, when lower temperatures slow tree growth and mite reproduction.

Trees may be infested with erinose mites from the time they are planted in the field if the air-layers are from infested parent trees. Otherwise, for infestation to occur, the mites must move directly between touching trees (Prasad and Singh, 1981), or be physically transported from tree to tree by human activity or other agents, or carried by the wind (Wen *et al.*, 1991). While wind transport may be the most common method of movement, they may also be transported by honeybees at flowering (Waite and McAlpine, 1992; Waite, 1999).

The mites attack the new vegetative growth, causing a felt-like erineum to be produced on the leaflets. This can take the form of small blisters, but if the infestation is severe, the erinose may eventually cover the entire leaflet, causing it to curl, and the foliage on whole terminals may be deformed. The erineum is at first silver-white, changing as it ages to light brown and then dark reddish brown. Very old erinose is almost black. The highest populations are present in light-brown, verging on dark-brown, erinose.

That development of erineum is a result of feeding by *A. litchii* has been questioned by Somchoudhury *et al.* (1989) and Sharma (1991), who proposed that the erineum does not arise from stimulated leaf cells, but is formed by the thalli of the alga, *Cephaleuros virescens* Kunze, with the alga and mite sharing a symbiotic relationship. On the basis of this, Somchoudhury *et al.* (1989) proposed that the name of *A. litchii* be changed to 'litchi algal mite'. Saha *et al.* (1996) studied the symptoms associated with the presence of *C. virescens* and *A. litchii* at a fundamental level and agreed that the alga was involved in the production of erinose. While *C. virescens* commonly grows on litchi trunks, branches and leaves, where it may be a problem in its own right, it is doubtful that it alone is responsible for the erinose, since chemical applications that kill the mites allow new foliage to develop without symptoms (Waite and Elder, 1996; Gupta *et al.*, 1997). In addition, symptoms are not present on litchis grown in countries that have never recorded the mite, but where the alga is present. The development of erinea and galls on leaves as a result of feeding by eriophyids is well documented (Westphal and Manson, 1996).

Many leaves may fall if mite infestations become severe. While established trees can tolerate substantial infestations, the growth of young trees may be restricted. If leaves immediately below an inflorescence are infested, the mites will attack the florets, preventing fruit set or producing malformed fruit. Even after fruit have set and have developed to half their final size, the mites can colonize them, producing erinose on the skin and making them unmarketable.

Numerous species of mites have been recorded in association with *A. litchii*, but not all have been proven to be predators. In India, Lall and Rahman (1975) reported that *Phytoseius intermedius* Evans and Macfarlane, *Phytoseius* spp., *Typhlodromus fleschneri* Chant, and *Cunaxa setirostris* were associated with the pest. Somchoudhury *et al.* (1987) added a further six species, namely *Amblyseius largoensis* Muma, *A. syzygii*, *A. herbicolus* (Chant), *Typhlodromus sonprayagenis*, *Typhlodromus homalii*, and *Agistemus* spp. Thakur and Sharma (1989) added *Amblyseius coccineae* Gupta, *A. finlandicus* Oudemans, *A. purni* Gupta, and *A. paraaerialis* Muma.

Wu *et al.* (1991) recorded the phytoseiid mites found on litchis in China, and the predator guilds associated with *A. litchii* in Australia and China were compared by Waite and Gerson (1994). They found that nine phytoseiid mites, *A. herbicolus* (Chant), *A. largoensis* (Muma), *A. barkeri* (Hughes), *Amblyseius nambourensis* Schicha, *A. neomarkwelli* Schicha, *Phytoseius hawaiiensis* Prasad, *P. rubiginosae* Schicha, *Okisieus morenoi* Schicha and *Typhlodromus haramotoi* Prasad, and six other mite species belonging to the Anystidae, Ascidae, Cheyletidae, Cunaxidae and Stigmaeidae, were associated with litchi erinose mite in Queensland. A cecidomyiid fly larva, *Arthrocnodax* spp., was also a common predator.

The limited census carried out in China over a shorter period recorded nine phytoseiid species: *Amblyseius eharai* Amitai and Swirski, *A. herbicolus*, *A. largoensis*, *A. ovalis* (Evans), *A. cantonensis* Schicha, *A. okinawanus* Ehara, *Phytoseius hawaiiensis*, *P. fujianensis* Wu and *Okisieus subtropicus* Ehara. Of these, *A. eharai* is considered to be the most effective predator of *A. litchii* in China. *Agistemus exsertus* Gonzalez (Stigmaeidae) has been used to control *A. litchii* in Guangdong, Guangxi and Fujian (Ren and Tian, 2000).

Frequent and regular monitoring of trees should be conducted to detect infestations of erinose mite. During the non-fruiting period, particularly during postharvest pruning, infested branches should be cut off and burned

(Anonymous, 1978; Wen *et al.*, 1991; Zhang *et al.*, 1997).

The critical time for treatment of erinose mite with chemicals is when the trees are about to flush (Pinese, 1981; Waite, 1992a; Waite and Elder, 1996). Monitoring for the presence of erinose on litchi trees requires regular inspections of the foliage to detect symptoms around the time that the trees are expected to flush, although not all the trees in an orchard will do so at the same time. The mites themselves are invisible to the naked eye, and they live amongst the erineum. To determine whether the erineum is active, the leaves should be picked and left for several hours to desiccate. As the leaves and erineum dry, the mites move to the surface, where they are easily visible with the aid of a hand lens or microscope. Populations in excess of 100,000 per leaflet have been recorded using this method in association with a washing and centrifuge technique (Waite, 1992a).

In Queensland, three sprays of dimethoate or wettable sulphur, applied at 2–3 week intervals during terminal emergence and leaf expansion, protect the new flush from infestation by mites migrating from infested leaves below. If this operation is carried out during the postharvest flush in late summer, the mite population on the tree will be minimal when inflorescences emerge in spring. If leaves below the emerging flower panicle remain infested, a similar series of chemical applications should be made to prevent mites from damaging the flowers and fruit. Sprays applied at times other than flushing often give poor control because of the protection provided to the mites by the erineum.

In India, Prasad and Bagle (1981) found that dicofol gave the best control of erinose mite, but that monocrotophos, carbaryl, cyhexatin and chlordimeform also performed satisfactorily. Sharma and Rahman (1982) confirmed that dicofol gave good control, but dimethoate also provided acceptable results. Sprays of wettable sulphur have been recommended in Australia (Waite, 1992b), China (Anonymous, 1978), Hawaii (Nishida and Holdaway, 1945) and Taiwan (Wen *et al.*, 1991). Other chemicals recommended in China include dichlorvos, dimethoate (Anonymous, 1978), dicofol, chlorpyrifos, omethoate and isocarbophos (Zhang *et al.*, 1997).

Longan erinose mite

The longan erinose mite, *Aceria longana* Boczek and Knihinicki, is recorded as a sporadic but major pest of longans in Thailand, severely affecting the terminals and flowers (Chai-ai and Visitpanich, 1997). *A. longana* is specific to longan. The mite is microscopic, creamy white, and lives amongst the erineum produced on the leaves, similar to *A. litchii*, or in the terminal buds. Boczek and Knihinicki (1998) describe this species as being very similar to *A. dimocarpi*.

Mites feeding on the growing points and flowers cause malformation and stunting, and severely infested trees may stop growing. Leaves may curl due to erinose development, and damage to buds produces a witches' broom effect. There is some uncertainty as to whether the witches' broom is always caused by the mite, or whether a phytoplasma is also involved, in which case it is suggested that the mites may be vectors (S. Dolsopon, Thailand, 1999, personal communication). Increased activity of the mite is noticed during hot, dry weather, which often prevails from February to April in Thailand. Similar symptoms occur on longans in China, where a different mite and phytoplasma have been implicated. No natural enemies are recorded for the mite, but it is assumed that there is a suite of predatory mites associated with it, similar to that found in association with the litchi erinose mite (Waite and Gerson, 1994).

Longan gall mite

The longan gall mite, *Aceria dimocarpi* (Kuang), is associated with longans in China, where it causes erineum on leaves (Kuang, 1997) and also witches' broom symptoms (He *et al.*, 2000). These symptoms have been reported from Fujian, Guangxi, Guangdong and Hainan as well as from Taiwan, although the cause has only recently been attributed to *A. dimocarpi* by He *et al.* (2000). As well as relating mite presence to witches' broom, they were able to reduce the incidence of symptoms in affected orchards through a combination of pruning and the application of miticides. Others claim that the symptoms are caused by a phytoplasma that could be transmitted by insect vectors such as litchi stinkbug and longan psylla, and via dodder (*Cuscuta tempestris*), and in budwood (Chen *et al.*, 2000). Infested young leaves roll up and do not expand. The shoots form compact clusters. Inflorescences are malformed and flowers fail to produce fruit.

Longan trees should be inspected regularly, and an appropriate miticide applied when witches' broom symptoms are found. In Guangdong, the application of a mixture of 20% omethoate, 40% dicofol and 50% colloidal sulphur reduced the incidence of affected shoots by up to 92% (He *et al.*, 2000).

Other mites

The tea red spider mite, *Oligonychus coffeae* (Nietner) (Tetranychidae), occasionally infests litchi leaves in Queensland, but is never a problem. In Queensland and other regions, *Polyphagotarsonemus latus* (Banks) (Tarsonemidae) may occasionally damage individual terminals on orchard trees. However, it is most often seen on nursery trees, where it can be easily controlled with dicofol, sulphur or endosulfan.

Gall midges

Yang and Luo (1999) described *Litchiomyia chinensis* Yang and Luo from specimens reared from leaf galls collected in Guangdong, with the suggestion that this species is the same as the litchi leaf midge, *Dasyneura* spp., which was never described morphologically. The larvae of the midge overwinter in the galls that form on the leaves as a result of their feeding. They pupate in the soil, the adult flies emerging in March–April to initiate the first of seven or eight overlapping generations for the year. The female midge prefers trees with dense foliage growing in moist environments in which to lay its eggs on young leaves. When the larvae hatch they mine the leaf, causing 'watery dots' to form, which become galls (Plate 48). The galls turn brown and eventually drop out, giving the leaf a 'shot-hole' appearance and reducing the leaf area.

In susceptible orchards, monitoring is not an option, and preventive procedures are adopted. As with the litchi erinose mite, infested leaves can be removed after harvest, and destroyed. In the spring, 75 kg/ha of 2.5% methyl parathion can be distributed on the ground under the tree, or isofenphos (0.001%) sprayed on the ground just prior to the expected emergence of the adult flies. In autumn, isocarbophos (0.001%) should be sprayed twice over a period of 14 days during a leaf flush (Zhang *et al.*, 1997).

Fruit flies

Ceratitis capitata (Weidemann) and *Ceratitis (Pterandrus) rosa* Karsch are recorded from litchi orchards in South Africa and Réunion, and *C. capitata*, *Bactrocera dorsalis* Hendel and *Bactrocera cucurbitae* (Coquillett) from Hawaii. In Queensland, *Bactrocera tryoni* (Froggatt) is occasionally found in litchis.

Female fruit flies lay their eggs through the skin of the fruit, often utilizing cracks and wounds made by other pests (Vayssieres, 1997). Although the eggs may hatch, the larvae rarely survive because of the amount of juice present in mature litchis, which drowns them (de Villiers, 1990a, 1992b; Vayssieres, 1997).

All of the above-mentioned species are capable of ovipositing through the sound skin of litchis, although in some cultivars the thickness of the skin may prevent successful oviposition. Most often, the presence of fruit fly eggs and the occasional larva in litchi fruit can be attributed to oviposition through prior damage inflicted by other pests such as *Cryptophlebia* spp. Only in South Africa are fruit flies considered a real problem, despite the relatively low damage levels recorded. *C. rosa* is thought to be responsible for most of the losses (Grové *et al.*, 1999a,b). The physical damage thus caused, as well as some damage inflicted on the flesh by early larval instars, initiates fruit rot and fermentation (de Villiers, 1992b). A study conducted by Gould *et al.* (1999) in Florida found that, despite the presence of large numbers of the Caribbean fruit fly, *Anastrepha suspensa* (Loew), in litchi and longan orchards, neither crop was attacked by the pest.

Numerous parasitoids, especially *Opius* spp. and *Biosteres* spp., have been recorded attacking the fruit fly species that damage litchis. Despite the activity of such parasitoids, fruit flies continue to flourish and often require chemical control in susceptible crops.

In South Africa, the use of pheromone-baited traps is recommended for monitoring fruit fly populations around litchi orchards. Control is achieved with bait sprays of protein hydrolysate mixed with trichlorfon or mercaptothion. Alternatively, panicles may be protected with paper bags applied just after the November fruit drop (de Villiers, 1990). Grové *et al.* (1999c) found that quarantine cold treatment killed *C. rosa* in litchi fruit. In other countries, no specific action is recommended because the problem is not serious, although quarantine protocols may change this in the future.

Bark borers

The larvae of *Salagena* spp. are dark brown and slightly hairy, and they feed on the bark and wood of litchi trees in the north-eastern Transvaal and Natal South Coast in South Africa. After hatching, the larvae bore into the wood, forming tunnels up to 70 mm long and 5 mm wide, usually in the crotches of branches. They cover the feeding site with frass held together by webbing. The larvae pupate in the tunnel, and moths emerge from November to January.

In India, *Indarbela quadrinotata* Walker and *Indarbela tetraonis* Moore are polyphagous, attacking aonla (*Emblica officinalis*), ber (*Zizyphus* spp.), citrus, falsa (*Grewia asiatica*), guava, jackfruit, loquat, mango, mulberry and pomegranate, as well as litchi. Eggs are laid in April and May, and the caterpillars are fully grown by December. Pupation is delayed until March or April, and the insect has one generation per year (Butani, 1977). *Arbela dea* Swinhoe attacks litchi in Guangdong. Large orchard trials showed that the application of the entomopathogenic nematode, *Steinernema carpocapsae*, gave 94% control (Xu and Xie, 1997).

The larvae of all these species feed on the bark and bore into the wood. If damage to the bark extends right around branches, they are ring-barked and die. The wounds may also allow infection by fungi, which can cause dieback.

During routine orchard operations, the frass-covered webbing associated with the borers can be seen on the branches. Unspecified insecticide applications, applied to areas of activity denoted by the presence of fresh frass, are effective against *Salagena* spp. (de Villiers, 1983a). In India, it was recommended that the frass and webbing be cleared away and the holes plugged with cotton wool soaked in petrol, chloroform or formalin. These chemicals are often too expensive and Shah (1946) proposed the use of hot water injected by syringe. Later, Khurana and Gupta (1972) suggested that dichlorvos, trichlorfon or endosulfan could be applied in the same way.

Comoritis albicapilla Moriuti is a new pest in Guangdong where it damages the bark of litchis. The larvae spin a double layer of webbing incorporating the faeces, making chemical control difficult. Entomopathogenic nematodes have provided variable levels of control, with the best results following treatment during rainy spring weather (Xu *et al.*, 2000).

Mealybugs

The citrus mealybug, *Planococcus citri* (Risso), is widely distributed throughout the world on numerous hosts. It sporadically attacks litchis in Taiwan, where it may contribute to the production of sooty mould.

Thrips

In India, *Dolichothrips indicus* Hood and *Megalurothrips distalis* Karny attacked the flowers and leaves of litchi, respectively (Ananthakrishnan, 1971). The damage caused by these species was not quantified, but phosphamidon and dimethoate were recommended for control (Butani, 1977).

The tea yellow thrips, *Scirtothrips dorsalis* Hood, infests litchis and longans in China. All life stages feed on the shoots, causing malformation of the new leaves. The crimped, yellow leaves have a mosaic appearance and eventually lose their sheen, and may fall. The thrips are most numerous from August to October, especially when it is dry. Sprays of isocarbophos, omethoate or dimethoate are recommended, along with orchard management procedures that encourage uniform flushing (Zhang *et al.*, 1997).

The plague thrips, *Thrips imaginis* Bagnall, often infests litchi flowers in Queensland. As its name implies, it may be present in extraordinary numbers, and it appears to feed on the florets. However, experiments have indicated that the insect has little, if any, effect on fruit set (Waite, 1992a).

Pollination

Du Toit and Swart (1995) concluded that litchi has a limited ability for self-pollination and that insects are necessary to ensure optimum fruit set. In India, Singh and Chopra (1998) recorded 16 species of bees, flies, wasps and other insects visiting litchi flowers. Most common were the honeybees, *A. cerana* Fabricius, *A. dorsata* Fabricius, *A. mellifera* Linnaeus and *A. florea* Fabricius, and the syrphid flies *Melanostoma univittatum* Wiedemann and *Episyrphus balteatus* (De Geer). Pollination by these insects resulted in a 387% increase in fruit set, and a 505% increase in fruit retention compared with panicles denied access by pollinators.

Of the bees, *A. dorsata* has been considered the most important, although *A. mellifera* is also a regular visitor to litchi flowers, and most probably has a significant impact on pollination (Kitroo and Abrol, 1996; Kumar *et al.*, 1996; Abrol, 1999). In Thailand, *A. mellifera* is preferred as a large-scale producer of honey and for the pollination of longans, while *A. cerana* is preferred for small-scale honey production and the pollination of litchis, rambutans and mangoes (Wongsiri and Chen, 1995). *Trigona iridipennis* Smith also contributes to pollination in both litchis and longans in India and Thailand (Boonithee *et al.*, 1991; Kumar *et al.*, 1996).

Eardley and Mansell (1996) concluded that of the 38 insect species visiting litchi flowers near Ofcolaco in South Africa, most were ineffectual as pollinators, and that honeybees and indigenous bees, such as *Plebeina denoiti* (Vachal), *Meliponula erythra junodi* (Friese), *Ctenoceratina moerenhouti* (Vachal), *Ctenoceratina rufigaster* (Cockerell) and *Braunsapis facialis* (Gerstaecker), contribute significantly to pollination. In contrast, litchis at Tzaneen were visited mostly by honeybees.

When litchi flowers are bagged to deny access to insects, fruit set is reduced. In India, Kumar *et al.* (1996) recorded an average of 1.4 fruit on panicles when insects were excluded, 8.9 on panicles caged with *A. mellifera*, and 14.9 on panicles with free access to all pollinators. An average of six fruit per panicle was set in South Africa when honeybees had access to flowers, compared with two per panicle when the panicles were bagged (du Toit, 1994).

Integrated pest management (IPM)

As with all crops, the ultimate aim in the protection of litchis and longans from insect pests is to implement a viable integrated pest management (IPM) system. This system will have been developed for each production region, accounting for the variations in geography and latitude that determine the timing of plant growth, and the local insect and mite fauna.

The tactics adopted against each pest depend on how well they fit in with the overall management strategy for an orchard. Some pests may be effectively controlled by natural enemies, and the options considered for the control of other pests must always take into account the possible side-effects on these, and the possible induction of problems caused by the injudicious use of insecticides. It should be noted, however, that when pest infestations start to cause economic losses, they should be controlled, but with consideration of the impact on the orchard ecosystem (see Figs 12.2–12.4 for the types of spray equipment used).

In China, where litchis have been cultivated for thousands of years, the need to control pests to produce high-quality fruit has been well recognized over the years. There is a strong belief in the concept of biological control combined with sound orchard management practices, especially the pruning of infested branches or leaves, as the basis for IPM. In addition, many studies have been conducted on the effect of orchard floor management on pest incidence. As well as providing a suitable habitat for natural enemies, suitable ground covers may improve the orchard micro-climate (Liang and Huang, 1994; Liu and Tan, 1999). When farm communes existed, such systems were widely used. However, since their demise and with the availability of more effective insecticides and spray equipment, adherence to this approach has waned.

For the Chinese crop, Zhang *et al.* (1997) divided the year into phenological stages so that the recommended management strategies can be implemented to match the crop's susceptibility to particular pests. A disease management system is also included. Although various biological controls are detailed in the preamble, the commercial recommendations listed do not include them.

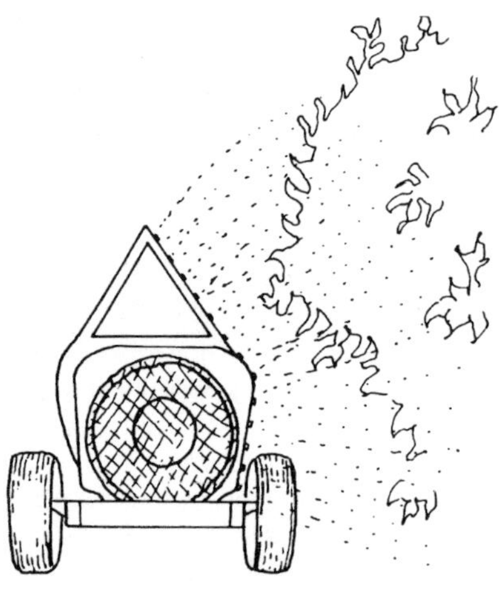

Fig. 12.2. Low profile air blast sprayer (drawing courtesy of Loraine Chapman).

During the winter flushing period, suppression of leaf flushes not only promotes flowering but also suppresses some of the overwintering insect and mite pests. In the spring flush/flowering period, trichlorfon, for the control of litchi stinkbug and suppression of erinose mite and leaf gall midge, is applied in combination with chlorbenside for downy blight disease. Two sprays of the latter chemical may be required during prolonged rainy periods at flowering. During the fruitlet period, pest targets are stem-end borer and litchi stinkbug, which are controlled with a mixture of cypermethrin or chlorpyrifos plus trichlorfon. Ridomil-MZ or Sandofan may be applied for downy blight control. The stem-end borer and downy blight are targeted in the maturing fruit period, with sprays as required. The autumn flush is protected from stem-end borer, erinose mite, leaf midge and various lepidopterous caterpillars by two sprays of isocarbophos or acephate, applied 10–15 days apart. Presumably, releases of *Anastatus* could be substituted for the stinkbug sprays, provided that they are correctly timed and integrated with the stem-end borer sprays.

The use of *A. japonicus* in combination with *Ooencytus phongi* has been adopted in Thailand, where the stinkbug is controlled on litchis and longans by mass releases of these egg parasitoids at flowering. If necessary, carbaryl can be applied at fruit set to control excessive

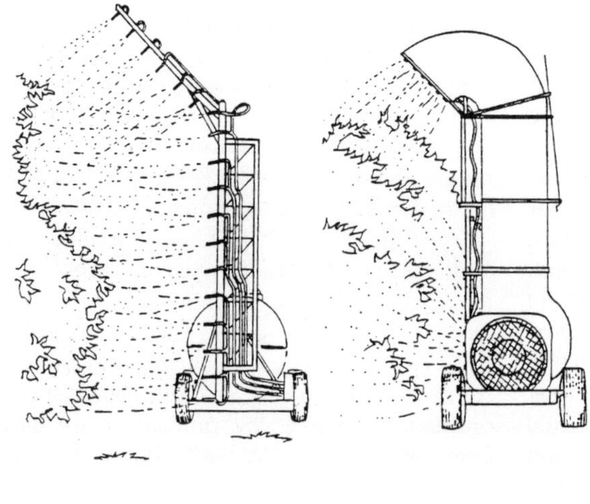

Fig. 12.3. Oscillating boom (left) and air blast spray unit with tower (drawing courtesy of Loraine Chapman).

Fig. 12.4. Hand spraying in China (photograph courtesy of Zhan Wei Zhang).

numbers of nymphs, if egg parasitism is low. The subsequent control of *Conopomorpha sinensis* using permethrin has no effect on biological control, since the oviposition period for the bugs, and hence the critical period of protection for the parasitoids, has passed by the time the small fruit become susceptible to the borer. In Thailand, pruning to remove infested leaves and fruit, and restriction of tree size so that sprays can be applied effectively, are recommended. Fruit bagging, while not always providing perfect pest control, provides the bonus of enhanced fruit colour in litchi and longan.

Waite (1992a) detailed the procedure that should be adopted for implementing IPM in Queensland litchis. This system follows the phenological cycle of the trees, and monitoring at least monthly during autumn and winter, and weekly during flowering and fruit production in spring and summer. Using this approach, the more important pests can be detected, and timely controls applied. In this way, conservation biological control is employed for pests such as the soft scales and erinose mite, although the latter may still require occasional chemical control. Leaf-eating loopers are tolerated until excessive foliage is consumed, and *Bacillus thuringiensis* is applied if necessary. In the past, carbaryl has been applied up to five times, or azinphos-methyl up to three times, during fruit growth to control the macadamia nut borer. Tebufenozide offers the opportunity for better control of this pest, with fewer sprays required and less disruption of natural enemies.

The complex of pests attacking litchis and longans throughout the world invariably includes at least one species of fruit borer. In the absence of completely effective natural enemies of these pests, chemicals will remain a necessary part of IPM. These will generally be the key determinants of the eventual fruit yield and quality, as well as the viability of the whole IPM system.

References

Abrol, D.P. (1999) Pollination importance value of bees. *Insect Environment* 5, 140–141.

Alam, M.Z. and Wadud, M.A. (1963) On the biology of litchi mite, *Aceria litchii* Keifer (Eriophyidae, Acarina) in East Pakistan. *Pakistan Journal of Science* 18, 232–240.

Ananthakrishnan, T.N. (1971) Thrips (Thysanoptera) in agriculture, horticulture and forestry: diagnosis, bionomics and control. *Journal of Scientific and Industrial Research* 30, 113–146.

Anonymous (1978) Lychee pest and disease control. In: Academy of Guangdong Agricultural Science (eds) *Monograph of Guangdong Lychee.* Guangdong Science and Technology Press, Guangzhou, China, pp. 119–146.

Banziger, H. (1982) Fruit-piercing moths (Lep., Noctuidae) in Thailand: a general survey and some new perspectives. *Mitteilungen der Schweizerischen Entomologiscchen Gesellschaft* 55, 213–240.

Boczek, J. and Knihinicki, D. (1998) Studies on eriophyoid mites (Acari: Eriophyoidea). XXVII. *Bulletin of the Polish Academy of Sciences, Biological Sciences* 46, 141–146.

Boonithee, A., Juntawong, N., Pechhacker, H. and Huttinger, E. (1991) Floral visits to select crops by four *Apis* species and *Trigona* sp. in Thailand. In: van Heemert, C. and de Ruijter, A. (eds) *Proceedings of the Sixth International Symposium on Pollination.* Research Centre for Insect Pollination and Beekeeping and International Society for Horticultural Science, Tilburg, The Netherlands, pp. 74–80.

Bradley, J.D. (1986) Identity of the south-east Asian cocoa moth, *Conopomorpha cramerella* (Snellen) (Lepidoptera: Gracillariidae), with descriptions of three allied new species. *Bulletin of Entomological Research* 76, 41–51.

Brown, J.W., Pena, J., Vazquez, T. and Baixeras, J. (2002) Description of a new tortricid pest (Lepidoptera: Tortricidae: Olethreutinae) of litchi (*Litchi chinensis*) in Florida, with a review of tortricid pests of litchi worldwide. *Proceedings of the Entomological Society of Washington* 104, 318–329.

Butani, D.K. (1977) Pests of litchi in India and their control. *Fruits* 32, 269–273.

Butcher, F.G. (1954) Insect problems in litchi production. In: *Florida Litchi Growers' Association Yearbook and Proceedings.* Second Annual Meeting, Winter Haven, Florida, pp. 15–16.

Chai-ai, P. and Visitpanich, J. (1997) Life cycle of four-legged mite, *Aceria longana* (Acarina: Eriophyidae), the major pest of longan. *Journal of Agriculture, Chiang Mai University* 13, 88–89.

Chen, J.Y., Chen, J. and Xu, X. (2000) Advances in the research of longan witches' broom disease. In: Huang, H.B and Huang, X.M (eds) *First International Symposium on Litchi and Longan.* South China Agricultural University, Guangzhou, China. Program and Abstracts E-03, p. 74.

de Villiers, E.A. (1983a) Bark borers in litchi trees. *Litchis H.2 Farming in South Africa.* 2pp.

de Villiers, E.A. (1983b) The litchi moth. *Litchis H.3 Farming in South Africa.* 2pp.

de Villiers, E.A. (1990a) Die beheer van vrugtevlieë by lietsjies. *Yearbook of the South African Litchi Growers' Association* 3, 29–30.

de Villiers, E.A. (1990b) Coconut bugs on avocados. *Farming in South Africa, Avocados* H.5. 2 pp.

de Villiers, E.A. (1992a) Control of the litchi moth, *Argyroploce peltastica* Meyr. *Yearbook of the South African Litchi Growers' Association* 4, 8.

de Villiers, E.A. (1992b) Fruit fly. In: *The Cultivation of Litchis.* The Agricultural Research Council of South Africa, Bulletin 425, 56–58.

de Villiers, E.A. and Stander, G.N. (1989) Proefresultate met trilumuron vir die beheer van die lietsjiemot. *Yearbook of the South African Litchi Growers' Association* 2, 20–21.

Dolsopon, S., Dasanoda, M. and Warapitirangsri, W. (1997a) Effectiveness of larval parasitoids to control litchi leafminer, *Conopomorpha litchiella* (Lepidoptera: Gracillariidae) in the field. *Journal of Thai Agricultural Research* 15, 35–44.

Dolsopon, S., Suphakamnuad, N. and Dasanoda, M. (1997b) Crop loss assessment caused by litchi fruit borer, *Conopomorpha sinensis* (Lepidoptera: Gracillariidae) and its parasitoids. *Chiangrai Horticultural Research Centre Annual Report.* Horticultural Research Institute, Department of Agriculture, Thailand, pp. 76–91.

du Toit, A.P. (1994) Pollination of avocados, mangoes and litchis. *Inligtingsbulletin Instituut vir Tropiese en Subtropiese Gewasse* 262, 7–8.

du Toit, A.P. and Swart, D.J. (1995) Pollination and fruit set in Bengal litchi during the 1993 flowering season at Politsi: second report. *Yearbook of the South African Litchi Growers' Association* 7, 31–32.

Eardley, C.D. and Mansell, M.W. (1996) The natural occurrence of insect pollinators in a litchi orchard. *Yearbook of the South African Litchi Growers' Association* 8, 27–29.

Fay, H.A.C. (2002) Fruitpiercing moths and fruit-spotting bugs: intractable pests of tree fruits in a reduced insecticide environment. *Acta Horticulturae* 575, 485–493.

Fay, H.A.C. and Halfpapp, K.H. (1993) Non-odorous characteristics of lychee (*Litchi chinensis*) and carambola (*Averrhoa carambola*) pertaining to fruitpiercing moth susceptibility. *Australian Journal of Experimental Agriculture* 33, 327–331.

Fay, H.A.C. and Halfpapp, K.H. (1999) Activity of fruitpiercing moths, *Eudocima* spp. (Lepidoptera: Noctuidae), in north Queensland: some effects of fruit type, locality and season. *Australian Journal of Entomology* 38, 16–22.

Fay, H.A.C. and Huwer, R.K. (1993) Egg parasitoids collected from *Amblypelta lutescens lutescens* (Distant) (Hemiptera: Coreidae) in north Queensland. *Journal of the Entomological Society of Australia* 32, 365–367.

Follett, P.A. and Lower, R.A. (2000) Irradiation to ensure security for *Cryptophlebia* spp. (Lepidoptera: Tortricidae) in sapindaceous fruits from Hawaii. *Journal of Economic Entomology* 93, 1848–1854.

Follett, P.A. and Sanxter, S.S. (2001) Hot water immersion to ensure quarantine security for *Cryptophlebia* spp. (Lepidoptera: Tortricidae) in lychee and longan exported from Hawaii. *Journal of Economic Entomology* 94, 1292–1295.

Gould, W.P., Hennessey, M.K., Pena, J., Castineiras, A., Nguyen, R. and Crane, J. (1999) Nonhost status of lychees and longans to Caribbean fruit fly (Diptera: Tephritidae). *Journal of Economic Entomology* 92, 1212–1216.

Grové, T., de Beer, M.S., Dreyer, S. and Steyn, W.P. (1999a) Monitoring fruit flies in litchi orchards. *Yearbook of the South African Litchi Growers' Association* 10, 34–36.

Grové, T., Steyn, W.P. and de Beer, M.S. (1999b) Exposure of different litchi cultivars to Mediterranean fruit fly, *Ceratitis capitata* (Weidemann) and the Natal fruit fly, *Ceratitis* rosa Karsch. *Yearbook of the South African Litchi Growers' Association* 10, 37–38.

Grové, T., Steyn, W.P. and de Beer, M.S. (1999c) Exposure of Natal fruit fly, *Ceratitis rosa* Karsch. to litchi export conditions. *Yearbook of the South African Litchi Growers' Association* 10, 39–40.

Gupta, D., Bhatia, R., Sharma, N.K., Chandel, J.S. and Sharma, R. (1997) Incidence and management of red rust of litchi in lower hills of Himachal Pradesh. *Pest Management in Horticultural Ecosystems* 3, 70–74.

He, D., Zeng, M., Zhou, B., Lin, S., Li, J. and Peng, J. (2000) The cause of witches' broom in longan: discovery of a new gall-mite species. In: Huang, H.B and Huang, X.M (eds) *First International Symposium on Litchi and Longan.* South China Agricultural University, Guangzhou, China. Program and Abstracts E-04, p.75.

Higgins, J.E. (1917) The litchi in Hawaii. *Hawaii Agricultural Experiment Station Bulletin* 44, 1–21.

Ho, D.P., Liang, H.W., Feng, Z.W. and Zhao, X.D. (1990) A study of the biology and control methods of the long horn beetle *Aristobia testudo* (Voet). *Natural Enemies of Insects* 12, 123–128.

Hsieh, F.K. and Hwang, J.S. (1981) The Indian lac insect in Taiwan, once a beneficial insect, now a nuisance. *Chinese Journal of Entomology* 1, 119–120.

Hsieh, F.K. and Hwang, J.S. (1983) Further studies on the control of the lac insect, *Kerria lacca* (Kerr.). *Plant Protection Bulletin* 25, 31–40.

Huang, B., Huang, J. and Chen, L. (2000) Studies on *Paradasynus longirostris* Hsiao and its control (Hemiptera: Coreidae). In: Huang, H.B and Huang, X.M. (eds) *First International Symposium on Litchi and Longan.* South China Agricultural University, Guangzhou, China. Program and Abstracts E-12, p. 79.

Huang, C.C., Chang, K.S. and Chu, Y.I. (1994) Damage and population fluctuation of the litchi fruit borer, *Conopomorpha sinensis* Bradley, in Chia-Nan district, Taiwan. *Plant Protection Bulletin Taipei* 36, 85–95.

Huang, M.D., Mai, S.H., Wu, W.N. and Poo, C.L. (1974) The bionomics of *Anastatus* sp. and its utilization for the control of litchi stink bug, *Tessaritoma papillosa* Drury. *Acta Entomologica Sinica* 17, 362–375.

Huang, T., Huang, K.W. and Horng, I.J. (1990) Two species of eriophyid mites injurious to litchi trees in Taiwan. *Chinese Journal of Entomology, Special Publication* 3, 57–64.

Hwang, J.S. and Hsieh, F.K. (1981) Bionomics of the lac insect in Taiwan. *Plant Protection Bulletin* 23, 103–115.

Jones, V.P. (1994) Oviposition patterns of koa seedworm and litchi fruit moth (Lepidoptera: Tortricidae) on macadamia and litchi in Hawaii. *Journal of Economic Entomology* 87, 1278–1284.

Khurana, A.D. and Gupta, O.P. (1972) Bark-eating caterpillars pose a serious threat to fruit trees. *Indian Farmers' Digest* 5, 51–52.

Kitroo, A. and Abrol, D.P. (1996) Studies on pollen carrying capacity and pollination efficiency of honey bees visiting litchi flowers. *Indian Bee Journal* 58, 55–57.

Kuang, H. (1997) Four new species of Eriophyidae (Acari: Eriophyoidea) from China. *Entomotaxonomia* 19, 74–78.

Kumar, R., Chaudhary, O.P. and Lenin, J.K. (1996) Studies on pollination of litchi (*Nephilium litchii* Camb.) in Dehradunn India. *Indian Bee Journal* 58, 9–12.

La Croix, E.A.S. and Thindwa, H.Z. (1986) Macadamia pests in Malawi. III. The major pests: the biology of bugs and borers. *Tropical Pest Management* 32, 11–20.

Lall, B.S. and Rahman, M.F. (1975) Studies on the bionomics and control of the erinose mite *Eriophyes litchii* Keifer (Acarina: Eriophyidae). *Pesticides* 9, 49–54.

Liang, W.G. and Huang, M.D. (1994) Influence of citrus orchard ground cover plants on arthropod communities in China. *Agriculture, Ecosystems and Environment* 50, 29–37.

Liu, D.G. and Tan, B.L. (1999) Studies on the parasitoid community in a complex litchi orchard. *Natural Enemies of Insects* 21, 134–139.

Liu, X.D. and Lai, C.Q. (1998) Experiment of control of litchi stink bug by using *Anastatus japonicus* Ashmead. *South China Fruits* 27 (3), 31.

Liu, Y.F. and Gu, D.X. (1998) Control of litchi stink bug by egg parasitoids. *Chinese Journal of Biological Control* 14, 145–147.

Nanta, P. (1992) *Biological Control of Insect Pests.* Biological Control Branch, Entomology and Zoology Division, Department of Agriculture, Bangkok, Thailand, 206 pp.

Newton, P.J. and Crause, C. (1990) Oviposition on *Litchi chinensis* by *Cryptophlebia* species (Lepidoptera: Tortricidae). *Phytophylactica* 22, 365–367.

Newton, P.J. and Odendaal, W.J. (1990) Commercial inundative releases of *Trichogrammatoidea cryptophlebiae* (Hym.: Trichogrammatidae) against *Cryptophlebia leucotreta* (Lep.: Tortricidae) in citrus. *Entomophaga* 35, 545–556.

Nishida, T. and Holdaway, F.G. (1945) *The Erinose Mite of Lychee*. Hawaii Agriculture Experiment Station Circular 48, 10 pp.

Pinese, B. (1981) Erinose mite: a serious litchi pest. *Queensland Agricultural Journal* 107, 79–81.

Prasad, V.G. and Bagle, B.G. (1981) Evaluation of acaricides for the control of litchi mite, *Aceria litchii* Keifer (Acarina: Eriophyidae). *Pesticides* 1, 22–23.

Prasad, V.G. and Singh, R.K. (1981) Prevalence and control of litchi mite, *Aceria litchii* Keifer in Bihar. *Indian Journal of Entomology* 43, 67–75.

Quilici, S. (1996) *Enquête sur les ravageurs et auxiliaires sur litchi à la Réunion*. Réport au Département des Productions Fruitières et Horticoles, CIRAD-FLHOR, Ile de la Réunion, 13 pp.

Quilici, S., Verbizier, B., Trahais, B. and Manikom, R. (1988) Note sur les ravageurs du litchi à la Réunion. *Fruits* 43, 459–464.

Ren, S. and Tian, M. (2000) Litchi pests and their natural enemies. In: Huang, H.B and Huang, X.M (eds) *First International Symposium on Litchi and Longan*. South China Agricultural University, Guangzhou, China. Program and Abstracts E-02, p. 74.

Rogers, D.J. and Blair, A.D. (1981) Assessment of insect damage to litchi fruit in northern Queensland. *Queensland Journal of Agricultural and Animal Sciences* 38, 191–194.

Saha, K., Somchoudhury, A.K. and Sarkar, P.K. (1996) Structure and ecology of *Cephaleuros virescens* Kunze and its relationship with *Aceria litchii* Keifer (Prostigmata: Acari) in forming litchi erineum. *Journal of Mycopathological Research* 34, 159–171.

Sahoo, A.K. and Maili, B. (1992) Control of litchi leaf roller. *Environment and Ecology* 10, 553–554.

Sánchez Soto, S. and Nakano, O. (2003) Ocorrência de imagos de *Proctophana tomentosa* Lacordaire (Coleoptera: Chrysomelidae) em lichia, no Brasil [Occurrence of adults of *Proctophana tomentosa* Lacordaire (Coleoptera: Chrysomelidae) on lychee, in Brazil]. *Neotropical Entomology* 32, 167–168.

Sands, D.P.A. and Schotz, M. (1989) Advances in research on fruit piercing moths of subtropical Australia. In: Batten D. (ed.) *Proceedings of the Fourth Australasian Conference on Tree and Nut Crops*. Lismore, NSW, Australia, pp. 378–383.

Shah, R. (1946) An effective and inexpensive method for the control of stem borers in fruit trees with special reference to santra trees in C.P. and Berar. *Current Science Bangalore* 15, 135.

Sharma, D.D. (1991) Occurrence of *Cephaleuros virescens*, a new record of leaf-curl in litchi (*Litchi chinensis*). *Indian Journal of Agricultural Sciences* 61, 446–448.

Sharma, D.D. and Rahman, M.F. (1982) Control of litchi mite *Aceria litchii* (Keifer) with particular reference to evaluation of pre-bloom and post-bloom application of insecticides. *Entomon* 7, 55–56.

Singh, R. and Chopra, S.K. (1998) Flower visitors of litchi (*Litchi chinensis* Sonn.) and their role in pollination and fruit production. *Pest Management and Economic Zoology* 6, 1–5.

Somchoudhury, A.K., Sarkar, P.K., Singh, P. and Mukherjee, A.B. (1987) Bioefficacy of some pesticides against *Aceria litchii* (Keifer) (Acari: Eriophyidae). *First National Seminar on Agricultural Acarology*. Kalyani, India, 48 pp.

Somchoudhury, A.K., Singh, P. and Mukherjee, A.B. (1989) Interrelationship between *Aceria litchii* (Acari: Eriophyidae) and *Cephaleuros virescens*, a parasitic alga in the formation of erineum-like structure on litchi leaf. *Progress in Acarology* 2, 147–152.

Storey, R.I. and Rogers, D.J. (1980) Lepidopterous pests of the litchi in North Queensland. *Queensland Journal of Agricultural and Animal Sciences* 37, 207–212.

Thakur, A.P. and Sharma, D.D. (1989) New records of predatory mites on mite pests of economic crops in Bihar. *Biojournal* 1, 155–156.

Unahawutti, C. (1990) Insect pests of fruit trees. In: *Proceedings of the Fifth Workshop on Insect Pests and Their Control*. Entomology and Zoology Division, Department of Agriculture, Bangkok, Thailand, 121 pp.

Vayssieres, J.F. (1997) *Enquête sur les arthropodes ravageurs et auxiliaires du litchi à la Réunion*. Réport au Département des Productions Fruitières et Horticoles, CIRAD-FLHOR, Ile de la Rèunion, 40 pp.

Waite, G.K. (1986) Pests of lychees in Australia. In: Menzel, C.M. and Greer, G.N. (eds) *Proceedings of the First National Lychee Seminar*. Australian Lychee Growers' Association, Nambour, Queensland, pp. 41–50.

Waite, G.K. (1988) Biological control of latania scale on avocados. *Queensland Journal of Agricultural and Animal Sciences* 45, 165–167.

Waite, G.K. (1990) *Amblypelta* spp. and green fruit drop in lychees. *Tropical Pest Management* 36, 353–355.

Waite, G.K. (1992a) *Pest Management in Lychees*. Australian Rural Industries Research and Development Corporation, Canberra, Australia, 42 pp.

Waite, G.K. (1992b) Integrated pest management in lychee. In: Fullelove, G.D. (ed.) *Proceedings of the Third National Lychee Seminar*. Australian Lychee Growers' Association, Bundaberg, Queensland, pp. 39–43.

Waite, G.K. (1999) New evidence further incriminates honey-bees as vectors of lychee erinose mite *Aceria litchii* (Acari: Eriophyidae). *Experimental and Applied Acarology* 23, 145–147.

Waite, G.K. and Elder, R.J. (1996) Lychee/longan insect pests and their control. In: Welch, T. and Ferguson, J. (eds) *Proceedings of the Fourth National Lychee Seminar Including Longans*. Australian Lychee Growers' Association, Yeppoon, Queensland, pp. 102–109.

Waite, G.K. and Gerson, U. (1994) The predator guild associated with *Aceria litchii* (Acari: Eriophyidae) in Australia and China. *Entomophaga* 39, 275–280.

Waite, G.K. and Huwer, R.K. (1998) Host plants and their role in the ecology of the fruitspotting bugs *Amblypelta nitida* Stål and *Amblypelta lutescens lutescens* (Distant) (Hemiptera: Coreidae). *Australian Journal of Entomology* 37, 340–349.

Waite, G.K. and McAlpine, J.D. (1992) Honey bees as carriers of lychee erinose mite *Eriophyes litchii* (Acari: Eriophyiidae). *Experimental and Applied Acarology* 15, 299–302.

Waite, G.K. and Petzl, N. (1997) A search for egg parasites of *Amblypelta* spp. (Hemiptera: Coreidae) in south-east Queensland. In: Menzel, C. and Crosswell, A. (eds) *Maroochy Research Station Report* 7, 13–14.

Wen, H.C., Lee, H.S. and Lin, C.C. (1991) Field studies on litchi erineum mite (*Eriophyes litchii* Keifer) in Southern Taiwan. *Journal of Agricultural Research of China* 40, 298–304.

Westphal, E. and Manson, D.C.M. (1996) Feeding effects on host plants: gall formation and other distortions. In: Lindquist, E.E., Sabelis, M.W. and Bruin, J. (eds) *Eriophyoid Mites: Their Biology, Natural Enemies and Control*. Elsevier Science, Amsterdam, pp. 231–242.

Wongsiri, S. and Chen, P. (1995) The effects of agricultural development on honey bees in Thailand. *Bee World* 76, 3–5.

Wu, W.N., Lan, W.M. and Liu, Y.H. (1991) Phytoseiid mites on litchis in China and their application. *Natural Enemies of Insects* 13, 82–91.

Xu, J.L. and Xie, R.C. (1997) Mass application of entomogenous nematode against litchi stem borer. *Acta Phytophylactica Sinica* 24, 150–154.

Xu, J.L., Han, R.H., Liy, X.L., Cao, L. and Yang, P. (1995) The application of codling moth nematode against the larvae of the litchi longicorn beetle. *Acta Phytophylactica Sinica* 22, 12–16.

Xu, J.L., Yang, P. and Xie, R.C. (2000) Studies on the applications of some entomopathogenic nematodes against litchi yponomeutid *Comoritis albicapilla* Moriuti. *Acta Phytophylactica Sinica* 27, 27–31.

Yang, C.K. and Luo, Q.H. (1999) A new genus and species of gall midge (Diptera: Cecidomyiidae) infesting litchi from China. *Entomotaxonomia* 21, 129–132.

Zeng, X.N., Deng, D. and Wang, J.M. (2001) Chlorpyrifos and cypermethrin for the control of litchi stink bug (*Tessaratoma papillosa*). *Acta Horticulturae* 558, 421–423.

Zhang, D.P. (1997) A study of the pericarp of litchi fruit and the injury caused by litchi stinkbug, *Tessaritoma papillosa* (Hem.: Pentatomidae). *Wuyu Science Journal* 13, 198–203.

Zhang, Z.W., Yuan, P.Y., Wang, B.Q. and Qui, Y.P. (1997) *Litchi Pictorial Narration of Cultivation*. Pomology Research Institute, Guangdong Academy of Agricultural Sciences, Guangzhou, China, 189 pp.

Zhou, Z.H. and Xian, X.Y. (1994) Investigation on the morphology and parasitism of *Ooencyrtus* sp. *Plant Protection* 20, 41–42.

13 Diseases

Lindy Coates[1], Erxun Zhou[2] and Chatree Sittigul[3]
[1]*Department of Primary Industries and Fisheries, 80 Meiers Road, Indooroopilly, Queensland 4068, Australia;* [2]*Department of Plant Pathology, College of Resources and Environment, South China Agricultural University, Guangzhou 510642, China;* [3]*Department of Plant Pathology, Chiang Mai University, Chiang Mai 50200, Thailand*

Introduction

Numerous diseases affect litchi and longan, although the biology of these has not been investigated extensively. Most leaf spots, flower blights and preharvest fruit diseases have little impact on commercial production, although downy blight is of significance in litchi in Asia. The causal agent of witches' broom, the most important disease of longan in Asia, has yet to be clearly established. In Thailand, a phytoplasma is thought to be responsible, whereas in China, a virus has been implicated. Although root rots and diebacks can result in serious losses in individual orchards, the causal agents are largely unknown. Postharvest diseases of fruit are common, but they can generally be managed with cool storage and careful handling. However, further studies are required on the aetiology of these diseases.

This chapter provides information on the major diseases of litchi and longan in terms of their importance, symptoms, and current status of causal agent/s, epidemiology, and management. We have drawn heavily on the recently published *Diseases of Tropical Fruit Crops* (Ploetz, 2003). Holcroft *et al.* (Chapter 14, this volume) provide further information on postharvest treatments that address the problems associated with disease.

Soil-borne diseases

Armillaria root rot

Armillaria or mushroom root rot is not very widespread, but can cause serious losses in individual litchi plantings in Florida (Cohen, 1955; McMillan, 1994b), South Africa (Darvas, 1992; Manicom, 1995) and Australia (Menzel *et al.*, 1988).

A loss of tree vigour and absence of new shoots are the first obvious symptoms of the disease (Darvas, 1992). The trees then begin to drop their leaves and may die. Young trees may collapse very suddenly, whereas older trees often decline over a number of years. The disease is distinguished from other forms of dieback by the presence of white to light-tan sheets of mycelium between the bark and the wood of the crown and large roots (McMillan, 1994b), particularly in young trees wilting suddenly (Darvas, 1992). These sheets are more difficult to distinguish in older trees. The wood underlying the sheets becomes discoloured, and after cool, wet weather, clumps of honey-coloured mushrooms may appear at the base of the trees.

The disease is caused by the basidiomycetes, *Armillaria mellea* in South Africa (Darvas, 1992) and *A. socialis* in Florida

(McMillan, 1994b). In Australia, an unidentified species, possibly *A. luteobubalina*, is involved. *A. luteobubalina*, a parasitic species attacking many native and introduced plants in Australia, has, in most cases, been misidentified as *A. mellea* (Grgurinovic, 1997).

The sporophores of *A. mellea* are ephemeral. The caps are honey-coloured, 5–30 cm in diameter and borne on stipes, 8–25 cm long. The mushroom has a ring, but no volva. Basidia are borne on the gills on the underside of the cap, and the dense basidiospores are white. Rhizomorphs that are cylindrical in cross-section (1–3 mm in diameter) are produced from colonized food bases. They are cream when young, but rapidly become black on the outside. *A. socialis* is distinguished from *A. mellea* by the absence of a ring around the stipe.

A. socialis may persist as a saprophyte on roots and plant debris in the soil for many years (Cohen, 1955). The most common sources of inoculum are infected tree stumps, large roots, and rhizomorphs in the soil. Infection occurs when the roots contact inoculum in the soil and, following penetration of a healthy root, the fungus grows up along the cambium. Eventually, the pathogen girdles the crown, and the tree dies (Cohen, 1955). The honey-coloured mushrooms that are occasionally produced at the base of a dying tree have no known role in infection (Shaw and Kile, 1991). In South Africa, it has been suggested that trees planted in poorly drained soils or weakened by nematodes may be more susceptible (Darvas, 1992).

Soil preparation is very important when new trees are planted into areas affected by *Armillaria*, including sites cleared of native vegetation. Tree stumps should be removed from the planting site, followed by deep cultivation to remove as many roots as possible. The use of methyl bromide following removal of infected plants has also been used, although penetration of the fumigant into large roots may not be effective. Replacement trees in established orchards should be planted some distance away from affected sites. When orchards are thinned, tree stumps should be removed or sealed, to limit cross-infection. There are currently no fungicides available for control.

Litchi decline

Litchi decline has been reported in many areas of northern Viet Nam (Trung, 1999; Trung *et al.*, 1999). Symptoms include leaf fall, loss of gloss in the leaves, and branch dieback. While symptoms are characteristic of *Phytophthora* infection, that fungus has not been isolated from affected trees. *Fusarium*, *Pythium*, *Cylindrocladium*, *Scopulariopsis* and *Rhizoctonia* have been isolated, but probably constitute secondary infections (Trung *et al.*, 1999). Poor drainage, deep planting, and inadequate nutrition favour the decline, but it can sometimes be slowed by the application of phosphonates (Trung, 1999).

In Australia, a branch dieback of unknown cause has been reported (Menzel *et al.*, 2002). The problem is rare and has only been observed in northern Queensland. A sudden death of litchi trees less than 3 years of age has also been reported in Queensland. While the cause is unknown, root damage at planting and poor soil drainage are contributing factors.

Longan decline

Longan decline is a serious disease in northern Thailand, affecting up to 40% of trees in some areas (Visitpanich *et al.*, 1999). Affected trees are unthrifty and dwarf (Plate 54). Trees may flower, but they set small, low-quality fruit. Trees also become more susceptible to insect pests and colonization by lichens and algae, and those growing in waterlogged soils often die (Visitpanich *et al.*, 1999).

While the cause of the disease is not known, there is a strong association between the syndrome and the nematode, *Rotylenchulus reniformis* (Visitpanich *et al.*, 1999). Both poor drainage and drought contribute to the problem. Improvements in soil drainage in lowland orchards, and irrigation during the dry season in upland areas, reduce the incidence of symptoms. Various soil amendments have also been trialled. Visitpanich *et al.* (1999) reported that cow manure and granular fertilizers reduced the severity of symptoms compared with untreated trees. More recently it was found that chicken manure plus urea increased leaf size in affected trees in pot trials, and dramatically reduced counts of *R. reniformis*.

Nematodes

Various nematodes have been associated with litchi dieback in South Africa (Milne *et al.*, 1971) and China (Yin *et al.*, 1994a,b). Affected trees have bare twigs and branches, yellow leaves with burnt tips, poor flowering, excessive fruit drop and erratic shoot growth (Milne *et al.*, 1971). The roots become 'stubby' and dark, with a smaller feeder-root mass, leading to decreased water and nutrient uptake, and tree dieback.

In South Africa, large numbers of the ring nematode, *Hemicriconemoides mangiferae*, and the dagger nematode, *Xiphinema brevicolle*, were associated with the disease (Milne *et al.*, 1971). In China, 32 species of nematode were identified in samples taken from litchi roots or the surrounding soil (Yin *et al.*, 1994a,b), although their role in the decline was not defined.

Pre-plant soil fumigation and the application of post-plant nematicides have shown considerable promise for controlling these nematodes in South Africa (Milne *et al.*, 1971; Cohn and Duncan, 1990). Because these nematodes can survive in the soil, the use of orchard soil for air-layering is not recommended.

Diseases in nurseries

A number of soil-borne diseases affect litchi and longan in nurseries. In Florida, a root rot (*Fusarium* spp.) occurs in nurseries where litchi air-layers have been planted in contaminated soil (McMillan, 1994b). *Fusarium* spp. are also possibly involved in Australia, where trees suddenly wilt and die (Coates *et al.*, 2003). Root rot of longan, caused by *F. solani*, has been reported in China, and usually occurs in seedlings from May to June. In some longan-growing regions, up to 10% of seedlings can be affected (Chi, 2000). Root rot (*Pythium* spp.) and stem rot (*Rhizoctonia solani*) also affect litchi stock in Florida (McMillan, 1994b).

Foliar, floral, fruit stem and preharvest diseases

Algal spot

Algal spot occurs in poorly managed orchards in the humid subtropics and tropics, and affects a wide range of important fruit, including litchi (Mishra *et al.*, 1974; Menzel *et al.*, 1988; McMillan, 1994b) and longan (Visarathanonth, 1999). The disease mainly manifests itself as circular, slightly raised patches of velvety, red-brown to orange algae on the surface of the leaves and branches (Plate 55). In the absence of sporangia, however, these spots remain green-brown (McMillan, 1994b).

The disease is caused by *Cephaleuros virescens*. The thallus produces erect cells, some of which enlarge to produce stalked, terminal or ovoid sporangia about 30 × 24 μm (Lim, 1994). Sporangia produce biflagellate zoospores, which are the main infective propagules. Sexual reproduction is by flask-shaped gametangia that form in the thallus and release 8–32 biflagellate gametes in the presence of free water. These gametes fuse, giving rise to dwarf sporophytes, which produce microsporangia. These then produce quadriflagellate zoospores, the role of which is unknown (Lim, 1994).

Algal spot is particularly prevalent in wet climates. Zoospores are dispersed by rain and wind, and infect leaves through the stomata, with chains of algal cells developing in the leaf tissue. Cultivars vary in their susceptibility (Menzel *et al.*, 1988), with the disease readily controlled by copper-based preparations (Menzel *et al.*, 1988; Gupta, 1992) and pruning, which improves tree ventilation.

Corky bark

Corky bark has recently been noted on litchi in Florida (Ploetz, 2003), and is similar to gall and scaly bark in mango. Affected trees have irregular patches of raised, cracked, and rough bark, which is friable and easily rubbed off. Branches die, and, eventually, so does the whole canopy. *Fusarium decemcellulare*, the same fungus as in mango, *Mangifera indica*, has been recovered from affected trees, although it has not been tested for its pathogenicity in litchi.

On mango, *F. decemcellulare* is a weak pathogen requiring a wound for infection. Armoured scales, *Andaspis punicae*, which are associated with lesions on litchi, possibly wound the fruit skin, although this relationship has not been confirmed. No pesticides are registered for its control. Affected branches and trees should be removed, and pruning equipment disinfected. Only healthy planting material should be used in new orchards, and air-layers should not be taken from affected trees (Ploetz, 2003).

Downy blight

Downy (brown) blight is the most serious disease of litchi in China (Chi *et al.*, 1984), and also occurs in Taiwan (Kao and Leu, 1980; Ann and Ko, 1984), Thailand and Viet Nam (Vien *et al.*, 2001).

Immature and ripe fruit, as well as leaves and inflorescences, are affected (Ann and Ko, 1984; Chi *et al.*, 1984). Symptoms appear as a brown blight, with the lesions covered by white hyphae, sporangia and sporangiophores during wet weather (Plates 50 and 51). Infected fruit may fall prematurely. The disease is caused by the fungus, *Peronophythora litchii* (Chen, 1961; Ko *et al.*, 1978), which produces colourless, aseptate mycelia, 4–6 μm wide, and which branch irregularly at right or acute angles (Hall, 1989). The thin-walled hyphal swellings are 30–35 μm in diameter, and the sporangia ovoid to obovoid, occasionally irregular, 28–33 μm × 19–22 μm, with distinct, flattened apical papillae. Sporangiophores are arborescent, one- to five-celled, 440–1325 μm × 4–6 μm, and branched dichotomously in the upper segment. Antheridia are paragynous, ellipsoidal and 9–13 μm, and the oogonia spherical to ellipsoidal, colourless, and 28–33 μm × 24–30 μm. Oospores are aplerotic, colourless, and 22–26 μm in diameter.

In northern Viet Nam, the disease is favoured by cool, wet weather during flowering and fruiting (Vien *et al.*, 2001). Primary infection is by direct penetration of the host by zoospores, with incubation less than 1 day at 25°C (Chi *et al.*, 1984). Dead branches and diseased leaves and fruit should be removed to reduce inoculum levels, and the canopy sprayed with copper oxychloride in winter (Li, 1997). Spraying the ground with copper sulphate, followed by the application of lime, is also recommended in China (Li, 1997). Metalaxyl, fosetyl-Al and mancozeb are effective when applied at flower budding, fruit set and just before harvest (Li, 1997; Ou *et al.*, 1999). Ann (2001) reported successful control in the field in Taiwan using phosphonates.

Lasiodiplodia branch blight and fruit rot

Lasiodiplodia branch blight and fruit rot are common on litchi in China (Chi, 2000). Brown spots appear on branches, followed by cracking of the bark. The lesions on the fruit spread rapidly, eventually covering large areas. The flesh becomes shrivelled and discoloured, with severely affected fruit falling. In advanced stages of infection, the affected areas are covered with numerous black pycnidia. Fruit may also break down after harvest (see section on 'Stem-end rot').

The fungus, *Lasiodiplodia theobromae*, causes the disease. When grown on media, the colonies are grey to black, fluffy, with abundant aerial mycelia. Pycnidia are simple or compound, often aggregated, stromatic, ostiolate, frequently setose, and up to 5 mm in diameter. Conidiophores are hyaline, simple, sometimes septate, rarely branched, cylindrical, and arise from the inner layers of cells lining the pycnidial cavity. Conidiogenous cells are hyaline, simple, cylindrical to sub-obpyriform, holoblastic and annelidic. Immature conidia are aseptate, hyaline, granulose, sub-ovoid to ellipsoid-oblong, thick-walled, with a truncate base, while mature conidia are 1-septate, cinnamon to fawn, often longitudinally striate, and 20–30 μm × 10–15 μm. Paraphyses when present are hyaline, cylindrical, sometimes septate, and up to 50 μm long. On host tissue, pycnidia are immersed, and later become erumpent, simple or grouped, 2–4 mm in diameter, ostiolate and frequently pilose, with conidia extruding in a black mass. A perfect state, *Botryosphaeria rhodina*, has been described, although the ascospores probably have only a minor role in infection.

The fungus survives in orchards on dead twigs, leaves, branches and infected fruit. Conidia are spread by water, with infection occurring through wounds or natural openings. Reductions in the amount of inoculum in an orchard, through pruning of dead wood and removal of infected fruit, help to reduce infection.

Pepper spot

Pepper spot affects litchi in Australia and has become a significant problem in recent years (Drew and Drew, 1999). While the disease only causes superficial blemishes, fruit are downgraded and unmarketable. Slightly raised, pinhead-sized dark spots develop (mostly) on the stem-end and shoulders of fruit (Bagshaw *et al.*, 1995). In severe cases, the spots may coalesce and darken the entire fruit surface (Plate 49). Spots also appear on the leaves.

Pepper spot is caused by *Colletotrichum gloeosporioides* (Yip, 1997; Cooke and Coates, 2002), the same fungus that causes a similar disease in avocado, *Persea americana* (Willingham *et al.*, 2000). *C. gloeosporioides* is also responsible for anthracnose in litchi, which leads to the fruit breaking down after harvest.

The morphology of *C. gloeosporioides* is highly variable. Colonies grown on media are grey-white to dark grey (Mordue, 1971). Conidia are hyaline, aseptate, uninucleate, 9–24 μm × 3–6 μm, cylindrical with obtuse ends, or sometimes slightly ellipsoid with a rounded apex and a narrow, truncate base. They are formed on unicellular, hyaline or faintly brown, cylindrical phialidic conidiophores. The teleomorph or sexual stage, *Glomerella cingulata*, sometimes occurs in culture, but is rare on the host.

Conidia produced on infected leaves and fruit, are the principal source of inoculum (Yip, 1997). Warm, wet weather favours infection and spread of the disease. The popular 'Kwai May Pink' is highly susceptible. A range of fungicides, including azoxystrobin, mancozeb, and copper, controlled the disease in northern Queensland (P. Langdon, South Johnstone, unpublished data, 2003).

Phomopsis leaf blight

Phomopsis leaf blight affects longan and litchi in parts of southern China (Chi, 2000). Initially, the infection appears as small spots on the leaf tips, progressing down the vein and forming characteristic 'V'-shaped lesions. In more advanced stages, pycnidia form on the surface of the lesions.

The disease is caused by *Phomopsis guiyuan* and *P. longanae* in longan, and by *P. longanae* in litchi (Chi, 2000). A similar leaf blight, caused by *Cladosporium oxysporum*, has also been reported in longan (Chen and Zhang, 2003).

Conidiomata of *P. longanae* are epiphyllous, eustromatic, globose-depressed or irregular, black, unilocular or bilocular, rarely trilocular, and 259–777 μm × 181–581 μm (Chi, 2000). Conidiophores are branched, septate and hyaline. Conidiogenous cells are phialidic, sub-cylindrical, hyaline and enteroblastic. The fungus produces two types of conidia: α-conidia are elliptical to fusiform, hyaline, unicellular, biguttulate, and 4–8 μm × 1.6–2.5 μm, while β-conidia are hyaline, filiform, unicellular, curvular, and 8–12 μm × 0.8 –1.2 μm.

Conidiomata of *P. guiyuan* are eustromatic, punctiform, ostiolate, immersed initially, erumpent when matured, gregarious, unilocular, triangle or globose-depressed, parietes 12–25 μm thick, dark brown and 64–170 μm × 60–144 μm (Chi, 2000). Conidiophores are hyaline, septate, branched and 12.5–50 μm × 2.0–4.8 μm. Conidiogenous cells are phialidic: α-conidia are hyaline, long-elliptical, sub-acute at two ends, biguttulate and 4.6–7.0 μm × 1.4–2.6 μm, while β-conidia are hyaline, hamate-filiform and 23–50 μm × 1–2 μm.

Infected and dead leaves are the principal sources of inoculum, and should be removed. It is thought that the conidia are dispersed primarily by air currents and rain splash. Fungicides such as chlorothalonil, thiophanate-methyl, and cardendazim control longan leaf blight in China.

Phytophthora-incited diseases

Phytophthora foliage blight and fruit rot are serious problems in areas of Thailand where longans set fruit in the off-season (Visitpanich *et al.*, 2000), particularly when cool weather follows 2–3 days of rain (Bhavakul *et al.*, 1998). Young shoots, panicles and fruit are mostly affected. Symptoms include a dark necrosis on young shoots, a brown blight on the inflorescences, irregular brown lesions on the fruit, and premature flower and fruit drop (Plate 57). During rainy weather, the fruit crack, with the lesions becoming covered in white sporangia and sporangiophores. Longan fruit are also affected in northern Viet Nam when weather conditions are favourable (A.W. Cooke, Australia, personal communication, 2003).

Phytophthora palmivora is the causal agent. According to Stamps (1985), the hyphae are fairly uniform, and rarely over 5 μm in diameter. Chlamydospores are abundant, formed early, and are 30–35 μm in diameter. Sporangiophores are narrow, developing as well-defined, simple sympodia. Sporangia are formed readily on solid media, and are ellipsoid or ovoid, and 35–60 μm × 20–40 μm, or up to 90 × 45 μm, caducous, with a short pedicel (up to 5 μm), broad and occluded, with a prominent papilla. Oogonia are rare in single cultures, but abundant when isolates of opposite compatibility are paired. The antheridia are amphigynous, spherical or oval, and 15 × 14 μm. The oospores nearly fill the oogonium, with a 2 μm-thick wall. The disease is favoured by moist conditions, since water is required for sporangium production, as well as for the release and movement of the zoospores. Infected fruit and plant material should be removed from affected orchards. Applications of fungicides such as metalaxyl may be necessary during rainy periods.

Sooty mould and black mildew

Sooty mould and black mildew (Plate 56) affect many tropical fruit, including litchi and longan. The surface of the fruit, leaves and twigs are covered with dense, black mats of mycelia. Although the damage is only superficial, affected fruit are downgraded due to their unsightly appearance. Leaf photosynthesis is also reduced due to restricted light penetration.

While the symptoms of sooty mould and black mildew are similar, the causal agents are different. The sooty mould fungi are saprophytes that do not penetrate the host, and require insect honeydew for colonization, whereas the black mildew fungi are parasitic (Lim and Sangchote, 2003). A number of genera within the Ascomycetes and Deuteromycetes have been listed as causal agents of sooty mould (Lim and Sangchote, 2003). In contrast, the black mildew fungi belong to the Meliolales, an order of the Ascomycetes (Lim and Khoo, 1985). In China, sooty mould of litchi is caused by *Capnodium* spp. (Li and Li, 1998), and black mildew by *Meliola capensis* and *M. commixta* (Hu, 1999). In Thailand, an unidentified species of *Meliola* affects longan.

The sooty mould fungi develop on honeydew that is excreted by sucking insects. Honeydew also encourages the growth of the black mildew fungi (Lim and Khoo, 1985). Effective suppression of these diseases can therefore be achieved by applying insecticides or oils, which control scale and other insects that produce the honeydew. Increasing air circulation in orchards, by pruning, weed control, and increased tree spacing, can also help.

Witches' broom

Witches' broom is the most important disease of longan in Asia, occurring in China (Menzel *et al.*, 1990), Hong Kong (So and Zee, 1972; Menzel *et al.*, 1990), Thailand (Menzel *et al.*, 1990; Sarindu, 1993) and Viet Nam (A.W. Cooke, Australia, personal communication, 2003). The disease has also been recorded on litchi in China (Chen *et al.*, 1996). Both inflorescences and leaves are affected. The inflorescences become deformed and shed their flowers prematurely, resulting in the characteristic 'broom-like' panicles (So and Zee, 1972; Menzel *et al.*, 1990). Fruit yield is greatly reduced as a result. Leaves are stunted, discoloured, deformed and blistered (Menzel *et al.*, 1990; Visitpanich *et al.*, 1996; Zhang, 1999) (see Plate 58).

There is some uncertainty regarding the causal agent in longan. A phytoplasma is responsible in Thailand (Visitpanich *et al.*,

1999), whereas a virus is involved in China (Chen *et al.*, 2000). Chen *et al.* (1996) also observed a virus associated with litchi witches' broom in China. In Thailand, the eriophyid mite, *Aceria dimocarpi*, is reported to transmit the phytoplasma (Visitpanich *et al.*, 1996). The virus in China is transmitted by the litchi stinkbug (*Tessaratoma papillosa*), the longan psylla (*Cornegenapsylla sinica*) and dodder (*Cuscuta campestris*), along with seed and budwood (Zhang, 1999; Chen *et al.*, 2000). A close association between the incidence of witches' broom and the gall mite, *Aceria* (*Eriophyes*) *dimocarpi*, has been reported from Guangdong (He *et al.*, 2001) (see also Waite, Chapter 12, this volume).

Strategies for disease management include the control of insect or mite vectors (Visitpanich *et al.*, 1996; Zhang, 1999; Chen *et al.*, 2000), the use of resistant cultivars, careful selection of propagating material, adoption of cultural practices that increase tree vigour, and the removal of infected branches, inflorescences and seedlings (Chen *et al.*, 2000). Strict quarantine is essential for all longan and litchi material that enters countries that are free of the disorder.

Minor foliar and stem diseases

There are a number of minor foliar and stem diseases. In China, leaf spot in litchi is caused by *Phyllosticta* spp., *Pestalotiopsis pauciseta* and *Coniothyrium litchi* (Chi, 2000). There are also reports of a leaf rust caused by *Uredo nephelii* (Sato, 1995). Leaf necrosis (*Colletotrichum gloeosporioides*), Gloeosporium leaf blight (*Gloeosporium* spp.), Phyllosticta leaf spot (*Phyllosticta* spp.), Phomopsis leaf spot, dieback caused by *Diplodia* sp., *Phomopsis* spp. and *Sphaeropsis* spp., and stem cankers caused by *Botryosphaeria* spp. and *Phomopsis* spp. occur in Florida (Alfieri *et al.*, 1994; McMillan, 1994b). In India, leaf blight and dieback in litchi is caused by *Pestalotiopsis mangiferae* (Kang and Singh, 1991). Leaf spot and blight in longan in China are associated with *Ascochyta longan*, *Leptosphaeria guiyuan* and *Marssonina euphoriae* (Chi, 2000).

Postharvest diseases

Anthracnose

Anthracnose is one of the most common diseases affecting litchi after harvest (Johnson, 1989; McMillan, 1994a; Chi, 2000), whereas in longan, it is only a minor foliage disease seen in poorly managed orchards. The disease can affect leaves, twigs, flowers and fruit in litchi. Although immature fruit can become infected and fall (Nakasone and Paull, 1998), it is more of a problem after harvest. Circular dark-brown to black lesions appear on the fruit, with salmon-coloured spore masses produced under humid conditions (Plate 52).

Anthracnose of litchi is predominantly caused by *Colletotrichum gloeosporioides*, although *C. acutatum* plays a minor role in Australia (Johnson *et al.*, 2002). An unidentified species of *Colletotrichum* causes the disease in longan. *C. gloeosporioides* has been described earlier. Colonies of *C. acutatum* are effuse, initially white, then pale orange, green-grey or black, and often pink or red-purple on the lower surface (Dyko and Mordue, 1979). Conidiophores are hyaline, septate, branched infrequently near the base, and smooth. Conidia are hyaline, aseptate, straight, smooth, fusiform, 8–16 μm × 2.5–4 μm and salmon-coloured in mass. Appressoria are few, mostly light to medium brown, clavate to obovate, and 6.5–11 μm × 4.5–7 μm, with smooth margins. This fungus differs from *C. gloeosporioides* in that the colony is orange-pink during the first few weeks of culture, and the conidia fusiform. Guerber and Correll (2001) have recently described a teleomorph or sexual stage, *Glomerella acutata*.

Infected leaves and stems are the principal sources of inoculum (McMillan, 1994a). Conidia are spread by water splash, with free water required for infection. Infection occurs by direct penetration of the fruit skin. Immature fruit are infected in the orchard, with the pathogen then remaining quiescent until they mature (McMillan, 1994b). High temperatures after harvest favour the disease.

Maintenance of orchard hygiene, through pruning of dead wood and removal of infected leaves, twigs and fruit, helps to reduce the incidence of the disease, while preharvest applications of fungicides such as mancozeb and benzimidazoles offer control (McMillan, 1994a; Crane *et al.*, 1997). Controlling insect pests to reduce alternative entry sites also helps. Treatment of harvested fruit with sulphur dioxide (Tongdee, 1986; Johnson, 1990), benomyl (Scott *et al.*, 1982; Huang and Scott, 1985;) and prochloraz (Brown *et al.*, 1984) offers control for rots of litchi, although there are restrictions on the use of these chemicals. Given current trends away from the use of chemicals in horticulture, storage at 5°C is the best method for suppressing rots, including anthracnose in litchi. Infected fruit, however, rot rapidly when returned to ambient temperatures.

Stem-end rot

Stem-end rot is a common postharvest disease in litchi and longan and is caused by a number of fungi. While symptoms vary with the causal agent, the rot generally appears as a browning of the skin at the stem-end of the fruit. Lesions expand rapidly, particularly those caused by *Lasiodiplodia theobromae*. Under humid conditions, the surface of the lesions may be dotted with numerous black pycnidia.

L. theobromae, *Phomopsis* spp., and anamorphs of *Botryosphaeria* spp. are the most common infective agents. *Colletotrichum* spp. has also been reported as a causal agent in litchi (Johnson *et al.*, 1998).

While no studies have established how the fungi infect litchi and longan, the main modes of infection are probably through the cut surface of the fruit pedicel or endophytic colonization of the fruit pedicel and stem-end. *Phomopsis*, *Botryosphaeria* and *Colletotrichum* have been isolated as endophytes from litchi and longan stems (Johnson *et al.*, 1998). Management practices outlined for the control of anthracnose, such as reductions in inoculum levels in the orchard, the application of fungicides, and cool storage, provide some control.

Other postharvest diseases

Although the incidence of most postharvest diseases can be minimized by correct handling and prompt marketing, there are a number that can cause significant losses at times (see Plate 53). In litchi, these include sour rot caused by *Geotrichum candidum* (Tandon and Tandon, 1975; Tsai and Hsieh, 1998) and *G. ludwigii* (Tsai and Hsieh, 1998), Rhizopus rot caused by *Rhizopus stolonifer* and *R. oryzae* (Sato, 1995), white mould caused by *Cylindrocladium scoparium* (Chi, 2000), and rots caused by *Alternaria* spp., *Aspergillus* spp., *Cladosporium* spp., *Curvularia* spp., *Fusarium* spp., *Mycosphaerella* spp., *Pestalotiopsis* spp., *Phoma epicoccina*, *Stemphylium* spp. and *Trichothecium* spp. (Prasad and Bilgrami, 1973; Snowdon, 1990; Coates *et al.*, 1994; Kooariyakul and Sardsud, 1998). Green and blue moulds initiated by *Penicillium* spp. can also be a problem in litchi following sulphur treatment. *Penicillium brevicompactum* (A.D. Hocking, Australia, personal communication, 1997) and *P. chrysogenum* (Chi, 2000) are associated with green mould in Australia and China, respectively.

In longan, the fungi commonly associated with decay include *Lasiodiplodia theobromae*, *Pestalotiopsis* spp., *Cladosporium* spp. and *Fusarium* spp., with *Aspergillus niger* isolated occasionally (Sardsud *et al.*, 1994). Yeasts can also be associated with decay in litchi (Roth, 1963; Fitzell and Coates, 1995) and longan (Tongdee, 1997).

Cool storage is essential for the effective management of postharvest diseases, although fruit will deteriorate rapidly when they are returned to ambient temperatures. Treatment with sulphur (Lonsdale and Kremer-Kohne, 1991; Coates *et al.*, 1993; Duvenhage, 1993; Tongdee, 1994, 1997) or fungicides (Scott *et al.*, 1982; Huang and Scott, 1985) controls some pathogens, although there are restrictions on their use in many countries. Furthermore, sulphur-treated fruit are more susceptible to colonization by *Penicillium* (Zauberman *et al.*, 1991; Coates *et al.*, 1993).

For pathogens that initially infect fruit in the field, maintenance of orchard hygiene and the application of preharvest fungicides, such as those used for anthracnose, offer some control. For pathogens that primarily infect after harvest through wounds in the fruit skin, careful handling to avoid injury is important.

Conclusions

Considerable scope exists for research into the cause, epidemiology and management of diseases in these two crops. Gaps exist in our knowledge of the major diseases in Asia, such as downy blight of litchi and witches' broom of longan. We also lack a basic understanding of the causal agents of the various diebacks and root rots. With improved understanding of pathogen biology, the formulation of more effective disease management strategies should be possible.

Options for disease management in the future include a range of strategies. We will probably see a shift towards the use of environmentally benign fungicides, coupled with more targeted chemical applications that use disease forecasting models and improved application technology. Studies of natural defence systems in these crops may lead to opportunities for enhancement or induction of antifungal compounds using defence elicitors. Nutrition has been shown to be important in disease resistance in other crops such as avocado (Willingham *et al.*, 2001), particularly concentrations of nitrogen and calcium in fruit tissue. The role of these nutrients in the development of fruit diseases in litchi and longan should be investigated. While there has been very little success in biological control of plant diseases, this strategy, when combined with other control methods, should not be discounted. The development of more disease-resistant cultivars through conventional breeding or biotechnology is another option for these industries.

Acknowledgements

We thank Ken Pegg and Roger Shivas for reviewing the manuscript.

References

Alfieri, S.A., Jr, Langdon, K.R., Kimbrough, J.W., El-Gholl, N.E. and Wehlburg, C. (1994) *Diseases and Disorders of Plants in Florida.* Bulletin No. 14, Division of Plant Industry, Florida Department of Agriculture USA, Consumer Services, Gainesville, Florida, 1114 pp.

Ann, P.J. (2001) Control of plant diseases with the non-pesticide compound: phosphorous acid. *Plant Pathology Bulletin* 10 (4), 147–154.

Ann, P.J. and Ko, W.H. (1984) Blossom blight of litchi in Taiwan caused by *Peronophythora litchii. Plant Disease* 68, 826.

Bagshaw, J.S., Underhill, S.J.R. and Fitzell, R.D. (1995) Lychees: disorders and injuries. In: Coates, L., Cooke, T., Persley, D., Beattie, B., Wade, N. and Ridgway, R. (eds) *Postharvest Diseases of Horticultural Crops*, Vol. 2, *Tropical Fruit.* Department of Primary Industries, Queensland, Australia, pp. 43–44.

Bhavakul, K., Tospol, M., Rakvitayasart, V. and Suwanketnikom, S. (1998) Studies on phytophthora leaf blight of longan: symptoms, causal organism and chemical control. In: *Proceedings of a Seminar and Workshop on Longan Producing Technology,* Chiang Mai, Thailand, pp. 62–73.

Brown, B.I., Scott, K.J. and Mayer, D.G. (1984) Control of fruit rots of guava, lychee and custard apple by prochloraz dips. *Singapore Journal of Primary Industries*, 12, 40–49.

Chen, C.C. (1961) A species of *Peronophythora* gen. nov. parasitic on litchi fruit in Taiwan. *Special Publication of the College of Agriculture, National Taiwan University* 10, 1–37.

Chen, J.Y., Li, K.B., Chen, J.Y. and Fan, G.C. (1996) A preliminary study on litchi witches' broom and its relation to longan witches' broom. *Acta Phytopathologica Sinica* 26, 331–335.

Chen, J.Y., Chen, J.Y. and Xu, X.D. (2000) Advances in the research of longan witches' broom disease. *Acta Horticulturae* 558, 413–416.

Chen, Y.S. and Zhang, S.S. (2003) Studies on the etiology and occurrence factors of longan leaf blight. *South China Fruits* 32 (2), 26–27.

Chi, P. (2000) *Records of Fungal Diseases of Fruits in Guangdong Province.* China Agricultural Press, Beijing, China, 241 pp.

Chi, P.K., Pang, S.P. and Liu, R. (1984) On downy blight of *Litchi chinensis* Sonn. 1. The pathogen and its infection process. *Acta Phytopathologica Sinica* 14, 113–119.

Coates, L.M., Doogan, V.J. and Gardiner, S. (1993) Postharvest disease control in lychees using sulphur dioxide treatment. In: *Proceedings of the Australasian Postharvest Conference.* The University of Queensland, Gatton College, Queensland, Australia, pp. 73–76.

Coates, L.M., Johnson, G.I., Sardsud, U. and Cooke, A.W. (1994) Postharvest diseases of lychee in Australia, and their control. In: Johnson, G.I. and Highley, E. (eds) *Development of Postharvest Handling Technology for Tropical Tree Fruits.* Australian Centre for International Agricultural Research Proceedings No. 58, pp. 68–69.

Coates, L.M., Sangchote, S., Johnson, G.I. and Sittigul, C. (2003) Diseases of longan, lychee and rambutan. In: Ploetz, R.C. (ed.) *Diseases of Tropical Fruit Crops.* CAB International, Wallingford, UK, pp. 307–325.

Cohen, M. (1955) Clitocybe rot of lychee trees. *Proceedings of the Florida State Horticultural Society* 68, 329–332.

Cohn, E. and Duncan, L.W. (1990) Nematode parasites of subtropical and tropical fruit trees. In: Luc, M., Sikora, R.A. and Bridge, J. (eds) *Plant Parasitic Nematodes in Subtropical and Tropical Agriculture.* CAB International, Wallingford, UK, pp. 347–362.

Cooke, A.W. and Coates, L.M. (2002) Pepper spot: a preharvest disease of lychee caused by *Colletotrichum gloeosporioides. Australasian Plant Pathology* 31, 303–304.

Crane, J.H., Sanford, R.E. and McMillan, R.T., Jr (1997) Control of lychee anthracnose by foliar applications of tebuconazole, mancozeb, and copper hydroxide on 'Mauritius' lychee fruit under south Florida conditions. *Proceedings of the Florida State Horticultural Society* 110, 149–152.

Darvas, J.M. (1992) Armillaria rot of litchi trees (*Litchi chinensis*) in South Africa. *Yearbook of the South African Litchi Growers' Association* 4, 2–4.

Drew, H. and Drew, J. (1999) Lychee pepper spot update: November 1999. In: Menzel, C.M., Waite, G.K. and Bronson, R. (eds) *Proceedings of the Fifth National Lychee Conference.* Australian Lychee Growers' Association, Twin Waters, Queensland, p. 24.

Duvenhage, J.A. (1993) Control of post-harvest decay and browning of litchi fruit by sodium metabisulphite and low pH dips. *Yearbook of the South African Litchi Growers' Association* 5, 31–32.

Dyko, B.J. and Mordue, J.E.M. (1979) Colletotrichum acutatum: *CMI Descriptions of Pathogenic Fungi and Bacteria* No. 630. Commonwealth Mycological Institute, Kew, UK.

Fitzell, R.D. and Coates, L.M. (1995) Lychees: diseases. In: Coates, L., Cooke, T., Persley, D., Beattie, B., Wade, N. and Ridgway, R. (eds) *Postharvest Diseases of Horticultural Crops,* Vol. 2, *Tropical Fruit.* Department of Primary Industries, Queensland, Australia, pp. 41–42.

Grgurinovic, C.A. (1997) *Larger Fungi of South Australia.* The Botanic Gardens of Adelaide and State Herbarium, Adelaide, Australia, 725 pp.

Guerber, J.C. and Correll, J.C. (2001) Morphological description of *Glomerella acutata,* the teleomorph of *Colletotrichum acutatum. Mycologia* 93, 216–229.

Gupta, J.H. (1992) Chemical control of algal rust of litchi caused by *Cephaleuros virescens* Kunze. *Progressive Horticulture* 24, 109–110.

Hall, G. (1989) Peronophythora litchii. *CMI Descriptions of Pathogenic Fungi and Bacteria* No. 974. Commonwealth Mycological Institute, Kew, UK, 2 pp.

He, D.P., Zeng, M.L., Zhou, B.P., Lin, S.X., Peng, J.X., Li, J.Y. and Huang, W.M. (2001) Occurrence, cause and control of longan witches' broom in Guangdong. *Acta Horticulturae* 558, 407–412.

Hu, Y. (1999) *Flora Fungorum Sinicorum,* Vol. 11: *Meliolales (II).* Science Press, Beijing, China, 252 pp.

Huang, P.Y. and Scott, K.J. (1985) Control of rotting and browning of litchi fruit after harvest at ambient temperatures in China. *Tropical Agriculture* 62, 2–4.

Johnson, G.I. (1989) Lychee disease control. In: Campbell, T. and Sing, N. (eds) *Proceedings of the Second National Lychee Seminar.* Australian Lychee Growers' Association, Cairns, Queensland, Australia, pp. 90–93.

Johnson, G.I. (1990) Postharvest disease assessment: rating systems. In: Beattie, B.B. (ed.) *Proceedings of the Australasian Postharvest Conference 1989.* Australian Institute of Agricultural Science, Occasional Publication No. 46, pp. 107–117.

Johnson, G.I., Joyce, D.C. and Gosbee, M.J. (1998) *Botryosphaeria* (anamorphs *Fusicoccum* and *Dothiorella*), *Diaporthe* (anamorphs *Phomopsis* spp.) and *Lasiodiplodia*: infection and defence. In: Johnson, G.I., Highley, E. and Joyce, D.C. (eds) *Disease Resistance in Fruit.* Australian Centre for International Agricultural Research Proceedings No. 80, pp. 46–52.

Johnson, G.I., Cooke, A.W. and Sardsud, U. (2002) Postharvest disease control in lychee. *Acta Horticulturae* 575, 705–715.

Kang, M.S. and Singh, I. (1991) Die-back and leaf blight of litchi incited by *Pestalotiopsis mangiferae* (Henn.) Steyaert. *Plant Disease Research* 6, 103–104.

Kao, C.W. and Leu, L.S. (1980) Sporangium germination of *Peronophythora litchii,* the causal organism of litchi downy blight. *Mycologia* 72, 737–748.

Ko, W.H., Chang, H.S., Su, H.J., Chen, C.C. and Leu, L.S. (1978) Peronophythoraceae, a new family of Peronosporales. *Mycologia* 70, 380–384.

Kooariyakul, S. and Sardsud, V. (1998) Bagging of lychee fruit to reduce postharvest disease. In: Coates, L.M., Hofman, P.J. and Johnson, G.I. (eds) *Disease Control and Storage Life Extension in Fruit.* Australian Centre for International Agricultural Research Proceedings No. 81, pp. 92–100.

Li, J.H. and Li, J. (1998) *New Cultivation Techniques for Litchi.* Fujian Science and Technology Press, Fuzhou, China, pp. 102–103.

Li, J.S. (1997) Diseases and pests and their control. In: Zhang, Z.W., Wang, B.Q., Li, J.S., Yuan, P.Y. and Qiu, Y.P. (eds) *Litchi Pictorial Narration of Cultivation.* Pomology Research Institute, Guangdong Academy of Agricultural Sciences, Guangzhou, China, pp. 140–161.

Lim, T.K. (1994) Algal leaf spot. In: Ploetz, R.C., Zentmyer, G.A., Nishijima, W.T., Rohrbach, K.G. and Ohr, H.D. (eds) *Compendium of Tropical Fruit Diseases.* American Phytopathological Society Press, St Paul, Minnesota, p. 41.

Lim, T.K. and Khoo, K.C. (1985) *Diseases and Disorders of Mango in Malaysia.* Tropical Press, Kuala Lumpur, Malaysia, 101 pp.

Lim, T.K. and Sangchote, S. (2003) Diseases of durian. In: Ploetz, R.C. (ed.) *Diseases of Tropical Fruit Crops.* CAB International, Wallingford, UK, pp. 241–251.

Lonsdale, J.H. and Kremer-Kohne, S. (1991) Maintaining market quality of fresh litchis during storage. Part 2. Control of postharvest decay. *Yearbook of the South African Litchi Growers' Association* 3, 18–20.

Manicom, B.Q. (1995) Litchi dieback. *Yearbook of the South African Litchi Growers' Association* 7, 3–4.

McMillan, R.T., Jr (1994a) Epidemiology and control of anthracnose of lychee. *Proceedings of the Florida State Horticultural Society* 107, 345–346.

McMillan, R.T., Jr (1994b) Diseases of *Litchi chinensis* in south Florida. *Proceedings of the Florida State Horticultural Society* 107, 360–362.

Menzel, C.M., Watson, B.J. and Simpson, D.R. (1988) The lychee in Australia. *Queensland Agricultural Journal* 114, 19–26.

Menzel, C.M., Watson, B.J. and Simpson, D.R. (1990) Longan. In: Bose, T.K. and Mitra, S.K. (eds) *Fruits: Tropical and Subtropical.* Naya Prokash, Calcutta, India, pp. 521–545.

Menzel, C., Bagshaw, J., Campbell, T., Greer, N., Noller, J., Olesen, T., Waite, G., Kernot, I., Chapman, L. and Rigden, P. (2002) *Lychee Information Kit.* Agrilink Series QAL9912, Department of Primary Industries, Queensland, Australia.

Milne, D.L., de Villiers, E.A. and Holtzhausen, L.C. (1971) Litchi tree decline caused by nematodes. *Phytophylactica* 2, 37–44.

Mishra, B., Prakash, O. and Misra, A.P. (1974) A serious leaf spot disease of litchi caused by *Cephaleuros virescens. Indian Journal of Mycology and Plant Pathology* 3, 219–220.

Mordue, J.E.M. (1971) Glomerella cingulata: *CMI Descriptions of Pathogenic Fungi and Bacteria* No. 315. Commonwealth Mycological Institute, Kew, UK. 2 pp.

Nakasone, H.Y. and Paull, R.E. (1998) *Tropical Fruits.* CAB International, Wallingford, UK, 445 pp.

Ou, Z.J., Deng, W.S. and Wu, C.T. (1999) Experiments to control litchi downy mildew disease using 80% mancozeb wettable powder. *China Fruits* 3, 32.

Ploetz, R.C. (2003) *Diseases of Tropical Fruit Crops.* CAB International, Wallingford, UK, 544 pp.

Prasad, S.S. and Bilgrami, R.S. (1973) Investigations on diseases of 'litchi'. III. Fruit rots and their control by post-harvest treatments. *Indian Phytopathology* 26, 523–527.

Roth, G. (1963) Post harvest decay of litchi fruit (*Litchi chinensis* Sonn.). *Technical Communications of the Citrus and Subtropical Fruit Research Institute, Nelspruit, South Africa* 11, 1–16.

Sardsud, V., Sardsud, U., Sittigul, C. and Chaiwangsri, T. (1994) Effects of postfumigation washing treatments and storage temperature on disease development in fresh longan. In: Johnson, G.I. and Highley, E. (eds) *Development of Postharvest Handling Technology for Tropical Tree Fruits.* Australian Centre for International Agricultural Research Proceedings No. 58, pp. 77–79.

Sarindu, N. (1993) Research on plant mycoplasma diseases in Thailand. In: *Plant Pathology and Microbiology.* Department of Agriculture, Bangkok, Thailand, pp. 123–143.

Sato, T. (1995) Litchi diseases. In: Anonymous (ed.) *Diseases of Tropical Trees.* Association for International Cooperation of Agriculture & Forestry (AICAF), Tokyo, Japan, pp. 95–96.

Scott, K.J., Brown, B.I., Chaplin, G.R., Wilcox, M.E. and Bain, J.M. (1982) The control of rotting and browning of litchi fruit by hot benomyl and plastic film. *Scientia Horticulturae* 16, 253–262.

Shaw, G.C. and Kile, G.A. (1991) *Armillaria Root Disease*. Agricultural Handbook No. 691. Forest Service, US Department of Agriculture, Washington, DC, 233 pp.

Snowdon, A.L. (1990) *A Colour Atlas of Post-Harvest Diseases and Disorders of Fruits and Vegetables*, Vol. 1: *General Introduction and Fruits*. Wolfe Scientific, London, 302 pp.

So, V. and Zee, S.Y. (1972) A new virus of longan (*Euphoria longana* Lam.) in Hong Kong. *PANS* 18 (3), 283–285.

Stamps, D.J. (1985) Phytophthora palmivora: *CMI Descriptions of Pathogenic Fungi and Bacteria No. 831*. Commonwealth Mycological Institute, Kew, UK, 2 pp.

Tandon, J.M.P. and Tandon, R.N. (1975) Rot of fruits of litchi (*Litchi chinensis*) in marketing processes. *Indian Phytopathology* 28, 530–531.

Tongdee, S.C. (1986) Sulfur dioxide fumigation in the storage of fresh litchi. *Thailand Institute for Scientific and Technological Research, Research News* (March).

Tongdee, S.C. (1994) Sulphur dioxide fumigation in postharvest handling of fresh longan and lychee for export. In: Champ, B.R., Highley, E. and Johnson, G.I. (eds) *Postharvest Handling of Tropical Fruits*. Australian Centre for International Agricultural Research Proceedings No. 50, pp. 186–195.

Tongdee, S.C. (1997) Longan. In: Mitra, S.K. (ed.) *Postharvest Physiology and Storage of Tropical and Subtropical Fruits*. CAB International, Wallingford, UK, pp. 335–345.

Trung, H.M. (1999) Lychee production in Vietnam: prospects and problems. In: Menzel, C.M., Waite, G.K. and Bronson, R. (eds) *Proceedings of the Fifth National Lychee Conference*. Australian Lychee Growers' Association, Twin Waters, Queensland, Australia, pp. 83–87.

Trung, H.M., Vien, N.V., Benyon, F.H.L., Nguyen, H.T., Summerell, B.A. and Burgess, L.W. (1999) Lychee decline in Northern Vietnam. *Handbook of the Twelfth Biennial Conference of the Australasian Plant Pathology Society*, Canberra, Australia, p. 52.

Tsai, J.N. and Hsieh, W.H. (1998) Occurrence of litchi sour rot and characteristics of the pathogens *Geotrichum candidum* and *G. ludwigii*. *Plant Pathology Bulletin* 7 (1), 10–18.

Vien, N.V., Benyon, F.H.L., Trung, H.M., Summerell, B.A., Van, N.K. and Burgess, L.W. (2001) First record of *Peronophythora litchii* on litchi fruit in Vietnam. *Australasian Plant Pathology* 30, 287–288.

Visarathanonth, N. (1999) *Diseases of Tropical Fruit and Their Control*, Vol. 2. J. Film Process Co., Bangkok, Thailand, 144 pp.

Visitpanich, J., Sittigul, C. and Sardsud, V. (1996) Longan leaf curl symptoms in Chiang Mai and Lam Phun. *Journal of Agriculture* 12 (3), 203–218.

Visitpanich, J., Sittigul, C., Sardsud, V., Chanbang, Y., Chansri, P. and Aksorntong, P. (1999) Determination of the causal agents of decline, witches' broom and sudden death symptoms of longan and their control. *Final Report, Thailand Research Fund Project*, Department of Plant Pathology, Chiang Mai University, Chiang Mai, Thailand, 288 pp.

Visitpanich, J., Sittigul, C. and Chanbang, Y. (2000) Longan leaf blight and fruit drop. *House Agricultural Magazine* (*Bangkok, Thailand*) 24 (1), 144–148.

Willingham, S.L., Cooke, A.W., Coates, L.M. and Pegg, K.G. (2000) Pepper spot: a new preharvest *Colletotrichum* disease of avocado cv. Hass. *Australasian Plant Pathology* 29, 151.

Willingham, S.L., Pegg, K.G., Cooke, A.W., Coates, L.M., Langdon, P.W.B. and Dean, J.R. (2001). Rootstock influences postharvest anthracnose development in 'Hass' avocado. *Australian Journal of Agricultural Research* 52, 1017–1022.

Yin, Y., Gao, X. and Feng, Z. (1994a) Investigations of parasitic nematodes on lychee in Guangdong province. *Journal of South China Agricultural University* 15 (3), 22–27.

Yin, Y., Yang, B. and Wang Q. (1994b) Identification of root-knot nematodes in thirty crops. *Journal of South China Agricultural University* 15 (1), 22–26.

Yip, H.Y. (1997) Lychee fruit pepper spot. *Living Lychee* 11, 6.

Zauberman, G., Ronen, R., Akerman, M., Weksler, A., Rot, I. and Fuchs, Y. (1991) Postharvest retention of the red colour of litchi fruit pericarp. *Scientia Horticulturae* 47, 89–97.

Zhang, Q. (1999) Investigation of the occurrence of longan witches' broom disease and its control. *South China Fruits* 28 (1), 24.

14 Harvesting and Storage

Deirdre M. Holcroft[1], Hetong Lin[2] and Saichol Ketsa[3]
[1]Dole Fresh Vegetables, PO Box 1759, Salinas, CA 93902, USA; [2]College of Food Science and Technology, Fujian Agriculture and Forestry University, Fuzhou 350002, China; [3]Department of Horticulture, Kasetsart University, Chatuchak, Bangkok 10900, Thailand

Introduction

Litchi and longan produce sweet, succulent fruit with a very aromatic aril or flesh. The indehiscent pericarp is orange-red at maturity in litchi, but yellow-brown in longan. Longan also tends to be smaller and rounder in shape, and the aril less juicy and acidic (Zee *et al.*, 1998). These fruit are highly prized in Asia, which accounts for more than 90% of the total world production. They are less well-known or appreciated elsewhere, despite their pleasant flavour (see Plates 69–72).

The perishability of the fruit limits marketing of these crops, and especially of exports to countries some distance from the main area of production. The skin of both litchi and longan turns brown soon after harvest, or rots after several days if the fruit are not handled correctly. In litchi, loss of the attractive red colour reduces its market appeal. Although the initial pericarp browning is primarily a cosmetic problem, it reduces the commercial value of the fruit (Joubert, 1986; Jiang, 1999a). Further browning is generally associated with the loss of eating quality.

In recent years, producing countries have identified new market niches for these distinctive fruit, transforming them from mainly domestic commodities to significant exports (Tongdee, 1997). Since the fruit are produced mainly in the tropical and subtropical regions of the world, far from these emerging markets, careful handling and transportation is required to ensure that they reach consumers in good condition. This has provided the impetus for research into improving postharvest handling technology.

Fruit morphology

Both litchi and longan produce conical, heart-shaped or spherical drupes. In litchi, the white, edible pulp or aril covers a smooth, brown seed and is surrounded by a thin, leathery, indehiscent pericarp. During fruit development, the pericarp is green and can photosynthesize. As the fruit mature, anthocyanins are synthesized in the pericarp, producing pink or red fruit depending on the stage of maturity and the cultivar (Joubert, 1986; Underhill and Critchley, 1992; Nacif *et al.*, 2001).

Longan fruit have a thin, leathery, smooth and indehiscent pericarp and contain a relatively large, black or brown seed at maturity (Wong and Ketsa, 1991). The edible portion of the fruit is a fleshy, translucent-white aril. The pericarp is green during initial development and yellow-brown, green-brown, russet, red-brown or black-brown at later stages, depending on the cultivar and the stage of maturity (Lin *et al.*, 2001c).

Respiration and ethylene production

Akamine and Goo (1973) monitored litchi during development and found that respiration and ethylene production declined continuously, typical of non-climacteric fruit. Immature green fruit produce more ethylene than mature red fruit (Chan *et al.*, 1998). Ethylene production by mature fruit is less than 0.5 μl/kg/h at 25°C (Akamine and Goo, 1973; Chan *et al.*, 1998) unless they start to decay (Tongdee *et al.*, 1982). Chan *et al.* (1998) showed that storage of fruit at 4 or 10°C stimulated ethylene production when they were subsequently transferred to 25°C, and that this effect was more pronounced in green than in pink or red fruit.

The wide range of published values for respiration rates of litchi can be attributed to cultivar differences (Tongdee *et al.*, 1982) as well as the continuing decline in respiration during storage. Paull and Chen (1987a) showed that respiration of 'Chenzi' ('Brewster') declined from 56 to 21 ml CO_2/kg/h (from 103 to 39 mg/kg/h) during 8 days of storage at 22°C, and Nagar (1994) showed that the respiration of 'Calcutta' declined from 20 to 10 ml CO_2/kg/h during 6 days of storage at 25°C. Zhang and Quantick (2000) measured an initial respiration rate of 107 ml CO_2/kg/h (200 mg CO_2/kg/h) in 'Huaizhi' ('Wai Chee'), but rates decreased during storage at 1 and 5°C. Kader (2000) reported generic respiration rates of 5–8 ml CO_2/kg/h at 5°C, 10–15 ml CO_2/kg/h at 10°C, and 25–40 ml CO_2/kg/h at 20°C.

Longans are also non-climacteric (Shi, 1990). The respiration rate of fruit stored at between 2 and 8°C declined continuously (Hong *et al.*, 1984; Shi, 1990; Zhou *et al.*, 1997; He *et al.*, 1997; Tongdee, 1997; Lin *et al.*, 1997, 2001b, 2002d,e; Li *et al.*, 1999). A later rapid increase in respiration was probably due to decay (Lin *et al.*, 2002d,e). Zhang *et al.* (2002) measured respiration rates of 8 mg CO_2/kg/h at 3°C, 12 mg CO_2/kg/h at 7°C, 54 mg CO_2/kg/h at 10°C, and 86 mg CO_2/kg/h at 25°C.

Longans produce relatively low concentrations of ethylene after harvest (< 2.3 μl/kg/h at 20°C); however, higher rates (28.3 μl/kg/h) are associated with decay (Shi, 1990; Zhou *et al.*, 1997). Low-temperature storage reduces ethylene production. Ethylene production of 'Shixia' fruit was 0.1 μl/kg/h at 0°C, 0.6 μl/kg/h at 4°C, and 1.1 μl/kg/h at 8°C (Zhou *et al.*, 1997). The highest ethylene production in 'Fuyan' fruit stored at 4 and 30°C was 2.3 and 3.9 μl/kg/h, respectively (Pan *et al.*, 1996).

Aril composition

In litchi, total soluble solids (TSS) increase during ripening, reaching 13–20% by harvest (Paull *et al.*, 1984; Paull and Chen 1987a; Nagar, 1994; Sonali *et al.*, 2001). The major sugars in ripe fruit are sucrose, fructose and glucose, with the relative ratios depending on the cultivar, stage of maturity and invertase activities (Chan *et al.*, 1975; Paull *et al.*, 1984). Titratable acidity (TA) decreases during development, while pH increases (Paull *et al.*, 1984). Succinic acid is the principal organic acid present during the greater part of fruit development, whereas malic acid dominates at maturity (Chan and Kwok, 1974; Paull *et al.*, 1984; Paull and Chen, 1987a). Fresh mature fruit are a significant source of ascorbic acid (40–100 mg/100 g) but concentrations decrease during storage, regardless of storage conditions (Singh and Abidi, 1986; Paull and Chen, 1987a). Paull *et al.* (1984) found that aril phenols decreased early in development and remained below 1 mg/100 g. Juice extracted from arils contained increasing amounts of acetaldehyde and ethanol when harvested later in the season, suggesting that fermentation can begin before the fruit are harvested (Pesis *et al.*, 2002).

Total soluble solids, total sugars and ascorbic acid contents increase, and TA decreases during longan fruit ripening (Chattopadhyay and Ghosh, 1991; Ke *et al.*, 1992; Huang, 1995; Lin, 2002). The main sugars present in the pulp are sucrose, fructose and glucose (Paull and Chen, 1987b), with sucrose concentrations in 'Fuyan' pulp being 2.3 times higher than those of the reducing sugars (Lin *et al.*, 2002e). The relative ratio of these sugars varies according to invertase activity, stage of maturity and cultivar (Huang, 1995; Lin, 2002). The main organic acids in the pulp are succinic, malic and citric acids (10:5:1) (Paull and Chen, 1987b). The pulp tends to have a relatively high sugar and low acid content (Han *et al.*, 2001; Lin *et al.*,

2002d). The TSS:TA ratios in 'Youtanben' and 'Fuyan' pulp at harvest were 399:1 and 88:1, respectively (Lin *et al.*, 2001b).

Total soluble solids of longan pulp decrease gradually after harvest until the later stages of storage, when the measure increases. This increase may be associated with the degradation of cell wall substances such as pectin, cellulose and other polysaccharides (Lin *et al.*, 2002d,e). Titratable acidity in the pulp usually decreases rapidly after harvest (Han *et al.*, 2001; Lin *et al.*, 2002d), although increases in fruit infected with pulp sour-rot disease (*Geotrichum candidum*) have been measured in 'Fuyan' during long-term storage at 20°C (Liu, 1991). Ascorbic acid concentrations of 69 mg/100 g have been measured in longan pulp (Tongdee, 1997) but these values decline during storage, even at 4°C (Lin *et al.*, 2002e). Increases in the activity of pectin methylesterase (PME) and polygalacturonase (PG) have been measured as the fruit mature and the pulp softens (Lu *et al.*, 1992; Lin, 2002). The pulp may harden in over-mature fruit as water is lost and lignification occurs (Lin, 2002).

Pericarp structure and composition

The mature litchi pericarp is 1–3 mm thick and consists of three distinct layers. The outermost epicarp has a continuous cuticle 1–3 μm thick, a single epidermal layer and sub-epidermal schlerenchyma. The mesocarp is composed of parenchyma and contains chlorophyll and most of the anthocyanins. The innermost endocarp is membranous and comprises small, thin-walled, unsuberized epidermal cells (Joubert and van Lelyveld, 1975; Underhill and Critchley, 1992; Nacif *et al.*, 2001). Rapid water loss occurs during storage due to the high permeability of the cuticle to water, damage to the cuticle, and the presence of lenticels. Underhill and Simons (1993) found that micro-cracks 20–100 μm wide in the pericarp surface were present at the time of harvest, but increased in density after 12 h and extended into the mesocarp.

The concentration of chlorophyll in the skin decreases at the beginning of litchi fruit maturation, coinciding with the synthesis of anthocyanin, which accounts for the red pigmentation of the pericarp. Cyanidin 3-rutinoside, cyanidin 3-glucoside, cyanidin 3-galactoside, malvidin 3-acetylglucoside, pelargonidin 3-glycosides and pelargonidin 3,5-diglucoside have been isolated from the pericarp (Prasad and Jha, 1978; Lee and Wicker, 1991a,b; Zhang *et al.*, 2000). Sarni-Manchado *et al.* (2000) used a combination of techniques including mass spectrometry and NMR spectroscopy to identify cyanidin 3-rutinoside as the major anthocyanin (0.43 mg/g), and cyanidin glucoside, probably cyanidin 3-O-glucoside (0.05 mg/g), as the minor pigment in 'Kwai Mi'.

The concentration of phenolics in the pericarp (1.4 mg/100 g) is higher than in the aril (0.5 mg/100 g), and varies with cultivar (Jaiswal *et al.*, 1986). Sarni-Manchado *et al.* (2000) identified several phenols in litchi pericarp. The flavonols quercetin 3-rutinoside (rutin) and quercetin glucoside had a combined concentration of 0.27 mg/g in 'Kwai Mi'. These compounds are ortho-diphenols and are therefore good substrates for the browning reaction. Epicatechin (1.72 mg/g), procyanidin A2 (0.68 mg/g), trimers of procyanidins (0.40 mg/g) and condensed tannins (4.02 mg/g) accounted for 90% of the phenols in 'Kwai Mi' litchi. These flavanols are the basis of the browning reaction (Sarni-Manchado *et al.*, 2000).

The mature longan pericarp consists of three layers: exocarp, mesocarp and endocarp (Pan, 1994; Qu *et al.*, 2001; Lin *et al.*, 2002a,c). The surface of the exocarp has a thin, suberized layer with many micro-cracks, lenticels, protuberances and setae (Lin *et al.*, 2002a,c). The mesocarp consists of schlerenchyma and parenchyma with large intercellular spaces (Pan, 1994; Qu *et al.*, 2001; Lin *et al.*, 2002a,c). The innermost endocarp is made up of small, thin-walled, unsuberized epidermal cells (Lin *et al.*, 2002a,c). Scanning election microscopy reveals that the surface of the endocarp has a thin cuticle and wax layer at maturity (Pan, 1994; Qu *et al.*, 2001; Lin *et al.*, 2002a). Degradation of chlorophyll and synthesis of carotenoids accounts for some of the changes in colour during longan fruit ripening (Ke *et al.*, 1992). The brown pericarp at maturity is also associated with the formation of a suberin layer in the exocarp (Lin *et al.*, 2002a,c).

Fruit of 'Fengliwei', 'Dongbi', 'Youtanben', 'Wulongling', 'Honghezi' and 'Jiaoyan' have thicker pericarps than those of 'Shuizhang',

'Chike', 'Fuyan' and 'Pumingan'. The thicker pericarps have many protuberances and setae in the exocarp surface, a thicker suberized layer in the exocarp, a higher proportion of closely grouped sclereid cells distributed in the mesocarp, well-developed vascular tissue in the mesocarp, and a thicker cuticle and wax layer in the endocarp. These characteristics were found to be associated with fewer diseases and better storage than cultivars with thin pericarps (Lin *et al.*, 2001a, 2002a,c,e).

Pericarp browning

Pericarp browning is one of the most important problems associated with both crops (Prapaipong and Rakariyatham, 1990; Underhill *et al.*, 1997; Jiang *et al.*, 2002). Browning has been attributed to desiccation (Scott *et al.*, 1982; Underhill and Simons, 1993; Lin *et al.*, 2002a,d), chilling (Tongdee *et al.*, 1982; McGuire, 1998), disease (Huang and Scott, 1985; Jiang *et al.*, 2002), heat stress (Wong *et al.*, 1991; Jiang *et al.*, 2002) and senescence (Huang and Wang, 1990). All these factors disrupt cellular compartmentation, allowing polyphenol oxidase (PPO) located in the chloroplasts and other plastids to react with phenolic substrates located in the vacuole, forming brown polymers (Macheix *et al.*, 1990; Underhill and Critchley, 1995; Lin *et al.*, 2002b). Peroxidase enzymes may also be involved in this process (Underhill and Critchley, 1995; Underhill *et al.*, 1997). Desiccation-induced browning begins on the tips of the protuberances of litchi and subsequently spreads across the entire fruit surface, with browning localized in the epicarp and upper mesocarp (Underhill and Critchley, 1995). Chilling injury also causes browning (Tongdee *et al.*, 1982).

Jiang and Fu (1999a) found that water loss from the litchi pericarp was more than 50% after 3 days of storage at 60% RH and 20°C, whereas there was minimal loss from the aril and seed. Hence, selective dehydration of the pericarp occurred with little movement of water between the aril and the pericarp. Eventually, the aril also loses water and the fruit becomes flaccid and bland (Underhill and Critchley, 1993; Underhill and Simons, 1993). The proportion of water lost from fruit increased as RH declined from 95% to 60% in storage at 20°C. Browning severity and loss of red colour were inversely related to RH. In addition, the activity of PPO increased at lower RHs and was associated with a decrease in the concentration of phenols (Jiang and Fu, 1999a,b).

Underhill and Critchley (1992, 1993) found that PPO activity was lower at maturity than during the first 62 days after anthesis and it progressively decreased in the 72 h after harvest at 25°C. Huang *et al.* (1990) found that PPO activity remained low during the first 29 days of storage at 4°C. Lin *et al.* (1988a) measured an increase in PPO during the first 2 days of storage, whereas Zauberman *et al.* (1991) observed few changes in PPO activity during the first 2 days of storage. Since peroxidase is also present in litchi pericarp and increases during storage, especially at low RHs (Lin *et al.*, 1988b; Huang *et al.*, 1990; Jiang and Fu, 1999a), browning is probably due to the activity of both enzymes (Underhill and Critchley, 1995).

While the disappearance of the red pigments occurs at the same time as browning, the pericarp of litchi contains many phenols, and these are better substrates for PPO than are anthocyanins (Sarni-Manchado *et al.*, 2000). However, when ortho-diphenolic compounds are oxidized to ortho-quinones, these quinones are then able to oxidize anthocyanins in a non-enzymatic reaction (Wesche-Ebeling and Montgomery, 1990). Jiang (2000) used an *in vitro* system to demonstrate that PPO extracted from litchi cannot oxidize an anthocyanin extract from the pericarp unless certain phenols are present. Thus, browning can be attributed to the action of PPO on the colourless phenolics, combined with the coupled oxidation of the red anthocyanins.

The structure and colour of anthocyanins is dependent on the environment, particularly cell pH. A rise in pH converts the red flavylium cations into the colourless carbinol or pseudobase. This in turn can form the pale-yellow chalcone, which is more susceptible to further degradation (Brouillard *et al.*, 1997). The colourless and yellow forms of anthocyanin allow expression of a brown background. Loss of cellular compartmentation and desiccation causes an increase in vacuole pH, which affects colour expression of the anthocyanins and their subsequent degradation. The fact that the red colour can be regenerated by acidification indicates that anthocyanins are still present in another form. Hence, not all anthocyanins are involved in the coupled oxidation reaction. The importance of intracellular pH in fixing anthocyanins and reversing the formation of colourless compounds has been recognized, and forms the basis of some postharvest practices (Zauberman *et al.*, 1990, 1991; Underhill *et al.*, 1992b).

Superficial browning of longan occurs within a day of harvest when fruit are stored at 25°C (Sardsud *et al.*, 1994), and it becomes obvious after 5–6 days at ambient temperatures (Cai, 1988). Since fresh longans are yellow-brown (Lin *et al.*, 2001c), the change from light to dark brown occurs more slowly and uniformly than in litchi. Browning severity is sometimes evaluated by measuring the extent of the total browned area on a fruit surface (Zhou *et al.*, 1997; Qu *et al.*, 2001; Lin *et al.*, 2002a,c,e). Mature fruit tend to brown more rapidly than immature fruit (Paull and Chen, 1987b). Pan (1994) observed that the mesocarp was the first to turn brown, followed by the endocarp. PPO activity in the pericarp of 'Shixia' longan held at 28–32°C increased gradually as browning occurred (Zhang *et al.*, 1999). At 4°C, PPO activity in 'Shixia' reached a peak after 30 days and was accompanied by severe browning of the pericarp (Wu *et al.*, 1999). Xu *et al.* (1999) recorded a gradual increase in PPO activity in 'Wulongling' held at 30°C for 6 days. This was followed by a rapid increase after a further 2 days. The PPO activity of these fruit was 6.8 times higher than in freshly harvested fruit.

Chilling injury

Litchis are susceptible to chilling but the symptoms are difficult to distinguish from those of desiccation (Tongdee *et al.*, 1982). Tongdee *et al.* (1982) reported damage to fruit stored for 30 days at 0 or 5°C when packed in punnets and wrapped in polyvinyl chloride (PVC) film. Uniform browning was observed at 0°C, while at 5°C irregular brown patches developed. Rapid breakdown occurred when these fruit were transferred to room temperature. Huang and Wang (1990) reported an increase in browning of fruit wrapped in polyethylene (PE) film and stored at 0–2°C. Different storage times, temperatures and cultivars used in this research prevent generalizations about the susceptibility of fruit to chilling.

Chilling injury is a serious problem in the cold storage of longan, with typical symptoms including brown, water-soaked areas on the pericarp (Zhou *et al.* 1997). Injury may occur in fruit stored at temperatures below 5–7°C for 20 days (Wang, 1998). Variations in susceptibility to chilling have been reported for different cultivars and production regions in China (Wang, 1998). O'Hare and Prasad (1991a) stored 'Biew Kiew' fruit at 0, 5, 10, 15 or 20°C for 12 days and found that fruit stored at 10°C had the best colour. Fruit stored at 5°C had good eating quality but developed mild chilling injury. 'Shixia' fruit stored at 0°C for 14 days developed slight water-soaking or scald spots in the inner pericarp (Zhou *et al.*, 1997). Fruit of 'Kohala' stored at 1.1°C for 15 days, followed by 3 days at 5°C, developed bronze patches on the pericarp, whereas fruit stored at 5°C for 3 days did not (McGuire, 1998). Longans stored at 0°C for 21 days or 1.1°C for 15 days maintained an acceptable texture despite browning of the pericarp (Zhou *et al.*, 1997; McGuire, 1998). Jiang (1999b) reported that the eating quality in 'Wuyan' was not adversely affected by a temperature of 1°C; however, O'Hare and Prasad (1991a) reported that the eating quality of fruit stored at 0°C for 12 days was poor.

Aril breakdown

The arils of litchi and longan break down and become bland when they lose turgidity and translucency, even if the pericarp is sound. Breakdown starts near the pericarp and is more prevalent at the distal end (Qu *et al.*, 2001). This disorder has been observed in stored fruit and in over-mature fruit on trees (R. Paull, Hawaii, 1996, personal communication; Shi, 1990; Pang and Zhang, 1999; Qu *et al.*, 2001; Han *et al.*, 2002). Breakdown of 'Shixia' longan was observed after 7 days at 25°C or 21 days at 8°C (Zhou *et al.*, 1997; Liang *et al.*, 1998). Longan cultivars vary in the severity of this disorder, with 'Chuliang', 'Fuyan' and 'Shuizhang' being more susceptible than 'Shixia', 'Dongbi', 'Wulongling' and 'Fengliwei' (Lin *et al.*, 2001a, 2002d,e; Han *et al.*, 2002).

Aril breakdown can be retarded by cold storage and controlled atmospheres (Zhou *et al.*, 1997; Liang *et al.*, 1998; Han *et al.*, 2002; Lin *et al.*, 2002d,e). It is probably associated with the degradation of the cell wall that is associated with natural fruit senescence (Shi, 1990; Han *et al.*, 2002; Lin *et al.*, 2002d) or pathogens (Liang *et al.*, 1998). However, the exact mechanism is unclear (Lin *et al.*, 2002d) and the relative roles of cell-wall metabolism, fruit senescence and infection require further investigation.

Postharvest pathology

Litchi and longan are very susceptible to postharvest decay caused by bacteria, yeasts and fungi (Roth, 1963; Liu, 1988; Sardsud *et al.*, 1994; Wang, 1998). The causal agents, symptoms and management of postharvest diseases are covered by Lindy Coates *et al.* (Chapter 13, this volume). Control measures include the use of fungicides, irradiation, heat, controlled atmospheres and biological agents (Underhill *et al.*, 1997; Jiang *et al.*, 2002).

Maturity indices

Litchis and longans should be harvested when they are fully mature and ready to eat, since they do not ripen after harvest. General guidelines for harvesting are difficult to establish because of the wide range of cultivars grown (Joubert, 1986; Jiang *et al.*, 2002). Most cultivars lose flavour if they are harvested when over-mature (Wong, 2000). Pulp hardening may also occur (Lin, 2002).

In litchi, maturity can be determined by fruit weight, colour, sugar or total soluble solids (TSS, %), titratable acidity (TA, %), sugar:acid ratio, flavour and days from anthesis. Batten (1989) found that TA and TSS:TA ratio (%) were good indicators of flavour, but that TSS was unreliable. Underhill and Wong (1990) recommended a TSS:TA ratio of 30–40. Fruit with a TSS:TA ratio of greater than 80:1 are considered to be over-mature (Joubert, 1970). In practice, maturity is usually assessed on characteristic fruit colour and taste for each cultivar.

In longan, maturity can be determined by fruit size, pericarp colour, TSS, TA, TSS:TA ratio, flavour and days from anthesis (Wara-Aswapati *et al.* 1994; Lin, 2002). In addition, the inner pericarp becomes lightly netted as the fruit mature (Wong and Ketsa, 1991). Wara-Aswapati *et al.* (1994) recommended TSS as a maturity index and established 15.5–16.0% as a minimum standard. An assessment of both TSS and TA provides a good indicator of flavour for most cultivars. Chattopadhyay and Ghosh (1991) reported that pulp TSS, total sugars and ascorbic acid in mature fruit were highest, while TA was lowest. As with litchi, maturity is usually assessed on the basis of fruit colour and taste (Jiang *et al.*, 2002; Lin, 2002).

Harvesting, grading and packing

Fruit are generally picked using ladders, or cherry pickers for larger trees (Figs 14.1 and 14.2). Fruit are harvested by breaking or cutting whole panicles, sometimes with a leafy branch attached, or individual fruit are cut or twisted from the panicles and placed in plastic crates

Fig. 14.1. Ladders are used to harvest litchis and longans growing on small trees (from Menzel *et al.*, 2002, with permission).

or bamboo baskets (Figs 14.3–14.6). Since mechanical damage to the fruit increases desiccation and decay, care should be taken during harvesting (Plates 59 and 60). Fruit should be harvested early in the day to minimize water loss and fruit heating from the sun. Harvesting

Fig. 14.2. Fruit growing on large trees need to be harvested using cherry pickers or similar devices. This increases the cost of harvesting the crop (from Menzel *et al.*, 2002, with permission).

Fig. 14.3. Picking litchis in Guangdong, southern China.

Fig. 14.4. Sorting litchis in southern China (photograph courtesy of Brian Brown).

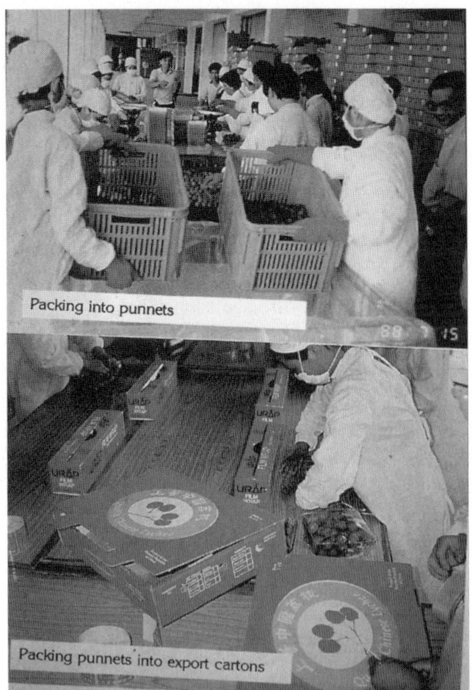

Fig. 14.5. Packing litchis in southern China (photograph courtesy of Brian Brown).

Fig. 14.6. Plastic crates are used to transport litchis from the field to the packing-house in Australia (photograph courtesy of Christopher Menzel).

during the hottest part of the day or on rainy days should be avoided. All fruit on a tree are usually harvested at once unless there is a great range in maturity. Shading of harvested fruit, rapid transfer of fruit from the orchard to the packing-house and minimal delays between harvesting and cooling all contribute to postharvest quality (Tongdee, 1997; Olesen, 2001; Bryant and Olesen, 2002; Lin, 2002).

Sorting and grading of fruit after harvest should take place in shaded and well-ventilated areas or in temperature-controlled packing-houses (Plates 61–63 and 65–68). The panicles are split into subpanicles and the leaves and twigs are removed. Fruit that are small, immature, overripe, misshapen, diseased, damaged by insects or split are discarded (Figs 14.7–14.9). Fruit attached to the stalks are reported to remain fresher than fruit that have been detached (Watson, 1984; Lin *et al.*, 2002b); however, Pornchaloempong *et al.* (1997) found that litchi fruit cut from panicles kept better than fruit that remained attached to panicles. Fruit should be graded to suit their intended markets (Lin *et al.*, 2001c). In addition to fruit weight, in Thailand, grading classifications also specify freshness, length of stalk, colour of the pericarp, freedom from fruit blemishes, rots or insect damage, and percentage of loose fruit (Tongdee, 1994). Fruit are packed in plastic crates, fibreboard cartons or polystyrene boxes lined with polythene for export or long-distance transportation (Figs 14.10 and 14.11). In Asia, bamboo baskets are widely used as packages for local markets but they are easily crushed, which causes damage to the fruit during storage and transport (Lin *et al.*, 2001c).

Fig. 14.7. Sorting and grading litchis in southern Queensland, Australia (photograph courtesy of Neil Greer).

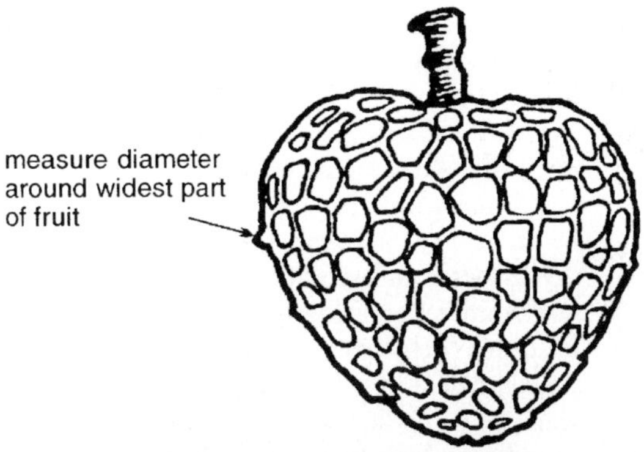

Fig. 14.8. Litchis in Australia must meet requirements for minimum fruit diameter (from Menzel *et al.*, 2002, with permission).

Fig. 14.9. Checking the quality of packed litchis in southern Queensland, Australia (photograph courtesy of Neil Greer).

Effects of sulphur fumigation on the fruit

Postharvest sulphur fumigation overcomes pericarp browning and reduces decay in litchi and longan fruit (Figs 14.12 and 14.13) (Tongdee, 1994). Browning is prevented since sulphur dioxide results in the formation of colourless quinone–sulphite complexes and it inhibits polyphenol oxidase activity, preventing the formation of quinones, which rapidly polymerize to form brown pigments (Macheix *et al.*, 1990). In addition, sulphur dioxide effectively suppresses surface fungi (Tongdee, 1994).

Sulphur treatment of red litchis results in a pliable but bleached pericarp, which turns a uniform pink after 3–5 days at 22°C (Zauberman *et al.*, 1990). If excess sulphur is used the fruit turn yellow or pale green, and fail to redden. Sulphite bleaching of anthocyanins occurs as the negative ions from the sulphurous acid react with the flavylium cation to form colourless chromen-4 (or 2)-sulphonic acid. This reaction is partially reversible as SO_2 is oxidized (Timberlake and Bridle, 1967; Bridle and Timberlake, 1997). Most of the litchis exported from South Africa to Europe are treated with SO_2.

SO_2 fumigation reduces decay of longan, allowing Thai exporters to successfully ship fruit to distant markets. In addition, sulphur inhibits browning and bleaches the fruit to a bright yellow (O'Hare and Prasad, 1991b; Tongdee, 1994). However, injury to the fruit may occur when they are subjected to excess rates or exposure times. Injury is manifested as irregular rusty-brown circles or lines on the underside of the pericarp a few days after fumigation (Tongdee, 1997). At very high rates the aril also turns dull white (Tongdee, 1994).

Gaseous SO_2 can be derived from burning sulphur powder, vaporizing liquid SO_2 held in a pressurized cylinder, or dissociation of sulphite compounds (Neamprem, 1992). Metabisulphite pads have been tested on longan, but when used at the high rate required to ensure an even colour, the pulp was tainted (O'Hare and Prasad, 1991b). In litchi, treatment with sodium metabisulphite is as effective as SO_2 if it is followed by an acid dip (Duvenhage *et al.*, 1995). While the use of liquid sulphur dioxide and sulphite compounds has some advantages, fumigation is usually achieved by burning 90% sulphur powder at ambient temperatures for 20 min, with no humidity control. Recommendations for different countries vary slightly (Zauberman *et al.*, 1990, 1991; Underhill *et al.*, 1992b; Fuchs *et al.*, 1993; Tongdee, 1994).

Fig. 14.10. Litchis packed in punnets in northern New South Wales, Australia (photograph courtesy of Christopher Menzel).

Fumigation of short duration and high concentration is used for longans. The amount of SO_2 required for a fumigation chamber depends on the quantity of fruit to be fumigated, the absorptivity of the fruit, the volume of free space in the chamber, the absorptivity of the containers, the type of packaging materials, and potential losses through leakage. Allowance is also made for the fruit stalks, which tend to be more absorbent than the fruit. The quantity of SO_2 needed is a combination of the space dosage and the commodity dosage, and can be calculated as:

$$\text{Weight of } SO_2 \text{ (g)} = (A \times B \times C) + (D \times E) \quad (1)$$

Fig. 14.11. Carton used to pack litchis in northern New South Wales, Australia (photograph courtesy of Christopher Menzel).

Fig. 14.12. Fresh litchis in Mauritius. The fruit turn brown very quickly at ambient temperature (photograph courtesy of Christopher Menzel).

Fig. 14.13. Fresh longans in Australia. The fruit turn brown or rot if not stored correctly (photograph courtesy of Christopher Menzel).

where A = the concentration of SO_2 required, B = free space in the chamber (litres), C = weight of 1 l of SO_2 at 30°C (2.574 g/l), D = weight of fruit (kg), E = SO_2 sorption of fruit (g/kg), and weight of sulphur to be burnt = weight of $SO_2/2$ (Tongdee, 1994, 1997).

A minimum fumigation period of 20 min is recommended and timing should begin only after the sulphur has been completely burnt. For litchi, fumigation was most effective when whole fruit sulphur residues were 200–350 mg/kg immediately after fumigation. Longan fumigated with 200–300 ml SO_2/kg fruit had SO_2 residues of 170–400 mg/kg in whole fruit and 1200–3000 mg/kg in the pericarp immediately after fumigation. Sulphur residues decrease rapidly in the first few days after fumigation; however, sulphur residues in the aril occasionally still exceed national and international limits. Progress has been made in achieving decreased SO_2 residues of 1–30 mg/kg in the aril (Tongdee, 1994; Ji *et al.*, 1999; Pan *et al.*, 1999; Xu *et al.*, 1999; Lin *et al.*, 2001b). However, there is increasing concern about these residues in the fruit, particularly for people who are sensitive to sulphites (Tongdee, 1994). Fumigation at the above rates results in about 15,000 mg/l (1.5%) SO_2 remaining in the chamber at the end of the fumigation period and this needs to be removed by a scrubbing system to safeguard workers and the environment (Tongdee, 1992, 1994).

Since importing countries are concerned about sulphur residues in fresh fruit, hazard analysis critical control point (HACCP) and good manufacturing practices (GMP) have been introduced as safety assurance systems (Rattanachai, 1997; Tongdee, 1997). In Thailand, exporting firms and suppliers are scored and granted a specific action status (Table 14.2), based on a fumigation performance rating developed by the Thailand Institute of Scientific and Technological Research (Table 14.1).

Acid treatments for retaining litchi pericarp colour

Dipping fruit in dilute hydrochloric acid (HCl) restored skin colour after SO_2 bleaching by converting the anthocyanin pigment to the red flavylium ion, which predominates at low pH. The colour faded during storage, possibly as a

Table 14.1. Fumigation performance rating and scoring system developed by the Thailand Institute of Scientific and Technological Research, for quality assurance of exported longans (Rattanachai, 1997).

Criteria	Activities to be assessed	Possible score
GMP and HACCP training	Management understanding and responsiveness (by interview or examination)	5
	Management training	5
	Department level training	5
	On the job training	5
Fumigation room audit	Structure and layout of fumigation room	5
	Fumigation process characteristics with and without fruit	10
	SO_2 scrubber system	5
Fumigation process control	Past performance	5
	Process flow design and equipment check	5
	Stable (validated and verifiable)	5
	Record of process monitoring	5
SO_2 residue report	Past record (2 years)	5
	SO_2 analysis of aril < 30 mg/kg	5
	SO_2 analysis of aril < 10 mg/kg	5
	Daily analysis of spot samples	5
Internal audit	Appointment of internal auditor	5
	Corrective action record and implementation	10
	Documentation	5
Total score		100

Table 14.2. Company rating system for export of sulphur-treated longans produced in Thailand, scored as outlined in Table 14.1 (Rattanachai, 1997).

Company rating	Score obtained	Action status
Premier	> 85	Shipment released on arrival, SO_2 analysis only in exceptional cases
Preferred	> 75	Shipment released on arrival, SO_2 analysis on spot samples, for example once a week
Approved	> 65	Shipment may be released on arrival, SO_2 analysis on 30–50% of samples or once every two or three shipments
Conditional	> 50	Shipment not released until SO_2 analysis
Candidate	> 20	Shipment not released until SO_2 analysis. Advise for further training

result of an increase in the pH of the cell sap (Zauberman *et al.*, 1990, 1991; Duvenhage *et al.*, 1995). SO_2 fumigation appeared to increase the permeability of the plasma membrane, which allowed the acid to reach the vacuoles, since fruit treated only with acid had a red colour with dark patches. However, there is a need to find effective alternatives, as there are increasing limitations on SO_2 use (Zauberman *et al.*, 1990, 1991; Ketsa and Leelawatana, 1992a; Underhill *et al.*, 1992b; Fuchs *et al.*, 1993). SO_2 fumigation can also taint the fruit.

Kaiser (1994) found that blanching fruit in water at 98°C for 30 s followed by hydro-cooling in a solution of distilled water adjusted to pH 0 for 5 min resulted in a pliable, red pericarp and no aftertaste. However, the aril suffered some damage and discolouration. A steam treatment for 2 s followed by immersion in water at pH 0 and then 1% Vaporgard, an anti-transpirant, resulted in red fruit with no discoloration of the pulp compared with sulphur-treated and control fruit, even after storage at 1°C for 28 days (Kaiser *et al.*, 1995). Peng and Cheng (2001) found that fruit that were treated at 98°C for 30 s or fumigated with sulphur and then dipped in acid had less decay and better colour at 4°C compared with fruit stored at 21°C. Hot-water brushing, where fruit are sprayed with hot water while being brushed in a revolving drum, followed by a hydrochloric acid and fungicide dip, has been developed in Israel (Lichter *et al.*, 2000).

Temperature management

Rapid cooling, followed by storage of fruit at low temperatures, reduces water loss and disease development. Low temperatures also decrease PPO activity, slow changes in TSS, TA and ascorbic acid concentrations, retard browning and aril breakdown, and thus prolong fruit life (Hong *et al.*, 1984; Paull and Chen, 1987b; Lin *et al.*, 2001a,b). However, despite the many advantages of cooling, longans and litchis are often stored at ambient temperatures because cold storage is not readily available in many producing areas. Storage at ambient temperature relies on the use of fungicides to reduce decay, and packing in plastic film to minimize browning. Under these conditions, fruit may last for 7–12 days (Lin *et al.*, 2001a).

Room cooling (3–5°C air temperature and 80–90% RH) took 13 h to reduce the temperature of litchis from 27 to 3°C; forced-air cooling (2.5 cm static pressure difference, 3–5°C air temperature and 80–90% RH) took 60 min to achieve the same result; while hydro-cooling (0–1°C water, 100 mg/l free chlorine and pH 6.8–7.3) took 12–15 min (Pornchaloempong *et al.*, 1997). Cooling takes longer if fruit are packed in plastic bags, and it can take up to 2 days to cool the pulp of packed fruit to 5°C with room cooling (Bagshaw *et al.*, 2000) (Fig. 14.14). While hydro-cooling is the most rapid cooling method (10–20 min) and has been used successfully on litchi (Fig. 14.15 and Plate 64), the fruit may decay if they are packed wet (Ketsa and Leelawatana, 1992b; Pornchaloempong *et al.*, 1997; Bagshaw *et al.*, 2000). Excess moisture should be drained from the fruit, with fans used to speed up drying (Bagshaw *et al.*, 2000). A potable water supply and chlorination with pH control are recommended (Thompson *et al.*, 1998).

Pornchaloempong *et al.* (1997) found that water loss was lower in fruit removed from panicles and hydro-cooled, compared with fruit that were stored loose in cartons for 15 days at room temperature or forced-air cooled at 3–5°C followed by 1 day at 20°C. Fruit without panicles stored in vented polystyrene clamshells under the same conditions lost more water regardless of the cooling method. Water loss was highest in fruit that were hydro-cooled and stored loose in cartons. Hydro-cooling litchis at 0–2°C helped retain colour, while fruit that were not hydro-cooled turned brown. Hydro-cooling at 0–2°C immediately after harvest, and packing in sealed polythene bags followed by cold storage, preserved colour for a month (Moreuil, 1973). Although vacuum cooling of litchi is rapid, the fruit lose water and turn brown (Chen *et al.*, 2001).

Room cooling, forced-air cooling, hydro-cooling and vacuum cooling have also been used in longan (Lin and Zhang, 1997; Chen and Lin, 2000). Room cooling is ineffective, with cooling times of 12 h recorded for the reduction of the temperature of 'Wulongling' and 'Youtanben' fruit from 30°C to 3–5°C (Lin, 2002). Forced-air cooling is three to four times faster

Fig. 14.14. Storage of packed litchis in a cool room in southern Queensland, Australia (photograph courtesy of Christopher Menzel).

Fig. 14.15. Hydro-cooling of litchis in southern Queensland, Australia (photograph courtesy of Neil Greer).

than room cooling. Cooling takes longer if the fruit are packed in plastic bags or in fibreboard cartons, so they should be cooled before packing. Container venting and stacking should be optimized to facilitate rapid cooling.

Hydro-cooling has been used widely in longan (Lin and Zhang, 1997; Chen *et al.*, 1998). In commercial practice, the fruit are immersed in water at a temperature of 0–6°C for 3–5 min, loaded into plastic crates and allowed to dry in a cold room before packing (Lin and Zhang, 1997; Chen *et al.*, 1998). Since hydro-cooling can increase the incidence of decay, sanitation of the water is essential. Hydro-cooling of SO_2-fumigated fruit can result in browning of the pulp near the fruit stalk. This damage can be prevented by holding the treated fruit at room temperature for 1–2 h before cooling.

Vacuum-cooling reduced the pulp temperature of 'Chike' longan from 28 to 7°C within 11 min (Chen and Lin, 2000). Wetting the fruit prior to vacuum-cooling reduced water loss and increased the speed of cooling (Lin, 2002); however, this technology is not used commercially because of its cost.

Litchis are usually stored at 5°C although recommendations vary (Table 14.3). Suggested storage temperatures range from 0 to 7°C, with Swarts and Anderson (1980) recommending

Table 14.3. Litchi fruit responses to postharvest handling procedures (Underhill *et al.*, 1997).

Variable	Conditions	Reference
Storage temperature		
Chilling temperature	Not defined (possibly 0.5–1°C)	–
Minimum temperature	1–2°C	Duvenhage *et al.* (1995)
Optimum temperature	5–7°C	Scott *et al.* (1982)
Heat damage	> 52°C for 2 min	Wong *et al.* (1991), Huang and Scott (1985)
Storage life		
At 25°C (open)	2–4 days	Datta *et al.* (1963)
At 25°C (plastic packaging)	Up to 7 days	Nip (1988)
At 5°C (plastic packaging)	Up to 28 days	Datta *et al.* (1963), Scott *et al.* (1982)

0–1°C for up to 30 days, and Tongdee *et al.* (1982) 7°C for up to 30 days. Zhang and Quantick (2000) found that fruit conditioned at 5°C for 5 days before transfer to 1°C had less browning than those stored at a constant 1 or 5°C.

In China, longans lasted for 30–40 days if they were treated with fungicides and stored at 1–5°C (Lin *et al.*, 2001a; Jiang *et al.*, 2002). However, the optimum temperature for different cultivars from different producing areas needs further investigation (Lin *et al.*, 2001a). In China, ice is used when transporting litchi or longan for 1–3 days over short distances in non-refrigerated vehicles. Fruit are packed in a polythene bag of 0.04–0.05 mm thickness and placed in polystyrene boxes. Ice is placed on top of the sealed bag (ratio of ice:fruit is 1:3–1:4) and the box is then closed (Chen *et al.*, 1998; Chen *et al.*, 2001). Chilling injury, manifest as darkening of the pericarp, occurs if the fruit are in contact with the ice.

Reducing water loss

Cold storage helps to reduce water loss because it lowers the vapour pressure difference between the fruit and the dry atmosphere (Thompson *et al.*, 1998). The maintenance of a relative humidity of 90–95% in the atmosphere surrounding stored fruit also reduces browning (Jiang and Fu, 1999a).

Litchi fruit treated with polysaccharide coatings such as Nature Seal developed less browning than untreated fruit, but were still unmarketable after 11 days (York, 1994). Chitosan coatings at 1% and 2% delayed browning in litchi compared with untreated control fruit, and reduced decay slightly (Zhang and Quantick, 1997a). The use of acid coatings has shown some potential in maintaining colour and preventing decay in litchi, but is still experimental (McGuire and Baldwin, 1998).

Pericarp coatings can extend the storage life of longans (Jiang *et al.*, 2002). Jiang and Li (2001) reported that the application of 2% chitosan extended the life of 'Shixia' from 30 to 40 days at 5°C, maintained fruit quality and delayed browning compared with untreated controls. Use of edible konjac glucomannan film in 'Wulongling' reduced water loss and respiration rate, inhibited pathogens, retarded browning, and extended storage and shelf-life compared with untreated fruit (Zou *et al.*, 2001).

Wrapping litchis in PE film reduced desiccation and browning compared with unwrapped fruit (Scott *et al.*, 1982; Paull and Chen, 1987a). Tongdee *et al.* (1982) found that fruit in punnets wrapped in polyvinylchloride (PVC) film and stored for 40 days lost 6.4% of their weight at 10°C and only 1.7% at 0°C. Litchis packed in unperforated PE (0.04–0.05 mm thick) bags at 20°C lost less than 2% of their fresh weight after 10 days, while unpacked control fruit lost 18–30%; however, the incidence of decay was higher in the packed fruit (Scott *et al.*, 1982). Treating litchis with hot benomyl dips prior to packing reduced decay. Fruit packed in plastic punnets wrapped in PVC (0.01 mm) had minimal condensation (Scott *et al.*, 1982; Huang and Scott, 1985).

Hot-benomyl-treated fruit packed in punnets with PVC and stored at 20°C remained in a good condition for 11 days, while untreated fruit turned brown after 1–2 days (Tongdee *et al.*, 1982). Fontes *et al.* (1999) compared low-density polyethylene (LDPE) with PVC and found that, while the reduction of water loss in litchis by LDPE was greater, condensation was higher. PVC reduced water loss without causing condensation, while browning was delayed for 36 days when fruit were stored at 5°C (Fontes *et al.*, 1999). Litchis turn brown after 4 weeks, even under high-humidity cool storage, due to natural senescence and the breakdown of cell compartments (Underhill *et al.*, 1992a).

Longans are very susceptible to water loss. Storage in plastic bags effectively reduced water loss and extended storage life (Lin *et al.*, 1997, 2001b, 2002a). 'Fuyan' fruit stored for 1, 2 or 6 days at 10°C and 50% RH lost 5.8%, 10.4% and 24.0% of their fresh weight, respectively, while fruit stored in PE film bags (0.015 mm) lost only 0.6% after 6 days (Lin *et al.*, 2002a). 'Fuyan' fruit packed in PE (0.015 mm) bags and held at 20°C lost 1.9% after 10 days, while fruit held at 4°C lost only 0.9%. Storage at 4°C also retarded browning (Lin *et al.*, 2002e). 'Pumingan' fruit packed in PE (0.010, 0.025, 0.040 or 0.060 mm) bags and stored at 29–33°C for 10 days, lost 0.5–4.0% of their fresh weight, while control fruit lost 11.3%. However, off-flavours tended to develop in bags thicker than 0.025 mm (Liu, 1988).

Controlled and modified atmospheres

Datta *et al.* (1963) found that an atmosphere enriched with 25% carbon dioxide (CO_2) at 4.4°C helped fruit to retain colour, flavour and texture, and reduced decay in litchi compared with an atmosphere of air. However, Vilasachandran *et al.* (1997) found that atmospheres of 15% CO_2 resulted in off-flavours and a dull, grey pulp in 'Mauritius' ('Tai So'). Controlled atmospheres (CA) of 3–5% O_2 and 3–7.5% CO_2 prolonged storage life and quality of litchi, reduced browning, reduced the loss of ascorbic acid, total soluble solids and titratable acidity, and reduced decay compared with air controls (Vilasachandran *et al.*, 1997; Pornchaloempong *et al.*, 1998; Jiang and Fu, 1999b). Jiang and Fu (1999b) found that the use of these atmospheres resulted in less water loss than did air controls, despite the fact that both were kept at 90% RH. Greater control of browning and decay may be achieved by using packages that reduce water loss and modify the atmosphere (Scott *et al.*, 1982).

Pesis *et al.* (2002) found that litchis packed in micro-perforated laminated polythene bags had less decay but poorer taste than fruit in macro-perforated bags. Carbon dioxide concentrations in the micro-perforated bags were about 5% at 2°C in fruit from the first harvest, but about 13% in fruit from the last harvest. When the bags were transferred to 20°C, CO_2 concentrations were as high as 26% in later-harvested fruit, resulting in higher acetaldehyde and ethanol concentrations, which contributed to off-flavours. Hence the success of modified-atmosphere packaging is dependent on harvesting fruit at the correct maturity and on the availability of cold-storage facilities.

Controlled and modified atmospheres have been reported to reduce browning and decay and extend the storage life of longan (Jiang, 1999b; Tian *et al.*, 2002). O'Hare and Prasad (1992) reported that CA storage for 12 days at 10°C did not affect pericarp colour, although an atmosphere with low oxygen (5% O_2) tended to maintain better eating quality compared with fruit stored at higher oxygen concentrations. Fruit of 'Shixia' held at atmospheres of 4–6% O_2 and 6–8% CO_2 at 1°C for 30 days had less decay and better eating quality than fruit kept in air, 21% O_2 and 4–6% CO_2, or 6–8% O_2 (Jiang, 1999b). Tian *et al.* (2002) found that 'Chuliang' fruit stored at 2°C under 4% O_2 and 5% or 15% CO_2 had less browning and decay than those stored in modified-atmosphere packaging (MAP), but they did not make a comparison with storage in air. The higher CO_2 concentration (15%) more effectively controlled decay than did the lower concentrations. Interestingly, Tian *et al.* (2002) found that fruit under atmospheres of 70% O_2, with the balance being nitrogen, had less decay and browning than fruit under 4% O_2 plus 5% or 15% CO_2.

Modified atmosphere packaging is used commercially in longan to reduce browning (Lin *et al.*, 2001a); however, the effects of packaging on humidity and water loss cannot be separated from the effects of O_2 and CO_2 (Jiang *et al.*, 2002). 'Shixia' fruit stored in 0.03 mm PE bags with an initial atmosphere of 3% O_2 and 5% CO_2 had less browning and higher pulp TSS and ascorbic acid concentrations compared with fruit stored in air (Zhang and Quantick, 1997b). 'Chieke' fruit packed in 0.06 mm-thick PE film bags flushed with N_2, and stored at 5°C had extended storage life, better fruit quality and reduced browning, compared with unpackaged fruit stored at the same temperature (Zhang *et al.*, 2002).

When CO_2 concentrations exceed 10–13%, the ethanol content in pulp increases and off-flavours develop (Shi, 1990; Lu *et al.*, 1992; Lin *et al.*, 2001a, 2002d). A rapid increase in ethanol concentration was measured in 'Chuliang' fruit packed in a PE bag (0.04 mm thick; 15–19% O_2 and 2–4% CO_2 measured every 10 days) after 30 days of storage at 2°C. This was probably due to decay, which was severe after this time (Tian *et al.*, 2002). 'Shixia' fruit stored in PE bags (0.03 mm thick; initial atmosphere of 1% O_2 and 5% CO_2) for 7 days at 25°C and 35 days at 4°C produced slight off-flavours (Zhang and Quantick, 1997b). 'Shixia' fruit packed in MAP, flushed with 75% N_2 plus 25% air and held for 7 days at 25°C produced an aril ethanol concentration of 6.4 mg/g (Liang *et al.*, 1998).

Disinfestation

Pest disinfestation, particularly for fruit fly, is required by many importing countries, especially the USA and Japan (Jacobi *et al.*, 1993; McGuire, 1998). Cold, heat and gamma irradiation treatments are acceptable for use in litchis and longans (Paull *et al.*, 1995; McGuire, 1997, 1998; Follett and Sanxter, 2001).

A vapour heat treatment (VHT) of 45°C for 30 min was sufficient to kill Queensland fruit fly, *Bactrocera tryoni*, without adversely affecting eating quality of litchi. However, the treatment caused the pericarp to brown, necessitating the use of an acid dip to correct this (Jacobi *et al.*, 1993). Japanese treatment protocols require that the pulp of VHT litchis reaches 46.2°C and remains at that temperature for 20 min before the fruit are cooled to 2°C and held at low temperature for a further 42 h (Armstrong, 1994). Storage at 1°C for 12–14 days or a hot-water treatment at 49°C for 20 min have been used on litchi, although both increase browning (Paull *et al.*, 1995). Sulphur dioxide, alone or with an acid dip, in conjunction with these disinfestation treatments, reduced browning, although some fruit decayed (Paull *et al.*, 1995), perhaps as a result of sulphur dioxide injury (Tongdee, 1994).

McGuire (1997) found that the loss of quality of litchi stored at 1.1°C for 15 days was minimal, although fruit were more susceptible to decay. Hot-water immersion for 20 min at 49°C used to control fruit flies in litchi and longan is also effective against nut borers, *Cryptophlebia* spp. (Follett and Sanxter, 2001). Hot-water treatment of 49°C for 15 min killed Chinese fly species without affecting the quality of longans (Lu *et al.*, 1992).

Gamma irradiation of 75 or 300 Gy kills the eggs and larvae of Queensland fruit fly without affecting the physical, chemical or organoleptic properties of the fruit, and is an effective disinfestation treatment for that species (McLauchlan *et al.*, 1992). McGuire (1998) found that 100–300 Gy of irradiation with ^{60}Co (cobalt) was a successful quarantine treatment, maintaining better quality of fresh longan than storage for 15 days at 1.1°C.

Transport and marketing

Fruit are transported by land, sea and air, depending on the market. Fruit sold in local markets are usually transported by truck to the distribution centres or wholesale markets. In Thailand, most vehicles used in transporting fruit to local markets are not refrigerated, while in China refrigerated trucks are used to transport fruit from the south of the country to the north. Longans produced in northern Thailand are shipped to Malaysia and Singapore in refrigerated trucks, while fruit exported overseas are usually shipped by sea in refrigerated containers. The high cost of air-freight precludes its use for the transport of these fruit in much of South-east Aisa.

Conclusions

Decay and browning are the most important factors affecting the shelf-life and quality of litchis and longans. Consequently, storage conditions and postharvest treatments are selected with the objective of reducing these problems. Sulphur is often used to reduce both decay and browning, with postharvest fungicides also providing good control of diseases. However, there is increasing pressure from consumers to find alternatives to chemicals while still satisfying the demand for sound, blemish-free fruit of good eating quality. Since low temperatures reduce both water loss and the incidence of disease, minimizing delays between harvesting and cooling, rapid cooling, refrigeration during transportation and cool storage remain the most promising means of limiting browning and decay. These treatments also maintain the red skin colour in litchi and aril quality in both species. The use of plastic films to reduce water loss, and modified atmospheres for reducing diseases, warrant further investigation.

Acknowledgements

Hetong Lin's work was supported by the State Natural Science Foundation of China (No. 30200192) and The Natural Science Foundation of Fujian Province (No. B9910016 and No. B0210021). We thank Yueming Jiang and Shiping Tian for providing their manuscripts and Trevor Olesen for commenting on drafts of this chapter.

References

Akamine, E.K. and Goo, T. (1973) Respiration and ethylene production during ontogeny of fruit. *Journal of the American Society for Horticultural Science* 93, 381–383.

Armstrong, J.W. (1994) Tropical and subtropical crops. In: Paull, R.E. and Armstrong, J.W. (eds) *Insect Pests and Fresh Horticultural Products: Treatments and Responses.* CAB International, Wallingford, UK, pp. 275–290.

Bagshaw, J., Underhill, S. and Dahler, J. (2000) Lychee hydrocooling. See http://www.dpi.qld.gov.au/horticulture/5439.html

Batten, D.J. (1989) Maturity criteria for litchis (lychees). *Food Quality Preferences* 1, 149–155.

Bridle, P. and Timberlake, C.F. (1997) Anthocyanins as natural food colours: selected aspects. *Food Chemistry* 58, 103–109.

Brouillard, R., Figueiredo, P., Elhabiri, M. and Dangles, O. (1997) Molecular interactions of phenolic compounds in relation to the colour of fruit and vegetables. In: Tomas-Barberan, F.A. and Robins, R.J. (eds) *Phytochemistry of Fruit and Vegetables. Proceedings of the Phytochemical Society of Europe*, Vol. 41. Clarendon Press, Oxford, UK, pp. 29–49.

Bryant, P. and Olesen, T. (2002) Harvesting and storage. In: Menzel, C.M. (ed.) *The Lychee Crop in Asia and the Pacific.* Food and Agriculture Organization of the United Nations, Bangkok, Thailand, pp. 84–94.

Cai, M.Z. (1988) A simple fumigation of SO_2 for storage of longan fruit. *Journal of Fujian Academy of Agricultural Science* 3 (1), 85–89.

Chan, H.T., Jr and Kwok, S.C.M. (1974) Nonvolatile acids in lychee. *Journal of Food Science* 39, 792–793.

Chan, H.T., Jr, Kwok, S.C.M. and Lee, C.W.O. (1975) Sugar composition and invertase activity in lychee. *Journal of Food Science* 40, 772–774.

Chan, Y.K., Yang, Y.H. and Li, N. (1998) Low-temperature storage elicits ethylene production in nonclimacteric lychee (*Litchi chinensis* Sonn.) fruit. *HortScience* 33, 1228–1230.

Chattopadhyay, P.K. and Ghosh, A. (1991) Studies on the physiology of fruit growth of longan. *Horticultural Journal* 4, 27–31.

Chen, J.Q. and Lin, H.T. (2000) Vacuum precooling model of longan fruit. *Transactions of the Chinese Society of Agricultural Engineering* 16 (Supplement), 108–110.

Chen, W., Wu, Z., Ji, Z. and Su, M. (2001) Postharvest research and handling of litchi in China: a review. *Acta Horticulturae* 558, 321–329.

Chen, Y.Z., Chen, M.H., Cui, J., Chen, K.G., Su, S.Y., Chen, M.D., Cui, T., Zhong, D.S. and Chen, Z.X. (1998) Major measurements of storage and transportation of longan fruit. *Guangdong Agricultural Science* 6, 20–21.

Datta, S.C., Sarkar, K.P. and Lodh, S.B. (1963) Storage behaviour of litchi (*Litchi chinensis* Sonner). *Science and Culture* 29, 405–406.

Duvenhage, J.A., Mostert, M.M. and Marais, J.J. (1995) Postharvest sulphuring and low pH treatment for retention of red skin colour of litchi fruit. *Yearbook of the South African Litchi Growers' Association* 7, 44–46.

Follett, P.A. and Sanxter, S.S. (2001) Hot water immersion to ensure quarantine security for *Cryptophlebia* spp. (Lepidoptera: Tortricidae) in lychee and longan exported from Hawaii. *Journal of Economic Entomology* 94, 1292–1295.

Fontes, V.L., de Moura, M.A., Vieira, G. and Finger, F.L. (1999) Influence of plastic films and temperature on postharvest pericarp browning in litchi (*Litchi chinensis*). *Revista Brasileira de Armazenamento* 24, 56–59.

Fuchs, Y., Zauberman, G., Ronen, R., Rot, I., Weksler, A. and Akerman, M. (1993) The physiological basis of litchi fruit pericarp color retention. *Acta Horticulturae* 343, 29–33.

Han, D.M., Cai, C.H. and Wu, Z.X. (2001) Postharvest physiology and storage characteristics of longan fruit. *Guangdong Agricultural Science* 6, 20–21.

Han, D.M., Wu, Z.X., Chen W.X. and Dai, H.F. (2002) Studies on the physiological and biochemical changes of longan (cv. Shixia and cv. Chuliang) pulp after harvest. *Journal of South China Agricultural University* 23, 20–23.

He, Z.G., Lu, X.M., Shen, C.G., Wei, Q., Cai, Z.J., Huang, T.J. and Bian, M.Y. (1997) Effect of pre-treatment on the respiration of stored longan fruit. *Journal of Fujian Academy of Agricultural Science* 12 (4), 32–34.

Hong, Q.Z., Sheng, H.Y. and Xie, Z.J. (1984) Study on storage of longan fruit. *Fujian Agricultural Science and Technology* 3, 50–51.

Huang, C.C. and Wang, Y.T. (1990) Effect of storage temperature on the color and quality of litchi fruit. *Acta Horticulturae* 269, 307.

Huang, H.B. (1995) Advances in fruit physiology of the arillate fruit of litchi and longan. *Annual Review of Horticultural Science* 1, 107–120.

Huang, P.Y. and Scott, K.J. (1985) Control of rotting and browning of litchi fruit after harvest at ambient temperatures in China. *Tropical Agriculture* 62, 2–4.

Huang, S., Hart, H., Lee, H. and Wicker, L. (1990) Enzymic and color changes during postharvest storage of lychee fruit. *Journal of Food Science* 55, 1762–1763.

Jacobi, K.K., Wong, L.S. and Giles, J.E. (1993) Lychee (*Litchi chinensis* Sonn.) fruit quality following vapour heat treatment and cool storage. *Postharvest Biology and Technology* 3, 111–119.

Jaiswal, B.P., Sah, N.L. and Prasad, U.S. (1986) Studies on phenolics of rind and aril during ripening and senescence of litchi fruit. *Plant Physiology and Biochemistry* 13, 40–45.

Ji, Z.L., Han, D.M. and Wu, Z.X. (1999) Effect of sulphur dioxide fumigation on postharvest longan fruit (cv. Shixia) and its residues. *Chinese Journal of Tropical Crops* 20 (3), 36–40.

Jiang, Y.M. (1999a) Purification and some properties of polyphenol oxidase of longan fruit. *Food Chemistry* 66, 75–79.

Jiang, Y.M. (1999b) Low temperature and controlled atmosphere storage of fruit of longan (*Dimocarpus longan* Lour.). *Tropical Science* 39, 98–101.

Jiang, Y.M. (2000) Role of anthocyanins, polyphenol oxidase and phenols in lychee pericarp browning. *Journal of the Science of Food and Agriculture* 80, 305–310.

Jiang, Y.M. and Fu, J.R. (1999a) Biochemical and physiological changes involved in browning of litchi fruit caused by water loss. *Journal of Horticultural Science and Biotechnology* 74, 43–46.

Jiang, Y.M. and Fu, J.R. (1999b) Postharvest browning of litchi fruit by water loss and its prevention by controlled atmosphere storage at high relative humidity. *Lebensmittel Wissenschaft und Technologie* 32, 278–283.

Jiang, Y.M. and Li, Y.B. (2001) Effects of chitosan coating on postharvest life and quality of longan fruit. *Food Chemistry* 73, 139–143.

Jiang, Y.M., Zhang, Z.Q., Joyce, D.C. and Ketsa, S. (2002) Postharvest biology and handling of longan fruit (*Dimocarpus longan* Lour.). *Postharvest Biology and Technology* 26, 241–252.

Joubert, A.J. (1970) The Litchi. *Bulletin of the Citrus and Subtropical Fruit Research Institute* Nelspruit, South Africa No. 389.

Joubert, A.J. (1986) Litchi. In: Monselise, S.P. (ed.) *CRC Handbook of Fruit Set and Development.* CRC Press, Boca Raton, Florida, pp. 233–246.

Joubert, A.J. and van Lelyveld, L.J. (1975) An investigation of pre-harvest browning of litchi peel. *Phytophylactica* 7, 9–14.

Kader, A.A. (2000) Lychee. See http://rics.ucdavis.edu/postharvest2/Produce/ProduceFacts/Fruit/lychees.html

Kaiser, C. (1994) Litchi (*Litchi chinensis* Sonn.) pericarp colour retention. *Journal of the Southern African Society for Horticultural Sciences* 4, 6–12.

Kaiser, C., Levin, J. and Wolstenholme, B.N. (1995) Vapour heat and low pH dips improve litchi (*Litchi chinensis* Sonn.) pericarp colour retention. *Journal of the Southern African Society for Horticultural Sciences* 5, 7–10.

Ke, G.W., Wang, C.C. and Huang, J.H. (1992) The aril initiation and ontogenesis of longan fruit. *Journal of Fujian Academy of Agricultural Science* 7 (1), 22–26.

Ketsa, S. and Leelawatana, K. (1992a) Effect of pre and poststorage acid dipping on browning of lychee fruit. *Acta Horticulturae* 321, 726–731.

Ketsa, S. and Leelawatana, K. (1992b) Effect of precooling and PE film liners in corrugated boxes on quality of lychee fruit. *Acta Horticulturae* 321, 742–746.

Lee, H.S. and Wicker, L. (1991a) Anthocyanin pigments in the skin of lychee fruit. *Journal of Food Science* 56, 466–468, 483.

Lee, H.S. and Wicker, L. (1991b) Quantitative changes in anthocyanin pigments of lychee fruit during refrigerated storage. *Food Chemistry* 40, 263–270.

Li, X.P., Pang, X.Q., Zhang, Z.Q., Ji, Z.L. and Li, T. (1999) Effect of SO_2 on cold storage and shelf-life of longan fruit. *Journal of South China Agricultural University* 20, 77–80.

Liang, H.H., Huang, X.Y. and Ji, Z.L. (1998) Modified atmosphere storage of longan fruit. *Chinese Journal of Tropical Crops* 19 (2), 49–54.

Lichter, A., Dvir, O., Rot, I., Akerman, M., Regev, R., Wiesblum, A., Fallik, E., Zauberman, G. and Fuchs, Y. (2000) Hot water brushing: an alternative method to SO_2 fumigation for color retention of litchi fruit. *Postharvest Biology and Technology* 18, 235–244.

Lin, H.T. (2002) Postharvest pericarp browning mechanism and postharvest handling of longan fruit. PhD dissertation, Zhejiang University, Hangzhou, China.

Lin, H.T. and Zhang, X.P. (1997) Regulations of technical operation in postharvest storage and transportation of longan fruit. *Applied Engineering Techniques in Rural Areas* No. 6, 25.

Lin, H.T., Zou, R.E., Hong, Q.Z. and Lin, Z.Y. (1997) Technology of prolonging the shelf-life of longan fruit after cold storage. *Journal of Fujian Agricultural University* 26, 113–118.

Lin, H.T., Chen, S.J., Chen, J.Q. and Hong, Q.Z. (2001a) Current situation and advances in postharvest storage and transportation technology of longan fruit. *Acta Horticulturae* 558, 343–351.

Lin, H.T., Chen, S.J. and Hong, Q.Z. (2001b) A study of the shelf-life of cold-stored longan fruit. *Acta Horticulturae* 558, 353–357.

Lin, H.T., Xi, Y.F. and Chen, S.J. (2001c) Main problems concerning storage, transportation, delivery and marketing of longan fruit and their countermeasures in China. *Research on Agricultural Modernization* 22 (Supplement), 150–152.

Lin, H.T., Chen, S.J., Xi, Y.F. and Guo, S.Z. (2002a) Observation on pericarp ultrastructure by scanning electron microscope and its relation to keeping quality of longan fruit. *Transactions of the Chinese Society of Agricultural Engineering* 18 (3), 95–99.

Lin, H.T., Xi, Y.F. and Chen, S.J. (2002b) A review of enzymatic browning in fruit during storage. *Journal of Fuzhou University* 30, 696–703.

Lin, H.T., Xi, Y.F., Chen, S.J. and Chen, J.Q. (2002c) A comparative observation on pericarp morphology and structure and its relationship to storability and transport of longan fruit. *Guihaia* 22, 413–419, 424.

Lin, H.T., Xi, Y.F., Chen, S.J., Chen, J.Q. and Hong, Q.Z. (2002d) Research advances of postharvest physiology, postharvest pathology and storage and transport technologies for longan fruit. *Transactions of the Chinese Society of Agricultural Engineering* 18, 185–190.

Lin, H.T., Xi, Y.F., Chen, S.J., Zhou, R.S., Wu, Y.J. and Li, W.L. (2002e) Effect of storage temperature on postharvest physiology and quality of longan fruit. *Transactions of the Chinese Society of Agricultural Engineering* 18 (supplement), 115–120.

Lin, Z.F., Li, S.S., Chang, D.L., Lin, G.Z., Li, Y.B., Liu, S.X. and Chen, M.D. (1988a) The changes of pigments, phenolics contents and activities of polyphenol oxidase and phenylalanine ammonia-lyase in pericarp of postharvest litchi fruit. *Acta Botanica Sinica* 30, 40–45.

Lin, Z.F., Li, S.S., Zhang, D.L., Liu, S.X., Li, Y.B., Li, G.Z. and Chen, M.D. (1988b) The changes of oxidation and peroxidation in postharvest litchi fruit. *Acta Botanica Sinica* 30, 383–387.

Liu, W.M. (1988) Studies on the postharvest physiology and fresh-keeping in storage of longan fruit. MSc thesis, Fujian Agricultural College, Fuzhou, China.

Liu, X.J. (1991) Postharvest diseases and its control measures on lychee and longan fruit. In: Leng, H.Q., Cao, R.P., Liu, X.J., Shen, Y.Z. and Rou, W.H. (eds) *Prevention and Controlling Diseases and Fresh-Keeping in Fruit Storage*. Shichuan Science and Technology Press, Chengdu, China, pp. 287–301.

Lu, R.X., Zhan, X.J., Wu, J.Z., Zhuang, R.F., Huang, W.N., Cui, L.X. and Huang, Z.M. (1992) Studies on storage of longan fruit. *Subtropical Plant Research Communications* 21 (2), 9–17.

Macheix, J.J., Fleuriet, A. and Billot, J. (1990) *Fruit Phenolics*. CRC Press, Boca Raton, Florida, 378 pp.

McGuire, R.G. (1997) Response of lychee fruit to cold and gamma irradiation treatments for quarantine eradication of exotic pests. *HortScience* 32, 1255–1257.

McGuire, R.G. (1998) The response of longan fruit to cold and gamma irradiation treatments for quarantine eradication of exotic pests. *Journal of Horticultural Science & Biotechnology* 73, 687–690.

McGuire, R.G. and Baldwin, E.A. (1998) Acidic fruit coatings for maintenance of color and decay prevention on lychee postharvest. *Proceedings of the Florida State Horticultural Society* 111, 243–247.

McLauchlan, R.L., Mitchell, G.E., Johnson, G.I., Nottingham, S.M. and Hammerton, K.M. (1992) Effects of disinfestation-dose irradiation on the physiology of Tai So lychee. *Postharvest Biology and Technology* 1, 273–281.

Menzel, C.M., Bagshaw, J., Campbell, T., Greer, N., Noller, J., Olesen, T. and Waite, G.K. (2002) *Lychee Information Kit.* Queensland Department of Primary Industries, Nambour, Australia.

Moreuil, C. (1973) Some observations and trials on the litchi. *Fruits* 28, 637–640.

Nacif, S.R., Paoli, A.A.S. and Salomao, L.C.C. (2001) Morphological and anatomical development of the litchi fruit (*Litchi chinensis* Sonn. cv. Brewster). *Fruits* 56, 225–233.

Nagar, P.K. (1994) Physiological and biochemical studies during fruit ripening in litchi (*Litchi chinensis* Sonn.). *Postharvest Biology and Technology* 4, 225–234.

Neamprem, S. (1992) SO_2 and determination of its residues in fresh longan. In: *Training on SO_2 Fumigation and Postharvest Handling of Longan for Export.* Thailand Institute of Scientific and Technological Research, Bangkok, Thailand, pp. 82–95.

Nip, W.K. (1988) Handling and preservation of lychee (*Litchi chinensis* Sonn.) with emphasis on colour retention. *Tropical Science* 28, 5–11.

O'Hare, T.J. and Prasad, A. (1991a) Cool temperature storage of longan fruit. *North Queensland Horticultural News*, p. 55.

O'Hare, T.J. and Prasad, A. (1991b) Longan storage in conjunction with metabisulphite pads. *North Queensland Horticultural News*, pp. 55–56.

O'Hare, T.J. and Prasad, A. (1992) Longan controlled atmosphere storage. In: Ledger, S., McLauchlan, R., Hofman, P. and Meiburg, G. (eds) *Horticulture Postharvest Group Biennial Review.* Queensland Department of Primary Industries, Brisbane, Australia, p. 21.

Olesen, T. (2001) Improved postharvest handling of lychee. *Living Lychee* 27, 25–30.

Pan, D.M., Guo, Z.X., Hong, Z.D., Pan, C.Z., Huang, Y.D. and Pan, J.G. (1999) Effects of SO_2 fumigation on the pericarp browning and mass fraction of SO_2 residue in longan fruit. *Journal of Fujian Agricultural University* 28, 425–427.

Pan, X.C. (1994) Study on relationship between preservation and microstructure of longan fruit. *Journal of Guangxi Agricultural University* 13, 185–188.

Pan, Y.S., Cai, X.D., Shun, L.N., Huang, Z.Z. and Yang, Z.G. (1996) The physiology of longan fruit in storage. *Journal of Fruit Science* 13, 19–22.

Pang, X.Q. and Zhang, Z.Q. (1999) Advances in studies of postharvest physiology, storage and fresh-keeping technologies of longan fruit. *Tropical Agricultural Science* 2, 56–59.

Paull, R.E. and Chen, N.J. (1987a) Effect of storage temperature and wrapping on quality characteristics of litchi fruit. *Scientia Horticulturae* 33, 223–236.

Paull, R.E. and Chen, N.J. (1987b) Changes in longan and rambutan during postharvest storage. *HortScience* 22, 1303–1304.

Paull, R.E., Chen, N.J., Deputy, J., Huang, H., Cheng, G. and Gao, F. (1984) Litchi growth and compositional changes during fruit development. *Journal of the American Society for Horticultural Science* 109, 817–821.

Paull, R.E., Reyes, M.E.Q. and Reyes, M.U. (1995) Litchi and rambutan insect disinfestation: treatments to minimize induced pericarp browning. *Postharvest Biology and Technology* 6, 139–148.

Peng, Y.H. and Cheng, W. (2001) Effect of postharvest handling on fruit quality, mass loss and respiration rate of litchi. *Acta Horticulturae* 558, 359–365.

Pesis, E., Dvir, O., Feygenberg, O., Ben-Arie, R., Ackerman, M. and Lichter, A. (2002) Production of acetaldehyde and ethanol during maturation and modified atmosphere storage of litchi fruit. *Postharvest Biology and Technology* 26, 157–165.

Pornchaloempong, P., Sargent, S.A. and Moretti, C.L. (1997) Cooling method and shipping container affect lychee fruit quality. *Proceedings of the Florida State Horticultural Society* 110, 197–200.

Pornchaloempong, P., Sargent, S.A. and Fox, A.J. (1998) Controlled atmosphere storage can extend postharvest quality of lychee fruit (cv. Mauritius). *Proceedings of the Florida State Horticultural Society* 111, 238–243.

Prapaipong, H. and Rakariyatham, N. (1990) Enzymic browning in longan. *Microbial Utilization of Renewable Resources* 7, 77–78.

Prasad, U.S. and Jha, O.P. (1978) Changes in pigmentation patterns during litchi ripening: flavonoid production. *Plant Biochemistry Journal* 5, 44–49.

Qu, H.X., Sun, G.C. and Jiang, Y.M. (2001) Study on the relationship between the peel structure and keeping quality of longan fruit. *Journal of Wuhan Botanical Research* 19, 83–85.

Rattanachai, Y. (1997) Set up HACCP for SO_2 fumigation of longan. In: *Proceedings of the Workshop on Application of GMP and HACCP in SO_2 Fumigation for Longan*, Bangkok, Thailand, pp. 75–113.

Roth, G. (1963) *Postharvest Decay of Litchi Fruit.* Technical Communication of the South African Department of Agriculture, No. 11, 16 pp.

Sardsud, V., Sittigul, C., Chaiwangsri, T. and Sardsud, U. (1994) Effects of post-fumigation washing treatments and storage temperature on disease development in fresh longan. In: Champ, B.R., Highley, E. and Johnson, G.I. (eds) *Proceedings of Postharvest Handling of Tropical Fruit.* Australian Centre for International Agricultural Research, Canberra, Australia, No. 50, pp. 77–79.

Sarni-Manchado, P., Le Roux, E., Le Guerneve, C., Lozano, Y. and Cheynier, V. (2000) Phenolic composition of litchi fruit pericarp. *Journal of Agricultural and Food Chemistry* 48, 5995–6002.

Scott, K.J., Brown, B.I., Chaplin, G.R., Wilcox, M.E. and Bain, J.M. (1982) The control of rotting and browning of litchi fruit by hot benomyl and plastic film. *Scientia Horticulturae* 16, 253–262.

Shi, Q. (1990) Studies on the characters of postharvest physiology and techniques of freshkeeping of longan fruit. *Fujian Fruit* 2, 1–4.

Singh, A. and Abidi, A.B. (1986) Level of carbohydrate fractions and ascorbic acid during ripening and storage of litchi (*Litchi chinensis* Sonn.) cultivars. *Indian Journal of Agricultural Chemistry* 19, 197–202.

Sonali, B., Das, B.C., Hasan, M.A. and Bhattacharjee, S. (2001) Studies on biochemical changes during fruit growth and development of litchi and their correlative relationship with peel anthocyanin content. *Research on Crops* 2, 41–46.

Swarts, D.H. and Anderson, T. (1980) Chemical control of mould growth on litchis during storage and sea shipment. *Subtropica* 1 (10), 13–15.

Thompson, J.F., Mitchell, F.G., Rumsey, T.R., Kasmire, R.F. and Crisosto, C.H. (1998) *Commercial Cooling of Fruit, Vegetables and Flowers.* University of California, DANR, Oakland, Publication No. 21567, 59 pp.

Tian, S.P., Xu, Y., Jiang, A.L. and Gong, Q.Q. (2002) Physiological and quality responses of longan fruit to high O_2 or high CO_2 atmospheres in storage. *Postharvest Biology and Technology* 24, 335–340.

Timberlake, C.F. and Bridle, P. (1967) Flavylium salts, anthocyanidins and anthocyanins. I. Reactions with sulphur dioxide. *Journal of the Science of Food and Agriculture* 18, 479–485.

Tongdee, S.C. (1992) Postharvest handling and SO_2 fumigation of longan for export. In: *Training of SO_2 Fumigation and Postharvest Handling of Longan for Export.* Thailand Institute of Scientific and Technological Research, Bangkok, Thailand, pp. 26–38.

Tongdee, S.C. (1994) Sulfur dioxide fumigation in postharvest handling of longan and lychee for export. In: Champ, B.R., Highley, E. and Johnson, G.I. (eds) *Proceedings of Postharvest Handling of Tropical Fruit.* Australian Centre for International Agricultural Research, Canberra, Australia No. 50, pp. 186–195.

Tongdee, S.C. (1997) Longan. In: Mitra, S. (ed.) *Postharvest Physiology and Storage of Tropical and Subtropical Fruit.* CAB International, Wallingford, UK, pp. 335–345.

Tongdee, S.C., Scott, K.J. and McGlasson, W.B. (1982) Packaging and cool storage of litchi fruit. *CSIRO Food Research Quarterly* 42, 25–28.

Underhill, S.J.R. and Critchley, C. (1992) The physiology and anatomy of lychee (*Litchi chinensis* Sonn.) pericarp during fruit development. *Journal of Horticultural Science* 67, 437–444.

Underhill, S.J.R. and Critchley, C. (1993) Physiological, biochemical and anatomical changes in lychee (*Litchi chinensis* Sonn.) pericarp during storage. *Journal of Horticultural Science* 68, 327–335.

Underhill, S.J.R. and Critchley, C. (1995) Cellular localisation of polyphenol oxidase and peroxidase activity in *Litchi chinensis* Sonn. pericarp. *Australian Journal of Plant Physiology* 22, 627–632.

Underhill, S.J.R. and Simons, D.H. (1993) Lychee (*Litchi chinensis* Sonn.) pericarp desiccation and the importance of postharvest microcracking. *Scientia Horticulturae* 54, 287–294.

Underhill, S.J.R. and Wong, L.S. (1990) A maturity standard for lychee (*Litchi chinensis* Sonn.). *Acta Horticulturae* 269, 181–187.

Underhill, S.J.R., Critchley, C. and Simons, D.H. (1992a) Postharvest pericarp browning of lychee (*Litchi chinensis* Sonn.). *Acta Horticulturae* 321, 718–725.

Underhill, S.J.R., Bagshaw, J., Prasad, A., Zauberman, G., Ronen, R. and Fuchs, Y. (1992b) The control of lychee (*Litchi chinensis* Sonn.) postharvest skin browning using sulphur dioxide and low pH. *Acta Horticulturae* 321, 732–741.

Underhill, S.J.R., Coates, L.M. and Saks, Y. (1997) Litchi. In: Mitra, S. (ed.) *Postharvest Physiology and Storage of Tropical and Subtropical Fruit.* CAB International, Wallingford, UK, pp. 191–208.

Vilasachandran, T., Sargent, S.A. and Maul, F. (1997) Controlled atmosphere storage shows potential for maintaining postharvest quality of fresh lychee fruit. In: Kader, A.A. (ed.)

Proceedings of the Seventh International Controlled Atmosphere Research Conference. Department of Pomology, University of California, Davis, USA. Postharvest Horticulture Series No. 17 (3), 83–89.

Wang, S.F. (1998) Longan. In: Hu, A.S. and Wang, S.F. (eds) *Fresh-Keeping and Commercial Treatment of Fruit.* China Agricultural Press, Beijing, China, pp. 153–160.

Wara-Aswapati, O., Sroymano, D., La-Ongsri, S. and Gomolmanee, S. (1994) Development of maturity indices for longan. In: Champ, B.R., Highley, E. and Johnson, G.I. (eds) *Proceedings of Postharvest Handling of Tropical Fruit.* Australian Centre for International Agricultural Research, Canberra, Australia No. 50, pp. 341–342.

Watson, B.J. (1984) Longan (*Euphoria longan* Lour.). In: Page, P.E. (ed.) *Tropical Tree Fruit for Australia.* Queensland Department of Primary Industries, Brisbane, Australia, pp. 192–197.

Wesche-Ebeling, P. and Montgomery, M.W. (1990) Strawberry polyphenoloxidase: its role in anthocyanin degradation. *Journal of Food Science* 55, 731–734.

Wong, K.C. (2000) *Longan Production in Asia.* Food and Agriculture Organization of the United Nations, Bangkok, Thailand, 44 pp.

Wong, K.C. and Ketsa, S. (1991) *Dimocarpus longan* Lour. In: Verheij, E.W.M. and Coronel, R.E. (eds) *Plant Resources of South-East Asia. No. 2: Edible Fruit and Nuts.* Pudoc, Wageningen, The Netherlands, pp. 146–151.

Wong, L.S., Jacobi, K.K. and Giles, J.E. (1991) The influence of hot benomyl dips on the appearance of cool stored lychee (*Litchi chinensis* Sonn.). *Scientia Horticulturae* 46, 245–251.

Wu, Z.X., Han, D.M., Ji, Z.L. and Chen, W.X. (1999) Effect of sulphur dioxide treatment on enzymatic browning of longan pericarp during storage. *Acta Horticulturae Sinica* 26, 91–95.

Xu, X.D., Huang, J.S., Zheng, S.Q., Xu, J.H. and Liu, H.Y. (1999) Effect of smouldering sulphur on browning of picked longan fruit and desulphurizatic effect. *Fujian Journal of Agricultural Science* 14 (1), 34–39.

York, G.M. (1994) An evaluation of two experimental polysaccharide Nature Seal coatings in delaying the post-harvest browning of the lychee pericarp. *Proceedings of the Florida State Horticultural Society* 107, 350–351.

Zauberman, G., Ronen, R., Akerman, M. and Fuchs, Y. (1990) Low pH treatment protects litchi fruit color. *Acta Horticulturae* 269, 309–314.

Zauberman, G., Ronen, R., Akerman, M., Weksler, S., Rot, I. and Fuchs, Y. (1991) Post-harvest retention of the red colour of litchi fruit pericarp. *Scientia Horticulturae* 47, 89–97.

Zee, F.T.P., Chan, H.T., Jr and Yen, C. (1998) Lychee, longan, rambutan and pulasan. In: Shaw, P.E., Chan, H.T., Jr and Nagy, S. (eds) *Tropical and Subtropical Fruit.* AgScience, Auburndale, Florida, pp. 290–335.

Zhang, D.L. and Quantick, P.C. (1997a) Effects of chitosan coating on enzymatic browning and decay during postharvest storage of litchi (*Litchi chinensis* Sonn.) fruit. *Postharvest Biology and Technology* 12, 195–202.

Zhang, D.L. and Quantick, P.C. (1997b) Preliminary study on effects of modified atmosphere packaging on postharvest storage of longan fruit. In: Kader, A.A. (ed.) *Proceedings of the Seventh International Controlled Atmosphere Research Conference.* Department of Pomology, University of California, Davis, USA. Postharvest Horticulture Series No. 17 (3), pp. 90–96.

Zhang, D.L. and Quantick, P.C. (2000) Effect of low-temperature hardening on postharvest storage of litchi fruit. *Acta Horticulturae* 518, 175–182.

Zhang, D.L., Quantick, P.C. and Grigor, J.M. (2000) Changes in phenolic compounds in litchi (*Litchi chinensis* Sonn.) fruit during postharvest storage. *Postharvest Biology and Technology* 19, 165–172.

Zhang, Q., Xie, J. and Xu, S.Q. (2002) Modified atmosphere packaging storage of 'Chike' longan fruit. *Food Science* 23 (4), 118–120.

Zhang, Z.Q., Pang, X.Q., Ji, Z.L. and Li, T. (1999) Effect of SO_2 release on storage and postharvest physiology of longan fruits. *Chinese Journal of Tropical Crops* 20 (1), 54–58.

Zhou, Y., Ji, Z.L. and Lin, W.Z. (1997) Study on the optimum storage temperature and chilling injury mechanism of longan fruit. *Acta Horticulturae Sinica* 24, 13–18.

Zou, S.Q., Pang, J., Lin, Q.X., Li, Y.X., Xie, J.H. and Zeng, J.H. (2001). A study of edible konjac glucomannan film on fresh-keeping and physiology of longan fruit. *Journal of Jiangxi Agricultural University* 23, 99–104.

Index